Springer Complexity

Springer Complexity is an interdisciplinary program publishing the best research and academic-level teaching on both fundamental and applied aspects of complex systems – cutting across all traditional disciplines of the natural and life sciences, engineering, economics, medicine, neuroscience, social and computer science.

Complex Systems are systems that comprise many interacting parts with the ability to generate a new quality of macroscopic collective behavior the manifestations of which are the spontaneous formation of distinctive temporal, spatial or functional structures. Models of such systems can be successfully mapped onto quite diverse "real-life" situations like the climate, the coherent emission of light from lasers, chemical reaction-diffusion systems, biological cellular networks, the dynamics of stock markets and of the internet, earthquake statistics and prediction, freeway traffic, the human brain, or the formation of opinions in social systems, to name just some of the popular applications.

Although their scope and methodologies overlap somewhat, one can distinguish the following main concepts and tools: self-organization, nonlinear dynamics, synergetics, turbulence, dynamical systems, catastrophes, instabilities, stochastic processes, chaos, graphs and networks, cellular automata, adaptive systems, genetic algorithms and computational intelligence.

The two major book publication platforms of the Springer Complexity program are the monograph series "Understanding Complex Systems" focusing on the various applications of complexity, and the "Springer Series in Synergetics", which is devoted to the quantitative theoretical and methodological foundations. In addition to the books in these two core series, the program also incorporates individual titles ranging from textbooks to major reference works.

Understanding Complex Systems

Founding Editor: J.A. Scott Kelso

Future scientific and technological developments in many fields will necessarily depend upon coming to grips with complex systems. Such systems are complex in both their composition – typically many different kinds of components interacting simultaneously and nonlinearly with each other and their environments on multiple levels – and in the rich diversity of behavior of which they are capable.

The Springer Series in Understanding Complex Systems series (UCS) promotes new strategies and paradigms for understanding and realizing applications of complex systems research in a wide variety of fields and endeavors. UCS is explicitly transdisciplinary. It has three main goals: First, to elaborate the concepts, methods and tools of complex systems at all levels of description and in all scientific fields, especially newly emerging areas within the life, social, behavioral, economic, neuro- and cognitive sciences (and derivatives thereof); second, to encourage novel applications of these ideas in various fields of engineering and computation such as robotics, nano-technology and informatics; third, to provide a single forum within which commonalities and differences in the workings of complex systems may be discerned, hence leading to deeper insight and understanding.

UCS will publish monographs, lecture notes and selected edited contributions aimed at communicating new findings to a large multidisciplinary audience.

Bernd Krauskopf · Hinke M. Osinga · Jorge Galán-Vioque (Eds.)

Numerical Continuation Methods for Dynamical Systems

Path following and boundary value problems

With 200 Figures

Dedicated to Eusebius J. Doedel for his 60th birthday

 Springer

Dr. Bernd Krauskopf
Dept of Engineering Mathematics
University of Bristol
Bristol
BS8 1TR
United Kingdom

Dr. Hinke M. Osinga
Dept of Engineering Mathematics
University of Bristol
Bristol
BS8 1TR
United Kingdom

Dr. Jorge Galán-Vioque
Applied Mathematics II
University of Sevilla
Escuela Superior de Ingenieros
Camino de los Descubrimientos, s/n,
41092 Sevilla
Spain

A C.I.P. Catalogue record for this book is available from the Library of Congress

ISSN 1860-0832

ISBN 978-1-4020-6355-8 (HB)

ISBN 978-1-4020-6356-5 (e-book)

Published by Springer,
P.O. Box 17, 3300 AA Dordrecht, The Netherlands
In association with
Canopus Publishing Limited,
27 Queen Square, Bristol BS1 4ND, UK

www.springer.com and www.canopusbooks.com

A Continuing Influence in Dynamics

A well-established method for studying a given dynamical system is to identify the compact invariant objects, such as equilibria, periodic orbits and invariant tori, and to consider the local behavior around them. This local information then needs to be assembled in a consistent way, frequently with the help of geometric and topological arguments, to obtain a unified global picture of the system. The aim is to find qualitative (and often also quantitative) representations of the different types of behavior that the system may exhibit in dependence of key parameters. The main result of such an effort is a bifurcation diagram, that is, information on the division of parameter space into regions of topologically different behavior together with representative phase portraits. The list of theoretical tools one may employ is long, well developed and dates back at least to the 19th century. However, even when one considers seemingly simple systems, theoretical tools need to be supplemented with numerical calculations.

Mainly for technological reasons, numerical methods have a shorter history than theoretical tools. The first and commonly used tool is numerical time integration, which allows one to explore the dynamics by solving a (possibly large) number of initial value problems. This approach is very practical for the representation of chaotic attractors, and especially their 'fingerprints' in a suitable Poincaré section. However, when it comes to the study of how the behavior changes as a function of parameters the tool of choice is numerical continuation — one also speaks of path following or homotopy methods. The basic idea is to compute an implicitly defined curve of a suitable system of equations that defines the dynamical object under consideration. In its basic form, numerical continuation implements the stability and bifurcation theory of equilibria of differential equations. More global objects, such as periodic and homoclinic orbits, and their bifurcations can be computed by setting up defining equations in the form of boundary value problems.

Path following in combination with boundary value problem solvers has emerged as a continuing and strong influence in the development of dynamical systems theory and its application in many diverse fields of science. It is widely acknowledged that the software package AUTO — developed by Eusebius J. Doedel about thirty years ago and further expanded and developed ever since — plays a central role in the brief history of numerical continuation. When we were thinking how best to present the origin and development of AUTO, a copy of the first edition of the Auto86 manual, with its authentic ochre Caltech cover, came in very handy. We quote from the preface, dated May 1986:

> *The* Auto *package was first written in 1979. It was based on a related program written in 1976 while the author was working with H.B Keller at the California Institute of Technology. A first publication referring to the package by its current name appeared in [22].*
>
> *Applications often revealed some inadequacy in the algorithms and resulted in changes. The applications also pointed to additional capabilities that would be useful to have integrated in the package, and effort was spent on making it easy to use. This explains the delay in publication of an extensive account of the algorithm implemented. Indeed, the difference in effort between a theoretical analysis of a new method and its implementation and integration appears to be considerable. We are confident, however, that the methods and software presented here will be of some use in the numerical exploration of nonlinear phenomena in ordinary differential equations.*

This quote not only highlights the intricate interplay at the very earliest stage between the development of the software and applications, but it also contains a major understatement: Auto has not just been "of some use", but it has been used by many hundreds of researchers from all around the world! To give a rough idea of its impact in the general scientific community, ISI Web of Knowledge reveals that the different versions of the Auto manual, which was never published other than as a Caltech preprint, has more than 700 citations. Similarly, the seminal reference [22] in the quote, the paper E.J. Doedel, Auto: A program for the automatic bifurcation analysis of autonomous systems, *Cong. Num.* 30 (Proc. 10th Manitoba Conf. Num. Math. and Comp.), 1981, 265–284, has more than 400 citations.

This book has been compiled on the occasion of Sebius Doedel's 60th birthday with the aim to illustrate the power and versatility of numerical continuation techniques. As is demonstrated in the chapters of this book, many recent developments build on the ideas of Sebius Doedel as implemented in the package Auto, whose core of path following routine and collocation boundary value problem solver is essentially still the same as when it was released in 1986. It lies in the nature of the subject and the versatility of Sebius Doedel's work that we had to make a choice about which topics to include. The emphasis of this book is on continuation methods for different types of systems and dynamical objects, and on examples of how numerical bifurcation analysis can be used in concrete applications. While recognizing that there are other topics that could have been included, we believe that this choice is in the spirit of the original motivation for the development of Auto as expressed in the above quote. In this way, we hope to give an impression of the continuing influence and future potential of these powerful numerical methods for the bifurcation analysis of different types of dynamical systems.

The book opens with an extended foreword by Herb Keller, who is widely recognized as the founding father of numerical continuation. Chapter 1 is an edited part of lecture notes that Sebius Doedel has been using in his own courses. It introduces the basic concepts of numerical bifurcation analysis and forms a basis for the remainder of the book. The other eleven chapters by leading experts focus on selected topics that have been influenced strongly by Sebius Doedel's work. In fact, at least half of the chapters discuss research in which he has been involved as a co-author. Chapter 2 by Willy Govaerts and Yuri Kuznetsov surveys recent developments of interactive continuation tools. Chapter 3 by Mike Henderson is concerned with higher-dimensional continuation, and Chap. 4 by Bernd Krauskopf and Hinke Osinga discusses the computation of invariant manifolds with a continuation approach. The next three chapters are devoted to applications. In Chap. 5 Don Aronson and Hans Othmer consider the dynamics of a SQUID consisting of two Josephson junctions. Chapter 6 by Sebastian Wieczorek discusses global bifurcations in laser systems, and Chap. 7 by Emilio Freire and Alejandro Rodríguez-Luis demonstrates the use of numerical bifurcation analysis for the study of electronic circuits. The remaining chapters deal with continuation for special types of dynamical systems. Chapter 8 by John Guckenheimer and Drew LaMar is concerned with slow-fast systems, and Chap. 9 by Jorge Galán-Vioque and André Vanderbauwhede with symmetric Hamiltonian systems. Spatially extended systems are the topic of Chap. 10 by Wolf-Jürgen Beyn and Vera Thümmler and of Chap. 11 by Alan Champneys and Björn Sandstede. Finally, in Chap. 12 Dirk Roose and Róbert Szalai survey numerical continuation techniques for systems with delay.

We are very grateful for the enthusiastic support from all who were involved in this book project. First of all, we thank all authors for their contributions and for making every effort to stay within the limits of a tight production schedule. We also thank Tom Spicer of Canopus Publishing Ltd for his support of this project from its conception to the final production of the book. Last, but not least, we would like to thank Sebius Doedel for his support over many years of collaboration, and for agreeing to the publication of Chap. 1 without knowing exactly what we were up to. Happy birthday, Sebius!

Bernd Krauskopf, Hinke Osinga and Jorge Galán-Vioque
Bristol and Sevilla, March 2007.

Foreword

Herbert B Keller

California Institute of Technology, Pasadena, USA, and University of California, San Diego, USA.

Sebius (diminutive for Eusebius) Doedel obtained his Ph.D. in Applied Mathematics from the University of British Columbia in 1976. His advisor was my friend and ex-colleague Jim Varah. As a consequence, I was able to employ Sebius as a Research Fellow in Applied Mathematics at Caltech in 1975. Over the next 26 years, he spent 13 of them at Caltech. However, he was also much appreciated at Concordia University, where he was employed in 1979 and rapidly rose to Professor of Computer Science, winning many awards which fortunately included several years on leave with pay!

We cycled together occasionally, even in Holland, where Sebius was born. I remember one ride in particular, when on a rather warm day we went east from Pasadena to Claremont, about 28 miles each way. Somewhere along the way, I became quite thirsty and so we stopped to get a cool drink. We sat outside, relaxed and, I thought, enjoyed our drinks. I started to wax poetic about how nice it was enjoying the outdoors and lovely California weather when Sebius said: "Herb, do you know where we are?" I said: "sure, near Claremont." He replied: "This is Pomona, the drive-by shooting capital of the world — I can't wait to get out of here." I have never again enjoyed going past that part of our ride.

Early during his first appointment at Caltech, Sebius became interested in bifurcation phenomena, two-point boundary value problems and numerical path-following or continuation methods, perhaps as a result of sitting in on my course in these areas. In 1976, I was writing the paper [17] in which pseudoarclength continuation was introduced and Sebius was willing to do some calculations to illustrate how these methods worked. Of course, he did a wonderful job producing all of the results in §7 of that paper, but more importantly, as a result, he essentially started working on AUTO at that time. The first publication on AUTO appeared in 1981 [7]. It has evolved dramatically since then, culminating in AUTO2000 [22], a fully parallel code in C++ with great graphics (available for free via http://sourceforge.net/projects/auto2000/) that was produced mainly by Randy Paffenroth, working at Caltech under Sebius's direction.

AUTO is without a doubt the most powerful and efficient tool for determining the bifurcation structure of nonlinear parameter-dependent systems of algebraic and ordinary differential equations. Influenced by his colleagues at the University of British Columbia, who developed COLSYS [5], Sebius has employed orthogonal collocation approximations and mesh refinement to obtain extremely high accuracy. The code is able to determine heteroclinic, homoclinic and periodic orbits, both stable and unstable, by means of a two-point boundary value problem formulation. Using a brilliant elimination procedure, the relatively sparse Jacobian is reduced to a low-dimensional dense matrix from which the Floquet multipliers are computed. The bifurcation structure at singular points is readily determined in this way.

The development of AUTO is but one of the main projects that Sebius has undertaken. In the course of this work powerful theoretical results have been produced, many with colleagues and his students, in the general areas of bifurcation theory, dynamical systems, periodic orbits, delay differential equations, collocation methods for nonlinear elliptic PDEs, coupled oscillator theory, control of bifurcation phenomena, continuation theory of manifolds, and numerous additional topics. However, there is no doubt that the AUTO software has had a tremendous impact on many applied mathematics areas and is, indeed, one of the leading tools in scientific computing. The code has been incorporated into many other large software systems that solve nonlinear problems involving continuation and bifurcation phenomena.

Essentially, all of the authors contributing to this volume have been coauthors with Sebius on papers related to the work presented here. However, I would like to point out a few of my favorite contributions made through these collaborations, not all of which have been fully appreciated yet. A brilliant contribution is contained in a paper by Wolf-Jürgen Beyn and Sebius Doedel [6], in which it is shown that a continuous nonlinear boundary value problem and the corresponding discretized problem have the same number of solutions for all sufficiently fine meshes.

Sebius introduced a very powerful technique to keep computed families of periodic solutions of autonomous differential equations in phase. The idea is simply to minimize the 'distance' between neighboring solutions with respect to a change in phase. That is, if $\mathbf{u}(t, \lambda)$ is the solution over the normalized period $0 \leq t \leq 1$ at parameter value λ, then the neighboring solution $\mathbf{u}(t, \lambda+\delta)$ with phase shift θ lies at distance

$$D^2(\theta) = \int_0^1 \| \mathbf{u}(t + \theta, \lambda + \delta) - \mathbf{u}(t, \lambda) \|^2 \ dt.$$

We seek to minimize this distance with respect to the phase shift θ. Standard calculus of variations near zero leads to the integral condition

$$\int_0^1 \dot{\mathbf{u}}^*(t, \lambda + \delta) \, \mathbf{u}(t, \lambda) \, dt = 0.$$

This is a generalization of the standard Poincaré phase or transversality condition that is applied only at one point on the orbit. However, the above global condition is much more robust in calculations, as has been shown in many examples [9, 10, 12] (the Poincaré condition remains preferable for analytical proofs). I am not sure when this global condition first appeared in the literature, but we have referred to it in [16] as having been introduced by Sebius in 1981 [7], which also happens to be his first publication on AUTO.

Many of Sebius's publications have to do with periodic solutions of dynamical systems. These arise in a great variety of applications starting with chemical reactors [23], then on to systems of oscillators [2, 21], heteroclinic orbits [9] in which the above phase condition is crucial, resonances in excitable systems [1] such as forced Fitzhugh-Nagumo systems, current biased and coupled Josephson junctions [3, 4], delay differential equations [11, 13, 14], modified Van der Pol oscillators [8], conservative and Hamiltonian systems [20], cardiac pacemakers [19], the circular restricted three-body problem and the figure-eight orbit of Chenciner and Montgomery [12, 15], and many more. A large number of these contributions are in the bio-physics area and, thus, it turns out that Sebius may be a closet biologist.

More recently, Sebius has returned with others to the important problem of computing higher-dimensional manifolds, either stable or unstable [18].

This brief account of some of Sebius's publications and obvious collaborations does not do justice to the impact he has had in the field of scientific computation. He has had numerous students, extremely well trained, and now making their own contributions. Furthermore, he has worked with many outstanding scientists and has invariably enhanced their ability to do significant scientific computations so much that it would be difficult to measure his tremendous influence in our field. Hopefully, he will continue as he reaches maturity.

H. B. Keller
Caltech / UCSD
November, 2006

References

1. J. C. Alexander, E. J. Doedel, and H. G. Othmer. On the resonance structure in a forced excitable system. *SIAM J. Appl. Math.* 50(5):1373–1418, 1990.
2. D. G. Aronson, E. J. Doedel, and H. G. Othmer. An analytical and numerical study of the bifurcations in a system of linearly-coupled oscillators. *Physica D* 25(1-3):20–104, 1987.
3. D. G. Aronson, E. J. Doedel, and H. G. Othmer. The dynamics of coupled current-biased Josephson junctions. II. *Internat. J. Bifur. Chaos Appl. Sci. Engrg.*, 1(1): 51–66, 1991.
4. D. G. Aronson, E. J. Doedel, and D. H. Terman. A codimension-two point associated with coupled Josephson junctions. *Nonlinearity* 10(5):1231–1255, 1997.
5. U. Ascher, J. Christiansen, and R. Russell. A collocation solver for mixed order systems of boundary value problems. *Math. Comp.*, 33(146):659–679, 1979.
6. W.-J. Beyn and E. J. Doedel. Stability and multiplicity of solutions to discretizations of nonlinear ordinary differential equations. *SIAM J. Sci. Comp.*, 2(1):107–120, 1981.
7. E. J. Doedel. AUTO: a program for the automatic bifurcation analysis of autonomous systems. *Congr. Numer.*, 30:265–384, 1981.
8. E. J. Doedel, E. Freire, E. Gamero, and A. J Rodríguez-Luis. An analytical and numerical study of a modified Van der Pol oscillator. *J. Sound Vibration*, 256(4):755–771, 2002.
9. E. J. Doedel and M. Friedman. Numerical computation of heteroclinic orbits. Continuation techniques and bifurcation problems. *J. Comput. Appl. Math.*, 26(1-2):155–170, 1989.
10. E. J. Doedel, H. B. Keller, and J.-P. Kernevez. Numerical analysis and control of bifurcation problems II: Bifurcation in infinite dimensions. *Internat. J. Bifur. Chaos Appl. Sci. Engrg.*, 1(4): 745–772, 1991.
11. E. J. Doedel and P. C. Leung. Numerical techniques for bifurcation problems in delay equations. *Congr. Numer.*, 34:225–237, 1982.
12. E. J. Doedel, R. C. Paffenroth, H. B. Keller, D. J. Dichmann, J. Galán-Vioque, and A. Vanderbauwhede. Computation of periodic solutions of conservative systems with applications to the 3-body problem. *Internat. J. Bifur. Chaos Appl. Sci. Engrg.*, 13(6): 1353–1381, 2003.
13. K. Engelborghs and E. J. Doedel. Convergence of a boundary value difference equation for computing periodic solutions of neutral delay differential equations. *J. Differ. Equations Appl.*, 7(6):927–940, 2001.
14. K. Engelborghs and E. J. Doedel. Stability of piecewise polynomial collocation for computing periodic solutions of delay differential equations. *Numer. Math.*, 91(4):627–648, 2002.
15. J. Galán-Vioque, F. J. Muñoz-Almaraz, E. Freire, E. J. Doedel, and A. Vanderbauwhede. Stability and bifurcations of the figure-8 solution of the three-body problem. *Phys. Rev. Lett.* 88(24):241101, 2002.
16. A. D. Jepson and H. B. Keller. Steady state and periodic solution paths: their bifurcations and computations. In T. Kupper and H. D. Mittleman, editors, *Numerical Methods for Bifurcation Problems*, pages 219–246. (Birkäuser Verlag, 1984).

17. H. B. Keller. Numerical solution of bifurcation and nonlinear eigenvalue problems. In P. H. Rabinowitz, editor, *Applications of Bifurcation Theory*, pages 359–384. (Academic Press, New York, 1978).

18. B. Krauskopf, H. M. Osinga, E. J. Doedel, M. E. Henderson, J. Guckenheimer, A. Vladimirsky, M. Dellnitz, and O. Junge. A survey of methods for computing (un)stable manifolds of vector fields. *Internat. J. Bifur. Chaos Appl. Sci. Engrg.*, 15(3):763–791, 2005.

19. T. Krogh-Madsen, L. Glass, E. J. Doedel, and M. R. Guevara. Apparent discontinuities in the phase-resetting response of cardiac pacemakers. *J. Theor. Biol.*, 230(4):499–519, 2004.

20. F. J. Muñoz-Almaraz, E. Freire, J. Galán, and E. J. Doedel. Continuation of periodic orbits in conservative and Hamiltonian systems. *Physica D*, 181(1-2):1–38, 2003.

21. H. G. Othmer, D. G. Aronson, and E. J. Doedel. Resonance and bistability in coupled oscillators. *Phys. Lett. A*, 113(7):349–354, 1986.

22. R. C. Paffenroth and E. J. Doedel. The AUTO2000 command line user interface. In *Ninth International Python Conference* (Long Beach, California, USA 2001), pages 233-241, 2001. Available via http://cmvl.cs.concordia.ca/.

23. A. B. Poore, E. J. Doedel, and J. E. Cermak. Dynamics of the Iwan-Blevins Wake Oscillator Model. *Internat. J. Non-Linear Mechanics*, 21(4):291–302, 1986.

Contents

List of Contributors

Don Aronson
School of Mathematics and
Institute for Mathematics
and its Applications
University of Minnesota
Minneapolis, MN 55455
USA
don@ima.umn.edu

Wolf-Jürgen Beyn
Department of Mathematics
University of Bielefeld
P.O. Box 100131
33501 Bielefeld
Germany
beyn@math.uni-bielefeld.de

Alan R Champneys
Department of Engineering
Mathematics
University of Bristol
Queen's Building
Bristol BS8 1TR
United Kingdom
A.R.Champneys@bristol.ac.uk

Eusebius J Doedel
Department of Computer Science
Concordia University
1455 De Maisonneuve Blvd. West
EV 3285

Montreal, QC H3G 1M8
Canada
doedel@cs.concordia.ca

Emilio Freire
Departamento de Matemática
Aplicada II
Escuela Superior de Ingenieros
Universidad de Sevilla
Camino de los Descubrimientos s/n
41092 Sevilla
Spain
efrem@us.es

Jorge Galán-Vioque
Departamento de Matemática
Aplicada II
Escuela Superior de Ingenieros
Universidad de Sevilla
Camino de los Descubrimientos s/n
41092 Sevilla
Spain
jgv@us.es

Willy Govaerts
Deptartment of Applied Mathematics and Computer Science
Ghent University
Krijgslaan 281 -S9
B-9000 Gent
Belgium
Willy.Govaerts@UGent.be

John Guckenheimer
Mathematics Department
Cornell University
Ithaca, NY 14853-2401
USA
Jmg16@cornell.edu

Michael E Henderson
IBM Watson Research Center
P.O. Box 218
33-215
Yorktown Heights, NY 10598
USA
mhender@watson.ibm.com

Herbert B Keller
Applied and Computational
Mathematics
California Institute of Technology
MC 217-50
Pasadena, CA 91125
USA
hbk@caltech.edu

Bernd Krauskopf
Department of Engineering
Mathematics
University of Bristol
Queen's Building
Bristol BS8 1TR
United Kingdom
B.Krauskopf@bristol.ac.uk

Yuri A Kuznetsov
Department of Mathematics
Budapestlaan 6
3584 CD Utrecht
The Netherlands
kuznet@math.uu.nl

M Drew LaMar
Mathematics Department
Cornell University
Ithaca, NY 14853-2401
USA
Mlamar@math.cornell.edu

Hinke M Osinga
Department of Engineering
Mathematics
University of Bristol
Queen's Building
Bristol BS8 1TR
United Kingdom
H.M.Osinga@bristol.ac.uk

Hans Othmer
School of Mathematics and
Digital Technology Center
University of Minnesota
Minneapolis, MN 55455
USA
othmer@math.umn.edu

Alejandro J Rodríguez-Luis
Departamento de Matemática
Aplicada II
Escuela Superior de Ingenieros
Universidad de Sevilla
Camino de los Descubrimientos s/n
41092 Sevilla
Spain
alejan@matinc.us.es

Dirk Roose
Department of Computer Science
KU Leuven
Celestijnenlaan 200A
B-3001 Heverlee - Leuven
Belgium
Dirk.Roose@cs.kuleuven.ac.be

Björn Sandstede
Department of Mathematics
University of Surrey
Guildford, GU2 7XH
United Kingdom
b.sandstede@surrey.ac.uk

Robert Szalai
Department of Engineering
Mathematics
University of Bristol
Queen's Building

Bristol BS8 1TR
United Kingdom
R.Szalai@bristol.ac.uk

Vera Thümmler
Department of Mathematics
University of Bielefeld
P.O. Box 100131
33501 Bielefeld
Germany
thuemmle@math.uni-bielefeld.de

André Vanderbauwhede
Department of Pure Mathematics
and Computer Algebra
Ghent University
Krijgslaan 281
B-9000 Gent
Belgium
avdb@cage.ugent.be

Sebastian M Wieczorek
Mathematics Research Institute
University of Exeter
Harrison Building
North Park Road
Exeter EX4 4QF
United Kingdom
S.M.Wieczorek@exeter.ac.uk

1

Lecture Notes on Numerical Analysis of Nonlinear Equations

Eusebius J Doedel

Department of Computer Science, Concordia University, Montreal, Canada

Numerical integrators can provide valuable insight into the transient behavior of a dynamical system. However, when the interest is in stationary and periodic solutions, their stability, and their transition to more complex behavior, then numerical continuation and bifurcation techniques are very powerful and efficient.

The objective of these notes is to make the reader familiar with the ideas behind some basic numerical continuation and bifurcation techniques. This will be useful, and is at times necessary, for the effective use of the software AUTO and other packages, such as XPPAUT [17], CONTENT [24], MATCONT [21], and DDE-BIFTOOL [16], which incorporate the same or closely related algorithms.

These lecture notes are an edited subset of material from graduate courses given by the author at the universities of Utah and Minnesota [9] and at Concordia University, and from short courses given at various institutions, including the Université Pierre et Marie Curie (Paris VI), the Centre de Recherches Mathématiques of the Université de Montréal, the Technische Universität Hamburg-Harburg, and the Benemérita Universidad Autónoma de Puebla.

1.1 The Implicit Function Theorem

Before starting our discussion of numerical continuation of solutions to nonlinear equations, it is important first to discuss under what conditions a solution will actually persist when problem parameters are changed. Therefore, we begin with an overview of the basic theory. The Implicit Function Theorem (IFT) is central to our analysis and we discuss some examples. The discussion in this section follows the viewpoint of Keller in graduate lectures at the California Institute of Technology, a subset of which was published in [23].

1.1.1 Basic Theory

Let $\mathcal{B}$ denote a Banach space, that is, a complete, normed vector space. In the presentation below it will be implicitly assumed that $\mathcal{B}$ is $\mathbb{R}^n$, although the results apply more generally. For $\mathbf{x}_0 \in \mathcal{B}$, we denote by $S_\rho(\mathbf{x}_0)$ the closed ball of radius ρ centered at $\mathbf{x}_0$, that is,

$$S_\rho(\mathbf{x}_0) = \{\mathbf{x} \in \mathcal{B} \mid \|\mathbf{x} - \mathbf{x}_0\| \le \rho\}.$$

Existence and uniqueness of solutions is obtained by using two theorems.

Theorem 1 (Contraction Theorem). *Consider a continuous function $F :$ $\mathcal{B} \to \mathcal{B}$ on a Banach space $\mathcal{B}$ and suppose that for some $\mathbf{x}_0 \in \mathcal{B}$, $\rho > 0$, and some K_0 with $0 \le K_0 < 1$, we have*

$$\|F(\mathbf{u}) - F(\mathbf{v})\| \le K_0 \|\mathbf{u} - \mathbf{v}\|, \text{ for all } \mathbf{u}, \mathbf{v} \in S_\rho(\mathbf{x}_0),$$
$$\|F(\mathbf{x}_0) - \mathbf{x}_0\| \le (1 - K_0)\,\rho.$$

Then the equation

$$\mathbf{x} = F(\mathbf{x}), \qquad \mathbf{x} \in \mathcal{B},$$

has one and only one solution $\mathbf{x}_* \in S_\rho(\mathbf{x}_0)$, and $\mathbf{x}_*$ is the limit of the sequence

$$\mathbf{x}_{k+1} = F(\mathbf{x}_k), \qquad k = 0, 1, 2, \ldots.$$

Proof. Let $\mathbf{x}_1 = F(\mathbf{x}_0)$. Then

$$\|\mathbf{x}_1 - \mathbf{x}_0\| = \|F(\mathbf{x}_0) - \mathbf{x}_0\| \le (1 - K_0)\,\rho \le \rho.$$

Thus, $\mathbf{x}_1 \in S_\rho(\mathbf{x}_0)$. Now assume inductively that $\mathbf{x}_0, \mathbf{x}_1, \cdots, \mathbf{x}_n \in S_\rho(\mathbf{x}_0)$. Then for $k \le n$ we have

$$
\begin{aligned}
\|\mathbf{x}_{k+1} - \mathbf{x}_k\| = \|F(\mathbf{x}_k) - F(\mathbf{x}_{k-1})\| &\le K_0 \|\mathbf{x}_k - \mathbf{x}_{k-1}\| \\
&= \qquad \cdots \qquad \le K_0^k \|\mathbf{x}_1 - \mathbf{x}_0\| \\
&\le \quad K_0^k (1 - K_0)\,\rho.
\end{aligned}
$$

Thus,

$$
\begin{aligned}
\|\mathbf{x}_{n+1} - \mathbf{x}_0\| &\le \|\mathbf{x}_{n+1} - \mathbf{x}_n\| + \|\mathbf{x}_n - \mathbf{x}_{n-1}\| + \cdots + \|\mathbf{x}_1 - \mathbf{x}_0\| \\
&\le (K_0^n + K_0^{n-1} + \cdots + 1)(1 - K_0)\,\rho \\
&= (1 - K_0^{n+1})\,\rho \\
&\le \rho.
\end{aligned}
$$

Hence $\mathbf{x}_{n+1} \in S_\rho(\mathbf{x}_0)$, and by induction $\mathbf{x}_k \in S_\rho(\mathbf{x}_0)$ for all k. We now show that $\{\mathbf{x}_k\}$ is a Cauchy sequence:

$$\|\mathbf{x}_{k+n} - \mathbf{x}_k\| \le \|\mathbf{x}_{k+n} - \mathbf{x}_{k+n-1}\| + \cdots + \|\mathbf{x}_{k+1} - \mathbf{x}_k\|$$
$$\le (K_0^{n-1} + K_0^{n-2} + \cdots + 1) K_0^k (1 - K_0) \rho$$
$$= (1 - K_0^n) K_0^k \rho$$
$$\le K_0^k \rho.$$

For given $\varepsilon > 0$, choose k such that $K_0^k \rho < \frac{1}{2} \varepsilon$. Then

$$\|\mathbf{x}_{k+\ell} - \mathbf{x}_{k+m}\| \le \|\mathbf{x}_{k+\ell} - \mathbf{x}_k\| + \|\mathbf{x}_{k+m} - \mathbf{x}_k\| \le 2K_0^k \rho < \varepsilon,$$

independently of ℓ and m. Hence, $\{\mathbf{x}_k\}$ is a Cauchy sequence and, therefore, converges to a unique limit $\lim \mathbf{x}_k = \mathbf{x}_*$, where $\mathbf{x}_* \in S_\rho(\mathbf{x}_0)$. Since we assumed that F is continuous, we have

$$\mathbf{x}_* = \lim \mathbf{x}_k = \lim F(\mathbf{x}_{k-1}) = F(\lim \mathbf{x}_{k-1}) = F(\lim \mathbf{x}_k) = F(\mathbf{x}_*).$$

This proves the existence of $\mathbf{x}_*$. We get uniqueness as follows. Suppose there are two solutions, say, $\mathbf{x}, \mathbf{y} \in S_\rho(\mathbf{x}_0)$ with $\mathbf{x} = F(\mathbf{x})$ and $\mathbf{y} = F(\mathbf{y})$. Then

$$\|\mathbf{x} - \mathbf{y}\| = \|F(\mathbf{x}) - F(\mathbf{y})\| \le K_0 \|\mathbf{x} - \mathbf{y}\|.$$

Since $K_0 < 1$, this is a contradiction. □

The second theorem ensures the parameter-dependent existence of a solution.

Theorem 2 (Implicit Function Theorem). *Let* $\mathbf{G} : \mathcal{B} \times \mathbb{R}^m \to \mathcal{B}$ *satisfy:*

- $\mathbf{G}(\mathbf{u}_0, \boldsymbol{\lambda}_0) = \mathbf{0}$ *for* $\mathbf{u}_0 \in \mathcal{B}$ *and* $\boldsymbol{\lambda}_0 \in \mathbb{R}^m$;
- $\mathbf{G}_{\mathbf{u}}(\mathbf{u}_0, \boldsymbol{\lambda}_0)$ *is nonsingular with bounded inverse,*

$$\|\mathbf{G}_{\mathbf{u}}(\mathbf{u}_0, \boldsymbol{\lambda}_0)^{-1}\| \le M$$

   for some $M > 0$;
- $\mathbf{G}$ *and* $\mathbf{G}_{\mathbf{u}}$ *are Lipschitz continuous, that is, for all* $\mathbf{u}, \mathbf{v} \in S_\rho(\mathbf{u}_0)$, *and for all* $\boldsymbol{\lambda}, \boldsymbol{\mu} \in S_\rho(\boldsymbol{\lambda}_0)$ *the following inequalities hold for some* $K_L > 0$:

$$\|\mathbf{G}(\mathbf{u}, \boldsymbol{\lambda}) - \mathbf{G}(\mathbf{v}, \boldsymbol{\mu})\| \le K_L (\|\mathbf{u} - \mathbf{v}\| + \|\boldsymbol{\lambda} - \boldsymbol{\mu}\|),$$
$$\|\mathbf{G}_{\mathbf{u}}(\mathbf{u}, \boldsymbol{\lambda}) - \mathbf{G}_{\mathbf{u}}(\mathbf{v}, \boldsymbol{\mu})\| \le K_L (\|\mathbf{u} - \mathbf{v}\| + \|\boldsymbol{\lambda} - \boldsymbol{\mu}\|).$$

Then there exists δ, *with* $0 < \delta \le \rho$, *and a unique function* $\mathbf{u}(\boldsymbol{\lambda})$ *that is continuous on* $S_\delta(\boldsymbol{\lambda}_0)$, *with* $\mathbf{u}(\boldsymbol{\lambda}_0) = \mathbf{u}_0$, *such that*

$$\mathbf{G}(\mathbf{u}(\boldsymbol{\lambda}), \boldsymbol{\lambda}) = \mathbf{0}, \text{ for all } \boldsymbol{\lambda} \in S_\delta(\boldsymbol{\lambda}_0).$$

If $\mathbf{G}(\mathbf{u}, \boldsymbol{\lambda}_0) = \mathbf{0}$ and if $\mathbf{G}_{\mathbf{u}}(\mathbf{u}_0, \boldsymbol{\lambda}_0)$ is invertible with bounded inverse, then $\mathbf{u}_0$ is called an *isolated solution* of $\mathbf{G}(\mathbf{u}, \boldsymbol{\lambda}_0) = \mathbf{0}$. Hence, the IFT states that isolation (plus Lipschitz continuity assumptions) implies the existence of a locally unique *solution family* (or *solution branch*) $\mathbf{u} = \mathbf{u}(\boldsymbol{\lambda})$, with $\mathbf{u}(\boldsymbol{\lambda}_0) = \mathbf{u}_0$.

Proof. We use the notation $\mathbf{G}_{\mathbf{u}}^0 = \mathbf{G}_{\mathbf{u}}(\mathbf{u}_0, \boldsymbol{\lambda}_0)$. Then we rewrite the problem as

$$\mathbf{G}(\mathbf{u}, \boldsymbol{\lambda}) = \mathbf{0} \Leftrightarrow \mathbf{G}_{\mathbf{u}}^0 \, \mathbf{u} = \mathbf{G}_{\mathbf{u}}^0 \, \mathbf{u} - \mathbf{G}(\mathbf{u}, \boldsymbol{\lambda})$$

$$\Leftrightarrow \mathbf{u} = \underbrace{\left(\mathbf{G}_{\mathbf{u}}^0\right)^{-1} \left[\mathbf{G}_{\mathbf{u}}^0 \, \mathbf{u} - \mathbf{G}(\mathbf{u}, \boldsymbol{\lambda})\right]}_{\equiv \mathbf{F}(\mathbf{u}, \boldsymbol{\lambda})}.$$

Hence, $\mathbf{G}(\mathbf{u}, \boldsymbol{\lambda}) = \mathbf{0}$ if and only if $\mathbf{u}$ is a fixed point of $\mathbf{F}(\cdot, \boldsymbol{\lambda})$. (Note that the corresponding fixed point iteration is, in fact, the Chord Method for solving $\mathbf{G}(\mathbf{u}, \boldsymbol{\lambda}) = \mathbf{0}$.) We must verify the conditions of the Contraction Theorem. Pick $\mathbf{u}, \mathbf{v} \in S_{\rho_1}(\mathbf{u}_0)$, and any *fixed* $\boldsymbol{\lambda} \in S_{\rho_1}(\boldsymbol{\lambda}_0)$, where ρ_1 is to be chosen later. Then

$$\mathbf{F}(\mathbf{u}, \boldsymbol{\lambda}) - \mathbf{F}(\mathbf{v}, \boldsymbol{\lambda}) = \left(\mathbf{G}_{\mathbf{u}}^0\right)^{-1} \left\{\mathbf{G}_{\mathbf{u}}^0 \left[\mathbf{u} - \mathbf{v}\right] - \left[\mathbf{G}(\mathbf{u}, \boldsymbol{\lambda}) - \mathbf{G}(\mathbf{v}, \boldsymbol{\lambda})\right]\right\}. \quad (1.1)$$

By the Fundamental Theorem of Calculus, we have

$$\mathbf{G}(\mathbf{u}, \boldsymbol{\lambda}) - \mathbf{G}(\mathbf{v}, \boldsymbol{\lambda}) = \int_0^1 \frac{d}{dt} \mathbf{G}(t\mathbf{u} + (1-t)\mathbf{v}, \boldsymbol{\lambda}) \, dt$$

$$= \int_0^1 \mathbf{G}_{\mathbf{u}}(t\mathbf{u} + (1-t)\mathbf{v}, \boldsymbol{\lambda}) \, dt \, [\mathbf{u} - \mathbf{v}]$$

$$= \hat{\mathbf{G}}_{\mathbf{u}}(\mathbf{u}, \mathbf{v}, \boldsymbol{\lambda}) \, [\mathbf{u} - \mathbf{v}],$$

where in the last step we used the Mean Value Theorem to get $\hat{\mathbf{G}}$. Then (1.1) becomes

$$\|\mathbf{F}(\mathbf{u}, \boldsymbol{\lambda}) - \mathbf{F}(\mathbf{v}, \boldsymbol{\lambda})\|$$

$$\leq M \, \|\mathbf{G}_{\mathbf{u}}^0 - \hat{\mathbf{G}}_{\mathbf{u}}(\mathbf{u}, \mathbf{v}, \boldsymbol{\lambda})\| \, \|\mathbf{u} - \mathbf{v}\|$$

$$= M \, \left\| \int_0^1 \mathbf{G}_{\mathbf{u}}(\mathbf{u}_0, \boldsymbol{\lambda}_0) - \mathbf{G}_{\mathbf{u}}(t\mathbf{u} + (1-t)\mathbf{v}, \boldsymbol{\lambda}) \, dt \right\| \, \|\mathbf{u} - \mathbf{v}\|$$

$$\leq M \int_0^1 \|\mathbf{G}_{\mathbf{u}}(\mathbf{u}_0, \boldsymbol{\lambda}_0) - \mathbf{G}_{\mathbf{u}}(t\mathbf{u} + (1-t)\mathbf{v}, \boldsymbol{\lambda})\| \, dt \, \|\mathbf{u} - \mathbf{v}\|$$

$$\leq M \int_0^1 K_L \left(\|\mathbf{u}_0 - \underbrace{(t\mathbf{u} + (1-t)\mathbf{v})}_{\in S_{\rho_1}(\mathbf{u}_0)}\| + \|\boldsymbol{\lambda}_0 - \boldsymbol{\lambda}\| \right) dt \, \|\mathbf{u} - \mathbf{v}\|$$

$$\leq \underbrace{M \, K_L \, 2\rho_1}_{\equiv K_0} \, \|\mathbf{u} - \mathbf{v}\|.$$

Therefore, if we take

$$\rho_1 < \frac{1}{2M \, K_L},$$

then $K_0 < 1$. The second condition of the Contraction Theorem is also satisfied, namely,

$$\| \mathbf{F}(\mathbf{u_0}, \boldsymbol{\lambda}) - \mathbf{u_0} \|$$
$$= \| \mathbf{F}(\mathbf{u_0}, \boldsymbol{\lambda}) - \mathbf{F}(\mathbf{u_0}, \boldsymbol{\lambda_0}) \|$$
$$= \| \left(\mathbf{G_u^0} \right)^{-1} [\mathbf{G_u^0 \, u_0} - \mathbf{G}(\mathbf{u_0}, \boldsymbol{\lambda})] - \left(\mathbf{G_u^0} \right)^{-1} [\mathbf{G_u^0 \, u_0} - \mathbf{G}(\mathbf{u_0}, \boldsymbol{\lambda_0})] \|$$
$$= \| \left(\mathbf{G_u^0} \right)^{-1} [\mathbf{G}(\mathbf{u_0}, \boldsymbol{\lambda_0}) - \mathbf{G}(\mathbf{u_0}, \boldsymbol{\lambda})] \|$$
$$\leq M \, K_L \, \| \boldsymbol{\lambda} - \boldsymbol{\lambda_0} \|$$
$$\leq M \, K_L \, \rho,$$

where ρ (with $0 < \rho \leq \rho_1$) is to be chosen. We want the above to be less than or equal to $(1 - K_0)\rho_1$, so we choose

$$\rho \leq \frac{(1 - K_0)\, \rho_1}{M \, K_L}.$$

Hence, for each $\boldsymbol{\lambda} \in S_\rho(\boldsymbol{\lambda_0})$ we have a unique solution $\mathbf{u}(\boldsymbol{\lambda})$. We now show that $\mathbf{u}(\boldsymbol{\lambda})$ is continuous in $\boldsymbol{\lambda}$. Let $\boldsymbol{\lambda_1}, \boldsymbol{\lambda_2} \in S_\rho(\boldsymbol{\lambda_0})$, with corresponding solutions $\mathbf{u}(\boldsymbol{\lambda_1})$ and $\mathbf{u}(\boldsymbol{\lambda_2})$. Then

$$\| \mathbf{u}(\boldsymbol{\lambda_1}) - \mathbf{u}(\boldsymbol{\lambda_2}) \|$$
$$= \| \mathbf{F}(\mathbf{u}(\boldsymbol{\lambda_1}), \boldsymbol{\lambda_1}) - \mathbf{F}(\mathbf{u}(\boldsymbol{\lambda_2}), \boldsymbol{\lambda_2}) \|$$
$$\leq \| \mathbf{F}(\mathbf{u}(\boldsymbol{\lambda_1}), \boldsymbol{\lambda_1}) - \mathbf{F}(\mathbf{u}(\boldsymbol{\lambda_2}), \boldsymbol{\lambda_1}) \| + \| \mathbf{F}(\mathbf{u}(\boldsymbol{\lambda_2}), \boldsymbol{\lambda_1}) - \mathbf{F}(\mathbf{u}(\boldsymbol{\lambda_1}), \boldsymbol{\lambda_2}) \|$$
$$\leq K_0 \, \| \mathbf{u}(\boldsymbol{\lambda_1}) - \mathbf{u}(\boldsymbol{\lambda_2}) \| +$$
$$\| \left(\mathbf{G_u^0} \right)^{-1} [\mathbf{G_u^0 \, u}(\boldsymbol{\lambda_2}) - \mathbf{G}(\mathbf{u}(\boldsymbol{\lambda_2}), \boldsymbol{\lambda_1})] - \left(\mathbf{G_u^0} \right)^{-1} [\mathbf{G_u^0 \, u}(\boldsymbol{\lambda_2}) - \mathbf{G}(\mathbf{u}(\boldsymbol{\lambda_2}), \boldsymbol{\lambda_2})] \|$$
$$\leq \underbrace{K_0}_{<1} \, \| \mathbf{u}(\boldsymbol{\lambda_1}) - \mathbf{u}(\boldsymbol{\lambda_2}) \| + M \, K_L \, \| \boldsymbol{\lambda_1} - \boldsymbol{\lambda_2} \| \, .$$

Hence,

$$\| \mathbf{u}(\boldsymbol{\lambda_1}) - \mathbf{u}(\boldsymbol{\lambda_2}) \| \leq \frac{M \, K_L}{1 - K_0} \, \| \boldsymbol{\lambda_1} - \boldsymbol{\lambda_2} \|,$$

which concludes the proof of the IFT. □

So far, under mild assumptions, we have shown that there exists a locally unique solution family $\mathbf{u}(\boldsymbol{\lambda})$. If we impose the condition that $\mathbf{F}(\mathbf{u}, \boldsymbol{\lambda})$ is continuously differentiable in $\boldsymbol{\lambda}$, then we can show that $\mathbf{u}(\boldsymbol{\lambda})$ is also continuously differentiable. To this end, the Banach Lemma is very useful.

Lemma 1 (Banach Lemma). *Let* $L : \mathcal{B} \rightarrow \mathcal{B}$ *be a linear operator with* $\| L \| < 1$. *Then* $(I + L)^{-1}$ *exists and*

$$\| (I + L)^{-1} \| \leq \frac{1}{1 - \| L \|}.$$

Proof. Suppose $I + L$ is not invertible. Then there exists $\mathbf{y} \in \mathcal{B}$, $\mathbf{y} \neq \mathbf{0}$, such that

$$(I + L)\, \mathbf{y} = \mathbf{0}.$$

Thus, $\mathbf{y} = -L\mathbf{y}$ and
$$\|\mathbf{y}\| = \|L\mathbf{y}\| \le \|L\| \|\mathbf{y}\| < \|\mathbf{y}\|,$$
which is a contradiction. Therefore, $(I+L)^{-1}$ exists. We can bound the inverse as follows:

$$(I+L)(I+L)^{-1} = I$$
$$\Leftrightarrow (I+L)^{-1} = I - L(I+L)^{-1}$$
$$\Leftrightarrow \|(I+L)^{-1}\| \le 1 + \|L\| \|(I+L)^{-1}\|$$
$$\Leftrightarrow \|(I+L)^{-1}\| \le \frac{1}{1-\|L\|}.$$

This proves the Banach Lemma. □

The Banach Lemma can be used to show the following.

Lemma 2. *Under the conditions of the IFT, there exists $M_1 > 0$ and $\delta > 0$ such that $\mathbf{G_u}(\mathbf{u}, \boldsymbol{\lambda})^{-1}$ exists and $\|\mathbf{G_u}(\mathbf{u}, \boldsymbol{\lambda})^{-1}\| \le M_1$ in $S_\delta(\mathbf{u}_0) \times S_\delta(\boldsymbol{\lambda}_0)$.*

Proof. Using again the notation $\mathbf{G_u^0} = \mathbf{G_u}(\mathbf{u}_0, \boldsymbol{\lambda}_0)$, we have

$$\mathbf{G_u}(\mathbf{u}, \boldsymbol{\lambda}) = \mathbf{G_u^0} + \mathbf{G_u}(\mathbf{u}, \boldsymbol{\lambda}) - \mathbf{G_u^0}$$
$$= \mathbf{G_u^0}\,[I + \underbrace{(\mathbf{G_u^0})^{-1}\,(\mathbf{G_u}(\mathbf{u}, \boldsymbol{\lambda}) - \mathbf{G_u^0})}_{\equiv L}].$$

Similar to how we verified the second condition of the Contraction Theorem in the proof of the IFT, we can show that

$$\|L\| \le M\,K_L\,(\|\mathbf{u} - \mathbf{u}_0\| + \|\boldsymbol{\lambda} - \boldsymbol{\lambda}_0\|) \le M\,K_L\,2\delta.$$

As for the IFT, we choose

$$\delta < \frac{1}{2M\,K_L},$$

and conclude that, therefore, $(I+L)^{-1}$ exists and

$$\|(I+L)^{-1}\| \le \frac{1}{1 - M\,K_L\,2\delta}.$$

Hence, $\mathbf{G_u}(\mathbf{u}, \boldsymbol{\lambda})^{-1}$ exists and

$$\|\mathbf{G_u}(\mathbf{u}, \boldsymbol{\lambda})^{-1}\| = \|(\mathbf{G_u^0})^{-1}\,(I+L)^{-1}\| \le \frac{M}{1 - M\,K_L\,2\delta} \equiv M_1,$$

as required. □

We are now ready to prove differentiability of the solution branch.

Theorem 3. *In addition to the assumptions of the IFT, assume that the derivative $\mathbf{G_\lambda}(\mathbf{u}, \boldsymbol{\lambda})$ is continuous in $S_\rho(\mathbf{u}_0) \times S_\rho(\boldsymbol{\lambda}_0)$. Then the solution branch $\mathbf{u}(\boldsymbol{\lambda})$ has a continuous derivative $\mathbf{u_\lambda}(\boldsymbol{\lambda})$ on $S_\delta(\mathbf{u}_0) \times S_\delta(\boldsymbol{\lambda}_0)$.*

Proof. Using the definition of (Fréchet) derivative, we are given that there exists $\mathbf{G_u}(\mathbf{u}, \boldsymbol{\lambda})$ such that $\mathbf{G}(\mathbf{u}, \boldsymbol{\lambda}) - \mathbf{G}(\mathbf{v}, \boldsymbol{\lambda}) = \mathbf{G_u}(\mathbf{u}, \boldsymbol{\lambda})(\mathbf{u} - \mathbf{v}) + R_1(\mathbf{u}, \mathbf{v}, \boldsymbol{\lambda})$, where $R_1(\mathbf{u}, \mathbf{v}, \boldsymbol{\lambda})$ is such that

$$\frac{\| R_1(\mathbf{u}, \mathbf{v}, \boldsymbol{\lambda}) \|}{\| \mathbf{u} - \mathbf{v} \|} \to 0 \text{ as } \| \mathbf{u} - \mathbf{v} \| \to 0. \tag{1.2}$$

Similarly, there exists $\mathbf{G}_{\boldsymbol{\lambda}}(\mathbf{u}, \boldsymbol{\lambda})$ such that $\mathbf{G}(\mathbf{u}, \boldsymbol{\lambda}) - \mathbf{G}(\mathbf{u}, \boldsymbol{\mu}) = \mathbf{G}_{\boldsymbol{\lambda}}(\mathbf{u}, \boldsymbol{\lambda})(\boldsymbol{\lambda} - \boldsymbol{\mu}) + R_2(\mathbf{u}, \boldsymbol{\lambda}, \boldsymbol{\mu})$, where $R_2(\mathbf{u}, \boldsymbol{\lambda}, \boldsymbol{\mu})$ satisfies

$$\frac{\| R_2(\mathbf{u}, \boldsymbol{\lambda}, \boldsymbol{\mu}) \|}{\| \boldsymbol{\lambda} - \boldsymbol{\mu} \|} \to 0 \text{ as } \| \boldsymbol{\lambda} - \boldsymbol{\mu} \| \to 0. \tag{1.3}$$

We must show that there exists $\mathbf{u}_{\boldsymbol{\lambda}}(\boldsymbol{\lambda})$ such that

$$\mathbf{u}(\boldsymbol{\lambda}) - \mathbf{u}(\boldsymbol{\mu}) = \mathbf{u}_{\boldsymbol{\lambda}}(\boldsymbol{\lambda})(\boldsymbol{\lambda} - \boldsymbol{\mu}) + r(\boldsymbol{\lambda}, \boldsymbol{\mu}),$$

with

$$\frac{\| r(\boldsymbol{\lambda}, \boldsymbol{\mu}) \|}{\| \boldsymbol{\lambda} - \boldsymbol{\mu} \|} \to 0 \text{ as } \| \boldsymbol{\lambda} - \boldsymbol{\mu} \| \to 0.$$

Now

$$\begin{aligned} 0 &= \mathbf{G}(\mathbf{u}(\boldsymbol{\lambda}), \boldsymbol{\lambda}) - \mathbf{G}(\mathbf{u}(\boldsymbol{\mu}), \boldsymbol{\mu}) \\ &= \mathbf{G}(\mathbf{u}(\boldsymbol{\lambda}), \boldsymbol{\lambda}) - \mathbf{G}(\mathbf{u}(\boldsymbol{\mu}), \boldsymbol{\lambda}) + \mathbf{G}(\mathbf{u}(\boldsymbol{\mu}), \boldsymbol{\lambda}) - \mathbf{G}(\mathbf{u}(\boldsymbol{\mu}), \boldsymbol{\mu}) \\ &= \mathbf{G_u}(\mathbf{u}(\boldsymbol{\lambda}), \boldsymbol{\lambda})(\mathbf{u}(\boldsymbol{\lambda}) - \mathbf{u}(\boldsymbol{\mu})) + R_1(\mathbf{u}(\boldsymbol{\lambda}), \mathbf{u}(\boldsymbol{\mu}), \boldsymbol{\lambda}) \\ &\quad + \mathbf{G}_{\boldsymbol{\lambda}}(\mathbf{u}(\boldsymbol{\mu}), \boldsymbol{\lambda})(\boldsymbol{\lambda} - \boldsymbol{\mu}) + R_2(\mathbf{u}(\boldsymbol{\mu}), \boldsymbol{\lambda}, \boldsymbol{\mu}). \end{aligned}$$

Lemma 2 guarantees the existence of $\mathbf{G_u}(\mathbf{u}(\boldsymbol{\lambda}), \boldsymbol{\lambda})^{-1}$, and we find

$$\begin{aligned} \mathbf{u}(\boldsymbol{\lambda}) - \mathbf{u}(\boldsymbol{\mu}) &= -\mathbf{G_u}(\mathbf{u}(\boldsymbol{\lambda}), \boldsymbol{\lambda})^{-1} [\mathbf{G}_{\boldsymbol{\lambda}}(\mathbf{u}(\boldsymbol{\mu}), \boldsymbol{\lambda})(\boldsymbol{\lambda} - \boldsymbol{\mu}) - (R_1 + R_2)] \\ &= -\mathbf{G_u}(\mathbf{u}(\boldsymbol{\lambda}), \boldsymbol{\lambda})^{-1} [\mathbf{G}_{\boldsymbol{\lambda}}(\mathbf{u}(\boldsymbol{\lambda}), \boldsymbol{\lambda})(\boldsymbol{\lambda} - \boldsymbol{\mu}) - r], \end{aligned}$$

where

$$r = [\mathbf{G}_{\boldsymbol{\lambda}}(\mathbf{u}(\boldsymbol{\lambda}), \boldsymbol{\lambda}) - \mathbf{G}_{\boldsymbol{\lambda}}(\mathbf{u}(\boldsymbol{\mu}), \boldsymbol{\lambda})](\boldsymbol{\lambda} - \boldsymbol{\mu}) + R_1 + R_2.$$

Let us, for the moment, ignore the harmless factor $\mathbf{G_u}(\mathbf{u}(\boldsymbol{\lambda}), \boldsymbol{\lambda})^{-1}$ and consider each term of r. Since $\mathbf{u}$ and $\mathbf{G}_{\boldsymbol{\lambda}}$ are continuous, we have

$$\frac{\| [\mathbf{G}_{\boldsymbol{\lambda}}(\mathbf{u}(\boldsymbol{\lambda}), \boldsymbol{\lambda}) - \mathbf{G}_{\boldsymbol{\lambda}}(\mathbf{u}(\boldsymbol{\mu}), \boldsymbol{\lambda})](\boldsymbol{\lambda} - \boldsymbol{\mu}) \|}{\| \boldsymbol{\lambda} - \boldsymbol{\mu} \|} \to 0 \text{ as } \| \boldsymbol{\lambda} - \boldsymbol{\mu} \| \to 0.$$

Also, the existence of $\mathbf{G}_{\boldsymbol{\lambda}}$ implies (1.3)

$$\frac{\| R_2(\mathbf{u}(\boldsymbol{\lambda}), \boldsymbol{\lambda}, \boldsymbol{\mu}) \|}{\| \boldsymbol{\lambda} - \boldsymbol{\mu} \|} \to 0 \text{ as } \| \boldsymbol{\lambda} - \boldsymbol{\mu} \| \to 0.$$

Using (1.2), we have

$$\frac{\|R_1(\mathbf{u}(\lambda), \mathbf{u}(\mu), \lambda)\|}{\|\lambda - \mu\|} = \frac{\|R_1(\mathbf{u}(\lambda), \mathbf{u}(\mu), \lambda)\|}{\|\mathbf{u}(\lambda) - \mathbf{u}(\mu)\|} \frac{\|\mathbf{u}(\lambda) - \mathbf{u}(\mu)\|}{\|\lambda - \mu\|} \to 0,$$

as $\|\lambda - \mu\| \to 0$ because the second factor is bounded due to continuity of $\mathbf{u}(\lambda)$ (see the end of the proof of the IFT). Thus,

$$\mathbf{u}_\lambda(\lambda) = -\mathbf{G_u}(\mathbf{u}(\lambda), \lambda)^{-1} \mathbf{G}_\lambda(\mathbf{u}(\lambda), \lambda).$$

To prove that $\mathbf{u}_\lambda$ is continuous it suffices to show that $\mathbf{G_u}(\mathbf{u}(\lambda), \lambda)^{-1}$ is continuous. Indeed,

$$
\begin{aligned}
&\|\mathbf{G_u}(\mathbf{u}(\lambda), \lambda)^{-1} - \mathbf{G_u}(\mathbf{u}(\mu), \mu)^{-1}\| \\
&= \|\mathbf{G_u}(\mathbf{u}(\lambda), \lambda)^{-1} \left[\mathbf{G_u}(\mathbf{u}(\mu), \mu) - \mathbf{G_u}(\mathbf{u}(\lambda), \lambda)\right] \mathbf{G_u}(\mathbf{u}(\mu), \mu)^{-1}\| \\
&\leq M_1^2 K_L(\|\mathbf{u}(\mu) - \mathbf{u}(\lambda)\| + \|\mu - \lambda\|),
\end{aligned}
$$

which concludes the proof of Theorem 3 □

Remark 1. In fact, if $\mathbf{G}_\lambda$ is Lipschitz continuous then $\mathbf{u}_\lambda$ is Lipschitz continuous (we already assume that $\mathbf{G_u}$ is Lipschitz continuous). More generally, it can be shown that $\mathbf{u}_\lambda$ is C^k if $\mathbf{G}$ is C^k, that is, $\mathbf{u}$ inherits the degree of continuity of $\mathbf{G}$.

We now give some examples where the IFT is used to show that a given solution persists, at least locally, when a problem parameter is changed. We also identify some cases where the conditions of the IFT are not satisfied.

1.1.2 A Predator-Prey Model

Our first example is that of a predator-prey model defined as

$$\begin{cases} u_1' = 3u_1(1 - u_1) - u_1 u_2 - \lambda(1 - e^{-5u_1}), \\ u_2' = -u_2 + 3u_1 u_2. \end{cases} \tag{1.4}$$

We can think of u_1 as 'fish' and u_2 as 'sharks', while the term $\lambda(1 - e^{-5u_1})$ represents 'fishing', with 'fishing-quota' λ. When $\lambda = 0$ the *stationary solutions* are

$$\left.\begin{array}{r} 3u_1(1 - u_1) - u_1 u_2 = 0 \\ -u_2 + 3u_1 u_2 = 0 \end{array}\right\} \Rightarrow (u_1, u_2) = (0, 0),\ (1, 0),\ (\tfrac{1}{3}, 2).$$

The Jacobian matrix is

$$J = \begin{pmatrix} 3 - 6u_1 - u_2 - 5\lambda e^{-5u_1} & -u_1 \\ u_2 & -1 + 3u_1 \end{pmatrix} = J(u_1, u_2; \lambda).$$

Hence, we have

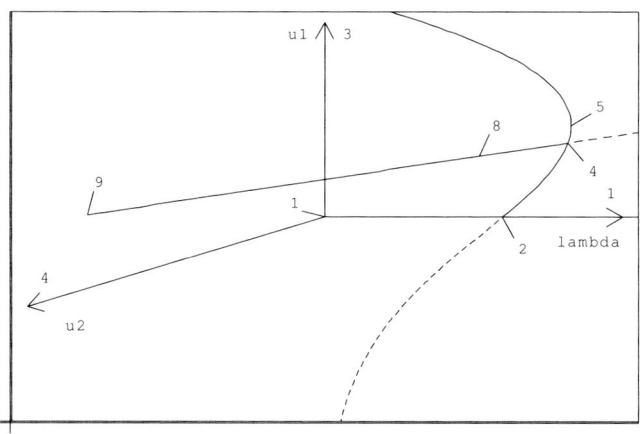

Fig. 1.1. Stationary solution branches of the predator-prey model (1.4). Solution 2 and solution 4 are branch points, while solution 8 is a Hopf bifurcation point.

$$J(0,0;0) = \begin{pmatrix} 3 & 0 \\ 0 & -1 \end{pmatrix}, \quad \text{eigenvalues } 3, -1 \text{ (unstable)};$$

$$J(1,0;0) = \begin{pmatrix} 3 & -1 \\ 0 & 2 \end{pmatrix}, \quad \text{eigenvalues } -3, 2 \text{ (unstable)};$$

$$J(\tfrac{1}{3},2;0) = \begin{pmatrix} -1 & -\tfrac{1}{3} \\ 6 & 0 \end{pmatrix}, \quad \text{eigenvalues} \begin{cases} (-1-\mu)(-\mu)+2 = 0 \Leftrightarrow \\ \mu^2 + \mu + 2 = 0 \Leftrightarrow \\ \mu_\pm = \frac{-1\pm\sqrt{-7}}{2}; \\ \text{Re}(\mu_\pm) < 0 \text{ (stable)}. \end{cases}$$

All three Jacobians at $\lambda = 0$ are nonsingular. Thus, by the IFT, all three stationary points persist for (small) $\lambda \neq 0$. In this problem we can *explicitly* find all solutions (see Fig. 1.1):

I: $(u_1, u_2) = (0,0)$.

II: $u_2 = 0$ and $\lambda = \dfrac{3u_1(1-u_1)}{1-e^{-5u_1}}$. (Note that $\lim\limits_{u_1 \to 0} \lambda = \lim\limits_{u_1 \to 0} \dfrac{3(1-2u_1)}{5e^{-5u_1}} = \dfrac{3}{5}$.)

III: $u_1 = \tfrac{1}{3}$ and $\tfrac{2}{3} - \tfrac{1}{3}u_2 - \lambda(1-e^{-5/3}) = 0 \Rightarrow u_2 = 2 - 3\lambda(1-e^{-5/3})$.

These solution families intersect at two *branch points*, one of which is $(u_1, u_2, \lambda) = (0,0,3/5)$.

The stability of Branch I follows from:

$$J(0,0;\lambda) = \begin{pmatrix} 3-5\lambda & 0 \\ 0 & -1 \end{pmatrix}, \quad \text{eigenvalues } 3-5\lambda, -1.$$

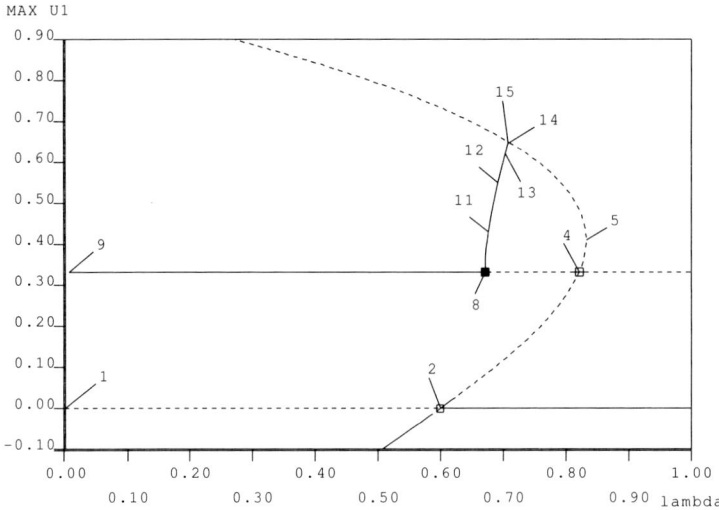

Fig. 1.2. Bifurcation diagram of the predator-prey model (1.4). The periodic solution branch is also shown. For stationary solutions the vertical axis is simply u_1, while for periodic solutions $\max(u_1)$ is plotted. Solid/dashed lines denote stable/unstable solutions. Open squares are branch points; the solid square is a Hopf bifurcation.

Hence, the trivial solution is unstable if $\lambda < 3/5$, and stable if $\lambda > 3/5$, as indicated in Fig. 1.2. Branch II has no stable positive solutions. At $\lambda_H \approx 0.67$ on Branch III (Solution 8 in Fig. 1.2) the complex eigenvalues cross the imaginary axis. This crossing is a *Hopf bifurcation*. Beyond λ_H there are *periodic solutions* whose period T increases as λ increases; see Fig. 1.3 for some representative periodic orbits. The period becomes infinite at $\lambda = \lambda_\infty \approx 0.7$. This final orbit is called a *heteroclinic cycle*.

From Fig. 1.2 we can deduce the solution behavior for increasing λ: Branch III is followed until λ_H; then the behavior becomes oscillatory due to the periodic solutions of increasing period until $\lambda = \lambda_\infty$; finally, the dynamics collapses to the trivial solution (Branch I).

1.1.3 The Gelfand-Bratu Problem

The IFT is not only useful in the context of solution branches of equilibria. The periodic orbits in Sect. 1.1.2 are also computed using the IFT principle. This section gives an example of a solution branch of a two-point boundary value problem. The Gelfand-Bratu problem [12] is defined as

$$\begin{cases} u''(x) + \lambda e^{u(x)} = 0, & \forall x \in [0,1], \\ u(0) = u(1) = 0. \end{cases} \tag{1.5}$$

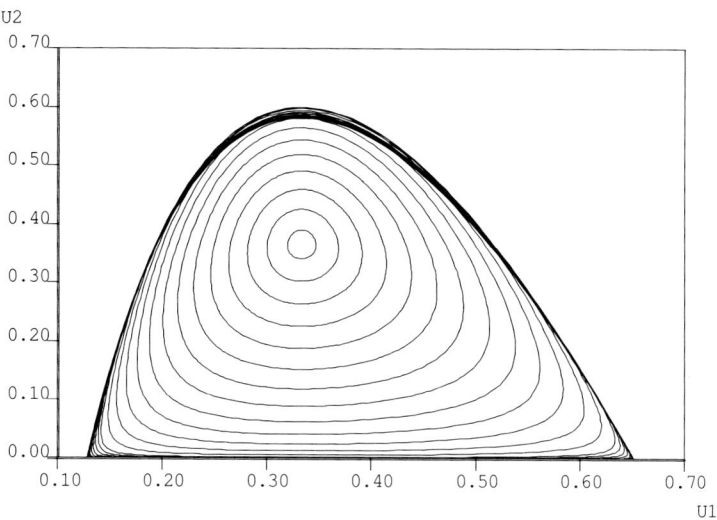

Fig. 1.3. Some periodic solutions of the predator-prey model (1.4). The final orbits are very close to a heteroclinic cycle.

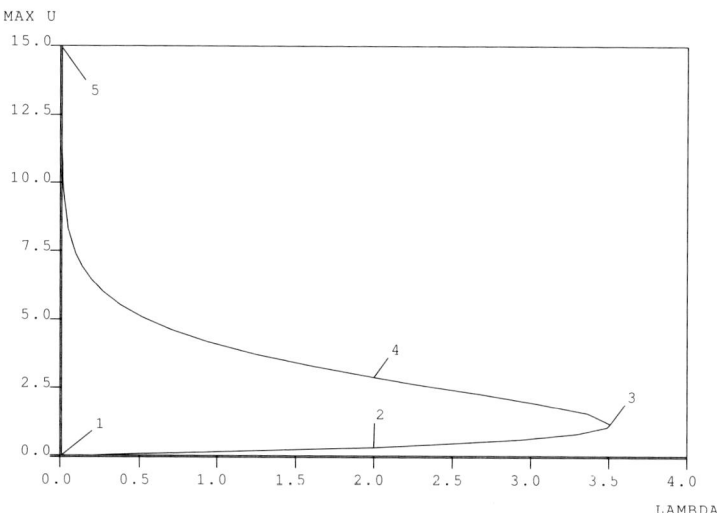

Fig. 1.4. Bifurcation diagram of the Gelfand-Bratu equation (1.5). Note that there are two solutions for $0 < \lambda < \lambda_C$, where $\lambda_C \approx 3.51$. There is one solution for $\lambda = \lambda_C$ and for $\lambda \leq 0$, and are no solutions for $\lambda > \lambda_C$.

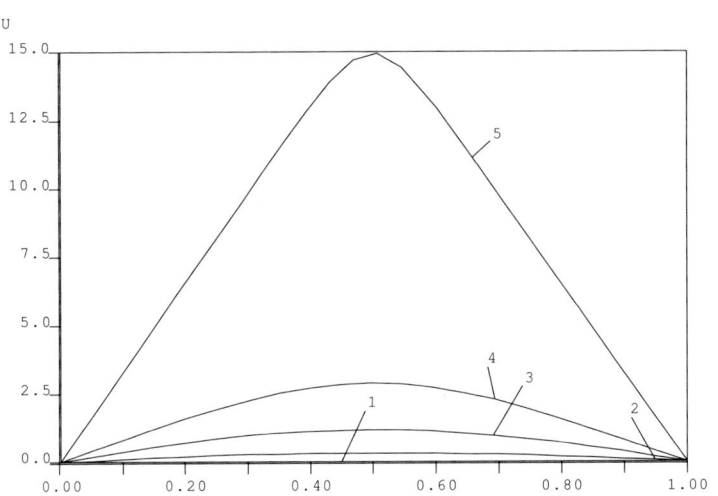

Fig. 1.5. Some solutions to the Gelfand-Bratu equation (1.5).

If $\lambda = 0$ then $u(x) \equiv 0$ is a solution. We show here that this solution is isolated, so that there is a continuation $u = \tilde{u}(\lambda)$, for $|\lambda|$ small. Consider

$$\left. \begin{array}{ll} u''(x) - \lambda e^{u(x)} = 0, \\ u(0) = 0, \qquad u'(0) = q, \end{array} \right\} \Rightarrow u = u(x; q, \lambda).$$

We want to solve $\underbrace{u(1; q, \lambda)}_{\equiv \mathbf{F}(q,\lambda)} = 0$, for $|\lambda|$ small. Here $\mathbf{F}(0,0) = 0$.

We must show (IFT) that $\mathbf{F}_q(0,0) \equiv u_q(1; 0, 0) \neq 0$:

$$\left. \begin{array}{ll} u_q''(x) - \lambda_0 e^{u_0(x)} u_q = 0, \\ u_q(0) = 0, \qquad u_q'(0) = 1, \end{array} \right\} \text{ where } u_0 \equiv 0.$$

Now $u_q(x; 0, 0)$ satisfies

$$\left\{ \begin{array}{l} u_q'' = 0, \\ u_q(0) = 0, \ u_q'(0) = 1. \end{array} \right.$$

Hence, $u_q(x; 0, 0) = x$, so that $u_q(1; 0, 0) = 1 \neq 0$.

1.1.4 A Nonlinear Eigenvalue Problem

The equations for column buckling (from nonlinear elasticity theory) [31] are given by

$$\begin{cases} u_1''(x) + \mu u_1(x) & = 0, \\ u_3'(x) + \frac{1}{2}u_1'(x)^2 + \mu\beta = 0, \end{cases}$$

for $x \in [0,1]$, with

$$\begin{cases} u_1(0) = & u_1(1) = 0, \\ u_3(0) = -u_3(1) = \lambda \quad (\lambda > 0), \end{cases}$$

where μ is a stress (to be determined) and β is another physical constant; take $\beta = 1$. Note that the boundary conditions are 'overspecified' (to determine μ). We rewrite the equations as the first order system

$$\begin{cases} u_1' = & u_2, & u_1(0) = & u_1(1) = 0, \\ u_2' = -\mu u_1, \\ u_3' = & -\frac{1}{2}u_2^2 - \mu, \ u_3(0) = -u_3(1) = \lambda. \end{cases} \tag{1.6}$$

Note that $u_1 \equiv u_2 \equiv 0$ implies $u_3' = -\mu$, so that $u_3(x) = \lambda - \mu x$, with $u_3(0) = \lambda$ and $u_3(1) = \lambda - \mu = -\lambda$. Thus, we must have $\mu = 2\lambda$, so that $u_3(x) = \lambda - 2\lambda x = \lambda(1 - 2x)$. Hence,

$$u_1 \equiv u_2 \equiv 0, \quad u_3(x) = \lambda(1 - 2x), \quad \mu = 2\lambda,$$

is a solution for all λ. Are these solutions isolated? In the formal set-up, consider

$$\begin{cases} u_1' = & u_2, & u_1(0) = 0, \\ u_2' = -\mu u_1, & u_2(0) = p, \\ u_3' = & -\frac{1}{2}u_2^2 - \mu, \ u_3(0) = \lambda. \end{cases} \Rightarrow \mathbf{u} = \mathbf{u}(x, p, \mu; \lambda).$$

We must have

$$\begin{rcases} u_1(1, p, \mu; \lambda) & = 0, \\ u_3(1, p, \mu; \lambda) + \lambda = 0, \end{rcases} \sim \mathbf{F}(p, \mu; \lambda) = 0,$$

with $\mathbf{F} : \mathbb{R}^2 \times \mathbb{R} \to \mathbb{R}^2$. So the question is: Is $(\mathbf{F}_p \mid \mathbf{F}_\mu)(\lambda)$ nonsingular along the basic solution branch?

To answer the above question quickly, we omit explicit construction of $\mathbf{F}$. We linearize (1.6) about $u_1, u_2, u_3, \mu,$ and λ, with respect to $u_1, u_2, u_3,$ and μ, acting on $v_1, v_2, v_3,$ and μ, to obtain the linearized homogeneous equations

$$\begin{cases} v_1' = & v_2, & v_1(0) = & v_1(1) = 0, \\ v_2' = & -\mu v_1 - \mu u_1, \\ v_3' = -u_2 v_2 - \mu, & v_3(0) = -v_3(1) = 0. \end{cases}$$

In particular, the linearized homogeneous equations about $u_1 \equiv u_2 \equiv 0$, $u_3(x) = \lambda(1 - 2x)$, and $\mu = 2\lambda$ are

$$\begin{rcases} v_1' = & v_2, & v_1(0) = & v_1(1) = 0, \\ v_2' = -2\lambda v_1, \\ v_3' = & -\mu, & v_3(0) = -v_3(1) = 0. \end{rcases} \Rightarrow \mu = 0, \ v_3 \equiv 0.$$

Now, if $2\lambda \neq k^2 \pi^2$, $k = 1, 2, 3, \ldots$, then

$$\begin{cases} v_1'' + 2\lambda v_1 = 0, \\ v_1(0) = v_1(1) = 0, \end{cases}$$

has the unique solution $v_1 \equiv 0$ and, hence, also $v_2 \equiv 0$. Thus, if $\lambda \neq \frac{1}{2}k^2\pi^2$ then the basic solution branch is locally unique. However, if $\lambda = \frac{1}{2}k^2\pi^2$ then the linearization is singular, and there may be bifurcations. (In fact, there are *buckled states*.)

1.1.5 The Pendulum Equation

The equation of a damped pendulum subject to a constant torque is given by

$$m R \phi''(t) + \underbrace{\varepsilon\, m\, \phi'(t)}_{\text{damping}} + m g \sin \phi(t) = \underbrace{I}_{\text{torque}} ,$$

that is,

$$\phi''(t) + \frac{\varepsilon}{R} \phi'(t) + \frac{g}{R} \sin \phi(t) = \frac{I}{m R}.$$

Scaling time as $s = c t$ we have $\phi' = \frac{d\phi}{dt} = \frac{d\phi}{ds}\frac{ds}{dt} = c\,\dot\phi$ and, similarly, $\phi'' = c^2\,\ddot\phi$, we obtain

$$c^2\, \ddot\phi(s) + \frac{\varepsilon c}{R} \dot\phi(s) + \frac{g}{R} \sin \phi(s) = \frac{I}{m R},$$

$$\ddot\phi + \frac{\varepsilon}{R c} \dot\phi + \frac{g}{R c^2} \sin \phi = \frac{I}{m R c^2}.$$

Choose c such that $\frac{g}{R c^2} = 1$, that is, $c = \sqrt{g/R}$, and set $\tilde\varepsilon = \varepsilon/(R c)$ and $\tilde I = I/(m R c^2)$. Then the equation becomes $\ddot\phi + \tilde\varepsilon\,\dot\phi + \sin\phi = \tilde I$, or, dropping the $\tilde{}$, and using $'$,

$$\phi'' + \varepsilon \phi' + \sin \phi = I. \tag{1.7}$$

We shall consider special solutions, called *rotations*, that satisfy $\phi(t + T) = \phi(t) + 2\pi$, for all t or, equivalently,

$$\phi(T) = \overbrace{\phi(0)}^{\equiv 0} + 2\pi \quad (= 2\pi),$$
$$\phi'(T) = \phi'(0),$$

where T is the *period*.

The Undamped Pendulum

First consider the undamped unforced pendulum

$$\phi'' + \sin\phi = 0,$$

that is, (1.7) with $\varepsilon = I = 0$. Suppose the initial data for a rotation are $\phi(0) = 0$, and $\phi'(0) = p > 0$. We have $\phi(T) = 2\pi$, and $\phi'(T) = \phi'(0) = p$. Integration gives

$$\int_0^t \phi'\,\phi''\,dt + \int_0^t \phi'\sin\phi\,dt = 0$$

$$\Leftrightarrow \qquad \tfrac{1}{2}\phi'^2\big|_0^t - \cos\phi\big|_0^t = 0$$

$$\Leftrightarrow \qquad \tfrac{1}{2}\phi'(t)^2 - \cos\phi(t) = \tfrac{1}{2}p^2 - 1$$

$$\Leftrightarrow \qquad \underbrace{\frac{1}{2}\phi'(t)^2}_{\text{kinetic energy}} + \underbrace{1 - \cos\phi(t)}_{\text{potential energy}} = \frac{1}{2}p^2.$$

Thus,

$$\phi'(t) = \tfrac{d\phi}{dt} = \sqrt{p^2 - 2 + 2\cos\phi(t)}$$

$$\Leftrightarrow \qquad \frac{dt}{d\phi} = \frac{1}{\sqrt{p^2 - 2 + 2\cos\phi}}$$

$$\Leftrightarrow \qquad \int_0^{2\pi} \tfrac{dt}{d\phi}\,d\phi = \int_0^{2\pi} \frac{1}{\sqrt{p^2 - 2 + 2\cos\phi}}\,d\phi$$

$$\Leftrightarrow \qquad T = \int_0^{2\pi} \frac{1}{\sqrt{p^2 - 2 + 2\cos\phi}}\,d\phi.$$

We see that

$$T \to 0 \quad \text{as} \quad p \to \infty,$$

and

$$T \to \int_0^{2\pi} \frac{1}{\sqrt{2 + 2\cos\phi}}\,d\phi = \infty \quad \text{as} \quad p \to 2.$$

In fact, rotations exists for all $p > 2$.

The Forced Damped Pendulum

We now consider the forced damped pendulum (1.7),

$$\phi'' + \varepsilon\,\phi' + \sin\phi = I,$$

with $\phi(0) = 0$ (which sets the phase) and $\phi'(0) = p$. We write the solution as $\phi = \phi(t; p, I, \varepsilon)$. Do there exist rotations, i.e., does there exist T such that $\phi(T; p, I, \varepsilon) = 2\pi$ and $\phi'(T; p, I, \varepsilon) = p$?

Theorem 4. *Let ϕ_0 be a rotation of the undamped unforced pendulum:*

$$\phi_0'' + \sin \phi_0 = 0,$$

$$\phi_0(0) = 0, \quad \phi_0'(0) = p_0,$$
$$\phi_0(T_0) = 2\pi, \; \phi_0'(T_0) = p_0.$$

Then there exist (smooth) functions $T = T(p, \varepsilon)$ and $I = I(p, \varepsilon)$, with $T(p_0, 0) = T_0$ and $I(p_0, 0) = 0$, such that $\phi(t; p, I(p, \varepsilon), \varepsilon)$ is a rotation of period $T(p, \varepsilon)$ of the damped forced pendulum

$$\phi_0'' + \varepsilon\, \phi' + \sin \phi_0 = I,$$

$$\phi(0; p, I(p, \varepsilon), \varepsilon) = 0, \qquad \phi'(0; p, I(p, \varepsilon), \varepsilon) = p,$$
$$\phi(T(p, \varepsilon); p, I(p, \varepsilon), \varepsilon) = 2\pi, \; \phi'(T(p, \varepsilon); p, I(p, \varepsilon), \varepsilon) = p.$$

for all (p, ε) sufficiently close to $(p_0, 0)$.

Proof. The Jacobian matrix with respect to T and I, of the algebraic system

$$\begin{cases} \phi(T; p, I, \varepsilon)\; -\; 2\pi = 0, \\ \phi'(T; p, I, \varepsilon)\; -\; \quad p = 0, \end{cases}$$

evaluated at $p = p_0$, $T = T_0$, and $I = \varepsilon = 0$, is

$$J_0 = \begin{pmatrix} \phi_0' & \phi_I^0 \\ \phi_0'' & \phi_I^{0'} \end{pmatrix}(T_0).$$

We must show that $\det J_0 \neq 0$. We have

$$\phi_0'' + \sin \phi_0 = 0 \Rightarrow \phi_0''(T_0) = -\sin(\phi_0(T_0)) = -\sin(2\pi) = 0,$$
$$\phi_0'(T_0) = p_0 \neq 0.$$

Thus, $\det J_0 \neq 0$ if $\phi_I^{0'}(T_0) \neq 0$. Here, ϕ_I satisfies

$$\phi_I'' + \varepsilon\, \phi_I' + \phi_I \cos \phi = 1, \qquad \phi_I(0) = \phi_I'(0) = 0.$$

In particular,

$$\phi_I^{0''} + \phi_I^0 \cos \phi_0 = 1, \qquad \phi_I^0(0) = \phi_I^{0'}(0) = 0.$$

From

$$\phi_I^{0''} \phi_0' + \phi_I^0 \cos \phi_0\, \phi_0' = \phi_0',$$

and

$$\phi_0'' + \sin \phi_0 = 0 \Rightarrow \phi_0''' + \cos \phi_0\, \phi_0' = 0,$$

we have

$$\phi_I^{0''} \phi_0' - \phi_I^0 \phi_0''' = \phi_0'.$$

Using integration, we find

$$\int_0^{T_0} \phi_I^{0''} \phi_0' - \int_0^{T_0} \phi_I^0 \phi_0''' = \int_0^{T_0} \phi_0' = 2\pi,$$

$$\phi_I^{0'} \phi_0' \Big|_0^{T_0} - \int_0^{T_0} \phi_I^{0'} \phi_0'' - \phi_I^0 \underbrace{\phi_0''}_{-\sin\phi_0}\Big|_0^{T_0} + \int_0^{T_0} \phi_I^{0'} \phi_0'' = 2\pi,$$

$$\phi_I^{0'}(T_0) \underbrace{\phi_0'(T_0)}_{p_0} - \underbrace{\phi_I^{0'}(0)\ \phi_0'(0)}_{0} = 2\pi.$$

Hence,

$$\phi_I^{0'}(T_0) = \frac{2\pi}{p_0} \neq 0.$$

□

A more general analysis of this type for *coupled* pendula can be found in [1] (see also Chap. 5).

1.2 Continuation of Solutions

As mentioned, the IFT plays an important role in the design of algorithms for computing families of solutions to nonlinear equations. Such continuation methods are applied in a parameter-dependent setting. Hence, we consider the equation

$$\mathbf{G}(\mathbf{u}, \lambda) = \mathbf{0}, \qquad \mathbf{u},\ \mathbf{G}(\cdot, \cdot) \in \mathbb{R}^n, \qquad \lambda \in \mathbb{R}.$$

Let $\mathbf{x} \equiv (\mathbf{u}, \lambda)$. Then the equation can be written as

$$\mathbf{G}(\mathbf{x}) = \mathbf{0}, \qquad \mathbf{G} : \mathbb{R}^{n+1} \to \mathbb{R}^n.$$

1.2.1 Regular Solutions

A solution $\mathbf{x}_0$ of $\mathbf{G}(\mathbf{x}) = \mathbf{0}$ is *regular* [22] if the n (rows) by $n + 1$ (columns) matrix $\mathbf{G}_\mathbf{x}^0 \equiv \mathbf{G}_\mathbf{x}(\mathbf{x}_0)$ has maximal rank, i.e., if $\mathrm{Rank}(\mathbf{G}_\mathbf{x}^0) = n$.

In the parameter formulation $\mathbf{G}(\mathbf{u}, \lambda) = \mathbf{0}$, we have

$$\mathrm{Rank}(\mathbf{G}_\mathbf{x}^0) = \mathrm{Rank}(\mathbf{G}_\mathbf{u}^0 \mid \mathbf{G}_\lambda^0) = n \Leftrightarrow \begin{cases} \text{(i) } \mathbf{G}_\mathbf{u}^0 \text{ is nonsingular,} \\ \text{or} \\ \text{(ii) } \begin{cases} \dim \mathcal{N}(\mathbf{G}_\mathbf{u}^0) = 1, \\ \text{and} \\ \mathbf{G}_\lambda^0 \notin \mathcal{R}(\mathbf{G}_\mathbf{u}^0). \end{cases} \end{cases}$$

Here, $\mathcal{N}(\mathbf{G}_\mathbf{u}^0)$ denotes the *null space* of $\mathbf{G}_\mathbf{u}^0$, and $\mathcal{R}(\mathbf{G}_\mathbf{u}^0)$ denotes the *range* of $\mathbf{G}_\mathbf{u}^0$, i.e., the linear space spanned by the n columns of $\mathbf{G}_\mathbf{u}^0$.

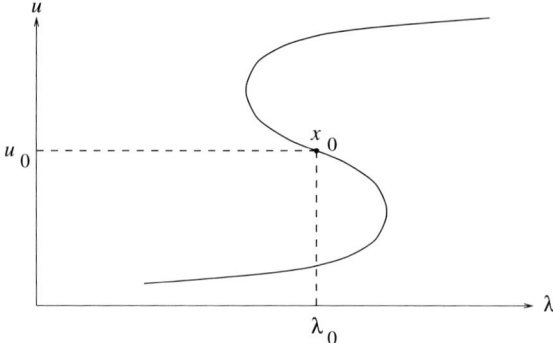

Fig. 1.6. A solution branch of $\mathbf{G}(\mathbf{u}, \lambda) = \mathbf{0}$; note the two folds.

Theorem 5. *Let* $\mathbf{x}_0 \equiv (\mathbf{u}_0, \lambda_0)$ *be a regular solution of* $\mathbf{G}(\mathbf{x}) = \mathbf{0}$. *Then, near* $\mathbf{x}_0$, *there exists a unique one-dimensional continuum of solutions* $\mathbf{x}(s)$, *called a solution family or a solution branch, with* $\mathbf{x}(0) = \mathbf{x}_0$.

Proof. Since $\text{Rank}(\mathbf{G}_\mathbf{x}^0) = \text{Rank}(\mathbf{G}_\mathbf{u}^0 \mid \mathbf{G}_\lambda^0) = n$, either $\mathbf{G}_\mathbf{u}^0$ is nonsingular and by the IFT we have $\mathbf{u} = \mathbf{u}(\lambda)$ near $\mathbf{x}_0$, or else we can interchange columns in the Jacobian $\mathbf{G}_\mathbf{x}^0$ to see that the solution can locally be parametrized by one of the components of $\mathbf{u}$. Thus, a unique solution family passes through a regular solution. $\square$

Remark 2. We remark here that the second case in the above proof is that of a *simple fold*; see also Fig. 1.6.

1.2.2 Parameter Continuation

In the parameter-dependent setting we assume that the continuation parameter is λ. Suppose we have a solution $(\mathbf{u}_0, \lambda_0)$ of

$$\mathbf{G}(\mathbf{u}, \lambda) = \mathbf{0},$$

as well as the direction vector $\dot{\mathbf{u}}_0 = d\mathbf{u}/d\lambda$, and we want to compute the solution $\mathbf{u}_1$ at $\lambda_1 = \lambda_0 + \Delta\lambda$; this is illustrated in Fig. 1.7.

To compute the solution $\mathbf{u}_1$ we use Newton's method

$$\begin{cases} \mathbf{G}_\mathbf{u}(\mathbf{u}_1^{(\nu)}, \lambda_1)\, \Delta\mathbf{u}_1^{(\nu)} = -\mathbf{G}(\mathbf{u}_1^{(\nu)}, \lambda_1), \\ \qquad\qquad \mathbf{u}_1^{(\nu+1)} = \mathbf{u}_1^{(\nu)} + \Delta\mathbf{u}_1^{(\nu)}, \qquad \nu = 0, 1, 2, \ldots. \end{cases} \tag{1.8}$$

As initial approximation, we use

$$\mathbf{u}_1^{(0)} = \mathbf{u}_0 + \Delta\lambda\, \dot{\mathbf{u}}_0.$$

If $\mathbf{G}_\mathbf{u}(\mathbf{u}_1, \lambda_1)$ is nonsingular and $\Delta\lambda$ is sufficiently small, then the convergence theory for Newton's method guarantees that this iteration will converge.

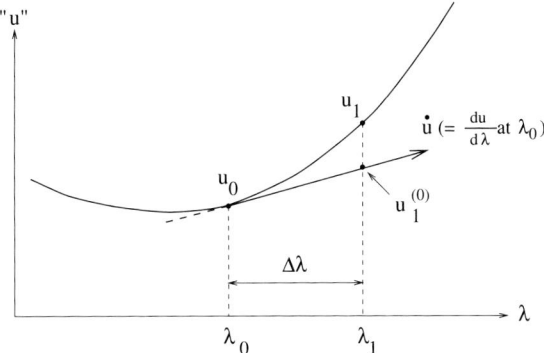

Fig. 1.7. Graphical interpretation of parameter continuation.

After convergence, the new direction vector $\dot{\mathbf{u}}_1$ can be computed by solving

$$\mathbf{G}_{\mathbf{u}}(\mathbf{u}_1, \lambda_1)\, \dot{\mathbf{u}}_1 = -\mathbf{G}_\lambda(\mathbf{u}_1, \lambda_1).$$

This equation follows from differentiating $\mathbf{G}(\mathbf{u}(\lambda), \lambda) = \mathbf{0}$ with respect to λ at $\lambda = \lambda_1$. Note that, in practice, the calculation of $\dot{\mathbf{u}}_1$ can be done without another LU-factorization of $\mathbf{G}_{\mathbf{u}}(\mathbf{u}_1, \lambda_1)$. Thus, the extra work to find $\dot{\mathbf{u}}_1$ is negligible.

As an example, consider again the Gelfand-Bratu problem of Sect. 1.1.3 given by

$$\begin{cases} u''(x) + \lambda e^{u(x)} = 0, & \forall x \in [0, 1], \\ \quad u(0) = u(1) = 0. \end{cases}$$

If $\lambda = 0$ then $u(x) \equiv 0$ is an isolated solution; see Sect. 1.1.3. We discretize this problem by introducing a mesh,

$$0 = x_0 < x_1 < \cdots < x_N = 1,$$
$$x_j - x_{j-1} = h, \quad 1 \leq j \leq N, \quad h = 1/N.$$

The discrete equations are:

$$\frac{u_{j+1} - 2u_j + u_{j-1}}{h^2} + \lambda e^{u_j} = 0, \qquad j = 1, \ldots, N - 1,$$

with $u_0 = u_N = 0$. (More accurate discretization is discussed in Sect. 1.3.1.) Let

$$\mathbf{u} \equiv \begin{pmatrix} u_1 \\ u_2 \\ \cdot \\ u_{N-1} \end{pmatrix}.$$

Then we can write the above as $\mathbf{G}(\mathbf{u}, \lambda) = \mathbf{0}$, where $\mathbf{G} : \mathbb{R}^n \times \mathbb{R} \to \mathbb{R}^n$, with $n = N - 1$.

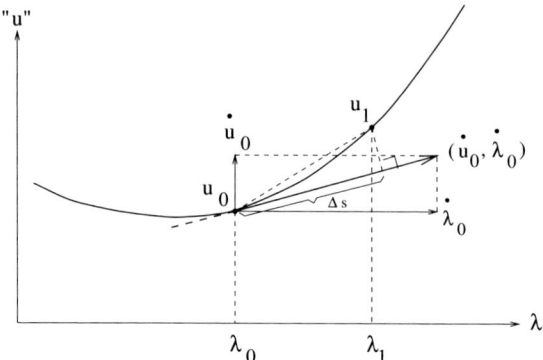

Fig. 1.8. Graphical interpretation of pseudo-arclength continuation.

For the parameter continuation we suppose that we know λ_0, $\mathbf{u}_0$, and $\dot{\mathbf{u}}_0$. Then we set $\lambda_1 = \lambda_0 + \Delta\lambda$ and apply Newton's method (1.8) with $\mathbf{u}_1^{(0)} = \mathbf{u}_0 + \Delta\lambda\,\dot{\mathbf{u}}_0$. After convergence find $\dot{\mathbf{u}}_1$ from

$$\mathbf{G_u}(\mathbf{u}_1, \lambda_1)\,\dot{\mathbf{u}}_1 = -\mathbf{G}_\lambda(\mathbf{u}_1, \lambda_1),$$

and repeat the above procedure to find $\mathbf{u}_2$, $\mathbf{u}_3$, and so on. Here,

$$\mathbf{G_u}(\mathbf{u}, \lambda) = \begin{pmatrix} -\frac{2}{h^2} + \lambda e^{u_1} & \frac{1}{h^2} & & & \\ \frac{1}{h^2} & -\frac{2}{h^2} + \lambda e^{u_2} & \frac{1}{h^2} & & \\ & . & . & . & \\ & & . & . & . \\ & & & \frac{1}{h^2} & -\frac{2}{h^2} + \lambda e^{u_{N-1}} \end{pmatrix}.$$

Hence, we must solve a tridiagonal system for each Newton iteration. The solution branch has a fold where the parameter-continuation method fails; see Figs. 1.4 and 1.5.

1.2.3 Keller's Pseudo-Arclength Continuation

In order to allow for continuation of a solution branch past a fold, AUTO [8, 11, 12] uses Keller's Pseudo-Arclength Continuation [22]. Suppose we have a solution $(\mathbf{u}_0, \lambda_0)$ of $\mathbf{G}(\mathbf{u}, \lambda) = \mathbf{0}$, as well as the direction vector $(\dot{\mathbf{u}}_0, \dot{\lambda}_0)$ of the solution branch. Pseudo-arclength continuation solves the following equations for $(\mathbf{u}_1, \lambda_1)$:

$$\begin{cases} \mathbf{G}(\mathbf{u}_1, \lambda_1) = \mathbf{0}, \\ (\mathbf{u}_1 - \mathbf{u}_0)^* \dot{\mathbf{u}}_0 + (\lambda_1 - \lambda_0)\,\dot{\lambda}_0 - \Delta s = 0. \end{cases} \tag{1.9}$$

Figure 1.8 shows a graphical interpretation of this continuation method. Newton's method for pseudo-arclength continuation becomes

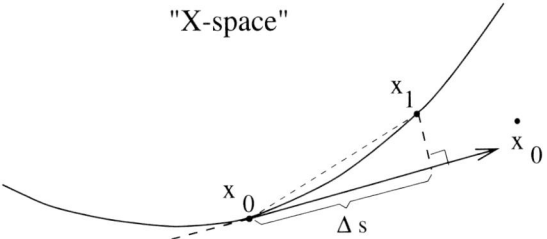

Fig. 1.9. Parameter-independent pseudo-arclength continuation.

$$\begin{pmatrix} (\mathbf{G_u^1})^{(\nu)} & (\mathbf{G_\lambda^1})^{(\nu)} \\ \dot{\mathbf{u}}_0^* & \dot{\lambda}_0 \end{pmatrix} \begin{pmatrix} \Delta\mathbf{u}_1^{(\nu)} \\ \Delta\lambda_1^{(\nu)} \end{pmatrix} = - \begin{pmatrix} \mathbf{G}(\mathbf{u}_1^{(\nu)}, \lambda_1^{(\nu)}) \\ (\mathbf{u}_1^{(\nu)} - \mathbf{u}_0)^* \dot{\mathbf{u}}_0 + (\lambda_1^{(\nu)} - \lambda_0) \dot{\lambda}_0 - \Delta s \end{pmatrix},$$

with the new direction vector defined as

$$\begin{pmatrix} \mathbf{G_u^1} & \mathbf{G_\lambda^1} \\ \dot{\mathbf{u}}_0^* & \dot{\lambda}_0 \end{pmatrix} \begin{pmatrix} \dot{\mathbf{u}}_1 \\ \dot{\lambda}_1 \end{pmatrix} = \begin{pmatrix} \mathbf{0} \\ 1 \end{pmatrix},$$

Note that

- In practice $(\dot{\mathbf{u}}_1, \dot{\lambda}_1)$ can be computed with one extra back-substitution;
- The orientation of the branch is preserved if Δs is sufficiently small;
- The direction vector must be rescaled, so that indeed $\|\dot{\mathbf{u}}_1\|^2 + \dot{\lambda}_1^2 = 1$.

Theorem 6. *The Jacobian of the pseudo-arclength system is nonsingular at a regular solution point.*

Proof. Let $\mathbf{x} = (\mathbf{u}, \lambda) \in \mathbb{R}^{n+1}$. Then pseudo-arclength continuation can be written as

$$\mathbf{G}(\mathbf{x}_1) = 0,$$
$$(\mathbf{x}_1 - \mathbf{x}_0)^* \dot{\mathbf{x}}_0 - \Delta s = 0, \qquad (\|\dot{\mathbf{x}}_0\| = 1).$$

Figure 1.9 shows a graphical interpretation. The matrix in Newton's method at $\Delta s = 0$ is $\begin{pmatrix} \mathbf{G_x^0} \\ \dot{\mathbf{x}}_0^* \end{pmatrix}$. At a regular solution we have $\mathcal{N}(\mathbf{G_x^0}) = \mathrm{Span}\{\dot{\mathbf{x}}_0\}$. We must show that $\begin{pmatrix} \mathbf{G_x^0} \\ \dot{\mathbf{x}}_0^* \end{pmatrix}$ is nonsingular at a regular solution. Suppose, on the contrary, that $\begin{pmatrix} \mathbf{G_x^0} \\ \dot{\mathbf{x}}_0^* \end{pmatrix}$ is singular. Then there exists some vector $\mathbf{z} \neq 0$ with

$$\mathbf{G_x^0} \mathbf{z} = 0 \quad \text{and} \quad \dot{\mathbf{x}}_0^* \mathbf{z} = 0.$$

Thus, $\mathbf{z} = c\dot{\mathbf{x}}_0$, for some constant c. But then

$$0 = \dot{\mathbf{x}}_0^* \mathbf{z} = c\dot{\mathbf{x}}_0^* \dot{\mathbf{x}}_0 = c \|\dot{\mathbf{x}}_0\|^2 = c,$$

so that $\mathbf{z} = \mathbf{0}$, which is a contradiction. $\qquad\square$

Consider pseudo-arclength continuation for the discretized Gelfand-Bratu problem of Sect. 1.1.3. Then the matrix

$$\begin{pmatrix} \mathbf{G_x} \\ \dot{\mathbf{x}}^* \end{pmatrix} = \begin{pmatrix} \mathbf{G_u} & \mathbf{G_\lambda} \\ \dot{\mathbf{u}}^* & \dot{\lambda} \end{pmatrix}$$

in Newton's method is a 'bordered tridiagonal' matrix of the form

$$\begin{pmatrix} \bullet & \bullet & & & & & & & \bullet \\ \bullet & \bullet & \bullet & & & & & & \bullet \\ & \bullet & \bullet & \bullet & & & & & \bullet \\ & & \bullet & \bullet & \bullet & & & & \bullet \\ & & & \bullet & \bullet & \bullet & & & \bullet \\ & & & & \bullet & \bullet & \bullet & & \bullet \\ & & & & & \bullet & \bullet & \bullet & \bullet \\ & & & & & & \bullet & \bullet & \bullet \\ \bullet & \bullet & \bullet & \bullet & \bullet & \bullet & \bullet & \bullet & \bullet \end{pmatrix}.$$

We now show how to solve such linear systems efficiently.

1.2.4 The Bordering Algorithm

The linear systems in Newton's method for pseudo-arclength continuation are of the form

$$\begin{pmatrix} A & \mathbf{c} \\ \mathbf{b}^* & d \end{pmatrix} \begin{pmatrix} \mathbf{x} \\ y \end{pmatrix} = \begin{pmatrix} \mathbf{f} \\ h \end{pmatrix}.$$

The special structure of this *extended system* can be exploited; a general presentation of the numerical linear algebra aspects of extended systems can be found in [20, 24]. If A is a sparse matrix whose LU-decomposition can be found relatively cheaply (e.g., if A is tridiagonal), then the following *bordered LU-decomposition* [22] will be efficient:

$$\begin{pmatrix} A & \mathbf{c} \\ \mathbf{b}^* & d \end{pmatrix} = \begin{pmatrix} L & \mathbf{0} \\ \boldsymbol{\beta}^* & 1 \end{pmatrix} \begin{pmatrix} U & \boldsymbol{\gamma} \\ \mathbf{0}^* & \delta \end{pmatrix}.$$

After decomposing $\boxed{A = LU}$ (which may require pivoting) we compute $\boldsymbol{\gamma}$, $\boldsymbol{\beta}$, and δ from

$$\boxed{L\boldsymbol{\gamma} = \mathbf{c},}$$

$$\boxed{U^*\boldsymbol{\beta} = \mathbf{b},}$$

$$\boxed{\delta = d - \boldsymbol{\beta}^*\boldsymbol{\gamma}.}$$

The linear system can then be written as

$$\begin{pmatrix} L & \mathbf{0} \\ \boldsymbol{\beta}^* & 1 \end{pmatrix} \underbrace{\begin{pmatrix} U & \boldsymbol{\gamma} \\ \mathbf{0}^* & \delta \end{pmatrix} \begin{pmatrix} \mathbf{x} \\ y \end{pmatrix}}_{} = \begin{pmatrix} \mathbf{f} \\ h \end{pmatrix},$$

$$\equiv \begin{pmatrix} \hat{\mathbf{f}} \\ \hat{h} \end{pmatrix}$$

and we can compute the solution $(\mathbf{x}, y)$ through the following steps:

$$\boxed{L\hat{\mathbf{f}} = \mathbf{f},}$$

$$\boxed{\hat{h} = h - \boldsymbol{\beta}^*\hat{\mathbf{f}},}$$

$$\boxed{y = \hat{h}/\delta,}$$

$$\boxed{U\mathbf{x} = \hat{\mathbf{f}} - y\boldsymbol{\gamma}.}$$

Theorem 7. *The bordering algorithm outlined above works if A and the full matrix $\mathcal{A} \equiv \begin{pmatrix} A & \mathbf{c} \\ \mathbf{b}^* & d \end{pmatrix}$ are nonsingular.*

In the proof of Theorem 7 we make use of the *Bordering Lemma* [22]

Lemma 3 (Bordering Lemma). *Let $\mathcal{A} \equiv \begin{pmatrix} A & \mathbf{c} \\ \mathbf{b}^* & d \end{pmatrix}$. Then*

(a) A nonsingular $\Rightarrow$ $\mathcal{A}$ nonsingular if and only if $d \neq \mathbf{b}^* A^{-1} \mathbf{c}$;

(b) $\dim \mathcal{N}(A) = \dim \mathcal{N}(A^*) = 1 \Rightarrow \mathcal{A}$ nonsingular if $\begin{cases} \mathbf{c} \notin \mathcal{R}(A), \\ \mathbf{b} \notin \mathcal{R}(A^*); \end{cases}$

(c) If $\dim \mathcal{N}(A) \geq 2$ then $\mathcal{A}$ is singular.

Proof.   (a) (A nonsingular)

In this case $\mathcal{A} = \begin{pmatrix} A & \mathbf{0} \\ \mathbf{b}^* & 1 \end{pmatrix} \begin{pmatrix} I & A^{-1}\mathbf{c} \\ \mathbf{0}^* & e \end{pmatrix}$, where $e = d - \mathbf{b}^* A^{-1} \mathbf{c}$. Clearly, $\mathcal{A}$ is nonsingular if and only if $e \neq 0$.

(b) ($\dim \mathcal{N}(A) = 1$)

Suppose $\mathcal{A}$ is singular in this case. Then there exist $\mathbf{z} \in \mathbb{R}^n$ and $\xi \in \mathbb{R}$, not both zero, such that

$$\mathcal{A} = \begin{pmatrix} A & \mathbf{c} \\ \mathbf{b}^* & d \end{pmatrix} \begin{pmatrix} \mathbf{z} \\ \xi \end{pmatrix} = \begin{pmatrix} A\mathbf{z} + \xi\mathbf{c} \\ \mathbf{b}^*\mathbf{z} + \xi d \end{pmatrix} = \begin{pmatrix} \mathbf{0} \\ 0 \end{pmatrix}.$$

We see that $\mathbf{c} \in \mathcal{R}(A)$ if $\xi \neq 0$, which contradicts the assumptions. On the other hand, if $\xi = 0$ and $\mathbf{z} \neq \mathbf{0}$ then

$$\mathcal{N}(A) = \text{Span}\{\mathbf{z}\} \quad \text{and} \quad \mathbf{b} \in \mathcal{N}(A)^{\perp}.$$

Since, in general, $\mathcal{N}(A)^\perp = \mathcal{R}(A^*)$, it follows that $\mathbf{b} \in \mathcal{R}(A^*)$, which also contradicts the assumptions.

(c) $(\dim \mathcal{N}(A) \geq 2)$

This case follows from a rank argument.   □

Proof (Theorem 7). The crucial step in the bordering algorithm is the computation of $z = \hat{h}/\delta$. Namely, we must have $\delta \neq 0$. Since δ is determined in the bordered LU-decomposition, we have

$$\begin{aligned}
\delta &= d - \boldsymbol{\beta}^* \boldsymbol{\gamma} && = d - (U^{*-1}\mathbf{b})^*(L^{-1}\mathbf{c}) \\
&= d - \mathbf{b}^* U^{-1} L^{-1} \mathbf{c} = d - \mathbf{b}^*(LU)^{-1}\mathbf{c} \\
&= d - \mathbf{b}^* A^{-1} \mathbf{c},
\end{aligned}$$

which is nonzero by Conclusion (a) of the Bordering Lemma.   □

Remark 3. In pseudo-arclength continuation we have

$$\mathcal{A} = \begin{pmatrix} A & \mathbf{c} \\ \mathbf{b}^* & d \end{pmatrix} = \begin{pmatrix} \mathbf{G}_u & \mathbf{G}_\lambda \\ \dot{\mathbf{u}}_0^* & \dot{\lambda}_0, \end{pmatrix}$$

that is, $A = \mathbf{G}_u$, which is singular at a fold. Therefore, the bordering algorithm will fail when it is used *exactly* at a fold. In practice, the method may still work. We consider another approach, used in AUTO, when discussing collocation methods in Sect. 1.3.1.

1.3 Boundary Value Problems

Consider the first-order system of ordinary differential equations

$$\mathbf{u}'(t) - \mathbf{f}(\mathbf{u}(t), \boldsymbol{\mu}, \lambda) = \mathbf{0}, \qquad t \in [0, 1],$$

where

$$\mathbf{u}(\cdot), \mathbf{f}(\cdot) \in \mathbb{R}^n, \qquad \lambda \in \mathbb{R}, \qquad \boldsymbol{\mu} \in \mathbb{R}^{n_\mu},$$

subject to boundary conditions

$$\mathbf{b}(\mathbf{u}(0), \mathbf{u}(1), \boldsymbol{\mu}, \lambda) = \mathbf{0}, \qquad \mathbf{b}(\cdot) \in \mathbb{R}^{n_b},$$

and integral constraints

$$\int_0^1 \mathbf{q}(\mathbf{u}(s), \boldsymbol{\mu}, \lambda) \, ds = \mathbf{0}, \qquad \mathbf{q}(\cdot) \in \mathbb{R}^{n_q}.$$

We want to solve this boundary value problem (BVP) for $\mathbf{u}(\cdot)$ and $\boldsymbol{\mu}$. In order for this problem to be formally well posed we require that

$$n_\mu = n_b + n_q - n \geq 0.$$

We can think of λ as the continuation parameter in which the solution $(\mathbf{u}, \boldsymbol{\mu})$ may be continued. A simple case is $n_q = 0$, $n_b = n$, for which $n_\mu = 0$.

1.3.1 Orthogonal Collocation

AUTO solves boundary value problems using the method of *orthogonal collocation with piecewise polynomials* [2, 7]. This method is very accurate, and allows adaptive mesh-selection. The set-up is as follows.

First, we introduce a mesh

$$\{0 = t_0 < t_1 < \cdots < t_N = 1\},$$

with

$$h_j = t_j - t_{j-1}, \qquad (1 \le j \le N).$$

Define the space of (vector-valued) *piecewise polynomials* $\mathcal{P}_h^m$ as

$$\mathcal{P}_h^m = \left\{ \mathbf{p}_h \in C[0,1] \mid \mathbf{p}_h|_{[t_{j-1},t_j]} \in \mathcal{P}^m \right\},$$

where $\mathcal{P}^m$ is the space of (vector-valued) polynomials of degree $\le m$. The orthogonal collocation method with piecewise polynomials [3] consists of finding $\mathbf{p}_h \in \mathcal{P}_h^m$ and $\boldsymbol{\mu} \in \mathbb{R}^{n_\mu}$, such that the following *collocation equations* are satisfied:

$$\mathbf{p}_h'(z_{j,i}) = \mathbf{f}(\mathbf{p}_h(z_{j,i}), \boldsymbol{\mu}, \lambda), \qquad j = 1, \ldots, N, \quad i = 1, \ldots, m,$$

and such that $\mathbf{p}_h$ satisfies the boundary and integral conditions. The *collocation points* $z_{j,i}$ in each subinterval $[t_{j-1}, t_j]$ are the (scaled) roots of the mth-degree orthogonal polynomial (*Gauss points*); see Fig. 1.10 for a graphical interpretation. Since each local polynomial is determined by $(m+1)n$, coefficients, the total number of degrees of freedom (considering λ as fixed) is $(m+1)nN + n_\mu$. This is matched by the total number of equations:

$$
\begin{aligned}
&\text{collocation:} \quad mnN, \\
&\text{continuity:} \quad (N-1)n, \\
&\text{constraints:} \quad n_b + n_q \quad (= n + n_\mu).
\end{aligned}
$$

If the solution $\mathbf{u}(t)$ of the BVP is sufficiently smooth then the order of accuracy of the orthogonal collocation method is m, i.e.,

$$\| \mathbf{p}_h - \mathbf{u} \|_\infty = \mathcal{O}(h^m).$$

At the main meshpoints t_j we have *superconvergence*:

$$\max_j |\mathbf{p}_h(t_j) - \mathbf{u}(t_j)| = \mathcal{O}(h^{2m}).$$

The scalar variables $\boldsymbol{\mu}$ are also superconvergent [7].

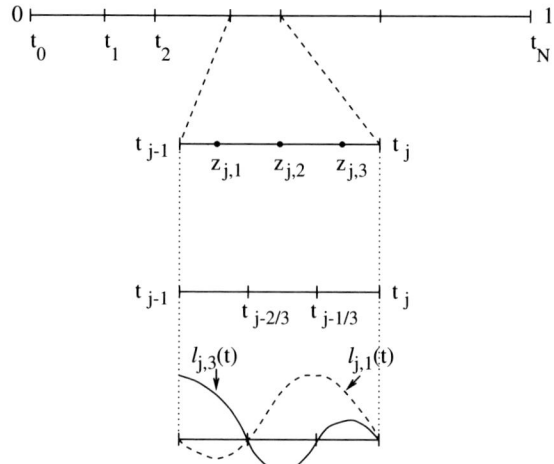

Fig. 1.10. The mesh $\{0 = t_0 < t_1 < \cdots < t_N = 1\}$. Collocation points and 'extended-mesh points' are shown for the case $m = 3$, in the jth mesh interval. Also shown are two of the four local Lagrange basis polynomials.

1.3.2 Implementation in AUTO

The implementation in AUTO [12] is done via the introduction of Lagrange basis polynomials for each subinterval $[t_{j-1}, t_j]$. Define

$$\{\ell_{j,i}(t)\}, \qquad j = 1, \ldots, N, \quad i = 0, 1, \ldots, m,$$

by

$$\ell_{j,i}(t) = \prod_{k=0, k \neq i}^{m} \frac{t - t_{j - \frac{k}{m}}}{t_{j - \frac{i}{m}} - t_{j - \frac{k}{m}}},$$

where

$$t_{j - \frac{i}{m}} = t_j - \frac{i}{m} h_j.$$

The local polynomials can then be written as

$$\mathbf{p}_j(t) = \sum_{i=0}^{m} \ell_{j,i}(t) \mathbf{u}_{j - \frac{i}{m}}.$$

With the above choice of basis

$$\mathbf{u}_j \text{ approximates } \mathbf{u}(t_j) \quad \text{and} \quad \mathbf{u}_{j - \frac{i}{m}} \text{ approximates } \mathbf{u}(t_{j - \frac{i}{m}}),$$

where $\mathbf{u}(t)$ is the solution of the continuous problem.

Then the collocation equations are

$$\mathbf{p}_j'(z_{j,i}) = \mathbf{f}(\mathbf{p}_j(z_{j,i}), \boldsymbol{\mu}, \lambda), \qquad i = 1, \ldots, m, \quad j = 1, \ldots, N,$$

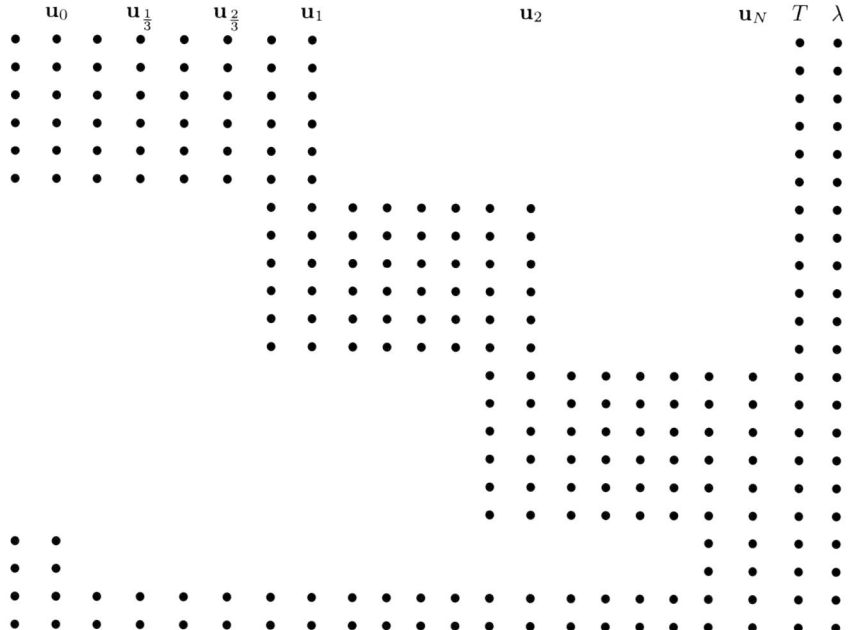

Fig. 1.11. Structure of the Jacobian for the case of $n = 2$ differential equations with the number of mesh intervals $N = 3$, the number of collocation points per mesh interval $m = 3$, the number of boundary conditions $n_b = 2$, and the number of integral constraints $n_q = 1$. The last row corresponds to the pseudo-arclength equation, which is not included in the $n_q = 1$ count. From E.J. Doedel, H.B. Keller, J.P. Kernévez, Numerical analysis and control of bifurcation problems (II): Bifurcation in infinite dimensions, *Internat. J. Bifur. Chaos Appl. Sci. Engrg.* 1(4) (1991) 745–772 ©1991 World Scientific Publishing; reproduced with permission.

the discrete boundary conditions are

$$b_i(\mathbf{u}_0, \mathbf{u}_N, \boldsymbol{\mu}, \lambda) = 0, \qquad i = 1, \dots, n_b,$$

and the integrals constraints can be discretized as

$$\sum_{j=1}^{N} \sum_{i=0}^{m} \omega_{j,i} q_k(\mathbf{u}_{j-\frac{i}{m}}, \boldsymbol{\mu}, \lambda) = 0, \qquad k = 1, \dots, n_q,$$

where the $\omega_{j,i}$ are the Lagrange quadrature coefficients.

The pseudo-arclength equation is

$$\int_0^1 (\mathbf{u}(t) - \mathbf{u}_0(t))^* \, \dot{\mathbf{u}}_0(t) \, dt + (\boldsymbol{\mu} - \boldsymbol{\mu}_0)^* \, \dot{\boldsymbol{\mu}}_0 + (\lambda - \lambda_0) \, \dot{\lambda}_0 - \varDelta s = 0,$$

where $(\mathbf{u}_0, \boldsymbol{\mu}_0, \lambda_0)$, is the previously computed point on the solution branch, and $(\dot{\mathbf{u}}_0, \dot{\boldsymbol{\mu}}_0, \dot{\lambda}_0)$, is the normalized direction of the branch at that point. The

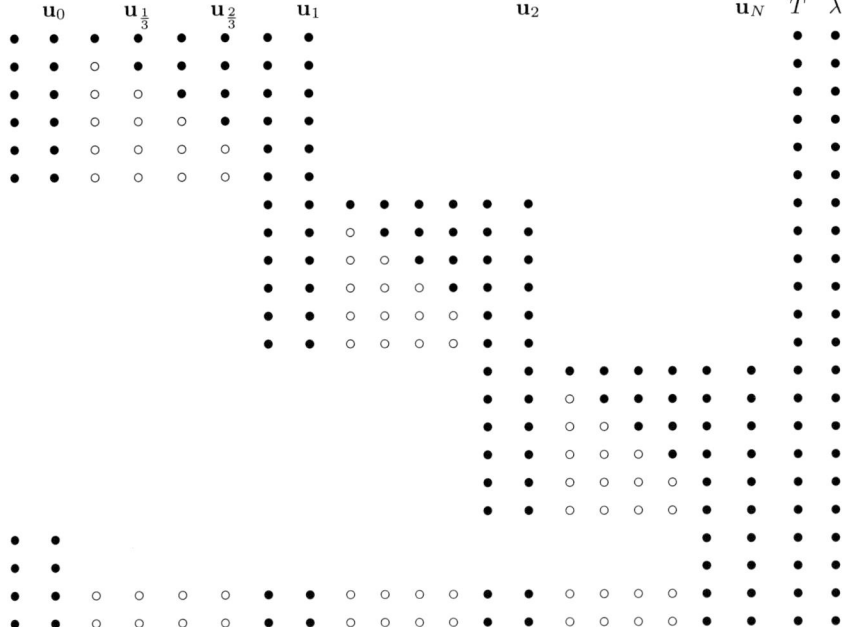

Fig. 1.12. The system after *condensation of parameters*. The entries ○ have been eliminated by Gauss elimination.

discretized pseudo-arclength equation is

$$\sum_{j=1}^{N}\sum_{i=0}^{m}\omega_{j,i}\left[\mathbf{u}_{j-\frac{i}{m}}-(\mathbf{u}_0)_{j-\frac{i}{m}}\right]^{*}(\dot{\mathbf{u}}_0)_{j-\frac{i}{m}}$$

$$+(\boldsymbol{\mu}-\boldsymbol{\mu}_0)^{*}\dot{\boldsymbol{\mu}}_0+(\lambda-\lambda_0)\dot{\lambda}_0-\Delta s=0.$$

The implementation in AUTO includes an efficient method to solve these linear systems [12]; this is illustrated in Figs. 1.12–1.15. Note that the figures only illustrate the matrix structure; the indicated operations are also carried out on the right-hand side, which is not shown in the figures. Figure 1.12 shows the system after *condensation of parameters*. The entries marked with ○ have been eliminated by Gauss elimination. These operations can be done in parallel [34]. The condensation of parameters leads to a system with a fully decoupled sub-system that can be solved separately. The decoupled sub-system is marked by ∗ in Fig. 1.13

1.3.3 Numerical Linear Algebra

The complete discretization consists of

$$mnN+n_b+n_q+1,$$

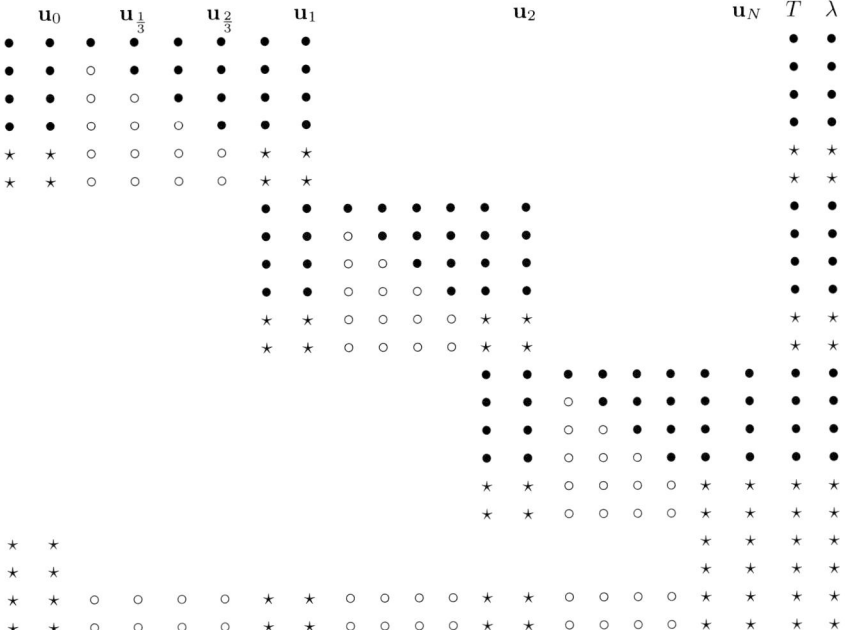

Fig. 1.13. This is the same matrix as in Fig. 1.12, except that some entries are now marked by a ⋆. The ⋆ sub-system is fully decoupled from the remaining equations and can, therefore, be solved separately. From E.J. Doedel, H.B. Keller, J.P. Kernévez, Numerical analysis and control of bifurcation problems (II): Bifurcation in infinite dimensions, *Internat. J. Bifur. Chaos Appl. Sci. Engrg.* 1(4) (1991) 745–772 ©1991 World Scientific Publishing; reproduced with permission.

nonlinear equations, in the unknowns

$$\{\mathbf{u}_{j-\frac{i}{m}}\} \in \mathbb{R}^{mnN+n}, \quad \boldsymbol{\mu} \in \mathbb{R}^{n_\mu}, \quad \lambda \in \mathbb{R}.$$

These equations can be solved by a Newton-Chord iteration. The structure of the associated Jacobian is illustrated in Fig. 1.11 for a system of $n = 2$ differential equations, with $N = 3$ mesh intervals, $m = 3$ collocation points per mesh interval, $n_b = 2$ boundary conditions, and $n_q = 1$ integral constraint. In a typical problem N will be larger, say, $N = 5$ for 'very easy' problems, and $N = 200$ for 'very difficult' problems. The 'standard' choice of the number of collocation points per mesh interval is $m = 4$.

The decoupled ⋆ sub-system can be solved by *nested dissection*. This procedure eliminates some of the ⋆-entries, but also introduces some new nonzero entries due to fill-in; see Fig. 1.14. However, the structure reveals a new decoupled sub-system that can be solved completely; this subsystem is highlighted in Fig. 1.15 with +. The + sub-system consists of two sub-matrices A_0 and A_1, as in Fig. 1.15. For periodic solutions, the Floquet multipliers are the eigenvalues of the matrix $-A_1^{-1}A_0$ [18].

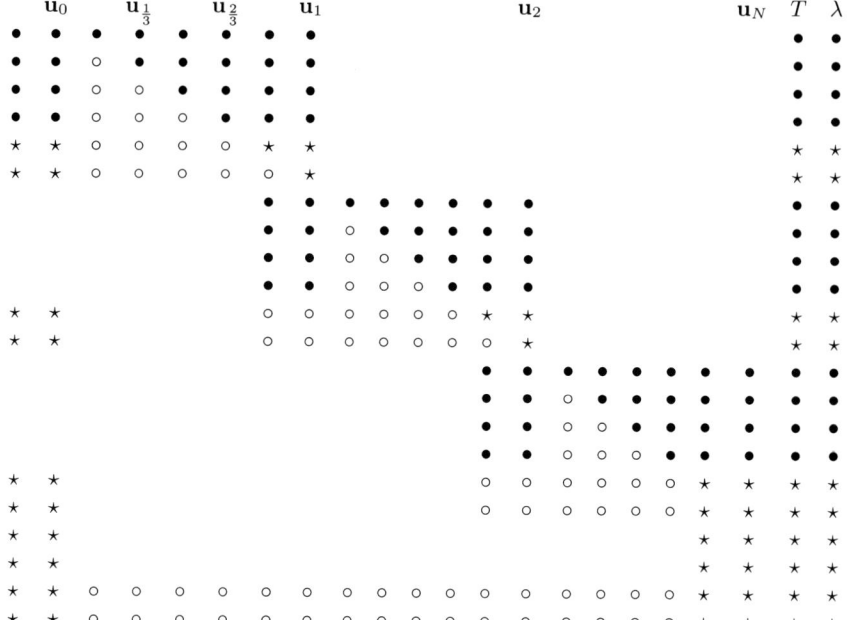

Fig. 1.14. The decoupled $\star$ sub-system solved by *nested dissection*. This procedure eliminates some of the $\star$-entries, but also introduces some new nonzero entries due to fill-in.

1.4 Computing Periodic Solutions

Periodic solutions can be computed very effectively by using a boundary value approach. This method also determines the period very accurately. Moreover, the technique allows asymptotically unstable periodic orbits to be computed as easily as asymptotically stable ones.

1.4.1 The BVP Approach.

Consider the first-order system

$$\mathbf{u}'(t) = \mathbf{f}(\mathbf{u}(t), \lambda), \qquad \mathbf{u}(\cdot), \mathbf{f}(\cdot) \in \mathbb{R}^n, \quad \lambda \in \mathbb{R}.$$

Fix the interval of periodicity by the transformation $t \mapsto \frac{t}{T}$. Then the equation becomes

$$\boxed{\mathbf{u}'(t) = T\,\mathbf{f}(\mathbf{u}(t), \lambda),} \qquad \mathbf{u}(\cdot), \mathbf{f}(\cdot) \in \mathbb{R}^n, \quad T, \lambda \in \mathbb{R}, \qquad (1.10)$$

and we seek solutions of period 1, i.e.,

$$\boxed{\mathbf{u}(0) = \mathbf{u}(1).} \qquad (1.11)$$

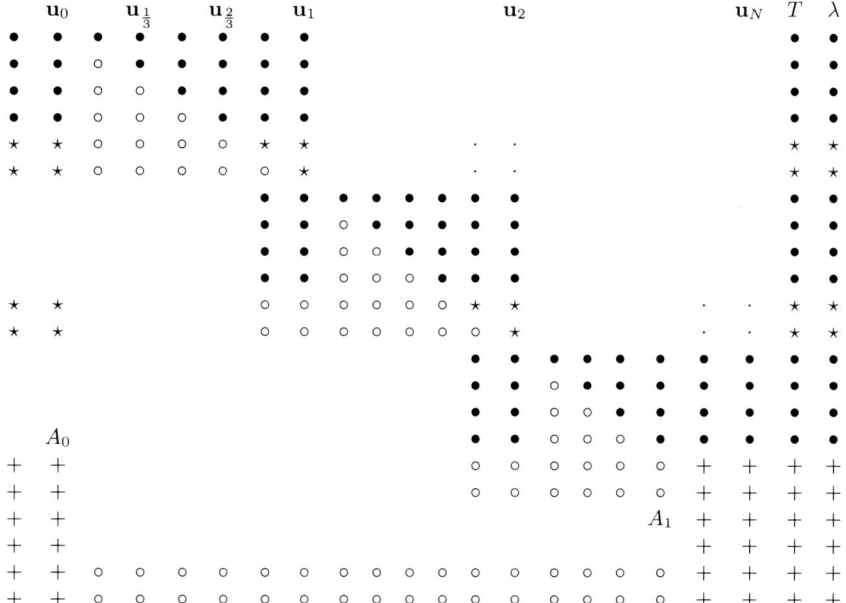

Fig. 1.15. The same matrix as in Fig. 1.14, except with some entries now marked by +. Note that the + sub-system is decoupled from the other equations, and can, therefore, be solved separately.

Note that the period T is one of the unknowns.

Equations (1.10)–(1.11) do not uniquely specify $\mathbf{u}$ and T. Assume that we have computed $(\mathbf{u}_{k-1}(\cdot), T_{k-1}, \lambda_{k-1})$ and we want to compute the next solution $(\mathbf{u}_k(\cdot), T_k, \lambda_k)$. Then $\mathbf{u}_k(t)$ can be translated freely in time: if $\mathbf{u}_k(t)$ is a periodic solution then so is $\mathbf{u}_k(t + \sigma)$ for any σ. Thus, a *phase condition* is needed. An example is the Poincaré orthogonality condition

$$\left(\mathbf{u}_k(0) - \mathbf{u}_{k-1}(0)\right)^* \mathbf{u}'_{k-1}(0) = 0,$$

where the phase of the next condition is fixed such that the difference at time $t = 0$ is perpendicular to the tangent vector of the current solution; this is illustrated in Fig. 1.16. In the next section we derive a numerically more suitable phase condition.

1.4.2 Integral Phase Condition

If $\tilde{\mathbf{u}}_k(t)$ is a solution then so is $\tilde{\mathbf{u}}_k(t + \sigma)$, for any σ. We want the solution that minimizes

$$D(\sigma) = \int_0^1 \|\tilde{\mathbf{u}}_k(t + \sigma) - \mathbf{u}_{k-1}(t)\|_2^2 \, dt.$$

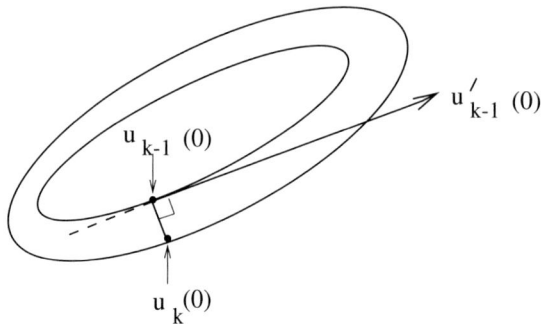

Fig. 1.16. Graphical interpretation of the Poincaré phase condition.

The optimal solution $\tilde{\mathbf{u}}_k(t+\hat{\sigma})$, must satisfy the necessary condition $D'(\hat{\sigma}) = 0$. Differentiation gives the necessary condition

$$\int_0^1 (\tilde{\mathbf{u}}_k(t+\hat{\sigma}) - \mathbf{u}_{k-1}(t))^* \, \tilde{\mathbf{u}}_k'(t+\hat{\sigma}) \, dt = 0.$$

Writing $\mathbf{u}_k(t) \equiv \tilde{\mathbf{u}}_k(t+\hat{\sigma})$, gives

$$\int_0^1 (\mathbf{u}_k(t) - \mathbf{u}_{k-1}(t))^* \, \mathbf{u}_k'(t) \, dt = 0.$$

Integration by parts, using periodicity, gives

$$\boxed{\int_0^1 \mathbf{u}_k(t)^* \, \mathbf{u}_{k-1}'(t) \, dt = 0.} \tag{1.12}$$

This is the *integral phase condition* [8].

1.4.3 Pseudo-Arclength Continuation

In practice, we use pseudo-arclength continuation to follow a family of periodic solutions; see Sect. 1.2.3. In particular, this allows calculation past folds along a family of periodic solutions. It also allows calculation of a 'vertical family' of periodic solutions, which has important applications to the computation of periodic solutions to conservative systems [14, 30] (see also Chap. 9). For periodic solutions the pseudo-arclength equation is

$$\boxed{\begin{aligned} &\int_0^1 (\mathbf{u}_k(t) - \mathbf{u}_{k-1}(t))^* \, \dot{\mathbf{u}}_{k-1}(t) \, dt \\ &+ (T_k - T_{k-1})^* \, \dot{T}_{k-1} + (\lambda_k - \lambda_{k-1}) \, \dot{\lambda}_{k-1} = \Delta s. \end{aligned}} \tag{1.13}$$

Equations (1.10)–(1.13) are the equations used in AUTO for the continuation of periodic solutions. In summary, given $\mathbf{u}_{k-1}, T_{k-1}$, and λ_{k-1}, we solve the system

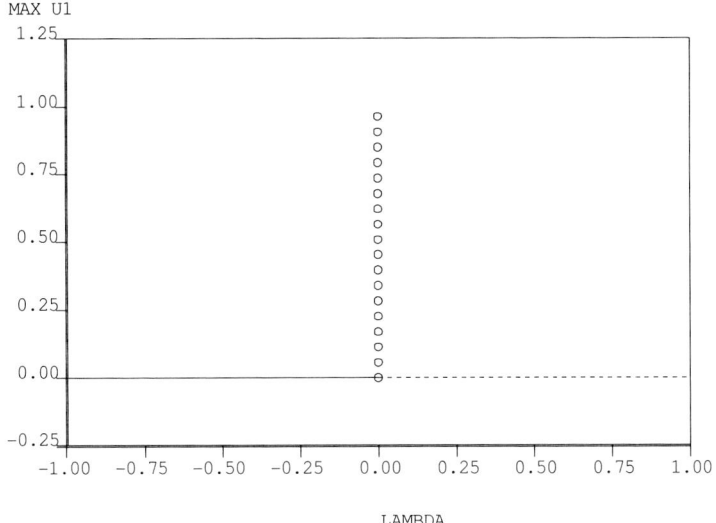

Fig. 1.17. Bifurcation diagram of the stationary solution $\mathbf{u}(t) \equiv \mathbf{0}$ of (1.14).

$$\mathbf{u}_k'(t) = T\,\mathbf{f}(\mathbf{u}_k(t), \lambda_k),$$
$$\mathbf{u}_k(0) = \mathbf{u}_k(1),$$

$$\int_0^1 \mathbf{u}_k(t)^* \, \mathbf{u}_{k-1}'(t)\, dt = 0,$$

$$\int_0^1 (\mathbf{u}_k(t) - \mathbf{u}_{k-1}(t))^* \, \dot{\mathbf{u}}_{k-1}(t)\, dt + (T_k - T_{k-1})\,\dot{T}_{k-1} + (\lambda_k - \lambda_{k-1})\,\dot{\lambda}_{k-1} = \Delta s,$$

where

$$\mathbf{u}(\cdot),\, \mathbf{f}(\cdot) \in \mathbb{R}^n, \qquad \lambda,\, T \in \mathbb{R}.$$

1.4.4 A Vertical Family of Periodic Orbits

Consider the system of equations

$$\begin{cases} u_1' = \lambda u_1 - u_2, \\ u_2' = u_1(1 - u_1). \end{cases} \tag{1.14}$$

Note that $\mathbf{u}(t) \equiv \mathbf{0}$ is a stationary solution for all λ. Another stationary solution is $\mathbf{u}(t) \equiv \begin{pmatrix} 1 \\ -\lambda \end{pmatrix}$.

The bifurcation diagram for $\mathbf{u}(t) \equiv \mathbf{0}$ is shown in Fig. 1.17, but we can also analyze the behavior analytically. The Jacobian along the solution family $\mathbf{u}(t) \equiv \mathbf{0}$ is

$$\begin{pmatrix} -\lambda & -1 \\ 1 & 0 \end{pmatrix},$$

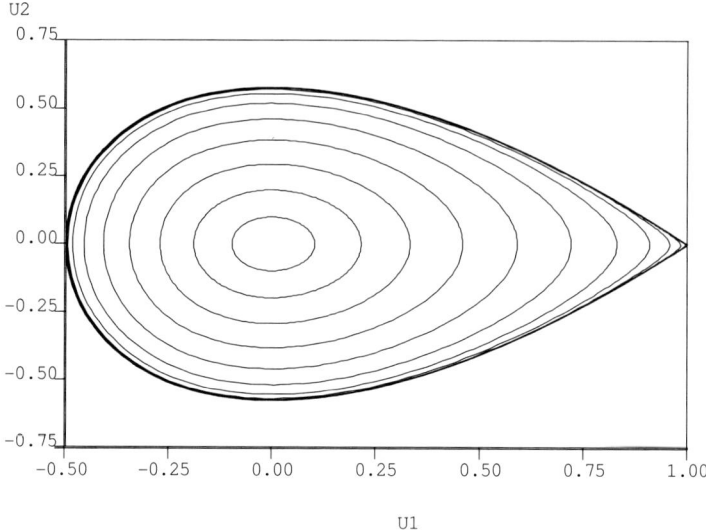

Fig. 1.18. A phase plot of some periodic solutions of (1.14).

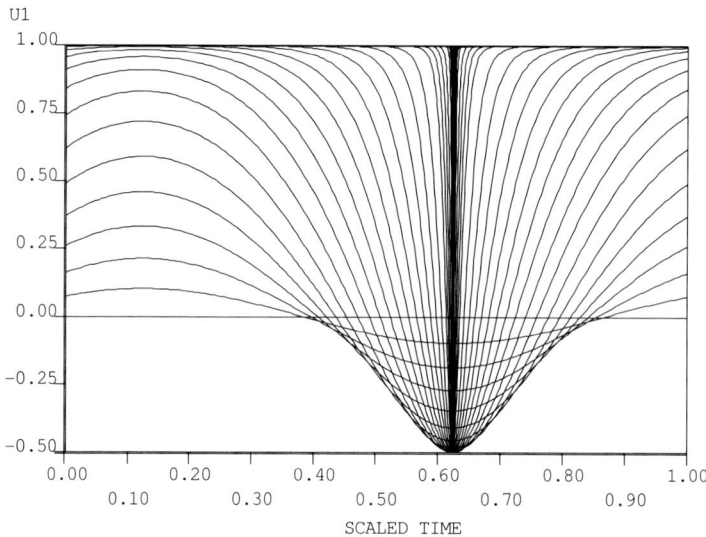

Fig. 1.19. Solution component u_1 of (1.14) as a function of the scaled time variable t.

with eigenvalues

$$\frac{-\lambda \pm \sqrt{\lambda^2 - 4}}{2}.$$

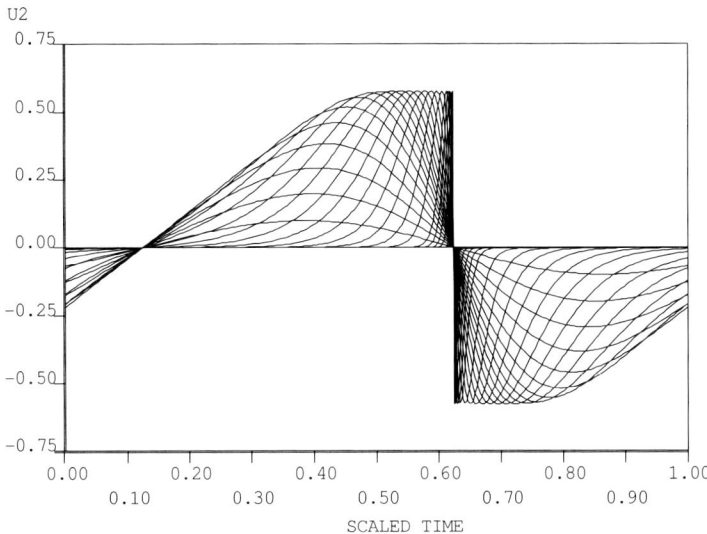

Fig. 1.20. Solution component u_2 of (1.14) as a function of the scaled time variable t.

Hence, the eigenvalues are complex for $\lambda \in (-2, 2)$. The eigenvalues cross the imaginary axis when λ passes through zero. Thus, there is a *Hopf bifurcation* along $\mathbf{u}(t) \equiv \mathbf{0}$ at $\lambda = 0$, and a family of periodic solutions bifurcates from $\mathbf{u}(t) \equiv \mathbf{0}$ at $\lambda = 0$. As shown in Fig. 1.17, the emanating family of periodic solutions is 'vertical'. Some periodic solutions are shown in Fig. 1.18 in the (u_1, u_2)-plane. These solutions are plotted versus time in Figs. 1.19 and 1.20.

Along this family the period tends to infinity. The final infinite-period orbit is *homoclinic* to $(u_1, u_2) = (1, 0)$. The time diagrams in Figs. 1.19 and 1.20 illustrate how the 'peak' in the solution remains in the same location. This is a result of the integral phase condition (1.12) and very advantageous for discretization methods.

1.4.5 FitzHugh-Nagumo Equations

The Fitzhugh-Nagumo equations of nerve-conduction are

$$\begin{cases} v' = c \left(v - \frac{1}{3} v^3 + w \right), \\ w' = -(v - a + bw)/c. \end{cases} \tag{1.15}$$

Let $b = 0.8$ and $c = 3$. Note that there is a stationary solution $(v(t), w(t)) = (0, 0)$ for $a = 0$.

We compute the solution family, starting at $(v(t), w(t)) = (0, 0)$ for $a = 0$, with AUTO. The bifurcation diagram is shown in Fig. 1.21. Note that the solution is unstable for a small and becomes stable after a Hopf bifurcation at $a \approx 0.4228$. Figure 1.21 also shows the emanating family of periodic solutions,

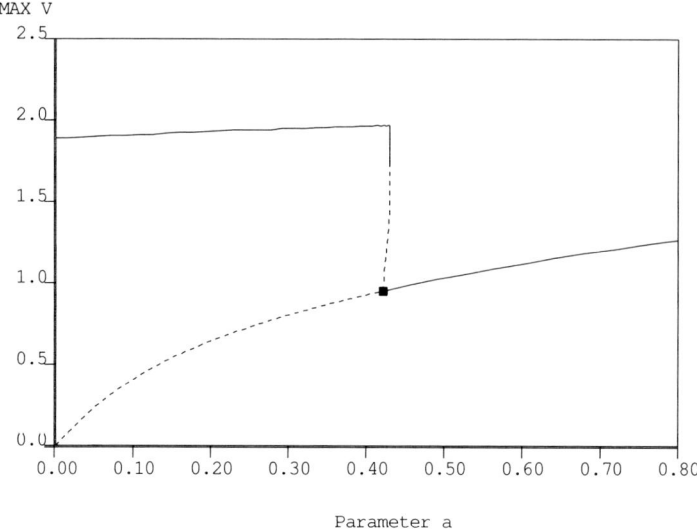

Fig. 1.21. Bifurcation diagram of the Fitzhugh-Nagumo equations (1.15).

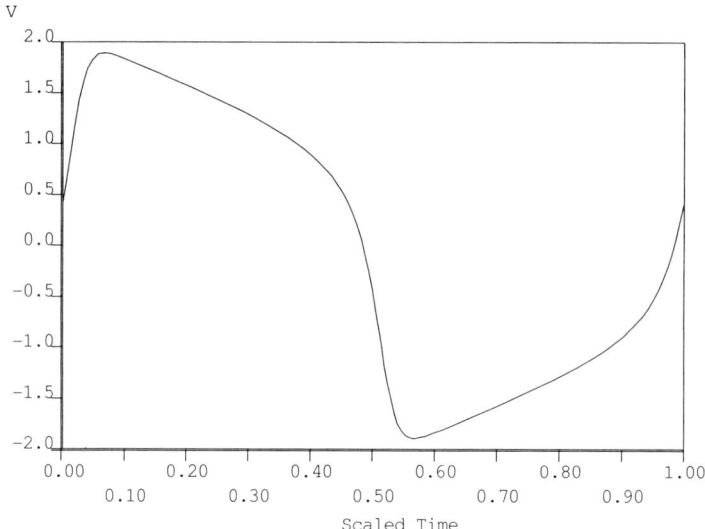

Fig. 1.22. The periodic solution of (1.15) at $a = 0$.

which turns back toward $a = 0$; the periodic solution at $a = 0$ is shown in Fig. 1.22.

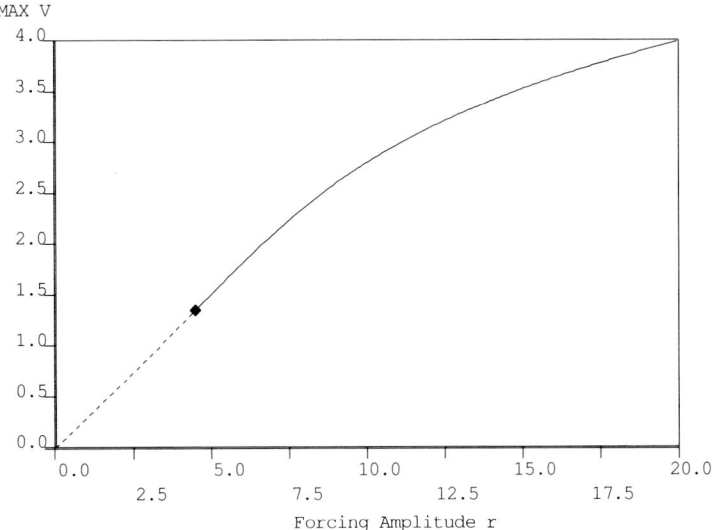

MAX V

Forcing Amplitude r

Fig. 1.23. Continuation of (1.16) from $r = 0$ to $r = 20$.

1.4.6 Periodically Forced and Non-Autonomous Systems

In this section we illustrate computing periodic solutions to non-autonomous systems. The classical example of a non-autonomous system is a periodically forced system. In AUTO periodic orbits of a periodically forced system can be computed by adding a nonlinear oscillator with the desired periodic forcing as one of its solution components. An example of such an oscillator is

$$\begin{cases} x' = \quad x + \beta y - x(x^2 + y^2), \\ y' = -\beta x + \quad y - y(x^2 + y^2), \end{cases}$$

which has the asymptotically stable solution

$$x(t) = \sin(\beta t), \quad y(t) = \cos(\beta t).$$

As an example, consider again the FitzHugh-Nagumo equations of Sect. 1.4.5, where we assume that the first component of the equations is periodically forced by $-r \cos \beta t$. Coupling the oscillator to the Fitzhugh-Nagumo equations gives:

$$\begin{cases} x' = \quad x + \quad \beta y - x(x^2 + y^2), \\ y' = -\beta x + \quad y - y(x^2 + y^2), \\ v' = \quad c(v - \frac{1}{3}v^3 + w - ry), \\ w' = -(v - \quad a + bw)/c, \end{cases} \tag{1.16}$$

where we take $b = 0.8$, $c = 3$, and $\beta = 10$. For $a = 0$ and $r = 0$ there exists the solution

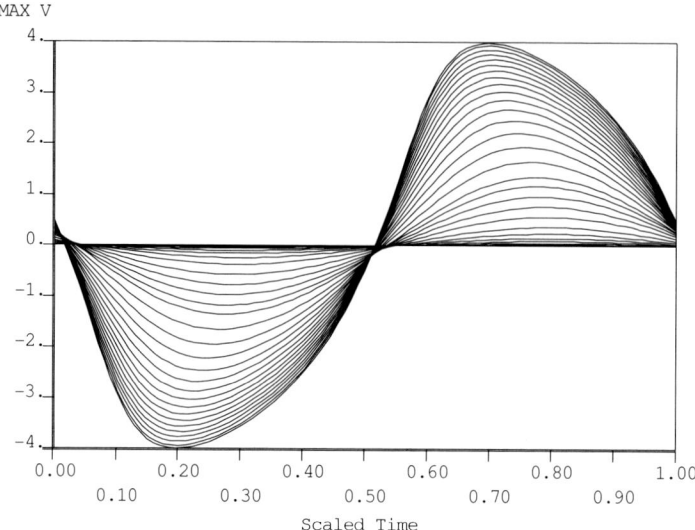

Fig. 1.24. Solutions along the continuation path of (1.16) from $r = 0$ to $r = 20$.

$$x(t) = \sin{(\beta t)}, \quad y(t) = \cos{(\beta t)}, v(t) \equiv 0, \quad w(t) \equiv 0.$$

We continue this solution in the forcing amplitude r, from $r = 0$ to, say, $r = 20$. The result is shown in Fig. 1.23, with some of the solutions along this family plotted versus time in Fig. 1.24.

If the forcing is not periodic, or difficult to model by an autonomous oscillator, then the equations can be rewritten in autonomous form as follows. The non-autonomous system

$$\begin{cases} \mathbf{u}'(t) = \mathbf{f}(t, \mathbf{u}(t)), & \mathbf{u}(\cdot), \mathbf{f}(\cdot) \in \mathbb{R}^n, \quad t \in [0, 1], \\ \mathbf{b}(\mathbf{u}(0), \mathbf{u}(1)) = \mathbf{0}, & \mathbf{b}(\cdot) \in \mathbb{R}^n, \end{cases}$$

can be transformed into

$$\begin{cases} \mathbf{u}'(t) = \mathbf{f}(v(t), \mathbf{u}(t)), & \mathbf{u}(\cdot), \mathbf{f}(\cdot) \in \mathbb{R}^n, \quad t \in [0, 1], \\ v'(t) = 1, & v(\cdot) \in \mathbb{R}, \\ \mathbf{b}(\mathbf{u}(0), \mathbf{u}(1)) = \mathbf{0}, & \mathbf{b}(\cdot) \in \mathbb{R}^n, \\ v(0) = 0, \end{cases}$$

which is autonomous, with $n + 1$ ODEs and $n + 1$ boundary conditions.

1.5 Computing Connecting Orbits

Orbits that connect fixed points of a vector field are important in many applications. A basic algorithm, which can be represented in various forms [25, 6, 19] consists of continuation of solutions to the equations

$$\mathbf{u}'(t) = T\,\mathbf{f}(\mathbf{u}(t), \boldsymbol{\lambda}), \qquad \mathbf{u}(\cdot),\, \mathbf{f}(\cdot, \cdot) \in \mathbb{R}^n, \quad \boldsymbol{\lambda} \in \mathbb{R}^{n_\lambda}, \tag{1.17}$$

$$\begin{cases} \mathbf{f}(\mathbf{w}_0, \boldsymbol{\lambda}) = \mathbf{0}, \\ \mathbf{f}(\mathbf{w}_1, \boldsymbol{\lambda}) = \mathbf{0}, \end{cases} \tag{1.18}$$

$$\begin{cases} \mathbf{f}_u(\mathbf{w}_0, \boldsymbol{\lambda})\mathbf{v}_{0i} = \mu_{0i}\mathbf{v}_{0i}, & i = 1, \ldots, n_0, \\ \mathbf{f}_u(\mathbf{w}_1, \boldsymbol{\lambda})\mathbf{v}_{1i} = \mu_{1i}\mathbf{v}_{1i}, & i = 1, \ldots, n_1, \end{cases} \tag{1.19}$$

$$\begin{cases} \mathbf{v}_{0i}^* \mathbf{v}_{0i} = 1, & i = 1, \ldots, n_0, \\ \mathbf{v}_{1i}^* \mathbf{v}_{1i} = 1, & i = 1, \ldots, n_1, \end{cases} \tag{1.20}$$

$$\int_0^1 (\mathbf{f}(\mathbf{u}, \boldsymbol{\lambda}) - \mathbf{f}(\hat{\mathbf{u}}, \hat{\boldsymbol{\lambda}}))^* \, \mathbf{f}_u(\hat{\mathbf{u}}, \hat{\boldsymbol{\lambda}})\, \mathbf{f}(\hat{\mathbf{u}}, \hat{\boldsymbol{\lambda}})\, dt = 0, \tag{1.21}$$

$$\begin{cases} \mathbf{u}(0) = \mathbf{w}_0 + \varepsilon_0 \sum_{i=1}^{n_0} c_{0i}\mathbf{v}_{0i}, & \sum_{i=1}^{n_0} c_{0i}^2 = 1, \\ \mathbf{u}(1) = \mathbf{w}_1 + \varepsilon_1 \sum_{i=1}^{n_1} c_{1i}\mathbf{v}_{1i}, & \sum_{i=1}^{n_1} c_{1i}^2 = 1. \end{cases} \tag{1.22}$$

Equation (1.17) is the ODE with independent variable t scaled to $[0, 1]$. Equation (1.18) defines two fixed points $\mathbf{w}_0$ and $\mathbf{w}_1$. We assume in (1.19) that $\mathbf{f}_u(\mathbf{w}_0, \boldsymbol{\lambda})$ has n_0 distinct real positive eigenvalues μ_{0i} with eigenvectors $\mathbf{v}_{0i}$, and $\mathbf{f}_u(\mathbf{w}_1, \boldsymbol{\lambda})$ has n_1 distinct real negative eigenvalues μ_{1i} with eigenvectors $\mathbf{v}_{1i}$. Equation (1.20) normalizes the eigenvectors. Equation (1.21) gives the *phase condition*, with *reference orbit* $\hat{\mathbf{u}}(t)$, which is a necessary condition for

$$D(\sigma) = \int_0^1 \| \mathbf{u}'(t + \sigma) - \hat{\mathbf{u}}'(t) \|^2 \, dt$$

to be minimized over σ; here we use $\mathbf{u}''(t) = \mathbf{f}_u(\mathbf{u}, \boldsymbol{\lambda})\, \mathbf{u}'(t) = \mathbf{f}_u(\mathbf{u}, \boldsymbol{\lambda})\, \mathbf{f}(\mathbf{u}, \boldsymbol{\lambda})$. Finally, (1.22) requires $\mathbf{u}(0)$ to lie in the tangent manifold $\mathbf{U}_0$ at 'distance' ε_0 from $\mathbf{w}_0$; similarly, $\mathbf{u}(1)$ must lie in $\mathbf{S}_1$ at distance ε_1 from $\mathbf{w}_1$.

Using (1.22) we can eliminate $\mathbf{w}_0$ and $\mathbf{w}_1$, to be left with n coupled differential equations subject to

$$n_c = 2n + (n + 1)(n_0 + n_1) + 3$$

constraints. In addition to $\mathbf{u}(t) \in \mathbb{R}^n$ we have scalar variables

$$\boldsymbol{\lambda} \in \mathbb{R}^{n_\lambda}, \quad \varepsilon_0, \varepsilon_1 \in \mathbb{R},$$
$$\mu_{0i}, c_{0i} \in \mathbb{R}, \quad \mathbf{v}_{0i} \in \mathbb{R}^n, \quad i = 1, \ldots, n_0,$$
$$\mu_{1i}, c_{1i} \in \mathbb{R}, \quad \mathbf{v}_{1i} \in \mathbb{R}^n, \quad i = 1, \ldots, n_1.$$

The total number of scalar variables equals

$$n_v = n_\lambda + (n + 2)(n_0 + n_1) + 2.$$

Formally, we need $n_v = n_c - n$ for a single heteroclinic connection; this gives $n_\lambda = n - (n_0 + n_1) + 1$. For a family of connecting orbits, we must use $n - (n_0 + n_1) + 2$ free parameters. Note that T is large and fixed in this continuation.

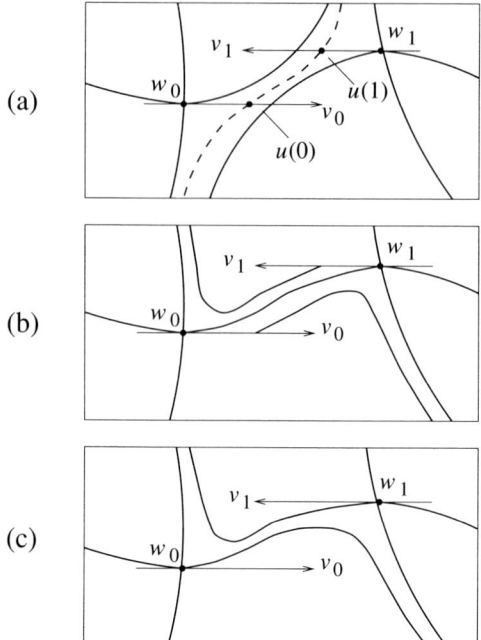

Fig. 1.25. Geometric interpretation of the equations for computing heteroclinic orbits in the special case $n = 2$, $n_0 = n_1 = 1$.

First consider the special case of a heteroclinic connection between two saddle points in $\mathbb{R}^2$, that is, $n = 2$, $n_0 = n_1 = 1$; a graphical illustration of this case is shown in Fig. 1.25. Then $n_\lambda + 1 = 2$, i.e., a branch of heteroclinic orbits requires two free problem parameters $\boldsymbol{\lambda} = (\lambda_1, \lambda_2)$. Consider λ_2 as fixed here.

For $\lambda_1 = \lambda_1^*$ we assume the existence of the heteroclinic connection in Fig. 1.25(b). Generically, perturbation of λ_1 will produce either Fig. 1.25(a) or Fig. 1.25(c), depending on the sign of the perturbation. If ε_0 and ε_1 are sufficiently small, then there exists a λ_1 close to λ_1^* for which (1.17)–(1.20) (and (1.22)) can be satisfied; here, this is satisfied for λ_1 as in Fig. 1.25(a). Furthermore, the radii ε_0 and ε_1 can be chosen such that the period of the orbit equals a given large value T, and such that the phase condition (1.21) is satisfied.

Some more particular cases are:

1. The connection of a saddle to a node in $\mathbb{R}^2$.
   Here $n = 2$, $n_0 = 1$, $n_1 = 2$, so $n_\lambda = 0$. A branch of connections requires one problem parameter;
2. If $n = 3$, $n_0 = 3$, $n_1 = 2$,
   then $n_\lambda = -1$, which means that a two-dimensional manifold of connecting orbits is already possible for fixed problem parameters;

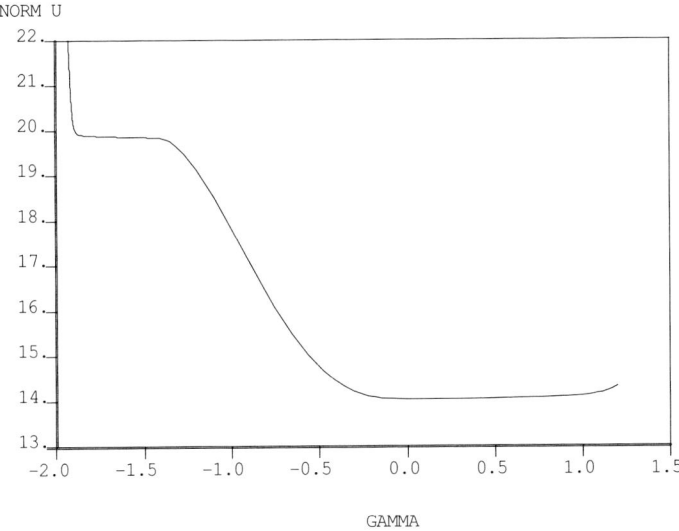

NORM U

GAMMA

Fig. 1.26. Bifurcation diagram of the singularly-perturbed BVP (1.23).

3. The *homoclinic orbit*.
   In this case $\mathbf{w}_0 = \mathbf{w}_1$ and $n_0 + n_1 = n$, so that $n_\lambda = 1$. Such orbits can also be computed as the limit of periodic orbits as the period $T \to \infty$.

1.6 Other Applications of BVP Continuation

We end this chapter with two examples where the boundary value continuation of AUTO is applied in special contexts.

Singularly Perturbed BVP

AUTO is well suited for computing solutions in systems with multiple timescales. The numerical sensitivity caused by the difference in timescales is dealt with by the orthogonal collocation solution technique with adaptive meshes. The pseudo-arclength continuation ensures detection of changes along the solution family. Consider the singularly perturbed system [26]

$$\varepsilon u''(x) = u(x)\, u'(x)\, (u(x)^2 - 1) + u(x),$$

with boundary conditions

$$u(0) = \frac{3}{2}, \quad u(1) = \gamma.$$

The computational formulation is in the form

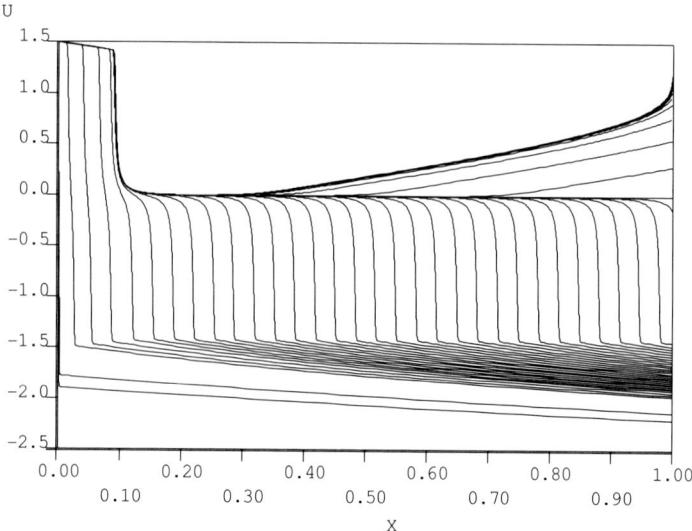

Fig. 1.27. Some solutions along the solution branch of (1.23).

$$\begin{cases} u_1' = u_2 \\ u_2' = \dfrac{\lambda}{\varepsilon}\left(u_1 u_2(u_1^2 - 1) + u_1\right), \end{cases} \qquad (1.23)$$

with boundary conditions

$$u_1(0) = \frac{3}{2}, \quad u_1(1) = \gamma.$$

The parameter λ is a homotopy parameter to locate a starting solution. In the first run λ varies from 0 to 1 and $\varepsilon = 1$ is fixed. In the second run ε is decreased by continuation to the desired value. We use $\varepsilon = 10^{-3}$.

Once a starting solution is obtained, we continue the solution for $\varepsilon = 10^{-3}$ in the parameter γ. This third run takes many continuation steps. Figure 1.26 shows the bifurcation diagram with the solution family obtained by continuation in γ. A selection of the solutions along the branch is shown in Fig. 1.27.

1.6.1 Orbit Continuation in IVP

One can also use continuation to compute solution families of initial value problems (IVP). Using continuation instead of integration of a large number of initial conditions has the advantage that the manifold described by the orbits is well covered, even in problems with very sensitive dependence on initial conditions. As an example, we consider the Lorenz equations given by

$$\begin{cases} x' = \sigma(y - x), \\ y' = \rho x - y - xz, \\ z' = xy - \beta z, \end{cases} \qquad (1.24)$$

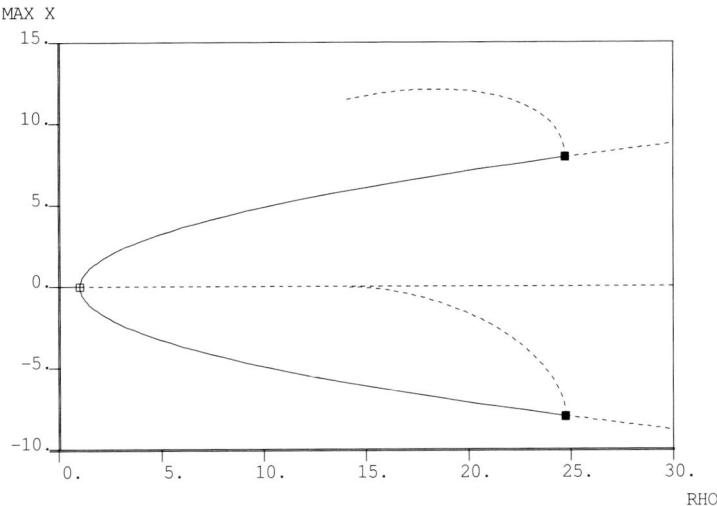

Fig. 1.28. Bifurcation diagram of the Lorenz equations (1.24).

with $\sigma = 10$ and $\beta = 8/3$.

Let us first analyze the stationary solutions of (1.24) as a function of ρ. A bifurcation diagram is shown in Fig. 1.28. The zero solution is unstable for $\rho > 1$. Two nonzero (symmetric) stationary solutions bifurcate at $\rho = 1$. These nonzero stationary solutions become unstable for $\rho > \rho_H \approx 24.7$. At ρ_H there are Hopf bifurcations, and a family of unstable periodic solutions emanates from each of the Hopf bifurcation points; only the maximal x-coordinate is shown in Fig. 1.28, and Fig. 1.29 shows some of these periodic orbits in the (x, y)-plane. The families of periodic solutions end in *homoclinic orbits* (infinite period) at $\rho \approx 13.9$.

Now let $\rho = 28$. For this parameter value the Lorenz equations have a *strange attractor*. Let

$$\mathbf{u} = \begin{pmatrix} x \\ y \\ z \end{pmatrix},$$

and write the Lorenz equations as

$$\mathbf{u}'(t) = \mathbf{f}(\mathbf{u}(t)).$$

The origin $\mathbf{0}$ is a saddle point, with eigenvalues $\mu_1 \approx -2.66$, $\mu_2 \approx -22.8$, $\mu_3 \approx 11.82$, and corresponding normalized eigenvectors $\mathbf{v}_1$, $\mathbf{v}_2$, and $\mathbf{v}_3$, respectively. We want to compute the *stable manifold* of the origin.

We compute an initial orbit $\mathbf{u}(t)$, for t from 0 to T (where $T < 0$), with $\mathbf{u}(0)$ close to $\mathbf{0}$ in the eigenspace spanned by $\mathbf{v}_1$ and $\mathbf{v}_2$, that is,

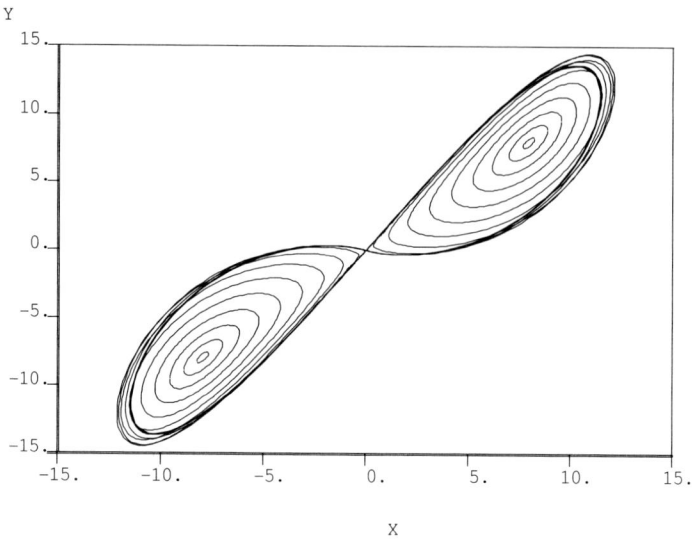

Fig. 1.29. Periodic orbits of the Lorenz equations (1.24).

$$\mathbf{u}(0) = \mathbf{0} + \varepsilon \left(\frac{\cos(\theta)}{|\mu_1|} \mathbf{v}_1 - \frac{\sin(\theta)}{|\mu_2|} \mathbf{v}_2 \right),$$

for, say, $\theta = 0$.

The IVP can be solved with AUTO as follows. Scale time $t \mapsto \frac{t}{T}$. Then the initial orbit satisfies

$$\mathbf{u}'(t) = T \mathbf{f}(\mathbf{u}(t)), \quad 0 \le t \le 1,$$

and

$$\mathbf{u}(0) = \frac{\varepsilon}{|\mu_1|} \mathbf{v}_1.$$

The initial orbit has length

$$L = T \int_0^1 \| \mathbf{f}(\mathbf{u}(s)) \| \; ds.$$

Thus the initial orbit is a solution of the equation $\mathbf{F}(\mathbf{X}) = \mathbf{0}$, where $\mathbf{X} = (\mathbf{u}(\cdot), \theta, T)$ (for given L and ε) and

$$\mathbf{F}(\mathbf{X}) = \begin{cases} \mathbf{u}'(t) - T \mathbf{f}(\mathbf{u}(t)), \\ \mathbf{u}(0) - \varepsilon \left(\dfrac{\cos(\theta)}{|\mu_1|} \mathbf{v}_1 - \dfrac{\sin(\theta)}{|\mu_2|} \mathbf{v}_2 \right), \\ T \int_0^1 \| \mathbf{f}(\mathbf{u}(s)) \| \; ds - L . \end{cases}$$

Once the initial orbit has been integrated up to a sufficiently long arclength L, we can use pseudo-arclength continuation to find a family of solution segments

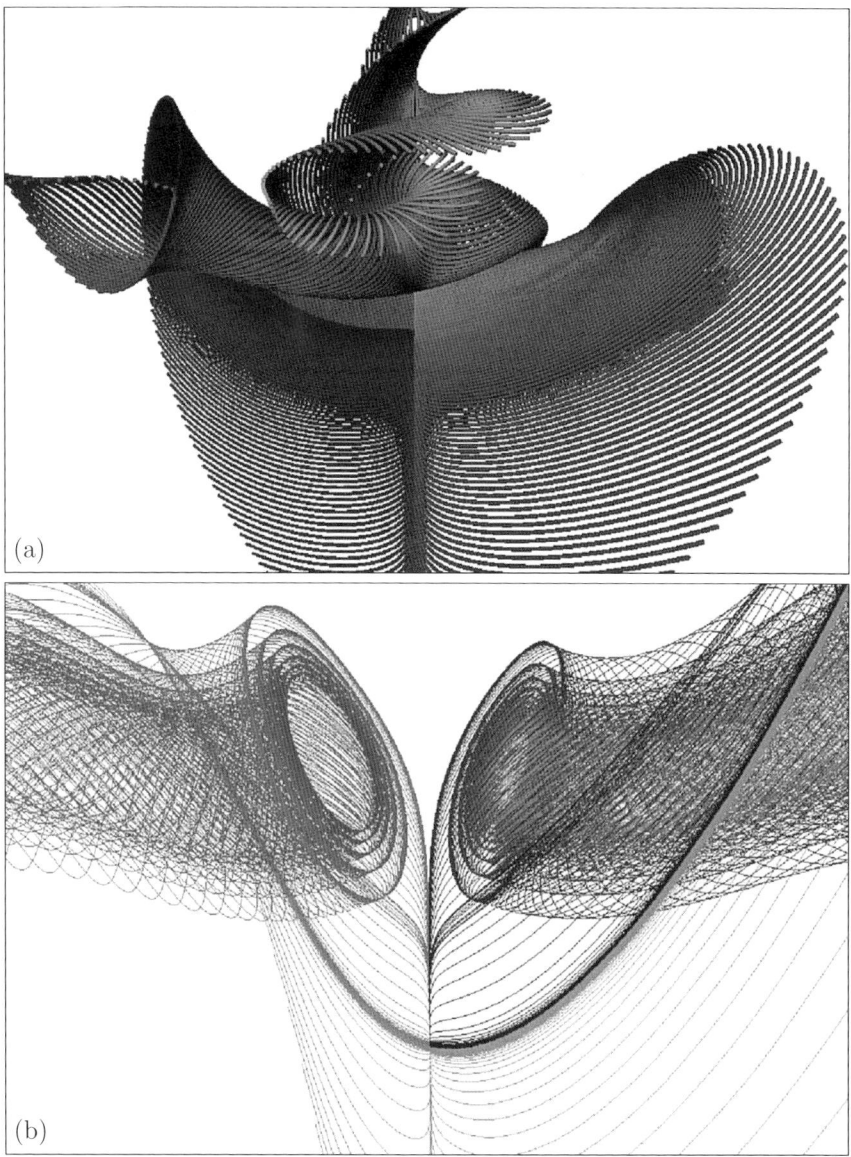

Fig. 1.30. The stable manifold of the origin in the Lorenz equations (1.24). Panel (a) shows the family of orbits that represent part of the manifold. Panel (b) shows another section of the Lorenz manifold.

that forms an approximation of the *Lorenz manifold*, the stable manifold of the origin. The set-up for pseudo-arclength continuation is now:

$$\mathbf{F}(\mathbf{X}_1) = 0,$$
$$(\mathbf{X}_1 - \mathbf{X}_0)^* \dot{\mathbf{X}}_0 - \Delta s = 0, \qquad (\| \dot{\mathbf{X}}_0 \| = 1),$$

with $\mathbf{X} = (\mathbf{u}(\cdot), \theta, T)$ and L and ε fixed. It is important to note here that we do not just change the initial point (i.e., the value of θ). The continuation stepsize Δs measures the change in $\mathbf{X}$. An impression of part of the computed Lorenz manifold is shown in Fig. 1.30. For more detailed results see [13].

1.7 Outlook

We discussed the set-up in AUTO for the numerical continuation of families of solutions to first-order systems of ordinary differential equations. AUTO uses Keller's pseudo-arclength continuation [22], which can equally well be applied to solution families of algebraic problems, e.g., families of stationary solutions. When applied to families of orbits, each continuation step involves solving a boundary value problem. AUTO uses piecewise polynomial collocation with Gauss-Legendre collocation points (orthogonal collocation) [7, 3], similar to COLSYS [2] and COLDAE [4], with adaptive mesh selection [32].

The basic objective behind the continuation methods of AUTO is the ability to perform a numerical bifurcation analysis. Such computational results give a deeper understanding of the solution behavior, stability, multiplicity, and bifurcations, and they often provide direct links to the underlying mathematical theories. We highlighted only the basic set-up in AUTO. For multi-parameter bifurcation analysis the system that implicitly defines the solution branch is extended to contain bifurcation conditions; see, for example, [11, 12]. By monitoring the appropriate bifurcation condition AUTO detects, say, a Hopf bifurcation when continuing a family of stationary solutions in one parameter. This bifurcation point can subsequently be continued by extending the set-up for pseudo-arclength continuation with extra equations (the bifurcation condition), and freeing a second parameter. For so-called *minimally extended systems* see [20, 24].

There is a need for further refinement of existing continuation algorithms and software for bifurcation analysis, and there is a need for their extension to new classes of problems. Probably the greatest challenges lie in the development of numerical continuation and bifurcation software for partial differential equations. There is such a package for scalar nonlinear elliptic PDEs on general domains in $\mathbb{R}^2$ [5], which is based on multigrid solution techniques; see also [27, 28, 29]. Good results have also been obtained with stabilized simple iteration schemes for computing stationary PDE solutions 'with mostly stable modes'; see, for example, [33]. There remains a need for general bifurcation software for *systems* of elliptic PDEs, subject to general boundary conditions and integral constraints. For the case of such systems on simple domains in $\mathbb{R}^2$, the generalization of the collocation method of Sect. 1.2.3 carries some promise. To become comparable in performance to current ODE bifurcation software it is necessary to use adaptive meshes. In this case the direct solution of the linear systems arising in Newton's method remains feasible, so that a high degree of robustness is possible. For developments in this directions, see [10, 15].

The chapters in this book also provide a wide range of examples of extensions and refinements of the continuation algorithms.

Acknowledgments

Although part of the material in these lecture notes is original, the author is greatly indebted to many collaborators, and above all to H.B. Keller of the California Institute of Technology, whose published and unpublished work is strongly present in these notes, and without whose inspiration and support since 1975 the algorithms and software described here would not exist in their current form. The author is grateful to World Scientific Publishing for permission to reproduce previously published material from [12].

References

1. D. G. Aronson, E. J. Doedel, and H. G. Othmer. The dynamics of coupled current-biased Josephson junctions II. *Internat. J. Bifur. Chaos Appl. Sci. Engrg.*, 1(1): 51–66, 1991.
2. U. M. Ascher, J. Christiansen, and R. D. Russell. Collocation software for boundary value ODEs. *ACM Trans. Math. Software*, 7:209–222, 1981.
3. U. M. Ascher, R. M. M. Mattheij, and R. D. Russell. *Numerical solution of boundary value problems for ordinary differential equations.* Prentice-Hall, 1988; SIAM, 1995.
4. U. M. Ascher and R. J. Spiteri. Collocation software for boundary value differential-algebraic equations. *SIAM J. Sci. Comput.*, 15:938–952, 1995.
5. R. E. Bank. *PLTMG, A Software Package for Solving Elliptic Partial Differential Equations.* (SIAM, Philadelphia, 1990).
6. W.-J. Beyn. The numerical computation of connecting orbits in dynamical systems. *IMA J. Num. Anal.*, 9:379–405, 1990.
7. C. de Boor and B. Swartz. Collocation at Gaussian points. *SIAM J. Numer. Anal.*, 10:582–606, 1973.
8. E. J. Doedel. AUTO: A program for the automatic bifurcation analysis of autonomous systems, *Cong. Num.* 30, 1981, 265–284. (Proc. 10th Manitoba Conf. on Num. Math. and Comp., Univ. of Manitoba, Winnipeg, Canada.)
9. E. J. Doedel. Numerical Analysis and Control of Bifurcation Problems. Report UMSI 89/17. University of Minnesota Supercomputer Institute. February 1989.
10. E. J. Doedel. On the construction of discretizations of elliptic partial differential equations. *J. Difference Equations and Applications*, 3:389–416, 1997.
11. E. J. Doedel, H. B. Keller, and J. P. Kernévez. Numerical analysis and control of bifurcation problems (I): Bifurcation in finite dimensions. *Internat. J. Bifur. Chaos Appl. Sci. Engrg.*, 1(3):493–520, 1991.

12. E. J. Doedel, H. B. Keller, and J. P. Kernévez. Numerical analysis and control of bifurcation problems (II): Bifurcation in infinite dimensions. *Internat. J. Bifur. Chaos Appl. Sci. Engrg.*, 1(4):745–772, 1991.

13. E. J. Doedel, B. Krauskopf, H. M. Osinga. Global bifurcations of the Lorenz manifold. *Nonlinearity*, 19(12):2947–2972, 2006.

14. E. J. Doedel, V. Romanov, R. C. Paffenroth, H. B. Keller. D. J. Dichmann, J. Galán, and A. Vanderbauwhede. Elemental periodic orbits associated with the libration points in the Circular Restricted 3-Body Problem. *Internat. J. Bifur. Chaos Appl. Sci. Engrg.*, 17(8), 2007 (in press).

15. E. J. Doedel and H. Sharifi. Collocation methods for continuation problems in nonlinear elliptic PDEs. In D. Henry and A. Bergeon, editors, *Issue on Continuation Methods in Fluid Mechanics*, Notes on Numer. Fluid. Mech., 74:105-118, (Vieweg, 2000).

16. K. Engelborghs, T. Luzyanina and D. Roose. Numerical bifurcation analysis of delay differential equations using DDE-BIFTOOL. *ACM Trans. Math. Software*, 28(1): 1–21, 2002. Available via http://www.cs.kuleuven.ac.be/cwis/research/twr/research/software/delay/ddebiftools.shtml.

17. B. Ermentrout. *Simulating, Analyzing, and Animating Dynamical Systems: A Guide to XPPAUT for Researchers and Students*, volume 14 of *Software, Environments, and Tools* (SIAM, Philadelphia, 2002). Available via http://www.math.pitt.edu/~bard/xpp/xpp.html.

18. T. F. Fairgrieve and A. D. Jepson. O.K. Floquet multipliers. *SIAM J. Numer. Anal.*, 28(5):1446–1462, 1991.

19. M. J. Friedman and E. J. Doedel. Numerical computation and continuation of invariant manifolds connecting fixed points. *SIAM J. Numer. Anal.*, 28:789–808, 1991.

20. W. Govaerts. *Numerical Methods for Bifurcations of Dynamical Equilibria.* (SIAM, Philadelphia, 2000).

21. W. Govaerts, Yu. A. Kuznetsov and A. Dhooge. Numerical continuation of bifurcations of limit cycles in Matlab. *SIAM J. Sci. Computing*, 27(1):231–252, 2005.

22. H. B. Keller. Numerical solution of bifurcation and nonlinear eigenvalue problems. In P. H. Rabinowitz, editor, *Applications of Bifurcation Theory.* (Academic Press, 1977), pages 359–384.

23. H. B. Keller. Lectures on Numerical Methods in Bifurcation Problems. Tata Institute of Fundamental Research. Bombay, 1987.

24. Yu. A. Kuznetsov. *Elements of Applied Bifurcation Theory.* (Springer-Verlag, New York, 2004).

25. M. Lentini and H. B. Keller. Boundary value problems over semi-infinite intervals and their numerical solution. *SIAM J. Numer. Anal.*, 17:557–604, 1980.

26. J. Lorenz. Nonlinear boundary value problems with turning points and properties of difference schemes. In W. Eckhaus and E. M. de Jager, editors, *Singular Perturbation Theory and Applications.* (Springer Verlag, New York, 1982).

27. K. Lust. PDECONT: A timestepper-based continuation code for large-scale systems. Available via http://www.dynamicalsystems.org/sw/sw/detail?item=8.

28. K. Lust, D. Roose, A. Spence, and A. R. Champneys. An adaptive Newton-Picard algorithm with subspace iteration for computing periodic solutions. *SIAM J. Sci. Computing*, 19(4):1188–1209, 1998.

29. K. Lust and D. Roose. Computation and bifurcation analysis of periodic solutions of large-scale systems. In E. J. Doedel and L. S. Tuckerman, editors, *Numerical Methods for Bifurcation Problems and Large-Scale Dynamical Systems*, IMA Vol. Math. Appl., volume 119. (Springer-Verlag, New York, 2000), pages 265–301.

30. F. J. Muñoz-Almaraz, E. Freire, J. Galán, E. J. Doedel, and A. Vanderbauwhede. Continuation of periodic orbits in conservative and Hamiltonian systems. *Physica D*, 181(1-2), 1–38, 2003.

31. E. L. Reiss. Column buckling — An elementary example of bifurcation. In J. B. Keller and S. Antman, editors, *Bifurcation Theory and Nonlinear Eigenvalue Problems*, pages 1–16. (W. A. Benjamin, Publishers, 1969).

32. R. D. Russell and J. Christiansen. Adaptive mesh selection strategies for solving boundary value problems. *SIAM J. Numer. Anal.*, 15:59–80, 1978.

33. G. M. Shroff and H. B. Keller. Stabilization of unstable procedures: The recursive projection method. *SIAM J. Numer. Anal.*, 30(4):1099–1120, 1993.

34. X.-J. Wang and E. J. Doedel. AUTO94P: An experimental parallel version of AUTO. Technical report, Center for Research on Parallel Computing, California Institute of Technology, Pasadena CA 91125. CRPC-95-3, 1995.

2

Interactive Continuation Tools

Willy Govaerts[1] and Yuri A Kuznetsov[2]

[1] Department of Applied Mathematics and Computer Science, Ghent University, Belgium
[2] Department of Mathematics, Utrecht University, The Netherlands

Systematic bifurcation analysis requires the repeated continuation of different phase objects in free parameters, the detection and analysis of their bifurcations, and branch switching. Such computations produce a lot of numerical data that must be analyzed and, finally, presented in graphical form. Thus, continuation programs should not only be efficient numerically but should allow for interactive management and have a user-friendly graphics interface. The development of such programs is progressing rapidly. Here we make an attempt to survey existing interactive continuation and bifurcation tools and outline their history and perspectives. This is followed by the presentation of a framework that organizes the different types of objects and bifurcating branches. We give a brief overview of how such a framework is implemented in the recent software environment MATCONT. In the final two sections we give a few examples that illustrate the use of MATCONT and indicate directions of future developments.

2.1 Overview of Existing Software

During the last decades, considerable efforts have been made to develop general-purpose software tools for bifurcation analysis. One may distinguish at least three types (generations) of such software:

1. Noninteractive packages and codes,
2. Interactive programs,
3. Software environments.

Since the development of numerical algorithms advances with each generation, these tools also differ in supported computations. We give here an overview for each of the above types, with emphasis on the best known packages.

2.1.1 Noninteractive Packages and Codes

Noninteractive packages and codes first appeared in the beginning of the 1980s and were written in FORTRAN. They allowed one to continue equilibria and limit cycles of ODEs, as well as detect and subsequently continue their basic bifurcations: limit point (saddle-node) bifurcation, Hopf bifurcation, and period-doubling bifurcation. The most widely used packages of this generation are AUTO86 [12] and LINLBF [24]. Although these two packages supported a similar level of bifurcation analysis, they employed very different numerical algorithms. For example, LINLBF bases its test functions to locate Hopf and Neimark-Sacker bifurcation points on Hurwitz determinants, while AUTO86 bases the detection and location of all local bifurcations on monitoring the eigenvalues (multipliers). Moreover, the continuation of equilibrium (fixed point) bifurcations in AUTO86 is done using extended augmented systems that include eigenvectors, while in LINLBF minimally augmented systems are used. The most essential difference, however, lies in the continuation of limit cycles and their codimension-one bifurcations. In LINLBF these tasks are performed via numerically constructed and differentiated Poincaré maps, while AUTO86 employs the discretization of the corresponding boundary value problems using piecewise-polynomial approximations and orthogonal collocation. The latter proved its superiority for more complex multi-dimensional ODEs.

Bifurcation theory relies on center manifold reduction, followed by transformation to a normal form. The computation of normal-form coefficients is an important aspect of bifurcation software; for background we refer to [32]. The explicit computation of normal-form coefficients is not supported by AUTO86, and LINLBF only computes normal-form coefficients for local codimension-one bifurcations. There were a few codes available in the 1980s for simple numerical normal-form and branching analysis, e.g. STUFF [5] and BIFOR2 [22], but switching to the computation of different bifurcating objects required manual restarting of the code with new initial data.

Several other noninteractive packages for the continuation of simplest bifurcations in ODEs also appeared around this time, e.g. BIFPACK by Seydel [38, 39], and development continues to date, in particular for large-scale dynamical systems with packages such as LOCA[3] by Salinger et al. [35, 36].

2.1.2 Interactive Programs

Interactive programs for the bifurcation analysis of ODEs appeared at the end of the 1980s, when workstations and IBM-PC compatible computers became widely available at universities and general research institutes. All programs of this generation have a simple Graphical User Interface (GUI) with buttons,

---

[3] LOCA is available via http://www.cs.sandia.gov/loca/.

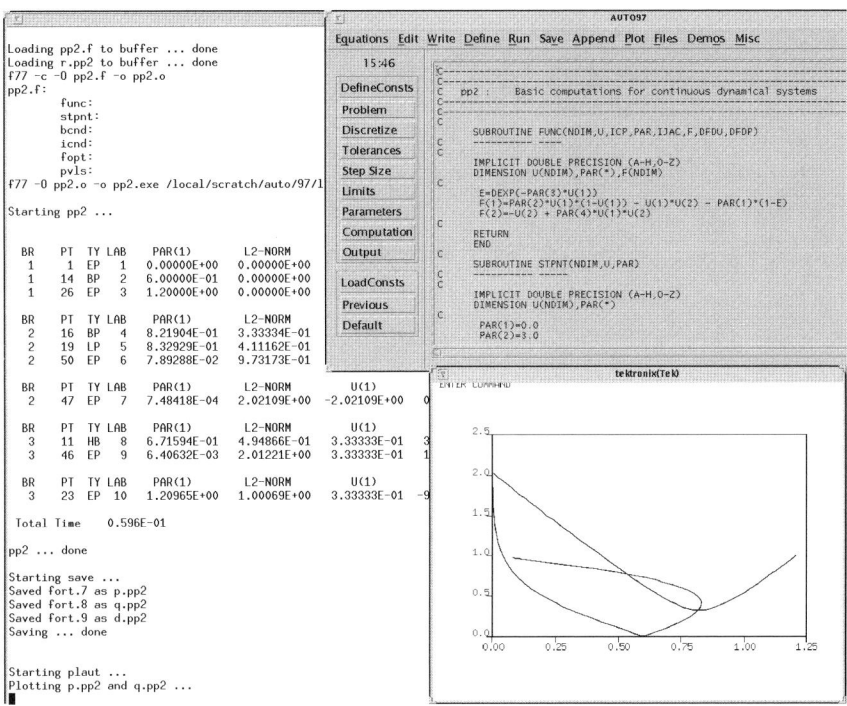

Fig. 2.1. Screen snapshot of Auto94/97.

windows, and pull-down menus, and support the on-line input of the right-hand side of ODEs (through compilation with a FORTRAN or C compiler). Computed curves could now be plotted directly in a graphics window.

The continuation code Auto86 mentioned above already comes with a simple interactive graphics program called `plaut` that allows for graphical presentation of computed data. There are versions of `plaut` for most of the widespread workstations, as well as a Matlab version `mplaut`[4], written by De Feo. There have been several attempts to improve the user interface of Auto86 and later versions of Auto. The first interactive version of Auto86 was developed at Princeton University by Taylor and Kevrekidis [41] for SGI workstations. Another example is XppAut[5] by Ermentrout [15] for workstations and PCs, which developed from combining the MS-DOS program PhasePlane with Auto. XppAut is also capable of simple phase-plane analysis, including the computation of one-dimensional global invariant manifolds of equilibria, as is Scigma [40]. Note that XppAut is still widely used and includes tools for the analysis of delay equations, functional equations, and stochastic equations. Doedel, Wang, and Fairgrieve also designed the interac-

---

[4] `mplaut` is available via `http://www.math.uu.nl/people/kuznet/cm`.
[5] XppAut is available via `http://www.math.pitt.edu/~bard/xpp/xpp.html`.

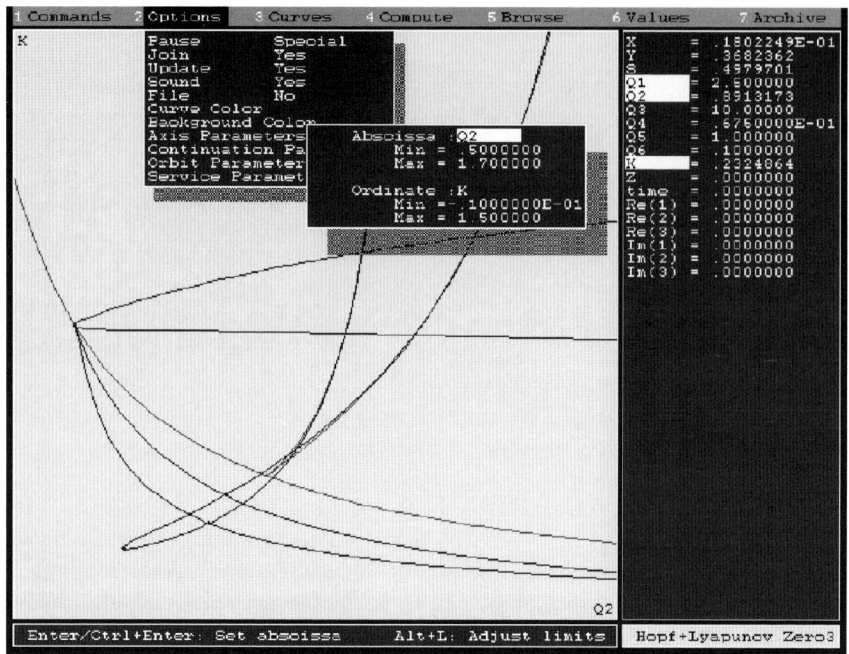

Fig. 2.2. Screen snapshot of LocBif 2.0.

tive version Auto94 for UNIX workstations with X-Windows; see Fig. 2.1. This version has extended numerical capabilities, including the continuation of all codimension-one bifurcations of limit cycles and fixed points. The software was upgraded in 1997 to support the continuation of homoclinic orbits using HomCont [9, 13]. This version is called Auto97[6] and there is also the C-version Auto2000 which has a new interactive graphics browser.

The major difficulty in using all versions of Auto is the analysis of detected bifurcation points and switching at these points to the continuation of other bifurcation curves, which requires browsing of several output files and a good understanding of their formats. However, due to the exceptional numerical efficiency of Auto, attempts to provide a better GUI for it continue to date; see, for example, Oscill8[7].

The first user-friendly interactive bifurcation program for bifurcation analysis was LocBif[8] developed for PCs under MS-DOS by Khibnik, Kuznetsov, Levitin, and Nikolaev [25]; a screen snapshot of version 2.0 is shown in Fig. 2.2. The numerical part of the program is based on the non-interactive code

[6] Auto is available via http://cmvl.cs.concordia.ca/.

[7] Oscill8 is available via http://oscill8.sourceforge.net/doc/.

[8] LocBif is available via http://www.math.uu.nl/people/kuznet/LOCBIF; LocBif works as a DOS-application under MS-Windows, but is no longer supported.

LINLBF and allows for continuation of equilibrium, fixed-point, and limit cycle bifurcations up to codimension three. The program allows for easy switching between the computation of various curves at detected bifurcation points. The user can manipulate individual computed bifurcation curves, which are stored separately in an archive. Version 1.0 of LOCBIF uses an external FORTRAN compiler and has a very simple keyboard-based interface, but version 2.0 can be driven by a mouse and has a special built-in compiler for the right-hand side. Neither version, however, has special tools to output the computed curves in a graphic format.

The program CANDYS/QA[9] by Feudel and Jansen [16] also belongs to this generation but is less widely used. All programs mentioned so far have closed architecture.

2.1.3 Software Environments

The first *software environments* for bifurcation analysis were DsTOOL[10] and CONTENT[11], developed in the 1990s. Both programs support the simulation of ODEs. The user can define/modify a dynamical model, perform a rather complete analysis, and export the results in a graphical form, all without leaving the program. Though hard, it is possible to extend them. The programs have an elaborate GUI and provide off- or on-line help and extensive documentation for users and developers.

DsTOOL [3] runs under UNIX or Linux. It performs simple phase-plane analysis and includes the computation of equilibria and associated one-dimensional stable and unstable manifolds, along with the continuation of equilibria and their codimension-one bifurcations, which is done by using parts of the LINLBF code.

The interactive software CONTENT was developed by Kuznetsov and Levitin with contributions by De Feo, Sijnave, Govaerts, Doedel, and Skovoroda; a screen snapshot of CONTENT 1.5 is shown in Fig. 2.3. The software runs on most popular workstations under UNIX and on PCs under Linux or MS-Windows and supports the continuation of equilibria and their bifurcations of codimension up to two. CONTENT uses minimal and extended augmented systems, as described in [19, 20], as well as the continuation of limit cycles using AUTO-like algorithms. Moreover, CONTENT supports the normal-form computations for many equilibrium bifurcations, taking advantage of internally generated symbolic derivatives of order up to three, and allows for branch switching by using algebraic branching equations. The software provides extensive storage, export and import facilities for computed curves and diagrams, including their numerical and PostScript formats. Switching between

---

[9] CANDYS/QA is available via `http://www.agnld.uni-potsdam.de/~wolfgang/candys.html`.

[10] DsTOOL is available via `http://www.cam.cornell.edu/~gucken/dstool`.

[11] CONTENTis available via `http://www.math.uu.nl/people/kuznet/CONTENT/`.

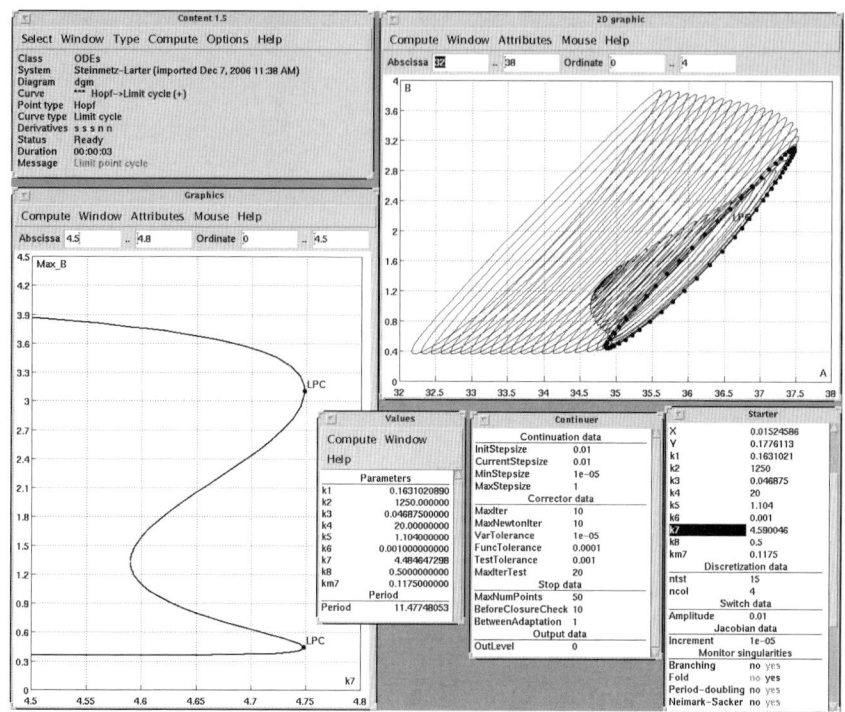

Fig. 2.3. Screen snapshot of CONTENT 1.5.

various bifurcating objects at special points is very easy and flexible in CON-TENT. The latest software project MATCONT[12] is a Matlab interactive toolbox for the continuation and bifurcation analysis of ODEs [11] that is based on experience in developing and using CONTENT.

To conclude this section, we mention that numerical bifurcation analysis of smooth iterated maps is also supported by existing software. Location, analysis, and continuation of fixed-point bifurcations are very similar to those for equilibria of ODEs, and are supported, for example, by AUTO, LOCBIF, and CONTENT [18]. Other problems, particularly the analysis of global bifurcations, require special algorithms. For example, one needs special algorithms for the computation of the one-dimensional stable and unstable invariant manifolds of fixed points of maps; implementations for one-dimensional manifolds already exist in DSTOOL [29, 30] and DYNAMICS [34, 43, 44]. Such algorithms are necessary for the continuation of homoclinic orbits and their tangencies [4], which is also implemented as an AUTO-driver by Yagasaki [42]. The computation of two- or higher-dimensional invariant manifolds, for example, global stable and unstable invariant manifolds and invariant tori, and their bifurcations both for ODEs and maps is much more difficult and only a few algorithms

---

[12] MATCONT is available via http://www.matcont.ugent.be/.

are available; see Chap. 4, and [28, 31] for global manifolds and [6, 14, 37] for invariant tori.

2.2 Bifurcation Objects and Their Relations

The application of bifurcation theory to multi-parameter dynamical systems requires a clear *continuation strategy* that should provide rules on how to increase and decrease the number of control parameters while studying objects of different codimension. In addition, this strategy should suggest how to switch between different bifurcations of the same codimension, keeping the number of control parameters constant. As was first mentioned in [25] from a theoretical point of view, this strategy must be based on *graphs of adjacency* [1] that describe relationships between bifurcations. Below, we present two graphs of adjacency, describing *detection* relationships and *branching* relationships. The next section identifies the bifurcation objects for ODEs; the equivalent for maps is done in Sect. 2.2.2.

2.2.1 Bifurcation Objects in ODEs

Tables 2.1 and 2.2 list the codimension-zero, -one, and -two objects that can be found in generic continuous dynamical systems, along with associated labels based on standard terminology [32]. Table 2.1 lists all objects related to equilibria and limit cycles, while Table 2.2 focusses on objects related to homoclinic orbits of equilibria. The relationships between these objects are complicated.

The detection relationships between the objects in Tables 2.1 and 2.2 are presented in Figs. 2.4 and 2.5, respectively. For example, the arrows from O to EP and LC mean that it is generically possible for a computed orbit (O) to converge to a (stable) equilibrium (EP) or to a (stable) limit cycle (LC). An arrow from an object A different from O to an object B means that the continuation of a one-parameter family of objects of type A can generically lead to the detection of an object of type B, either because object B is a special case of object A or because it is a limiting case when the parameter tends to a special value. An example of the first situation is a Hopf bifurcation point (H) on a curve of equilibria (EP); an example of the second situation is a homoclinic orbit of a hyperbolic saddle (HHS), because it is the limit of a branch of periodic orbits (LC) when the period tends to infinity. We do not distinguish between the two situations, because the difference depends somewhat on the definition of a family of objects and it may depend on the implementation of the defining system that is used in the computation of the branch (e.g. an H point on a family of LC objects). Note that any computation normally starts with a point P as an initial condition to generate an orbit O; this first step does not feature in Fig. 2.4. We are interested in generic detection relationships, which is why the arrows always connect objects from

Table 2.1. Objects and associated labels related to equilibria and limit cycles of ODEs

Type of object	Label
Point	P
Orbit	O
Equilibrium	EP
Limit cycle	LC
Limit Point (fold) bifurcation	LP
Hopf bifurcation	H
Limit Point bifurcation of cycles	LPC
Neimark-Sacker (torus) bifurcation	NS
Period Doubling (flip) bifurcation	PD
Branch Point	BP
Cusp bifurcation	CP
Bogdanov-Takens bifurcation	BT
Zero-Hopf bifurcation	ZH
Double Hopf bifurcation	HH
Generalized Hopf (Bautin) bifurcation	GH
Branch Point of Cycles	BPC
Cusp bifurcation of Cycles	CPC
Generalized Period Doubling	GPD
Chenciner (generalized Neimark-Sacker) bifurcation	CH
1:1 Resonance	R1
1:2 Resonance	R2
1:3 Resonance	R3
1:4 Resonance	R4
Fold–Neimark-Sacker bifurcation	LPNS
Flip–Neimark-Sacker bifurcation	PDNS
Fold-flip	LPPD
Double Neimark-Sacker	NSNS

one codimension level down to objects on the next codimension level. The only two exceptions are the arrows from EP to BP and from LC to BPC, which jump over two codimension levels. In fact, these situations are non-generic, but they are so often found in systems with equivariance or invariant subspaces that most software packages support their detection.

The branching relationships between the objects in Tables 2.1 and 2.2 can be obtained directly from Figs. 2.4 and 2.5, respectively. In general, if there is an arrow in Fig. 2.4 or 2.5 from an object A different from O to an object B then for each object of type B there is a unique one-parameter family of objects of type A that branches off B, provided a total of $k+1$ free variables is chosen, where k is the codimension level of A. There are only four exceptions:

1. The arrows from EP to BP and from LC to BPC: there are generically two codimension-zero curves emanating from the codimension-two points.

Table 2.2. Objects and associated labels related to homoclinic orbits of equilibria of ODEs

Type of object	Label
Limit cycle	LC
Homoclinic orbit of a Hyperbolic Saddle	HHS
Homoclinic orbit of a Saddle-Node	HSN
Neutral saddle	NSS
Neutral saddle-focus	NSF
Neutral Bi-Focus	NFF
Shilnikov-Hopf	SH
Double Real Stable leading eigenvalue	DRS
Double Real Unstable leading eigenvalue	DRU
Neutrally-Divergent saddle-focus (Stable)	NDS
Neutrally-Divergent saddle-focus (Unstable)	NDU
Three Leading eigenvalues (Stable)	TLS
Three Leading eigenvalues (Unstable)	TLU
Orbit-Flip with respect to the Stable manifold	OFS
Orbit-Flip with respect to the Unstable manifold	OFU
Inclination-Flip with respect to the Stable manifold	IFS
Inclination-Flip with respect to the Unstable manifold	IFU
Non-Central Homoclinic to saddle-node	NCH

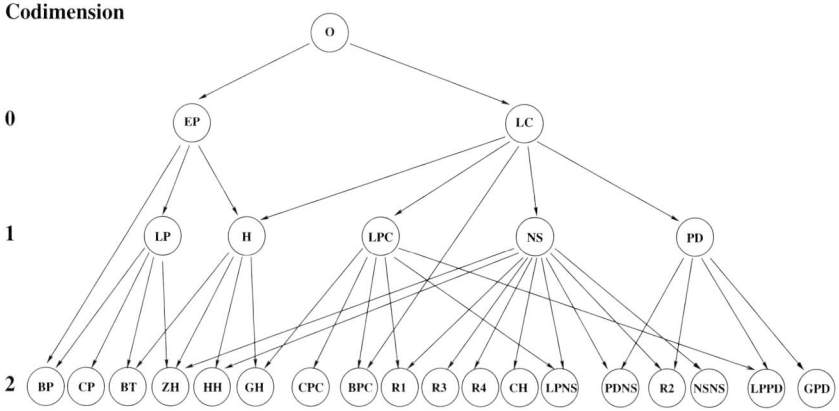

Fig. 2.4. Detection relationships between bifurcations of equilibria and limit cycles of ODEs; the branching relationships are found by following the arrows in the opposite directions, with four exceptions as discussed in the text.

2. The arrows from H to HH and from NS to NSNS: there are generically two codimension-one curves emanating from the codimension-two points.
3. The arrow from NS to ZH: the existence of the NS curve rooted in the ZH point is subject to an inequality constraint.

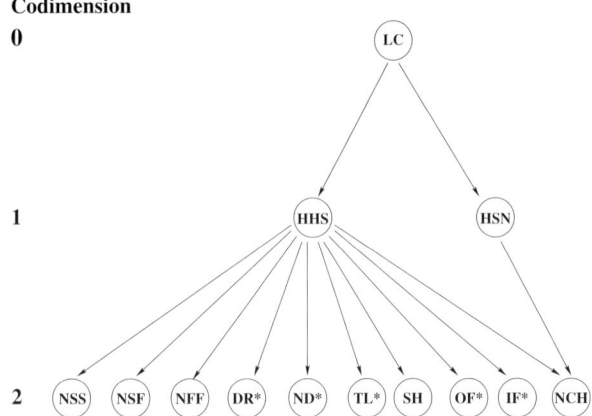

Fig. 2.5. Detection relationships between homoclinic bifurcations of ODEs; here *
stands for S or U; the branching relationships are found by following the arrows in
the opposite directions.

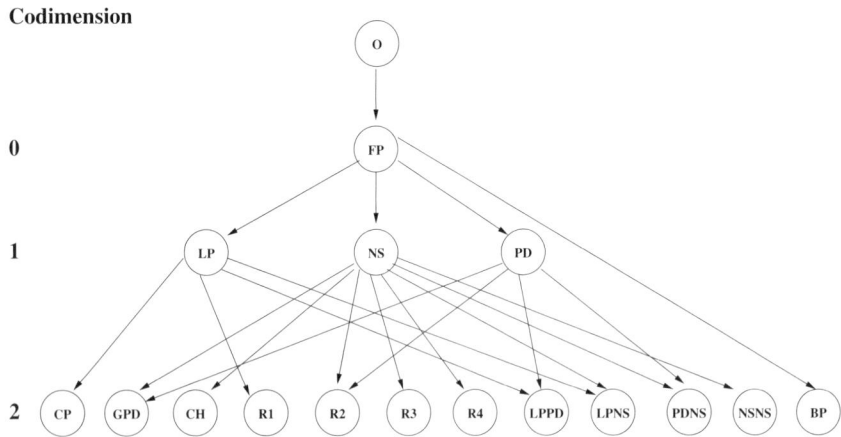

Fig. 2.6. Detection relationships between dynamical objects for maps.

4. The arrow from NS to HH: there are generically two NS curves emanating
   from an HH point.

We note that generically a curve of HHS orbits emanates from a Bogdanov
Takens point (BT), as well as two such curves from a Zero-Hopf bifurcation
point (ZH). These are not indicated in Figs. 2.4 and 2.5.

2.2.2 Bifurcation Objects for Cycles of Maps

In this section we present the equivalent objects and relationships for maps.
Table 2.3 lists the codimension-zero, -one, and -two objects that can be found

Table 2.3. Objects and associated labels related to equilibria and cycles of maps

Type of object	Label
Point	P
Orbit	O
Fixed Point	FP
Limit Point of cycle bifurcation	LP
Period Doubling Point of cycles	PD
Neimark-Sacker bifurcation	NS
Branch Point	BP
Cusp bifurcation	CP
Generalized Period Doubling	GPD
Chenciner (generalized Neimark-Sacker) bifurcation	CH
1:1 Resonance	R1
1:2 Resonance	R2
1:3 Resonance	R3
1:4 Resonance	R4
Fold–Neimark-Sacker bifurcation	LPNS
Flip–Neimark-Sacker bifurcation	PDNS
Fold-flip	LPPD
Double Neimark-Sacker	NSNS

in generic maps, together with the associated labels [32]. The detection relationships between them are presented in Fig. 2.6. The precise meaning of the arrows is simpler than in the case of ODEs: if we exclude O then an arrow from an object A to an object B indicates that object B can generically be found as a regular point on a branch of objects of type A. The only exception is the arrow from FP to BP which is again not generic but found in many examples that exhibit a form of equivariance or have invariant subspaces.

The branching diagram for maps, on the other hand, is far more complicated than for ODEs; this is largely due to the fact that one needs to consider different iterates of the underlying maps, which causes an additional complication. For reasons of clarity we, therefore, present two branching diagrams; see Figs. 2.7 and 2.8. As before, the arrows indicate the type of object to which one can generically switch from a given codimension-one or -two bifurcation point. If the arrow is dashed then this switching is subject to additional constraints. Furthermore, several switches to branches of lower codimension lead to curves with double, triple or quadruple iteration number, which is indicated by the symbols $\times 2$, $\times 3$, and $\times 4$, respectively.

2.3 The Implementation in MATCONT

The framework described in the previous section has been implemented in the recent Matlab-based software environment MATCONT [11]. At present

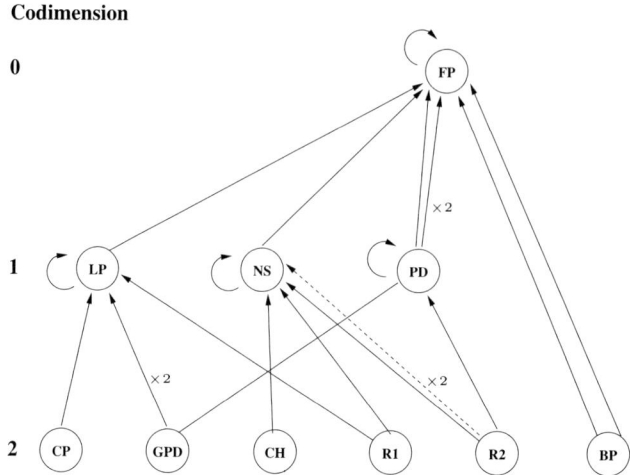

Fig. 2.7. Partial branching relationships for maps; see also Fig. 2.8. Dashed lines indicate switching subject to constraints and ×2 indicates switching to a curve with twice the period.

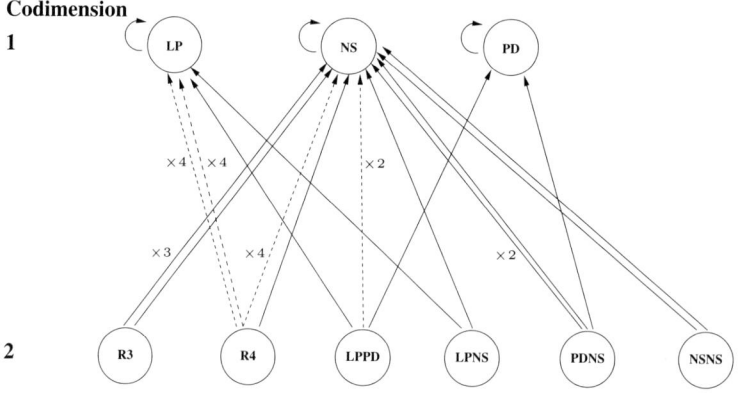

Fig. 2.8. Partial branching relationships for maps; see also Fig. 2.7. Dashed lines indicate switching subject to constraints and ×2 (×3, ×4) indicates switching to a curve with twice (three times, four times) the period.

three related Matlab packages are distributed, namely a command-line version CL_MATCONT and a GUI version MATCONT for ODEs, and a command-line version CL_MATCONTM for Maps[13]. As in AUTO and CONTENT, limit cycles are computed by an approach based on the discretization via piecewise-polynomial approximation with orthogonal collocation of the corresponding boundary value problem. However, MATCONT uses sparse Matlab solvers

---

[13] CL_MATCONT, MATCONT, and CL_MATCONTM are all available via http://www.matcont.ugent.be/.

instead of the original AUTO algorithm (a special block elimination; see Chap. 1). The same approach is applied for homoclinic orbits, in combination with the continuation of invariant subspaces for the equilibrium end point of the homoclinic orbit; for details on this method we refer to [10] and for its implementation in MATCONT to [17].

Nearly all functionalities described in Sect. 2.2 are supported. Remaining functionalities (now under construction) are:

- Branch switching at HH (equilibria), NSNS (limit cycles) and NSNS (cycle of maps) to the secondary branch of type H, NS (limit cycle) or NS (cycle of maps), respectively.
- Branch switching from ZH (equilibria) to NS (limit cycles).
- Branch switching from HH (equilibria) to NS (limit cycles).

A computationally more difficult problem is branch switching from ZH to HHS [8]. It is also planned to have a GUI version for CL_MATCONTM and to introduce automatic differentiation routines for the computation of the normal-form coefficients, which are now computed either via numerical directional derivatives or using a user-supplied code. Preliminary evidence indicates that finite difference approximations are not reliable for these computations. Also, for high-order iterates of maps the normal-form computations are much faster when using automatic differentiation compared to symbolically generated derivatives. The computation of normal-form coefficients for codimension-two bifurcations of limit cycles is not yet supported in MATCONT and is another topic for further development.

In some cases normal-form coefficients are not very informative. For example, for a limit point of equilibria (LP) the quadratic normal-form coefficient a is defined up to a nonzero multiple and the LP point is nondegenerate if $a \neq 0$ and degenerate (CP) if $a = 0$. However, because of truncation and round-off errors the value computed for a will always be nonzero. Therefore, the value of a reported at LP points is not very useful. However, provided its continuity along the LP-branch is ensured, this value is important for the detection of the CP-points.

In other cases the normal-form coefficients are very useful for the user, because their values determine the number and type of branches of new objects that emanate from the bifurcation points and whether these objects are stable or not. In the following two sections we discuss the cases that are of particular interest.

2.3.1 Normal-Form Coefficients for Bifurcations of ODEs as Given in MATCONT

The implementation in MATCONT provides normal-form coefficients for all codimension-one and -two bifurcations of equilibria, and periodic normal-form coefficients for all codimension-one bifurcations of limit cycles; see [32] and [33] for further details and notation used. We discuss here four cases:

1. The Hopf bifurcation H of an equilibrium, where the equilibrium has a pair of purely imaginary eigenvalues, is determined by the first Lyapunov coefficient l_1, which is the real part of the third-order coefficient in the complex normal form. If $l_1 < 0$ then the Hopf bifurcation is supercritical, i.e., unstable fixed points coexist with stable periodic orbits on one side of the bifurcation point in the center manifold. If $l_1 > 0$ then the Hopf bifurcation is subcritical, i.e., stable fixed points coexist with unstable periodic orbits on one side of the bifurcation point in the center manifold.

2. The Zero-Hopf bifurcation ZH, also called saddle-node Hopf, fold-Hopf or zero-pair bifurcation [32], is a codimension-two bifurcation where an equilibrium has one zero eigenvalue together with a pair of purely imaginary eigenvalues. The normal form involves quadratic coefficients denoted s and θ; see [32, Lemma 8.11]. An NS curve emanates from the ZH point only if $s\theta < 0$. The implementation in MATCONT also computes a relatively technical term $E(0)$. If $E(0) < 0$ then time has to be reversed in the unfolding analysis in [32], i.e., stable becomes unstable, and vice versa.

The above two cases deal with bifurcation of equilibria. For limit cycles we have:

3. The period-doubling or flip bifurcation PD, where the limit cycle has one Floquet multiplier at -1, involves the coefficient c in the periodic normal form that determines the bifurcation. If $c < 0$ then the flip bifurcation is supercritical, i.e., unstable periodic orbits coexist with stable double-period orbits on one side of the bifurcation point in the center manifold. If $c > 0$ then the flip bifurcation is subcritical, i.e. stable periodic orbits coexist with unstable double-period orbits on one side of the bifurcation point in the center manifold.

4. At a Neimark-Sacker bifurcation NS the limit cycle has a pair of complex conjugate Floquet multipliers on the unit circle. The bifurcation is determined by the cubic coefficient $\text{Re}(d)$ of the periodic normal form. If $\text{Re}(d) < 0$ then the NS bifurcation is supercritical, i.e., unstable limit cycles coexist with stable invariant tori on one side of the bifurcation point in the center manifold. If $\text{Re}(d) > 0$ then the NS bifurcation is subcritical, i.e., stable limit cycles coexist with unstable invariant tori on one side of the bifurcation point in the center manifold.

2.3.2 Normal-Form Coefficients for Bifurcations of Maps as Given in MATCONTM

The implementation in MATCONTM provides normal-form coefficients for all codimension-one and -two bifurcations of fixed points; see [32] for details. The codimension-one cases are very similar to the corresponding bifurcations of limit cycles listed in the previous section. The user should, in particular, be aware of:

1. The period-doubling or flip bifurcation PD, where the fixed point has an eigenvalue -1. The sign of the cubic normal-form coefficient b_1 determines whether the bifurcation is supercritical ($b_>0$) or subcritical ($b_1 < 0$) as before.
2. The Neimark–Sacker bifurcation NS, where the fixed point has a pair of complex conjugate eigenvalues on the unit circle. As before, the NS bifurcation is supercritical (subcritical) if the cubic normal-form coefficient $c_1 = \mathrm{Re}(d_1)$ is negative (positive) and no strong resonances (1:1, 1:2, 1:3, 1:4) are present.

For codimension-two bifurcation points the user should pay particular attention to:

1. At a 1:2 Resonance point R2 the fixed point has a pair of complex conjugate eigenvalues on the unit circle that are both at -1. The normal form contains two cubic coefficients C_1 and D_1 that determine this bifurcation. If $C_1 < 0$, then an NS curve of double-period cycles emanates from the R2 point. The MATCONTM output is $[c, d] = [4C_1, -2D_1 - 6C_1]$.
2. At a 1:4 Resonance point R4 the fixed point has eigenvalues $\pm i$. This bifurcation is determined by a complex normal-form coefficient $A_0 = a + ib$. If $a^2 + b^2 - 1 > 0$ then two half lines $l_{1,2}$ of limit points of quadruple-period cycles emanate from the R4 point. If $|b| > (1 + a^2)/\sqrt{1 - a^2}$ then there is a curve of quadruple-period cycles that contains an NS bifurcation point.
3. At a fold-flip bifurcation LPPD the fixed point has eigenvalues 1 and -1. MATCONTM computes normal-form coefficients $\frac{a}{2e}$ and $\frac{be}{2}$; see [32] for details. If $be > 0$ then an NS curve of double period emanates from the LPPD point. In this case, MATCONTM also reports an approximation of the corresponding first Lyapunov coefficient. The NS points of the second iterate are stable in the center manifold if this coefficient is negative; they are unstable if it is positive.

2.4 Examples and Applications

We end this chapter with two examples that illustrate how MATCONT is used in practice. In the next section we describe the process of a continuation strategy for a vector field. We use the model of a Van der Pol–Duffing oscillator that is also used in Chap. 4. Section 2.4.2 illustrates the use of MATCONTM for a discrete model of a production strategy involving two competing firms.

2.4.1 The Koper Model

In [27] Koper introduced the following model to describe a three-dimensional Van der Pol–Duffing oscillator:

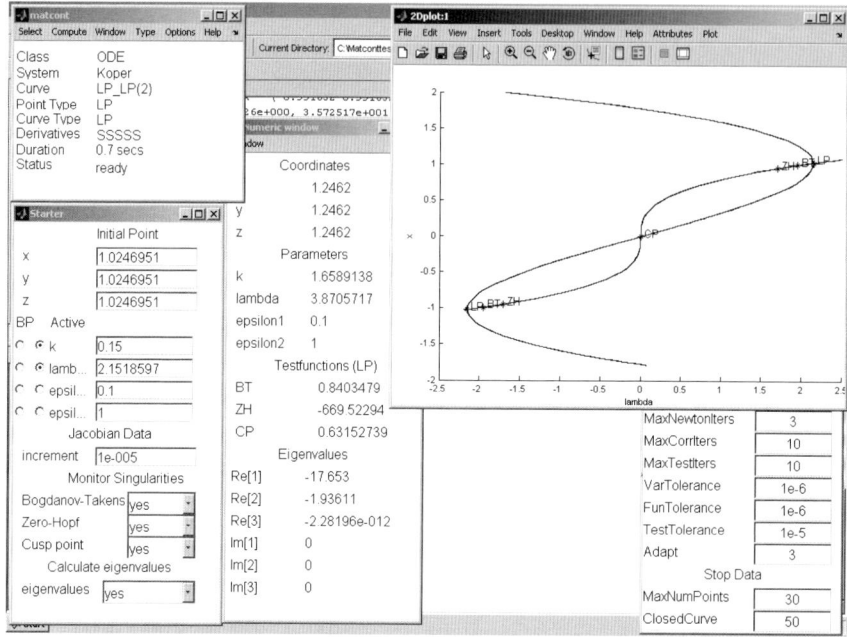

Fig. 2.9. Screen snapshot of MATCONT with the computed equilibrium and LP curves of the Van der Pol–Duffing oscillator (2.1).

$$\begin{cases} \dot{x} = (ky - x^3 + 3x - \lambda)/\varepsilon_1, \\ \dot{y} = x - 2y + z, \\ \dot{z} = \varepsilon_2(y - z). \end{cases} \qquad (2.1)$$

As in [27] we use $\varepsilon_1 = 0.1$ and $\varepsilon_2 = 1$. We note that if $(x(t), y(t), z(t))$ is a solution of (2.1) for a particular value of λ, then $(-x(t), -y(t), -z(t))$ is a solution for $-\lambda$. Therefore, bifurcation diagrams in which λ is represented usually have some symmetry.

We begin the analysis of (2.1) by determining the equilibria. Note that an equilibrium solution (x_0, y_0, z_0) satisfies $x_0 = y_0 = z_0$, which must be a solution of

$$kx - x^3 + 3x - \lambda = 0. \qquad (2.2)$$

In particular, for $\lambda = 0$ and $k = 0.15$ the equilibria are $(0, 0, 0)$ and (x_0, y_0, z_0) with $x_0 = y_0 = z_0 = \pm\sqrt{3.15} \approx \pm 1.77482393492988$. By selecting one of these latter two points in MATCONT we compute by numerical continuation the solution of (2.2) as a function of λ; the cubic solution curve is visualized in the two-dimensional graphics window of the screen snapshot of MATCONT in Fig. 2.9. On the equilibrium curve MATCONT detects two limit points LP at $\lambda = \pm 2.151860$ and reports for both points the critical normal-form coefficient $a = -4.437060$. We select one of the LP points, set both k and λ as free parameters, and compute a curve of LP points that connects

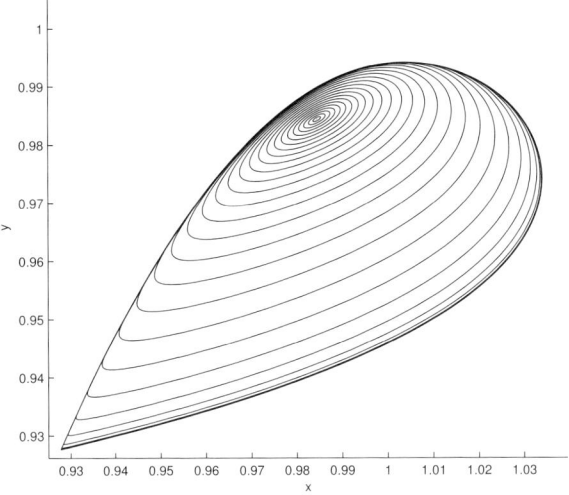

Fig. 2.10. Limit cycles of (2.1) started from a Hopf point and converging to a homoclinic orbit.

the two LP points for $k = 0.15$; this curve is also shown in Fig. 2.9. During the computation of the LP curve MATCONT detects two BT points at $(k, \lambda) = (-0.050000, 1.950209)$ and $(k, \lambda) = (-0.050000, -1.950209)$, with normal-form coefficients $(a, b) = (6.870226e + 000, 3.572517e + 001)$ and $(a, b) = (-6.870226e + 000, -3.572517e + 001)$, respectively; two Zero-Neutral Saddle points (formally ZH) at $(k, \lambda) = (-0.300000, 1.707630)$ and $(k, \lambda) = (-0.300000, -1.707630)$; and a cusp point CP at $(k, \lambda) = (3.000000, 0.000000)$ with normal-form coefficient $c = 5.649718e - 002$. These codimension-two points are also shown in Fig. 2.9.

Starting from the BT point at $\lambda = 1.950209$ we can compute a Hopf curve in the two free parameters k and λ. We stop, fairly arbitrarily, at the Hopf point with $x_0 = y_0 = z_0 = 0.98460576$, $k = -0.25185549$, and $\lambda = 1.7513143$. Starting from this point we keep k fixed and compute a curve of limit cycles (LC) as a function of λ; see Fig. 2.10. It is visually clear that the limit cycles converge to a homoclinic orbit; this can also be inferred from the fact that the parameters change very slowly at the end of the continuation, while the period increases rapidly.

When computing limit cycles, MATCONT allows for the computation and visualization of their *phase response curves* (PRC) [21] as well as the time derivatives of these phase response curves (dPRC). The study of such curves is an important subject in the theory of weakly connected neural networks [23]. In particular, it is well known that they take very specific shapes in the neighborhoods of bifurcations of limit cycles [7]. We demonstrate this by presenting

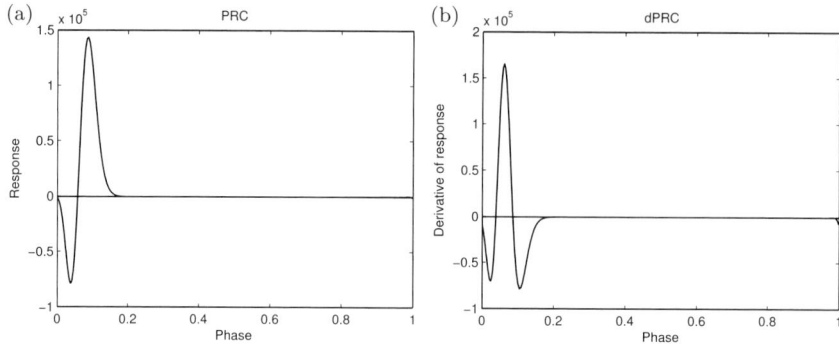

Fig. 2.11. The phase response curve PRC (a) and its derivative dPRC (b) of a limit cycle of (2.1) close to a homoclinic orbit.

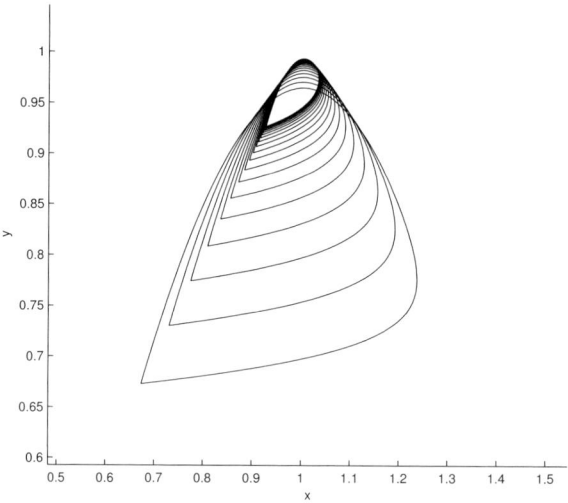

Fig. 2.12. Continuation of an orbit of (2.1) that is homoclinic to a hyperbolic saddle, starting from a limit cycle with large period.

the curves PRC and dPRC in Fig. 2.11 for the limit cycle of the above continuation at $\lambda = 1.7510571$. This limit cycle has period 46.799011, that is, it is close to a homoclinic orbit.

It is possible in MATCONT to start a continuation of homoclinic orbits in two parameters from a limit cycle close to a homoclinic orbit; see Fig. 2.5. An example is presented in Fig. 2.12, where we start from the last limit cycle computed in the previous run, declare it to be of type **HHS**, and choose k and λ as the two free parameters. The time length of the discretized part of the

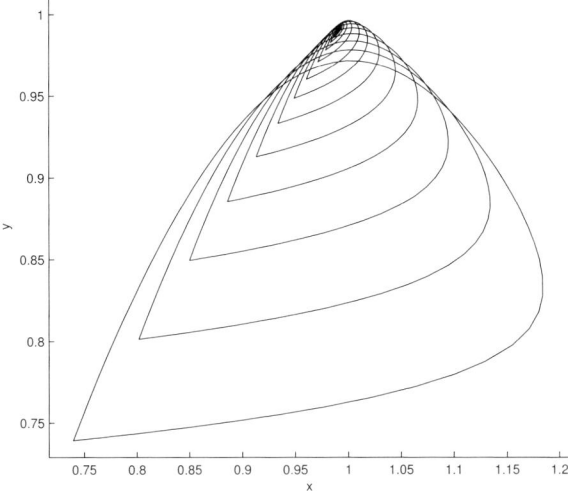

Fig. 2.13. Continuation of an orbit of (2.1) that is homoclinic to a hyperbolic saddle, starting from a Bogdanov-Takens point.

orbit is kept fixed at the period of the original limit cycle while the distances from the end points in the stable and unstable directions are free.

It is also possible to start the continuation of a curve of homoclinic orbits from a Bogdanov-Takens point BT; cf. Sect. 2.2.1. An example of such a continuation is presented in Fig. 2.13. Here we started from the BT point at $\lambda = 1.950209$ that is shown in Fig. 2.9. In this case the distance from the end point in the unstable direction was fixed.

2.4.2 The Duopoly Model

We demonstrate the use of MATCONTM for an example of two competing firms that decide on annual production quantities in a duopoly environment. The two firms are homogeneous with regard to forming their expectation and the action effect on each other. The model that we use is the two-dimensional map

$$F : \begin{cases} x_1(t+1) = (1 - \rho)x_1(t) + \rho\mu x_2(t)(1 - x_2(t)), \\ x_2(t+1) = (1 - \rho)x_2(t) + \rho\mu x_1(t)(1 - x_1(t)), \end{cases} \quad (2.3)$$

described in [2, 26]. The duopoly model assumes that at each discrete time t the two firms produce the quantities $x_1(t)$ and $x_2(t)$, respectively, and decide their productions $x_1(t+1)$ and $x_2(t+1)$ for the next period. The parameter $\mu > 0$ measures the intensity of the effect that one firm's actions has on the other firm. The parameter ρ, which is typically in $[0, 1]$, has an averaging effect.

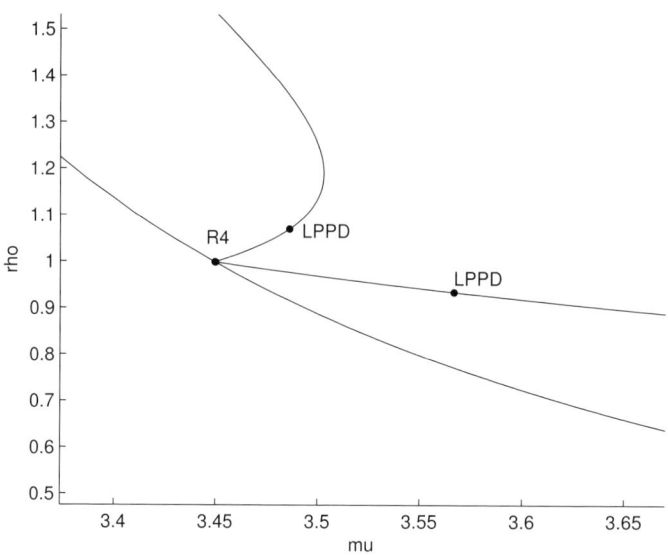

Fig. 2.14. An R4 point on an NS curve of (2.3) with emanating branches of fold curves of period-four cycles.

We start with parameter values $\mu = 3.5$ and $\rho = 0.1$. It is checked easily that F has a fixed point

$$(x_1, x_2) = \left(\frac{\mu + 1 + \sqrt{(\mu + 1)(\mu - 3)}}{2\mu}, \; \frac{\mu + 1 - \sqrt{(\mu + 1)(\mu - 3)}}{2\mu} \right).$$

We now perform a continuation of fixed points of F with free parameter ρ and find an NS point at $(x_1, x_2) = (0.857143, 0.428571)$ for $\rho = 0.888889$. The normal-form coefficient is $-6.273434\mathrm{e}+001$. Since it is negative, the NS bifurcation is supercritical.

Starting from this NS point we can now compute a curve of NS points in the two free parameters μ and ρ. On this curve we find a 1:4 resonance point R4 at $(x_1, x_2) = (0.849938, 0.439960)$ for $\rho = 1.000000$ and $\mu = 3.449490$. It is worthwhile to note that this R4 point lies precisely on the boundary of the region where $\rho \leq 1$, i.e., the region that is relevant from the application's point of view. The normal-form coefficient is $A_0 = (-3.000000\mathrm{e}+000 - 9.231411\mathrm{e}{-}017\, i)$. Since $|A_0| > 1$, two cycles of period four are born at the R4 point. Furthermore, near the R4 point their region of existence is bounded by two fold curves of period-four cycles that emanate from the R4 point. We can start the continuation of these curves from the codimension-two point R4 in MATCONTM. Interestingly, on each of these two curves an LPPD point (of the fourth iterate) is found. A picture of this situa-

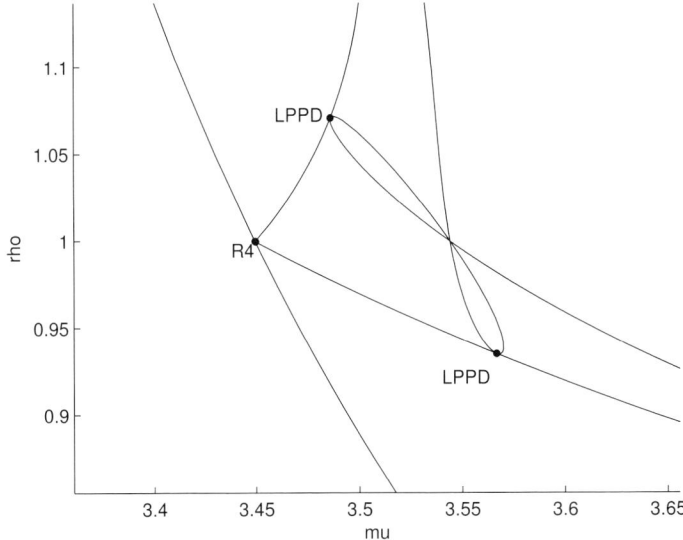

Fig. 2.15. The diamond-shaped region bounded by period-four LP and PD curves near the R4 point on the NS curve of (2.3) contains stable period-four cycles.

tion is presented in Fig. 2.14. The lower LPPD point is detected at $(x_1, x_2) = (0.841586, 0.354516)$ for $\rho = 0.935299$ and $\mu = 3.566686$; its normal-form coefficients are $a/(2e) = 2.574002e+000$ and $be/2 = -5.829597e+001$. The upper LPPD point is detected at $(x_1, x_2) = (0.836428, 0.522216)$ for $\rho = 1.071080$ and $\mu = 3.486079$, and has normal-form coefficients $a/(2e) = 1.733856e+000$ and $be/2 = -2.471512e+001$. We note that the lower LPPD point is in the region relevant to applications while the upper one is not.

It is further interesting to compute the PD curves that emanate from the LPPD points and they are presented in Fig. 2.15. The stable period-four cycles exist in the diamond-shaped region bounded by curves LP and PD of the fourth iterate F^4.

2.5 Directions for Future Development

We presented an overview of software tools for bifurcation analysis. At present, the state of the art is a software environment that provides a clear continuation strategy as implemented in MATCONT. We showed two examples of how to use MATCONT for ODEs and maps. MATCONT has the advantage that it is implemented in Matlab, which is standard in many applied fields, particularly in engineering. Furthermore, its numerical capabilities include the computation of normal-form coefficients and automatic branch switching.

The development of MATCONT is ongoing. In the near future, we hope to implement the remaining functionalities listed in Sect. 2.3. Furthermore, we plan to include algorithms for the computation of invariant manifolds and develop facilities to analyze global bifurcations. Other directions for further development would be the provision of higher-codimension bifurcations and the corresponding detection and branching relationships. Moreover, it would be of interest to generalize the functionalities to other classes of systems, for example, systems with symmetry or preserved quantities; see also Chap. 9.

Acknowledgments

The authors thank Reza Khoshsiar Ghaziani (Ghent), Hil Meijer (Utrecht) and Bart Sautois (Ghent) for help in preparing the figures and several helpful comments and suggestions.

References

1. D. V. Anosov, S. Kh. Aranson, V. I. Arnold, I. U. Bronshtein, V. Z. Grines, and Yu. S. Il'yashenko. Ordinary differential equations and smooth dynamical systems. In *Dynamical Systems I*, volume 1 of *Encyclopaedia Math. Sci.* (Springer-Verlag, Berlin, 1988).
2. H. N. Agiza. On the analysis of stability, bifurcation, chaos and chaos control of Kopel map. *Chaos, Solitons & Fractals*, 10(11): 1909–1916, 1999.
3. A. Back, J. Guckenheimer, M. R. Myers, F. J. Wicklin, and P. A. Worfolk. DsTool: Computer assisted exploration of dynamical systems. *Notices Amer. Math. Soc.*, 39(4): 303–309, 1992. Available via http://www.cam.cornell.edu/~gucken/dstool.
4. W.-J. Beyn and J.-M. Kleinkauf. The numerical computation of homoclinic orbits for maps. *SIAM J. Numer. Anal.*, 34(3): 1207–1236, 1997.
5. R. M. Borisyuk. Stationary solutions of a system of ordinary differential equations depending upon a parameter. FORTRAN Software Series 6, Research Computing Centre, USSR Academy of Sciences, Pushchino, Moscow Region, 1981 (In Russian).
6. H. W. Broer, H. M. Osinga, and G. Vegter. Algorithms for computing normally hyperbolic invariant manifolds. *Z. Angew. Math. Phys.*, 48(3): 480–524, 1997.
7. E. Brown, J. Moehlis and P. Holmes. On the phase reduction and response dynamics of neural oscillator populations. *Neural Comput.*, 16: 673–715, 2004.
8. A. R. Champneys and V. Kirk. The entwined wiggling of homoclinic curves emerging from saddle-node/Hopf instabilities. *Physica D*, 195: 77–105, 2004.
9. A. R. Champneys, Yu. A. Kuznetsov, and B. Sandstede. A numerical toolbox for homoclinic bifurcation analysis. *Internat. J. Bifur. Chaos Appl. Sci. Engrg.*, 6(5): 867-887, 1996.
10. J. W. Demmel, L. Dieci, and M. J. Friedman. Computing connecting orbits via an improved algorithm for continuing invariant subspaces. *SIAM J. Sci. Computing*, 22(1): 81–94, 2001.

11. A. Dhooge, W. Govaerts, and Yu. A. Kuznetsov. MATCONT: A Matlab package for numerical bifurcation analysis of ODEs. *ACM Trans. Math. Software* 29(2): 141–164, 2003. Available via http://www.matcont.ugent.be/.
12. E. J. Doedel and J.-P. Kernévez. AUTO: Software for continuation problems in ordinary differential equations with applications, Applied Mathematics, California Institute of Technology, Pasadena, 1986.
13. E. J. Doedel, A. R. Champneys, T. F. Fairgrieve, Yu. A. Kuznetsov, B. Sandstede, and X.-J. Wang. AUTO97: Continuation and bifurcation software for ordinary differential equations (with HOMCONT). Computer Science, Concordia University, Montreal, 1997. Available via http://cmvl.cs.concordia.ca/
14. K. D. Edoh, and J. Lorenz. Computation of Lyapunov-type numbers for invariant curves of planar maps. em SIAM J. Comput., 23(4): 1113–1134, 2001.
15. B. Ermentrout. *Simulating, Analyzing, and Animating Dynamical Systems: A Guide to XPPAUT for Researchers and Students*, volume 14 of *Software, Environments, and Tools* SIAM, Philadelphia, 2002. Available via http://www.math.pitt.edu/~bard/xpp/xpp.html.
16. U. Feudel and W. Jansen. CANDYS/QA - a software system for the qualitative analysis of nonlinear dynamical systems. *Internat. J. Bifur. Chaos Appl. Sci. Engrg.*, 2(4): 773-794, 1992. Available via http://www.agnld.uni-potsdam.de/~wolfgang/candys.html.
17. M. Friedman, W. Govaerts, Yu. A. Kuznetsov, and B. Sautois. Continuation of homoclinic orbits in Matlab. In V. S. Sunderam, G. D. van Albada, P. M. A. Sloot, and J. J. Dongarra, editors. *Computational Science – ICCS 2005, Atlanta*, LNCS 3514, pages 263–270. Springer, 2005.
18. W. Govaerts, Yu. A. Kuznetsov, and B. Sijnave. Bifurcations of maps in the software package CONTENT. In V. G. Ganzha, E. W. Mayr, and E. V. Vorozhtsov, editors. *Computer Algebra in Scientific Computing—CASC'99*, pages 191–206. Springer, Berlin, 1999. Available via http://www.math.uu.nl/people/kuznet/CONTENT/.
19. W. Govaerts, Yu. A. Kuznetsov, and B. Sijnave. Continuation of codimension-2 equilibrium bifurcations in CONTENT. In E. J. Doedel and L. S. Tuckerman, editors. *Numerical Methods for Bifurcation Problems and Large-Scale Dynamical Systems*, volume 119 of *IMA Vol. Math. Appl.*, pages 163–184. Springer, New York, 2000. Available via http://www.math.uu.nl/people/kuznet/CONTENT/.
20. W. Govaerts, Yu. A. Kuznetsov, and B. Sijnave. Numerical methods for the generalized Hopf bifurcation. *SIAM J. Numer. Anal.*, 38(1): 329–346, 2000.
21. W. Govaerts and B. Sautois. Computation of the phase response curve: a direct numerical approach. *Neural Comput.*, 18(4): 817–847, 2006.
22. B. D. Hassard, N. D. Kazarinoff, and Y.-H. Wan. *Theory and Applications of Hopf Bifurcation*. In volume 41 of *London Mathematical Society Lecture Note Series*. Cambridge University Press, London, 1981.
23. F. C. Hoppensteadt and E. M. Izhikevich. *Weakly Connected Neural Networks*. Springer, Berlin Heidelberg New York, 1997.
24. A. I. Khibnik. LINLBF: A program for continuation and bifurcation analysis of equilibria up to codimension three. In D. Roose, B. De Dier, and A. Spence, editors. *Continuation and Bifurcations: Numerical Techniques and Applications*, volume 313 of *NATO Adv. Sci. Inst. Ser. C Math. Phys. Sci.*, pages 283–296. Dordrecht, 1990.

25. A. I. Khibnik, Yu. A. Kuznetsov, V. V. Levitin, and E. V. Nikolaev. Continuation techniques and interactive software for bifurcation analysis of ODEs and iterated maps. *Physica D*, 62(1-4): 360–371, 1993.

26. M. Kopel. Simple and complex adjustment dynamics in Cournot duopoly models. *Chaos, Solitons & Fractals*, 7(12): 2031–2048, 1996.

27. M. Koper Bifurcations of mixed-mode oscillations in a three-variable autonomous Van der Pol-Duffing model with a cross-shaped phase diagram. *Physica D*, 80(1-2): 72–94, 1995.

28. B. Krauskopf and H. M. Osinga. Globalizing two-dimensional unstable manifolds of maps. *Internat. J. Bifur. Chaos Appl. Sci. Engrg.*, 8(3): 483-503, 1998.

29. B. Krauskopf and H. M. Osinga. Growing 1D and quasi-2D unstable manifolds of maps. *J. Comput. Phys.*, 146(1): 404–419, 1998.

30. B. Krauskopf and H. M. Osinga. Investigating torus bifurcations in the forced Van der Pol oscillator. In E. J. Doedel and L. S. Tuckerman, editors. *Numerical Methods for Bifurcation Problems and Large-Scale Dynamical Systems*, volume 119 of *IMA Vol. Math. Appl.*, pages 199–208. Springer, New York, 2000.

31. B. Krauskopf, H. M. Osinga, E. J. Doedel, M. E. Henderson, J. Guckenheimer, A. Vladimirsky, M. Dellnitz, and O. Junge. A survey of methods for computing (un)stable manifolds of vector fields. *Internat. J. Bifur. Chaos Appl. Sci. Engrg.*, 15(3): 763–791, 2005.

32. Yu. A. Kuznetsov. *Elements of Applied Bifurcation Theory*, 3rd edition. (Springer, Berlin Heidelberg New York, 2004).

33. Yu. A. Kuznetsov, W. Govaerts, E. J. Doedel, and A. Dhooge. Numerical periodic normalization for codim 1 bifurcations of limit cycles. *SIAM J. Numer. Anal.*, 43(4): 1407–1435, 2005.

34. H. E. Nusse and J. A. Yorke. *Dynamics: Numerical Explorations*, 2nd edition. (Springer-Verlag, New York, 1998).

35. A. G. Salinger, N. M. Bou-Rabee, E. A. Burroughs, R. P. Pawlowski, R. B. Lehoucq, L. A. Romero, and E. D. Wilkes. Loca 1.0 library of continuation algorithms: Theory and implementation manual. Sandia National Laboratories Technical Report, SAND2002-0396, 2002. available via http://www.cs.sandia.gov/loca/.

36. A. G. Salinger, E. A. Burroughs, R. P. Pawlowski, E. T. Phipps, and L. A. Romero. Bifurcation tracking algorithms and software for large scale applications.. *Internat. J. Bifur. Chaos Appl. Sci. Engrg.*, 15(3): 1015–1032, 2005.

37. F. Schilder, H. M. Osinga, and W. Vogt. Continuation of quasi-periodic invariant tori. *SIAM J. Appl. Dyn. Sys.*, 4(3): 459-488, 2005.

38. R. Seydel. *From Equilibrium to Chaos: Practical Bifurcation and Stability Analysis*. (Elsevier, New York, 1988).

39. R. Seydel. Tutorial on continuation. *Internat. J. Bifur. Chaos Appl. Sci. Engrg.*, 1(1): 3–11, 1991.

40. M. A. Taylor, M. S. Jolly, and I. G. Kevrekidis. Scigma: Stability computations and interactive graphics for invariant manifold analysis Technical report, Dept. of Chem. Eng. Princeton University, 1989.

41. M. A. Taylor and I. G. Kevrekidis. Interactive Auto: A graphical interface for Auto86. Technical report, Dept. of Chem. Eng. Princeton University, 1990.

42. K. Yagasaki. Numerical detection and continuation of homoclinic points and their bifurcations for maps and periodically forced systems. *Internat. J. Bifur. Chaos Appl. Sci. Engrg.*, 8(7): 1617–1627, 1998.

43. Y. Zhiping, E. J. Kostelich, and J. A. Yorke. Calculating stable and unstable manifolds. *Internat. J. Bifur. Chaos Appl. Sci. Engrg.*, 1(3): 605–623, 1991.
44. Y. Zhiping, E. J. Kostelich, and J. A. Yorke. Erratum: "Calculating stable and unstable manifolds". *Internat. J. Bifur. Chaos Appl. Sci. Engrg.*, 2(1): 215, 1992.

3

Higher-Dimensional Continuation

Michael E Henderson

IBM T.J. Watson Research Center, Yorktown Heights, NY, USA

Numerical continuation in one dimension produces a sequence of points $(\mathbf{u}_i, \lambda_i)$, $i \in [0, N)$, along a set of arcs that are connected at their endpoints. Each point is the solution of an equation

$$F(\mathbf{u}_i, \lambda_i) = 0, \qquad F : \mathbb{R}^n \times \mathbb{R} \to \mathbb{R}^n.$$

A *regular point* is a solution $F(\mathbf{u}_0, \lambda_0) = 0$ where the Jacobian $F_\mathbf{u}(\mathbf{u}_0, \lambda_0)$ is nonsingular. According to the Implicit Function Theorem (IFT) there is a unique curve $(\mathbf{u}(s), \lambda(s))$ of regular points in some small neighborhood $|s| < \varepsilon$ of a regular point. By extending this small piece of the solution curve, detecting singular points and switching branches at singular points, the continuation method produces an approximation of the *connected component* $\Gamma_{(\mathbf{u}_0, \lambda_0)}$ of solutions of $F(\mathbf{u}, \lambda) = 0$ containing $(\mathbf{u}_0, \lambda_0)$; see Fig. 3.1(a).

Most physical problems depend on more than one parameter. While it can be useful to study how the solution depends on each parameter separately, what is really wanted is an understanding of the behavior over some finite piece of parameter space. This is the higher-dimensional continuation problem. Instead of solution curves, we wish to find *solution manifolds* of

$$F(\mathbf{u}, \lambda) = 0, \qquad F : \mathbb{R}^n \times \mathbb{R}^k \to \mathbb{R}^n.$$

Using the IFT again, at a regular point there is a unique manifold of regular solutions $F(\mathbf{u}(s), \lambda(s)) = 0$, which exists in some small neighborhood $|\mathbf{s}| < \varepsilon$, $\mathbf{s} \in \mathbb{R}^k$. The connected component $\Gamma_{(\mathbf{u}_0, \lambda_0)}$ is as defined above, but it is a branched manifold (the solution manifold), rather than a branched curve; see Fig. 3.1 (b). Singular points on the manifold where the Jacobian has a rank deficiency of 1 form the boundary of each branch. These singular manifolds are generically $(k - 1)$-dimensional submanifolds.

Numerical continuation methods use a local analysis, usually a computational version of the IFT, which approximates the solution manifold in a neighborhood of a regular point $\mathcal{N}_i(\mathbf{u}_i)$, and aggregate n of these neighborhoods into a global approximation of the solution manifold M. The geometric problem of maintaining the aggregate of neighborhoods, and merging a

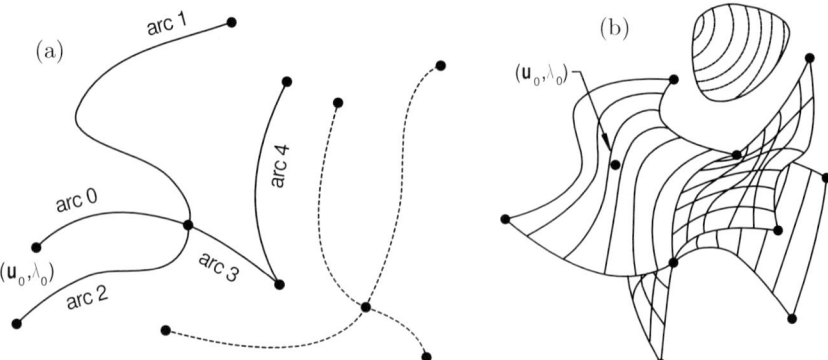

Fig. 3.1. Panel (a) shows connected components of the solution set of $F(\mathbf{u}, \lambda) = 0$; each component consists of a set of smooth arcs that meet at singular points. Panel (b) shows a branched two-dimensional manifold that consists of a set of smooth manifolds that meet along shared one-dimensional boundary manifolds.

new neighborhood into the aggregate, is roughly equivalent to advancing-front mesh generation on a surface. This is not an easy problem, especially in higher dimensions n and for manifolds of dimension $k > 3$. In Sect. 3.1 we give a brief survey of the methods that can be used to represent and manipulate *complexes*, which are generalizations of meshes and are used to represent the branched manifold that approximates M. In Sect. 3.2 we describe five algorithms that are in the literature, and attempt to classify them in terms of the representations of the manifold as discussed in Sect. 3.1. Finally, in Sect. 3.3 we compare the results of the five algorithms when applied to a sphere.

3.1 Mathematical Setting and Background

In computational and pure mathematics it is necessary to choose a representation or notation for the object being computed or manipulated. The parallel between notation and computational representation is quite close, although the computer is able to deal with expressions that are far too complicated to be dealt with by hand. A good notation makes manipulations easier and reduces errors, and a good computational representation makes operations easier to implement, thus reducing coding errors. The choice of representation is often not made explicitly, especially when the objects are simple.

The choice of representation also serves to distinguish between algorithms. Indeed, once the choice of a representation is made there are often fewer choices of how to proceed. We use this approach in Sect. 3.2 to analyze five algorithms for higher-dimensional continuation that are in the literature.

For one-dimensional continuation many issues are quite straightforward, since curves are easily represented as lists of points. However, representing surfaces and manifolds is more challenging. In Sect. 3.1.1 we discuss simplicial and cell complexes, which are general meshes that are commonly used in algebraic topology. Then we define manifolds and manifolds with boundary, which are the analogues of curves and arcs in one dimension.

Section 3.1.2 describes two fundamental geometric abstractions, the Voronoi and Delaunay tessellations, and a particular generalization of the Voronoi tessellation that represents the boundary of a union of spherical balls. In Sect. 3.1.3 we discuss how these topological and geometrical ideas have been used to represent manifolds and, finally, how to construct the neighborhood of a regular point.

3.1.1 Cell and Simplicial Complexes, and Manifolds

There are only a few ways to deal with general surfaces, and most are variations on meshes. In this section we discuss representations of general meshes in higher dimensions, called cell and simplicial complexes. The Voronoi diagram and Delaunay triangulation are two instances of complexes with special properties. The Voronoi diagram or tessellation contains information about neighborhoods of a set of points. There is a variant, the Laguerre-Voronoi diagram, that provides an efficient means of determining the boundary of a union of spherical balls. The Delaunay triangulation is related to the Voronoi diagram, and is frequently used to generate meshes in two and three dimensions.

This subject can be presented in a very opaque, but abstract way. The author would recommend the books [14, 35] and the paper [16] for reasonably clear expositions of complexes. Hopefully, the descriptions that follow are as clear.

A *cell complex* of dimension k is a set of cells of dimension 0 to k. The 0-cells (vertices) may (or may not) be identified with points in an n-dimensional embedding space. If they are not embedded, the complex represents purely topological information. For our purposes the vertices will be points on or near the solution manifold. Each p-cell (except the 0-cells) has a set of faces, which are $(p-1)$-cells in the complex. Cells must be *compatible* with each other: the intersection of two cells is either empty, or it is a cell in the complex that both cells have as a common face. This compatibility condition turns out to create major difficulties for some of the algorithms described in Sect. 3.2 below. Figure 3.2 shows some cells that are not compatible.

A *complex* is a set of lists: of cells of each dimension, the faces of each cell with orientations (the boundary), and a list of cells of which the cell is a face (the co-boundary). Some of this information is redundant, and when storage is an issue it is necessary to avoid storing redundant information. Guibas and Stolfi [18] describe a minimal data structure for representing two-dimensional cell complexes and show that, with operations which are fairly simple in this representation, a Delaunay triangulation algorithm can be built;

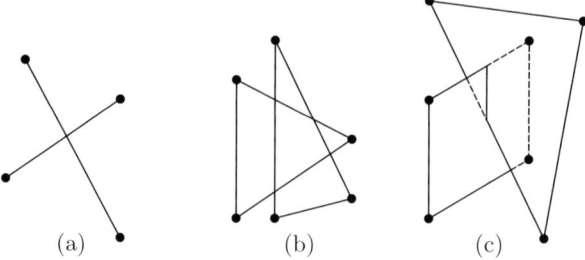

Fig. 3.2. Cells that are not compatible, namely 1-cells embedded in $\mathbb{R}^2$ (a), 2-cells embedded in $\mathbb{R}^2$ (b), and 2-cells embedded in $\mathbb{R}^3$ (c).

see Sect. 3.1.3. For three-dimensional complexes there are similar representations, e.g., the 'winged edge'; see, for example, [13, 19].

   The basic geometrical object in higher-dimensional continuation is a k-dimensional branched manifold. (Recall that k is the number of parameters.) A branched manifold is a set of *manifolds with boundaries* (the branches), glued together along common boundaries. The boundaries are sets of singular points, which are generically $(k-1)$-dimensional manifolds. Branch switching is the process of finding the branches that meet at a singular surface.

Convex Polyhedral Cells

Convex polyhedral cells have a particularly nice structure. The polyhedron is formed by intersecting half-spaces whose planar boundaries contain the faces of the polyhedron. The cells can be represented by the list of planes that they lie on. So for example, the set of linear inequalities in three dimensions

$$P = \begin{cases} (\ \hat{\mathbf{e}}_0, -1).(\mathbf{v}, 1) \leq 0 \\ (-\hat{\mathbf{e}}_0, \ \ 1).(\mathbf{v}, 1) \leq 0 \\ (\ \hat{\mathbf{e}}_1, -1).(\mathbf{v}, 1) \leq 0 \\ (-\hat{\mathbf{e}}_1, \ \ 1).(\mathbf{v}, 1) \leq 0 \\ (\ \hat{\mathbf{e}}_2, -1).(\mathbf{v}, 1) \leq 0 \\ (-\hat{\mathbf{e}}_2, \ \ 1).(\mathbf{v}, 1) \leq 0 \end{cases}$$

defines a polyhedron (a cube), as is shown in Fig. 3.3(a). Given a vertex $\mathbf{v}_0$ (a 0-cell) with representation $(0, 2, 4)$, this indicates that $\mathbf{v}_0$ satisfies the linear system

$$(\hat{\mathbf{e}}_0, -1).(\mathbf{v}_0, 1) = 0,$$
$$(\hat{\mathbf{e}}_1, -1).(\mathbf{v}_0, 1) = 0,$$
$$(\hat{\mathbf{e}}_2, -1).(\mathbf{v}_0, 1) = 0.$$

   The representations of the vertices is sufficient to define the rest of the cells. In a k-dimensional polyhedron each vertex must lie on k or more planes.

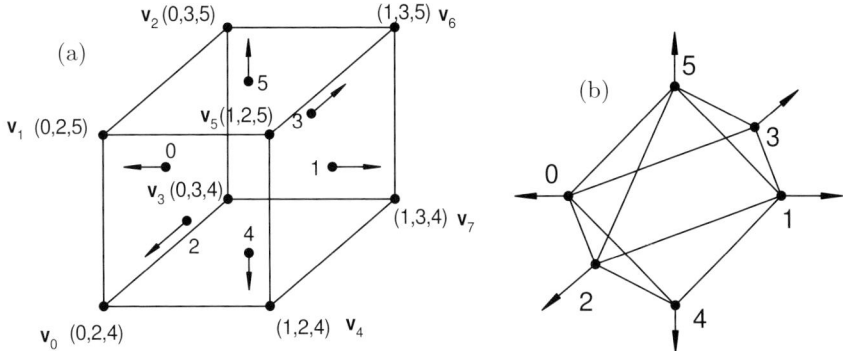

Fig. 3.3. A cube with the face and vertex labels (a) and its dual (b), which is an octahedron.

Cells of dimension $p > 0$ must lie on exactly $k - p$ faces, and so have $k - p$ indices. There is an edge between two vertices if the vertices have $k - 1$ indices in common. For example, for the cube in Fig. 3.3

$$\mathbf{v}_0 \cap \mathbf{v}_1 = (0, 2, 4) \cap (0, 2, 5) = (0, 2) \,,$$
$$\mathbf{v}_3 \cap \mathbf{v}_4 = (0, 3, 4) \cap (1, 2, 4) = (4) \,.$$

So there is an edge with endpoints $\mathbf{v}_0$ and $\mathbf{v}_1$, but not with endpoints $\mathbf{v}_3$ and $\mathbf{v}_4$. The same holds for cells of higher dimension. A p-cell exists that contains a set of vertices if, and only if, the intersection of the index sets of the vertices contains $k - p$ indices. This works because each plane corresponds to a single linear constraint, and a set of indices is just a linear system. For example, in $\mathbb{R}^k$, $k - 1$ linear equations define a line. The notation used here is introduced (in a different guise) in [17, Sec. 2.6]; for more information on polyhedra see, for example, [17] or [12].

Subtracting a Half-Space from a Convex Polyhedron

One operation on convex polyhedra requires mention, namely, that of intersecting the polyhedron with a half-space. This adds a plane to the polyhedron, and the vertices that are the 'wrong' side of the plane and any cell containing those vertices must be removed. If the convex polyhedron is represented by a list of inequalities and a list of vertices, then the algorithm described in [12] may be used.

This algorithm is simple to describe. First the half-space is added to the list of inequalities, then the vertices are tested for inclusion in the half-space. Those that are not in the half-space are removed and a new vertex is added on each edge containing the deleted vertex. The index of these new points is the index of the edge plus the index of the new inequality. The performance of the algorithm can be improved by storing not just the vertices but the edges as well, as is described in [12].

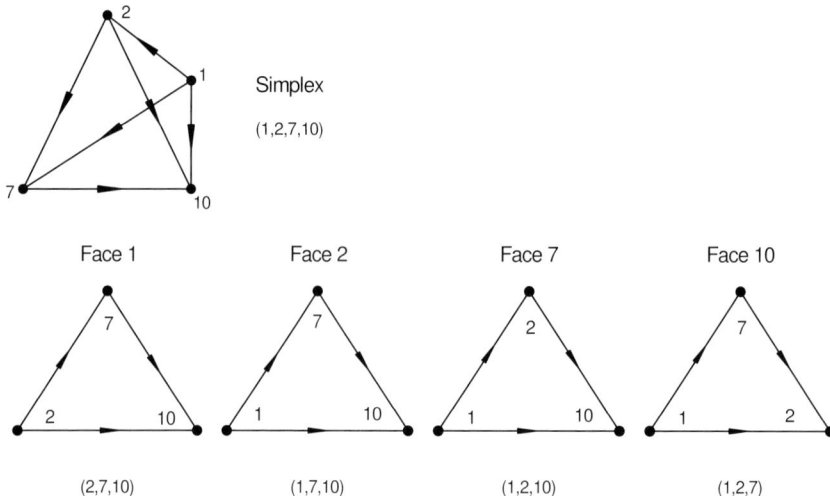

Fig. 3.4. The simplex $(1, 2, 7, 10)$ and its faces. The faces can be enumerated by removing from the label of the simplex the index of each vertex in turn. Removing a vertex index yields the label of the face opposite that vertex. The arrows indicate the positive ordering of the vertices on the edges.

Simplices

A *simplex* of dimension k is a set of $k + 1$ vertices that do not lie in a linear subspace of $\mathbb{R}^k$. A *simplicial complex* is a cell complex whose cells are all simplices. Polyhedra are represented in terms of their faces, but a simplex is represented in terms of its vertices. The faces of a simplex can be obtained by simple operations on the list of vertices, so do not have to be tabulated. A two-dimensional simplex (a triangle) with vertices 1, 4 and 5 is represented as $(1, 4, 5)$. The natural ordering of the integers determines the order in which the vertices are listed.

The faces of a k-dimensional simplex are simplices whose representation is the same as that of the 'parent' simplex, but with one vertex dropped. The face is the one opposite the vertex that is dropped; see Fig. 3.4. This enumeration holds for $(k - 1)$-cells down to 0-cells, which have a single index.

Orientation, the Boundary Operator, Duality and the Co-Boundary Operator

Cells in a complex can have an *orientation* of ± 1. For simplicial cells the sign can be computed directly from the indices. The orientations of the faces of a cell are chosen so that the boundary of the boundary of a cell is empty; see Fig. 3.5. The sets of cells of the various dimensions in the complex are connected by boundary operators: cells of dimension p are connected to cells

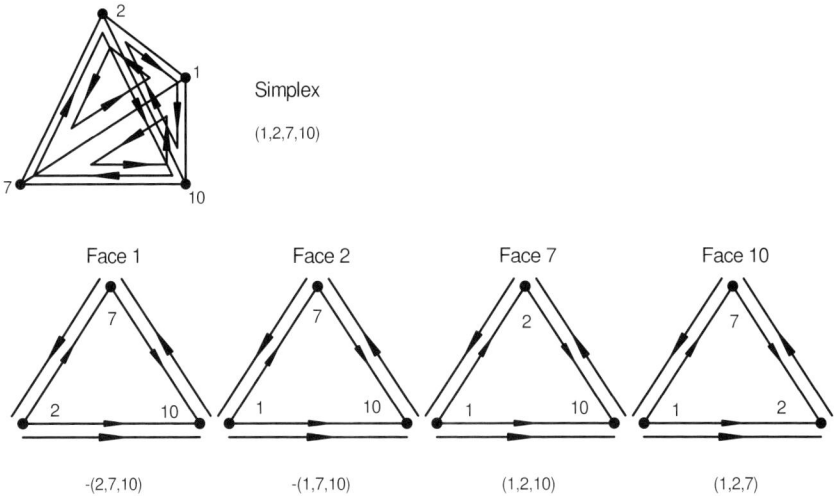

Fig. 3.5. The Simplex $(1, 2, 7, 10)$ and its faces, oriented so that the boundary of its boundary is empty.

of dimension $p - 1$ by the boundary operator d_p, and p-cells are connected to $p + 1$ cells by the co-boundary operator d^p, as defined below.

Exchanging two adjacent vertices changes the orientation of the simplex, and the simplex with vertices in numerical order is assigned an orientation of $+1$. So for an arbitrary list of vertices the sign of the simplex is the parity of the permutation that orders the vertex list. As a result there is a simple expression for the boundary operator:

$$d_p(v_0, \ldots, v_i, \ldots, v_p) = \sum_{i=0}^{p} (-1)^i (v_0, \ldots, v_{i-1}, v_{i+1}, \ldots v_p).$$

This is illustrated in Fig. 3.5. For example, the face $(2, 7, 10)$ in the simplex $(1, 2, 7, 10)$ (a tetrahedron) has boundary

$$d_2(2, 7, 10) = (7, 10) - (2, 10) + (2, 7),$$

$$d_1 d_2(2, 7, 10) = \{(10) - (7)\} - \{(10) - (2)\} + \{(7) - (2)\} = 0.$$

Missing from the representation of a k-dimensional simplicial complex as a list of the vertices in each simplex is information about how the simplices are connected. Note, however, that if the complex is the surface of a single $(k+1)$-dimensional simplex then the complex is completely defined by the list of vertices, as is the case in the complex shown in Figs. 3.4 and 3.5. Figure 3.6 illustrates why the orientation is important. The sum $(1, 2, 7, 10) + (1, 2, 9, 10)$ does not contain the simplex $(1, 2, 10)$, but $(1, 2, 7, 10) + (0, 1, 2, 10)$ does. Note

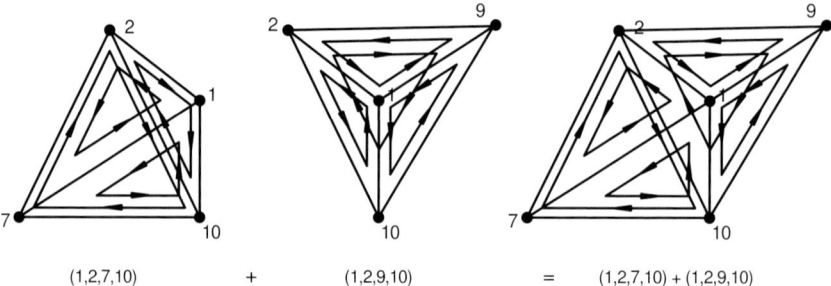

(1,2,7,10)   +   (1,2,9,10)   =   (1,2,7,10) + (1,2,9,10)

Fig. 3.6. The sum of the boundaries of two simplices $(1, 2, 7, 10)$ and $(1, 2, 9, 10)$.

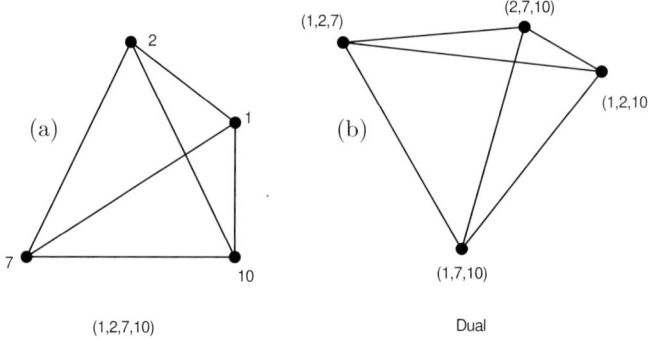

Fig. 3.7. The simplex $(1, 2, 7, 10)$ (a) and its dual (b).

that, if a cell appears twice in a sum it is only counted once, the addition is like a signed union of the sets of cells.

For polyhedral cells the boundary operator is usually tabulated, because it is not expressed easily. For the cube in Fig. 3.3(a) the boundary of face 0 is

$$d_2(0) = (0, 2) + (2, 5) - (1, 2) - (2, 4),$$
$$d_1 d_2(0) = (0, 2, 5) - (0, 2, 4) + (1, 2, 5) - (0, 2, 5) + (1, 2, 4) - (1, 2, 5)$$
$$+ (0, 2, 4) - (1, 2, 4) = 0.$$

Figure 3.3(b) shows the dual of the cube with respect to the faces of the cube, and Fig. 3.7(b) shows the dual of the simplex. Duality is a general relationship, and like the duality between a vector space and its adjoint space, the dual of a cell is a linear functional. The dual of $\sigma_p[i]$ maps the sum of p-cells onto the values $\{-1, 0, 1\}$. The dual cell $\sigma^p[i]$ is the functional that maps the sum to 1 if $\sigma_p[i]$ occurs in the sum with a positive orientation, to -1 if it occurs with a negative orientation, and to 0 if $\sigma_p[i]$ is not present in the sum. That is,

p	cell	dual	p	cell	dual	p	cell	dual
0	(0,2,4)	$(0,2,4)^T$	1	(0,1)	(0,2)	2	(0)	(0,2,4)
0	(0,2,5)	(1)	1	(0,3)	(0,4)	2	(1)	(0,2,5)
0	(0,4,5)	(2)	1	(2,3)	(0,3)	2	(2)	(0,3,5)
0	(0,3,4)	(3)	1	(1,2)	(0,5)	2	(3)	(0,3,4)
0	(1,2,4)	(4)	1	(4,5)	(1,2)	2	(4)	(1,2,4)
0	(1,2,5)	(5)	1	(4,7)	(1,4)	2	(5)	(1,2,5)
0	(1,3,5)		1	(5,6)	(1,5)	2	(6)	(1,3,5)
0	(1,3,4)		1	(0,4)	(2,4)	2	(7)	(1,3,4)
			1	(1,5)	(2,5)			
			1	(3,7)	(3,4)			
			1	(2,6)	(3,5)			

Table 3.1. The cells of the cube and of the dual octahedron from Fig. 3.3(b); all of the indices refer to the faces of the cube.

$$\sigma^p[i]\sigma_p[j] = \delta_{ij},$$

where δ_{ij} is the Kronecker delta. There is an obvious identification of $\sigma^p[i]$ with $\sigma_p[i]$, which is what is meant by the dual of a cell.

The *co-boundary operator* d^p is the adjoint of the boundary operator d_p, defined by the relation

$$(d^p\sigma^p[i])\,\sigma_{p+1}[j] = \sigma^p[i]d_{p+1}\sigma_{p+1}[j].$$

The boundary operator reduces the dimension of a simplex, and the co-boundary operator increases the dimension of a dual cell. The boundary operator on the right-hand side of this equation is d_{p+1} because $\sigma^p[i]$ acts on p-cells and the boundary operator d_{p+1} produces simplices of dimension p. If σ_{p+1} is a $(p+1)$-cell then

$$(d^p\sigma^p[i])\sigma_{p+1}[j] = \sigma^p[i](d_{p+1}\sigma_{p+1}[j]) = \sum_k (-1)^{l_k}\sigma^p[i]\sigma_p[j_k].$$

Here the boundary of the jth $(p+1)$-dimensional cell is given by the list of simplices $j_0, j_1, \cdots$ with orientations $(-1)^{l_0}, (-1)^{l_1}, \cdots$. The co-boundary boundary operator applied to $\sigma_{p+1}[j]$ is zero unless $\sigma_p[i]$ is a face of $\sigma_{p+1}[j]$, the co-boundary of $\sigma^p[i]$ is therefore a signed sum of the duals of the $p+1$-cells that are incident on $\sigma_p[i]$.

For a simplex there is an expression for the boundary operator d_{p+1}: the list of simplices containing a face can be found by adding each vertex in turn to the face. The co-boundary operator for a simplicial complex consisting of a single simplex can be written explicitly as

$$d^p f^p = \sum_{i=0}^{p+1} (-1)^i f^p(v_0, .., v_{i-1}, w_i, v_i, \ldots v_p),$$

p	σ_p	σ^{k-p+1}	$d_p\sigma_p$	$d_{k-p+1}\sigma^{k-p+1}$	$d^p\sigma_p$
0	(10)	(1,2,7)		$(2,7)-(1,7)+(1,2)$	(1,10)-(2,10)+(7,10)
0	(7)	(1,2,10)		$(2,10)-(1,10)+(1,2)$	(1,7)-(2,7)+(7,10)
0	(2)	(1,7,10)		$(7,10)-(1,10)+(1,7)$	(1,2)-(1,7)+(1,10)
0	(1)	(2,7,10)		$(7,10)-(2,10)+(2,7)$	(1,2)-(2,7)+(2,10)
1	(7,10)	(1,2)	(10)-(7)	$(2)-(1)$	(1,7,10)-(2,7,10)
1	(2,10)	(1,7)	(10)-(2)	$(7)-(1)$	(1,2,10)-(2,7,10)
1	(2,7)	(1,10)	(7)-(2)	$(10)-(1)$	(1,2,7)-(2,7,10)
1	(1,10)	(2,7)	(10)-(1)	$(7)-(2)$	(1,2,10)-(1,7,10)
1	(1,7)	(2,10)	(7)-(1)	$(10)-(2)$	(1,2,7)-(1,7,10)
1	(1,2)	(7,10)	(2)-(1)	$(10)-(7)$	(1,2,7)-(1,2,10)
2	(2,7,10)	(1)	(7,10)-(2,10)+(2,7)		
2	(1,7,10)	(2)	(7,10)-(1,10)+(1,7)		
2	(1,2,10)	(7)	(2,10)-(1,10)+(1,2)		
2	(1,2,7)	(10)	(2,7)-(1,7)+(1,2)		

Table 3.2. The simplicial complex that forms the surface of the simplex $(1,2,7,10)$; compare with Figs. 3.4 and 3.5.

where w_i is in turn each of the vertices that are missing from the index of the cell. For cell complexes the boundary and co-boundary need to be tabulated, although there is redundant information in the lists. Table 3.1 shows the cells in the cube and their duals. Note that, in general, we would write the dual of 1-cell $(0,1)$ as $(0,1)^T$. The surface of the simplex $(1,2,7,10)$ from Figs. 3.4 and 3.5 is the simplicial complex shown in Table 3.2.

One of the advantages of this notation is that two simplicial complexes can be added. The example in Fig. 3.6 removes the face they share from the boundary. The result is still a simplicial complex, but it is the boundary of an octahedron. This makes it possible to perform such operations as adding a 'handle' to a cell complex [28]. For simplices there is no need to represent the faces, since they can be generated from the vertex list. For a simplicial complex the dual edges to each face of a simplex must be stored (i.e., the pair of simplices that share the faces). For a cell complex there needs to be a data structure for storing the lists associated with each cell.

3.1.2 Manifolds

A *manifold* is the generalization of a surface, but it has a lot of the features of a cell complex. A k-dimensional manifold is a set of k-dimensional neighborhoods of the origin that are isomorphic to the k-dimensional unit ball $B_1(0) = \{x \mid |x| < 1\}$, called *charts*, along with adjacency relations indicating which charts 'overlap'. Two adjacent charts must agree on some common non-empty sub-region, which means that there is a one-to-one and onto mapping from the subregion of one chart to the corresponding subregion of the other; see Fig. 3.8. The neighborhoods are called *chart domains*, and the collection

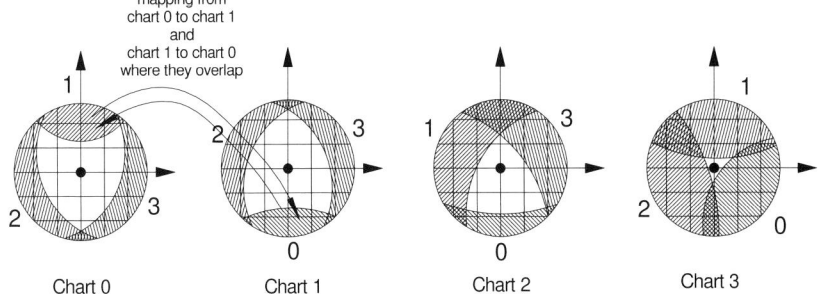

Fig. 3.8. A manifold is an atlas of overlapping charts, with an identification of overlapping subregions and one-to-one and onto mappings between them.

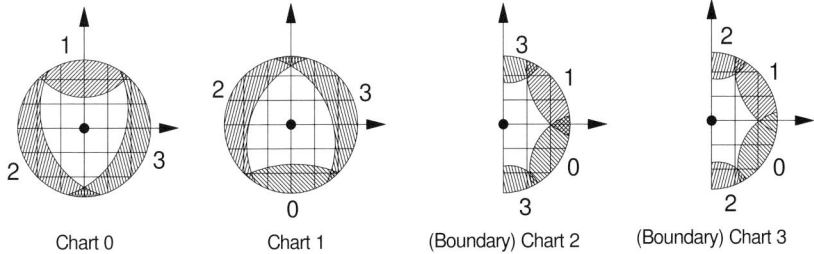

Fig. 3.9. A manifold with boundary is a manifold but, in addition to charts with domains that are full neighborhoods, there are boundary charts with domains that are the intersection of a full neighborhood with half-spaces containing the origin.

of charts is called an *atlas*. This is in reference to navigational charts, which cover a small piece of the globe and are bound together into an atlas. In this analogy, the overlaps are needed to move from one chart to the next, and the adjacency is usually indicated by the number of the neighboring chart in the margin; see Fig. 3.8.

The relation between a manifold and a cell complex is straightforward. A chart is a k-cell, the overlap between two charts is a $k-1$ cell and a common face of each chart, and so on. With this interpretation the manifold shown in Fig. 3.8 is a tetrahedron.

Manifolds with Boundary

A manifold with boundary has special charts called *boundary charts*. Instead of a full neighborhood of the origin, the domain of a boundary chart is isomorphic to the half-ball $B_1^{1/2}(0) = \{x = (x^0, x^1, \ldots) \mid |x| \leq 1, x^0 \leq 0\}$. The restriction to the ball $D_1(0) = \{x = (x^0, x^1, \ldots) \mid |x| \leq 1, x^0 = 0\}$, a full ball of one dimension less, is a chart on one of the *boundary manifolds*. The boundaries themselves can have boundaries (just as the edges of a square have end

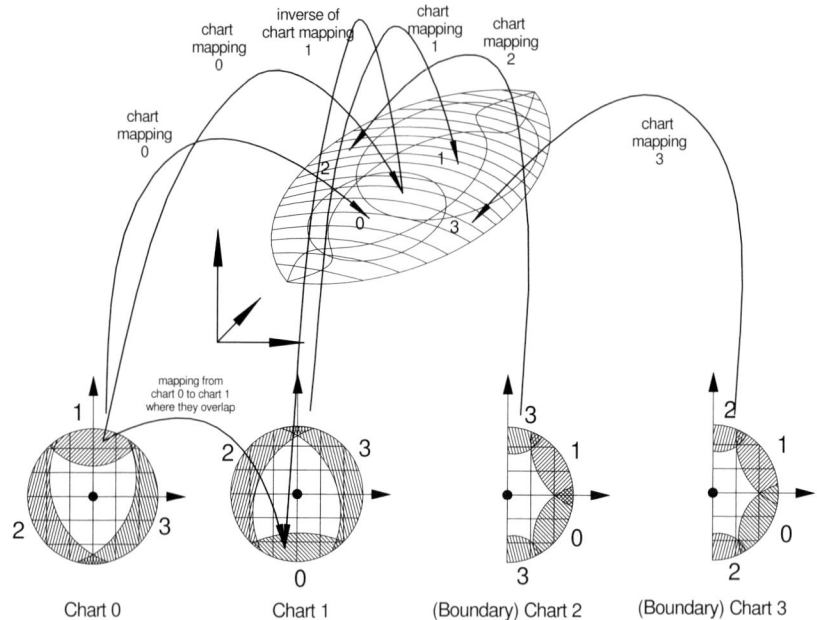

Fig. 3.10. An embedded manifold with boundary.

points), so instead of a half-ball we really mean a ball that is restricted so that some set of the coordinates are non-negative; see Fig. 3.9.

This is a second way that manifolds with boundary are like a cell complex. The manifold itself is a cell, and the faces of the cell are the boundary manifolds.

Embedding a Manifold

The solution manifold M of $F(\mathbf{u}, \lambda)$ is a set of points in $\mathbb{R}^n$, but so far the definition of a manifold has been in terms of neighborhoods of the origin in $\mathbb{R}^k$. The relation between a k-dimensional manifold and the solution space $\mathbb{R}^n$ is an embedding of the manifold. (Here we assume that n is large enough so that the manifold does not self-intersect.) Each chart is assigned a one-to-one mapping to $\mathbb{R}^n$, called the *chart mapping*, which must map points in the overlap of two charts to the same points in the embedding space. With an embedding the mapping between the overlap of two charts can be expressed using the chart mapping of one and the inverse of the chart mapping of the other. For a manifold defined by a system of equations we considered branches that consist of regular points, and the IFT gives the chart mapping. The singular boundary manifolds are where this embedding fails; see Fig. 3.10.

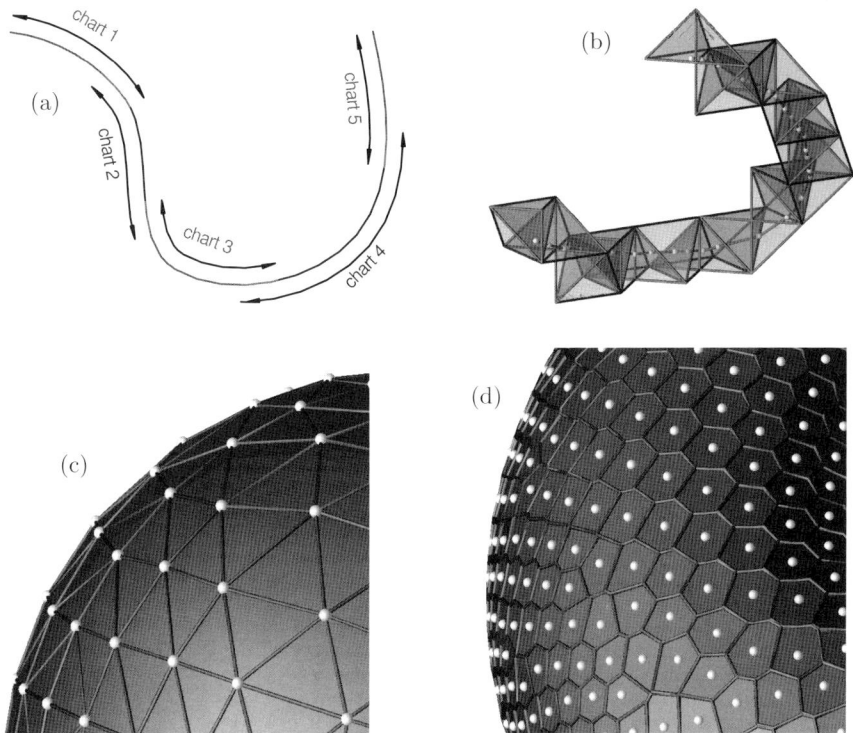

Fig. 3.11. The four main ways of representing a manifold. Panel (a) illustrates the definition for $k = 1$ and $n = 2$, namely a list of overlapping chart mappings from $\mathbb{R}^k$ to $\mathbb{R}^n$. Panel (b) shows a covering or containment in solution space for $k = 1$ and $n = 3$, panel (c) a triangulation for $k = 2$ and $n = 3$, and panel (d) a polygonal tiling for $k = 2$ and $n = 3$.

Representing a Manifold

The algorithms described in Sect. 3.2 use different representations of the solution manifold. Allgower and Schmidt's algorithm [3] represents the solution manifold as a set of simplices in the embedding space $\mathbb{R}^n$, each of which contains a piece of the manifold. Rheinboldt's moving-frame algorithm [31], Brodzik's algorithm [9], and Melville and Mackey's boundary representation [29] use instead a simplicial complex whose vertices are points on the solution manifold. The author's algorithm [20] represents the solution manifold as a complex with convex polyhedral cells. These three different representations, as used by the five algorithms described in Sect. 3.3, are illustrated in Fig. 3.11.

3.1.3 Basic Computational Geometry

The representations of manifolds that are used for higher-dimensional continuation are all based on cell or simplicial complexes. In computational geometry

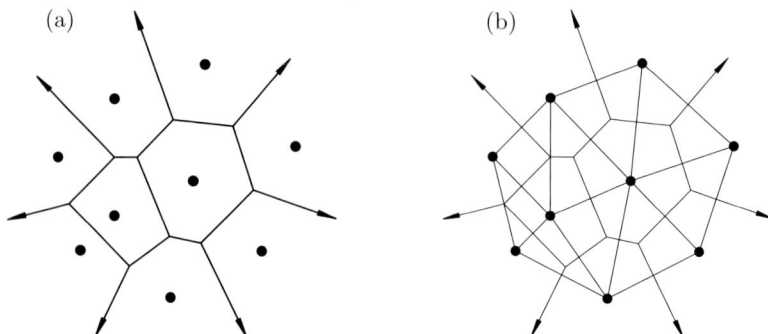

Fig. 3.12. The Voronoi diagram (a) of a set of points in the plane, where each region is the set of points closest to a particular point, and the dual Voronoi and Delaunay diagrams (b).

two of the most frequently used complexes are the Delaunay triangulation of a set of points and the dual Voronoi diagram. The Delaunay triangulation in higher dimensions is in fact not a triangulation, but it is still referred to as a triangulation. Generically, the Delaunay triangulation is a simplicial complex, and the Voronoi diagram is a complex with convex polyhedral cells. There is a large literature on both, and the reader may wish to consult [7, 8, 11, 15, 34], or any introductory text on computational geometry.

The Delaunay triangulation has good properties for mesh cells, namely it creates 'fat' simplices. In two dimensions it has been proved [30] that, over all triangulations of a fixed set of points, the Delaunay triangulation is the one that maximizes the smallest angle in any of the triangles. The Voronoi diagram contains information about 'nearest neighbors', so is used in many pattern matching applications.

Voronoi Diagrams

Given a set of points $\mathbf{u}_i$ in $\mathbb{R}^n$, the Voronoi diagram of the points is a decomposition of $\mathbb{R}^n$ into n-cells, each associated with one of the points. The Voronoi cell V_i of $\mathbf{u}_i$ is

$$V_i = \{\mathbf{u} \in \mathbb{R}^n \mid |\mathbf{u} - \mathbf{u}_i| < |\mathbf{u} - \mathbf{u}_j| \text{ for all } j \neq i\}.$$

Here $|\cdot|$ is the Euclidean 2-norm. Figure 3.12(a) shows an example of a Voronoi diagram for a set of points in the plane.

Each Voronoi region is a 'domain of influence' of the respective point. A Voronoi region also gives information about the nearest neighbors of each point through $(n-1)$-dimensional faces that separate the Voronoi regions of two nearby points.

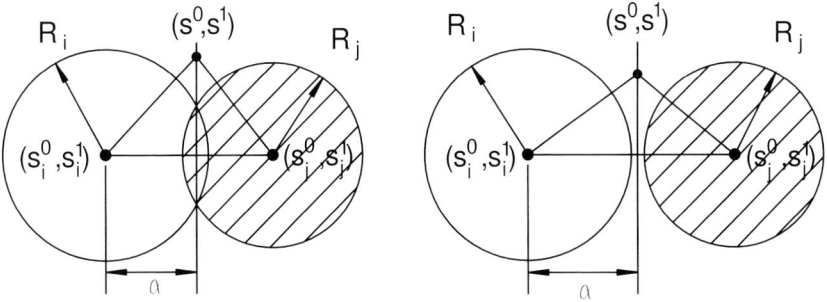

$dB_i - B_j$:

$$
\begin{array}{ccc}
(s^0 - s_i^0)^2 + (s^1)^2 = R_i^2 & & (s^0 - s_i^0)^2 + (s^1)^2 = R_i^2 \\
(s^0 - s_j^0)^2 + (s^1)^2 \le R_j^2 & \equiv & 2(s_i^0 - s_j^0)s^0 + (s_j^0)^2 - (s_i^0)^2 \le R_j^2 - R_i^2
\end{array}
$$

Fig. 3.13. The Laguerre-Voronoi face when $|s_i^0 - s_j^0| \le R_i + R_j$ (a), and when $|s_i^0 - s_j^0| > R_i + R_j$ (b).

The Laguerre-Voronoi Diagram and the Boundary of a Union of Spherical Balls

There are several generalizations of the Voronoi diagram [24]. One of them, the Laguerre-Voronoi diagram or power diagram [4, 5, 6, 22], can be used to find points on the boundary of a union of spherical balls, an operation which will be referred to in Sect. 3.2. In the Laguerre-Voronoi diagram each point is given a weight, which is the radius of the spherical ball about the point.

The face between two Laguerre-Voronoi cells is defined by the equation

$$
|\mathbf{s} - \mathbf{s}_i|^2 - R_i^2 = |\mathbf{s} - \mathbf{s}_j|^2 - R_j^2 ,
$$

which is easily solved. We find that points $\mathbf{s}$ on this Voronoi face are solutions of

$$
2(\mathbf{s}_j - \mathbf{s}_i).\mathbf{s} = R_i^2 - R_j^2 + |\mathbf{s}_i|^2 + |\mathbf{s}_j|^2 ,
$$

which is a plane orthogonal to the line connecting the centers of the two cells. So the Laguerre-Voronoi diagram has cells with planar faces. Substituting $\mathbf{s} = \mathbf{s}_i + \alpha(\mathbf{s}_j - \mathbf{s}_i)$, we find that the plane intersects the line between the two points at

$$
\alpha = \frac{R_i^2 - R_j^2 + |\mathbf{s}_i - \mathbf{s}_j|^2}{2\,|\mathbf{s}_i - \mathbf{s_j}|^2} .
$$

If $|\mathbf{s}_i - \mathbf{s}_j| \le R_i + R_j$ then the plane contains the intersection of the two spheres, and it divides the part of the first sphere that lies outside the second from the part that lies inside the second (and vice versa); see Fig. 3.13. The

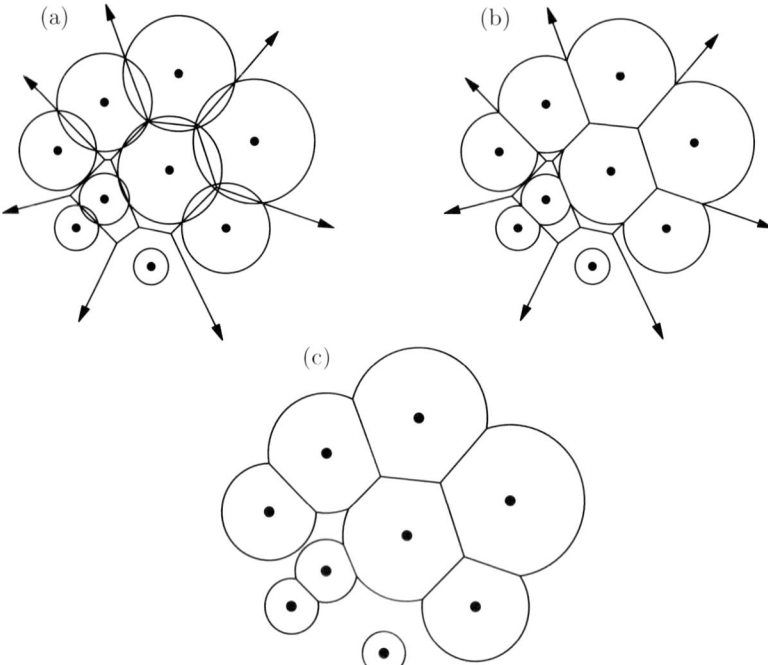

Fig. 3.14. Panel (a) shows the Laguerre-Voronoi diagram of a set of points in the plane, where each region is the set of points closest to a particular point for the distance $|\mathbf{s} - \mathbf{s}_i|^2 - R_i^2$. Panel (b) shows that the Laguerre-Voronoi diagram contains information about the boundary of the union of balls, namely, the boundary of the union of balls is the union of the part of the boundary of each ball that lies within its Voronoi region. Panel (c) illustrates that only the part of the Laguerre-Voronoi diagram is needed that corresponds to overlapping balls; it is called the restricted Laguerre-Voronoi diagram.

n-cell is the intersection of the half-spaces defined by each pair of points, so it is a convex polyhedron, and the boundary of the union of balls which define the diagram is the union of the parts of each sphere that lie inside the corresponding Laguerre-Voronoi n-cell.

The boundary of a union of spherical balls is the union of the parts of the spherical boundaries that do not lie inside another ball. Therefore, the boundary of the union consists of the parts of the spheres that lie inside their restricted-Laguerre Voronoi diagram; see Fig. 3.14.

The Delaunay Triangulation

The Delaunay triangulation is the dual of the Voronoi diagram in the sense discussed in Sect. 3.1.1 for complexes; see Fig. 3.15. In two dimensions it has

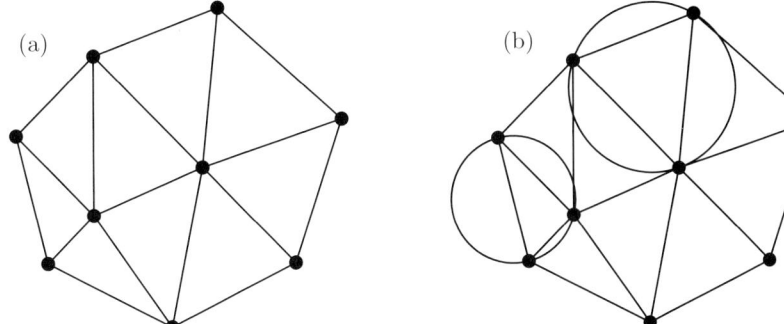

Fig. 3.15. Panel (a) shows the Delaunay tessellation of a set of points in the plane. Panel (b) illustrates that each simplex in the tessellation satisfies the 'in circle' test, that is, no point lies interior to a Euclidean sphere passing through the vertices of the simplex.

been shown [30] that Delaunay triangulations, among all possible triangulations of a set of points, maximize the smallest angle in any triangle. In three dimensions there appears to be a similar property, but it is not clear exactly what maximum principle there might be. Because of this property of having 'fat' cells, the Delaunay triangulation is widely used for mesh generation. For a more complete explanation of the properties of Delaunay triangulations, and how they are constructed, see [7, 15, 18, 33].

The Delaunay tessellation can be defined independently of the Voronoi diagram by means of the 'in-circle' test. Each n-cell of the triangulation is such that no point is inside the Euclidean sphere that contains the vertices of the n-cell; see Fig. 3.15.

The Coxeter-Freudenthal-Kuhn Triangulation

The Coxeter-Freudenthal-Kuhn triangulation (or tessellation) [2] is a simplicial decomposition of $\mathbb{R}^n$. (It is also described in [3], but with quite a few typographic errors.) The tessellation is defined in terms of an initial n-dimensional simplex with $n + 1$ vertices labeled v_i for $i = 0$ to n. An $(n - 1)$-dimensional face separates two n-dimensional simplices that share the face. If one of the simplices has vertices $(v_0, v_1, \ldots, v_n)$ then in this triangulation the simplex across the face v_i is defined as the simplex with vertices

$$P_{v_i}(v_0, v_1, \ldots, v_n) := (v_0, v_1, \ldots, v_{i-1}, \tilde{v}_i, v_{i+1}, \ldots, v_n),$$

where

$$\tilde{v}_i = \begin{cases} v_1 + v_n - v_0, & i = 0, \\ v_{i+1} + v_{i-1} - v_i, & 0 < i < n, \\ v_{n-1} + v_0 - v_n, & i = n. \end{cases}$$

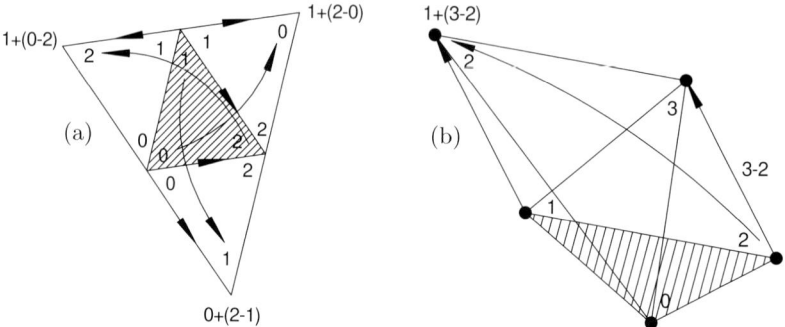

Fig. 3.16. A pivot in the two-dimensional Kuhn triangulation (a), and in the three-dimensional Kuhn triangulation (b).

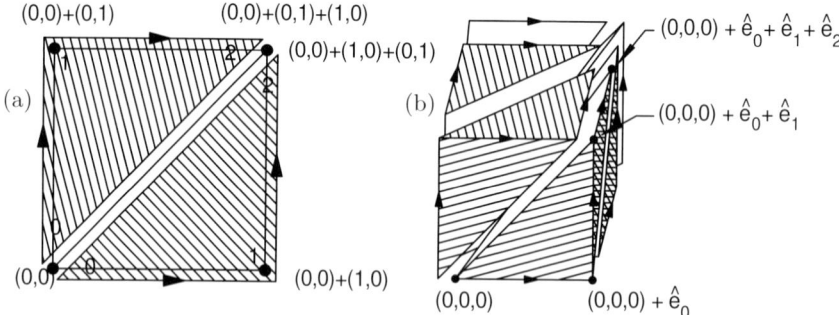

Fig. 3.17. A path simplex decomposition of a square (a) and of a cube (b). The simplices for the n-dimensional cube are generated by paths with n segments, each parallel to a coordinate direction, with no direction repeated.

This operation of moving across a face to an adjacent simplex is called a *pivot*. Pivoting works for any initial simplex, but one particular choice has an explicit representation and is well suited to computations. The initial simplex is a *path simplex*. A path simplex has vertices that are defined by the end points of the segments of a piecewise linear path; see Fig. 3.16. In particular, if the coordinate directions are used as segments, and no coordinate direction is allowed to appear twice, then the set of path simplices decomposes the interior of the unit cube; see Fig. 3.17 and [25] where the idea is attributed to Tucker [26, Problem 3, p. 140]). It is trivial to tile $\mathbb{R}^n$ with cubes and, hence, this decomposition of the cube also gives a simplicial decomposition of $\mathbb{R}^n$. This is not true for an arbitrary decomposition of the cube, since translation in a coordinate direction will generate incompatible faces unless opposite faces of the cube have the same decomposition.

If the coordinate of the 'lower left' corner of a cube — the corner that is closest to the origin in Fig. 3.17(b) — is $\mathbf{v}^0$, and the side of the cube is length δ then the simplices in this decomposition can be represented by a translation vector $\mathbf{z} \in \mathbb{N}^n$, and a permutation $\pi \in S_n$ from the group of permutations of the symbols $\{1, \cdots, n\}$. The permutation identifies the simplex in the decomposition of the cube. The vertices of the simplex with representation $(\mathbf{z}, \pi)$ are given by the recursion

$$\mathbf{v}_0(\mathbf{z}, \pi) = \mathbf{v}^0 + \delta \mathbf{z},$$

$$\mathbf{v}_{j+1} = \mathbf{v}_j + \delta \hat{\mathbf{e}}_{\pi_{j+1}}.$$

The effect of the pivot across face j ($\mathcal{P}_j$) on the permutation π is

$$\mathcal{P}_j \pi = \begin{cases} \{\pi_0, \ldots, \pi_{j-1}, \pi_{j+1}, \pi_j, \pi_{j+2}, \ldots, \pi_{n-1}\}, & 0 < j < n, \\ \{\pi_1, \ldots, \pi_{n-1}, \pi_0\}, & j = 0, \\ \{\pi_{n-1}, \pi_0, \ldots, \pi_{n-2}\}, & j = n, \end{cases}$$

and the action on $\mathbf{z}$ is

$$\mathcal{P}_j \mathbf{z} = \begin{cases} \mathbf{z}, & 0 < j < n, \\ \mathbf{z} + \hat{\mathbf{e}}_{\pi_0}, & j = 0, \\ \mathbf{z} - \hat{\mathbf{e}}_{\pi_{n-1}}, & j = n. \end{cases}$$

3.2 Five Algorithms

We now have the language and tools for describing algorithms for higher-dimensional continuation. Cell and simplicial complexes are used to represent the solution manifold, Delaunay triangulations to decompose the manifold, the restricted Laguerre-Voronoi tessellation to obtain information about the boundary of a collection of balls, and the Coxeter-Kuhn-Freudenthal tessellation for finding a set of simplices that encloses the solution manifold.

Specifically, we describe five algorithms:

- Allgower and Schmidt's pattern algorithm [3];
- Rheinboldt's moving-frame algorithm [31];
- Rheinboldt and Brodzik's [10] tiling algorithm for $k = 2$ and Brodzik's [9] generalization to any dimension;
- Melville and Mackey's $k = 2$ boundary representation algorithm [29];
- and the author's covering algorithm [20].

All of these algorithms can be viewed as an iterative application of three basic steps to a representation M_i of the manifold:

1. find a point $\mathbf{u}_i$ on the boundary of M_i.
2. build a neighborhood $\mathcal{N}_i$ of $\mathbf{u}_i$.
3. merge $\mathcal{N}_i$ into M_i to obtain M_{i+1}.

The algorithms differ in how M_i is represented, and how the three operations are performed.

Continuation of higher-dimensional manifolds is conceptually different from that of one-dimensional ones. In fact, there are only two ways to represent a 1-manifold.

The first corresponds to the approach of pseudo-arclength continuation [23], and it uses a set of polygonal arcs to represent M. Pseudo-arclength continuation exploits the fact that for 1-manifolds the boundary is a set of points, so that the merge operation is just a matter of discarding one of the two intervals that make up the neighborhood of a point on M. However, if a simplicial approximation of M is used in higher dimensions then the boundary of M_i is a $(k-1)$-dimensional simplicial complex. Constructing a simplicial neighborhood of a boundary point so that the simplices are compatible with M_i seems tractable, but turns out to be a difficult problem.

The second corresponds to simplicial or piecewise-linear continuation [1], and it uses n-dimensional simplices in $\mathbb{R}^n$ to cover M. For 1-manifolds in $\mathbb{R}^{n+1}$ simplicial continuation uses an $(n+1)$-dimensional Coxeter-Kuhn-Freundenthal simplicial complex, and only requires pivots across n-cells. This is because a boundary 'point' is the face of an $(n+1)$-dimensional simplex, and with $k=1$, n-dimensional faces intersect M at a point (or not at all). There are exactly two simplices that contain this boundary face. One is the simplex on the boundary and the other can be found with the pivot operation. In higher dimensions the intersection of M with an $(n+k)$-dimensional simplex is a k-cell, and there are more than two simplices containing an n-dimensional simplex that intersects M at a point.

It is easier to understand an algorithm when the manifold is flat, so we will make use of natural parameter continuation to present the five algorithms.

3.2.1 Natural Parameter Continuation

Let us begin with the case that there is a unique solution at each point in parameter space, and consider the generalization of natural parameter continuation; see Fig. 3.18. Natural parameter continuation is a fairly obvious means of adapting an iterative solver for $F(\mathbf{u}, \lambda)$ at a fixed parameter value to map out the solution manifold. Iterative methods require an initial guess, and rather than start with the same initial guess at each new parameter value, the solution at a nearby parameter value is used.

If a set mesh is used, natural parameter continuation selects a point from the mesh at which the solution is known, and which has a neighbor at which the solution is not yet known; see Fig. 3.18(a) and (b). If the points in parameter space are chosen adaptively, the method becomes a generalization of advancing-front mesh generation, which is a challenging problem in higher dimensions. The main consideration is that the new point lies on the edge of the previously computed points, but near enough so that one of the known

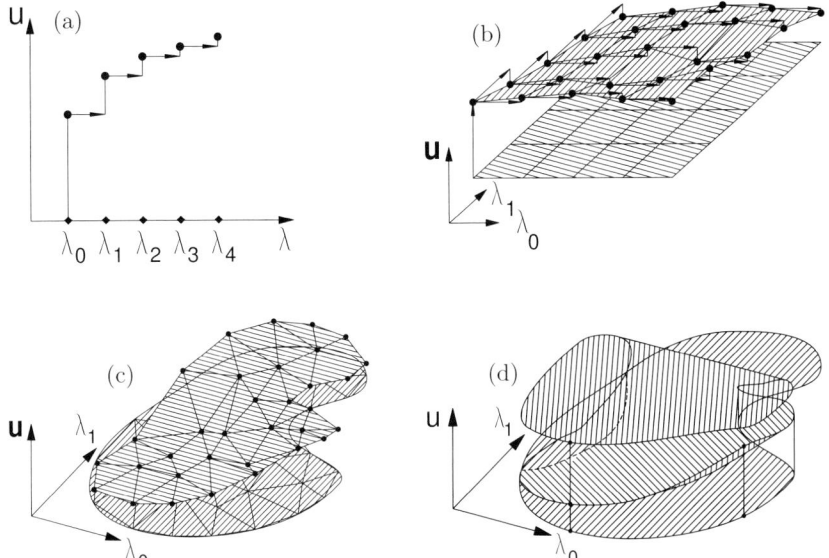

Fig. 3.18. Panel (a) illustrates natural parameter continuation in one parameter; the solution at the point to the left is used as an initial guess for an iterative method at the next point. Panel (b) illustrates natural parameter continuation in two parameters, making use of a rectangular grid on parameter space; the technique is the same as for one parameter, but the grid vertices are traveled in a predetermined order. Panel (c) illustrates natural parameter continuation in two parameters, but with a triangular grid on the region of interest in parameter space. Panel (d) shows a case where natural parameter continuation fails.

points provides a good initial guess. This makes the boundary of the meshed region an important object; see Fig. 3.18(c).

3.2.2 Solution Space Continuation

For many interesting problems the solution manifold cannot be expressed as a function of the parameter. Fig. 3.18(d) shows such a case for $k = 2$. Natural parameter continuation would find the lower or upper sheet, depending on the initial guess, and the iteration would fail to converge beyond the fold.

When $k = 1$ there are two choices of algorithm. If n (the dimension of the solution space) is small then simplicial continuation can be used. On the other hand, pseudo-arclength continuation may be used for any n.

Allgower and Georg's $k = 1$ simplicial continuation begins with a simplex in $\mathbb{R}^{n+1}$ that contains a point on the solution manifold, which is a branched curve. If the simplex is sufficiently small and the point is a regular point then the curve enters this simplex through one face and leaves through another. With these assumptions a linear approximation of F over the simplex results

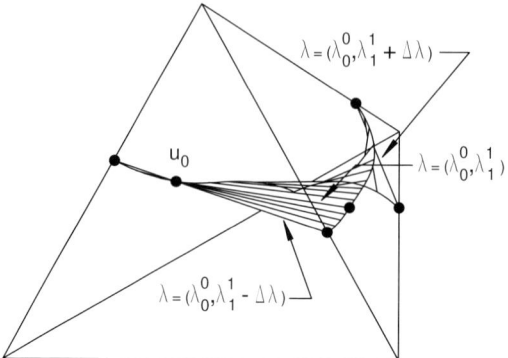

Fig. 3.19. The solution of $F = 0$ for a piecewise-linear interpolant with the second parameter fixed. As the second parameter is allowed to change, the entry and exit points move and the line segment connecting them sweeps out a ruled surface.

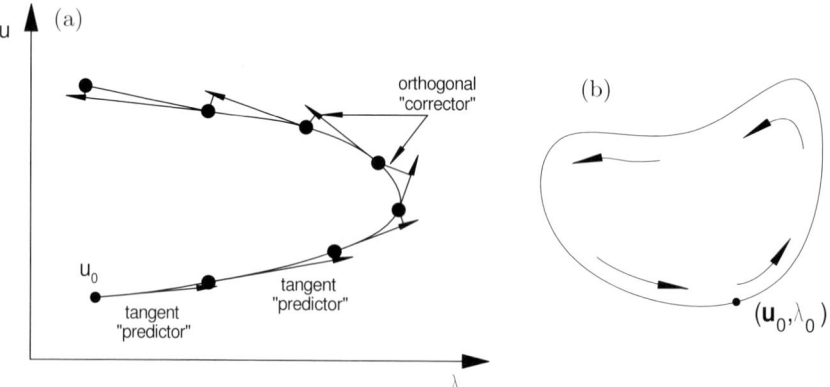

Fig. 3.20. Pseudo arclength continuation (a), and an isola (b) that is repeatedly traced.

in a good approximation to these two points on the $(n-1)$-dimensional faces of the simplex. It is important that simplices are used, so that there is a unique, continuous piecewise-linear interpolant. This also produces a piecewise-linear solution curve. Figure 3.19 shows the effect of a second parameter on the piecewise-linear solution of $F = 0$.

Pseudo-arclength continuation (PSALC) [23] uses the tangent of the solution curve to define a new parameter. This is done by appending a constraint that the projection of a solution point $(\mathbf{u}(s), \lambda(s))$ onto the tangent vector is s. The curve of solutions is represented as a piecewise-linear arc, and is extended by using one of the end points and a linear extrapolation for an initial guess to Newton's method for the system plus the pseudo-arclength

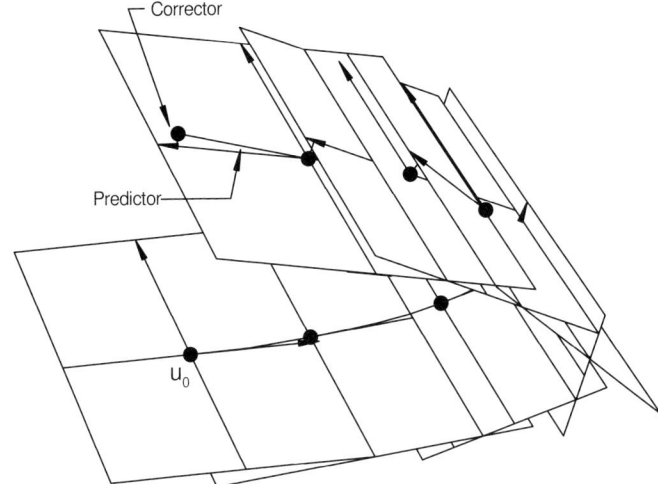

Fig. 3.21. Pseudo-arclength continuation in higher dimensions where one parameter is chosen to follow the corresponding curve on the solution manifold M .

constraint; see Fig. 3.20(a) and Chap. 1. When the solution set is a closed curve, or isola, the PSALC algorithm will repeatedly trace the isola, as shown in Fig. 3.20(b). Usually an upper limit on the arclength is included in the conditions that terminate the algorithm. In higher dimensions we will call this the *self-intersection* problem, and it is related to the problem of incompatible simplices.

Figure 3.21 shows the effect of adding an additional parameter to PSALC. The tangent space is no longer one-dimensional, and it is possible to make a step in more than one direction. The continuation algorithm must choose in which direction to step while ensuring that the resulting points sample the solution manifold uniformly.

3.2.3 Local Analysis

To build a neighborhood of M at a point $\mathbf{u}_0 \in \mathbb{R}^n$ a local analysis is required. The operations used by the five algorithms are

- Find a basis for the k-dimensional tangent space of M at a point $\mathbf{u}$,
- Project a point from the tangent space onto M,
- Estimate the size of a ball in the tangent space so that points within the ball project uniquely onto M.
- Determine if an n-dimensional simplex contains a point on a piecewise-linear approximation to M.

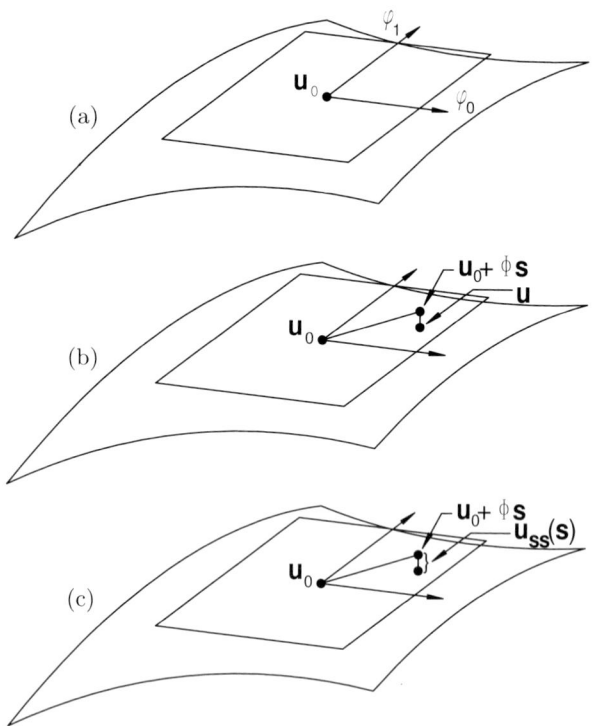

Fig. 3.22. The tangent space of the solution manifold M at $\mathbf{u}_0$ spanned by the orthonormal basis $\Phi = [\varphi_0, \varphi_1]$ (a), projecting a point $\mathbf{s}$ in the tangent space orthogonally to obtain a point $\mathbf{u}$ on M (b), and the curvature $\mathbf{u}_{ss}(\mathbf{s})$ of M at $\mathbf{u}$ (c).

Finding a Basis for the Tangent Space

The manifold M is defined by the equation $F(\mathbf{u}) = 0$. If $\mathbf{u}$ is a point on M then the tangent space of M at $\mathbf{u}$ satisfies

$$F_{\mathbf{u}}(\mathbf{u})\Phi = 0,$$

$$\Phi^T \Phi = I.$$

Finding the tangent space means finding a basis for the nullspace of the Jacobian. At a regular point F^i_j is full rank, which is rank n, so the nullspace is k-dimensional; see Fig. 3.22(a). If n is moderately small a singular value decomposition or QR decomposition might be used. If the problem is large, one approach is to use the tangent space at a nearby point. This replaces the first Φ in the normalization, and results in the linear system

$$\begin{bmatrix} F_{\mathbf{u}}(\mathbf{u}) \\ \Phi_0^T \end{bmatrix} \Phi = \begin{bmatrix} 0 \\ I \end{bmatrix}.$$

The columns of $\varPhi_0$ are the orthonormal basis for the nearby tangent space. The columns of $\varPhi$ will not be orthonormal, but Gram-Schmidt can easily be used to find an orthonormal basis from $\varPhi$. This linear system aligns the new tangent space as much as possible with the old one, which is used in Rheinboldt's wrapping algorithm.

Whichever system is used, any structure in F_u should be exploited. For example, if F is a two point boundary value problem and collocation is used, then the collocation points can be eliminated to yield a smaller system.

Projecting a Point in the Tangent Space onto M

The projection of a point $\mathbf{s} \in \mathbb{R}^k$ in the tangent space of M at $\mathbf{u}_0$ (with orthonormal basis $\varPhi_0$) orthogonally to the tangent space is the solution of the system

$$F(\mathbf{u}) = 0 \,,$$
$$\varPhi^T (\mathbf{u} - \mathbf{u}_0) = \mathbf{s} \,.$$

This is illustrated in Fig. 3.22(b). If $\mathbf{u}_0$ is a regular point of M and $\mathbf{s}$ is small enough then the Jacobian of this system is nonsingular. Modified Newton's method for the nonlinear system results in linear systems with the same matrix as for the tangent space, but with different right-hand sides.

Estimating the Size of a Ball in the Tangent Space

The IFT in finite-dimensional spaces is a modified Newton's method, and there are bounds on the size of the ball in terms of norms of the Jacobian, its inverse and Lipschitz bounds on the Jacobian. Experience indicates that local estimates of these quantities are expensive to compute and provide a very conservative radius. Global bounds can sometimes be found, but then the estimated radius is even more conservative.

A different approach is to impose a maximum number of Newton iterations required for the projection, and to reduce the radius by a fixed factor if the number exceeds the maximum allowed. This is an estimate that comes after the fact, and the way we posed the continuation method above requires an estimate before the projection is performed.

Here we present a method from [20] that chooses the radius of the ball so that the distance from the tangent space to the manifold is roughly constant; see Fig. 3.22(c). If the rate of convergence of Newton's method is constant over the manifold then this method is equivalent to limiting the number of Newton steps.

Using Taylor's remainder theorem, we have

$$|\mathbf{u}(\mathbf{s}) - \mathbf{u}_0 - \varPhi\mathbf{s}| \leq \frac{1}{2} |\mathbf{u}_{\mathbf{ss}}(\xi)s^2| \,,$$

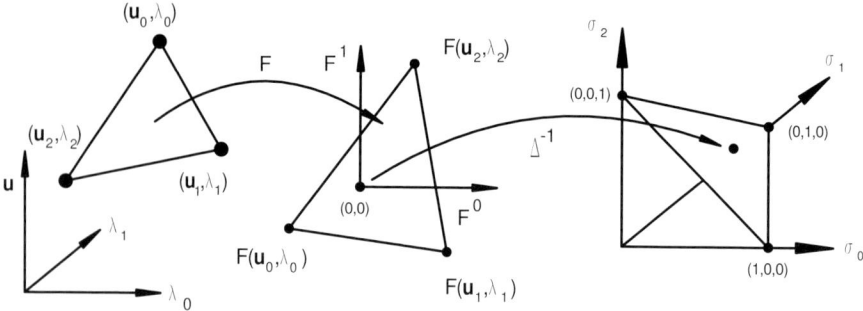

Fig. 3.23. The image of an n-dimensional face in a simplicial complex under F (with the mapping applied only to the vertices). This is equivalent to a global piecewise linear approximation of F over the complex. The question of whether an n-dimensional face in a simplicial decomposition of $\mathbb{R}^{n+k}$ contains a zero of the interpolant is equivalent to whether 0 lies in the image of the face.

where ξ is some point in the ball $\xi \leq |\mathbf{s}|$. If a tolerance ε is given on the error then $\mathbf{s}$ must lie in a ball of the radius of the ball in which the error is less than that tolerance, that is

$$R(\mathbf{s}) = \sqrt{\frac{\varepsilon}{2|\mathbf{u}_{ss}|}} \, .$$

The second derivative of $\mathbf{u}$ is the solution of the system

$$F_{\mathbf{u}}(\mathbf{u}_0)\mathbf{u}_{ss} = -F_{\mathbf{uu}}\varPhi\varPhi \, ,$$

$$\varPhi^T \mathbf{u}_{ss} = 0 \, .$$

It is, therefore, possible to find the second derivative (which is a $k \times k$ matrix whose entries are vectors in $\mathbb{R}^n$), and estimate its norm.

Determining Whether an n-Dimensional Simplex Crosses M

The mapping $F(\mathbf{u})$ takes a simplicial complex with vertices $\mathbf{v}_i \in \mathbb{R}^{n+k}$ and assigns new coordinates $F(\mathbf{v}_i)$. In particular, F maps an n-dimensional simplex with vertices in $\mathbb{R}^{n+k}$ into an n-dimensional simplex in $\mathbb{R}^n$ (provided the vertices are regular points and the simplices are small enough); see Fig. 3.23.

If an n-dimensional simplex contains a point on the solution manifold then it will generically be a single point. Using barycentric coordinates α, the system for the point is

$$F\left(\sum_0^{n+k} \alpha_i \mathbf{v}_i\right) = 0 \, , \qquad \sum_0^{n-k} \alpha_i = 1 \, .$$

This square system for the unknowns α_i is a mapping from a reference simplex in $\mathbb{R}^{n+1}$ to the simplex with vertices $F(\mathbf{v}_i)$. In practice, a piecewise-linear approximation to F is used. Again using barycentric coordinates, the approximation within a p-cell is

$$F(\sum_0^p \alpha_i \mathbf{v}_i) \sim \sum_0^p F(\mathbf{v}_i)\alpha_i .$$

This expression only depends on the vertices of the p-cell, so the approximation is continuous between adjacent simplices. The point where the interpolant is zero is the solution of the $(n+1) \times (n+1)$ linear system

$$\Delta\alpha \equiv \begin{bmatrix} F^0(\mathbf{v}_0) & \cdots & F^0(\mathbf{v}_n) \\ \vdots & & \vdots \\ F^{n-1}(\mathbf{v}_n) & \cdots & F^{n-1}(\mathbf{v}_n) \\ 1 & \cdots & 1 \end{bmatrix} \begin{bmatrix} \alpha_0 \\ \vdots \\ \alpha_{n-1} \\ \alpha_n \end{bmatrix} = \begin{bmatrix} 0 \\ \vdots \\ 0 \\ 1 \end{bmatrix} .$$

This is one step of Newton's method when using differences for the Jacobian. If, say, α_0 is eliminated using the last row then we have that $\alpha_0 = 1 - \alpha_1 - \ldots - \alpha_n$, and the system for the remaining α_i is

$$\begin{bmatrix} F^0(\mathbf{v}_1) - F^0(\mathbf{v}_0) & \cdots & F^0(\mathbf{v}_n) - F^0(\mathbf{v}_n) \\ \vdots & & \vdots \\ F^{n-1}(\mathbf{v}_n) - F^0(\mathbf{v}_0) & \cdots & F^{n-1}(\mathbf{v}_n) - F^0(\mathbf{v}_n) \end{bmatrix} \begin{bmatrix} \alpha_0 \\ \vdots \\ \alpha_n \end{bmatrix} = \begin{bmatrix} -F^j(\mathbf{v}_0) \\ \vdots \\ -F^j(\mathbf{v}_n) \end{bmatrix} .$$

3.2.4 Allgower and Schmidt's Pattern Algorithm

This algorithm described in [3] extends Allgower and Georg's one-dimensional simplicial continuation [1] to arbitrary dimension. It produces a list of $(n+k)$-dimensional simplices and points at which a piecewise-linear approximation to F is zero.

The solution space $\mathbb{R}^{n+k}$ is decomposed as a simplicial complex by using the Kuhn-Freudenthal triangulation; cf. Sect. 3.1.3. The initial solution $(\mathbf{u}_0, \lambda_0)$ is used to choose the origin z so that the simplex (z, π) with π the identity permutation contains the initial solution. For example, the expression for the vertices of the simplex can be used to choose z so that the point with barycentric coordinates $(1, \ldots, 1)/\sqrt{n+k}$ is the initial point. A list of the indices (z, π) of the simplices which will be output is kept, starting with the initial simplex.

Iteratively, the n-dimensional faces of the simplices in the aggregate that lie on the boundary are tested to find those that cross M; cf. Sect. 3.2.3. The candidate faces are kept in a list, which is updated as simplices are added to the aggregate. The n-faces of the initial simplex are found by removing all combinations (without order) of k vertices from the index of the simplex. The n-dimensional simplices that are found to cross M play the role of points on

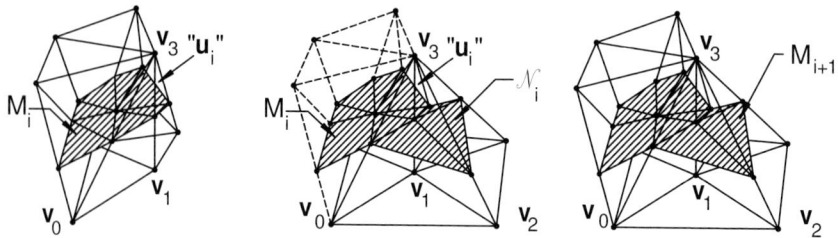

Fig. 3.24. Allgower and Schmidt's algorithm: a transverse $(n-k)$-dimensional face (an edge since $n-k=1$) and the simplices on which it lies. This collection of simplices is the neighborhood of the transverse face. The merge is easy, since the simplices are all drawn from a larger simplicial complex.

the boundary. The points on the piecewise-linear interpolant can be stored, but are not needed by the algorithm.

With a simplex that crosses M, all of the $(n+k)$-dimensional simplices that have this simplex as a face (i.e. the co-boundary) are checked. Any that are not on the list are added. This is a process of adding k vertices to the index of the face, with vertices that lie across any $(n+k-1)$-dimensional face that contains the boundary simplex. This set of simplices is the neighborhood of the boundary; see Fig. 3.24.

Finally, the list of boundary points is updated, by removing those that lie on an $(n+k-1)$-dimensional face for which both cells on either side of the face are in the list, and by adding those n-dimensional faces of the newly added simplices that lie on $(n+k-1)$-dimensional simplices with only one cell on the list. Finding the index of the simplex on the opposite side of an $(n+k-1)$-dimensional face is a pivot, and the action of the pivot on the index is known.

3.2.5 Rheinboldt's Moving-Frame Algorithm

Rheinboldt's moving-frame algorithm [32] represents M as a k-dimensional mesh with vertices on M. Any fixed mesh can be used, but there must be an ordering on the mesh points s_i and a way to find a point $s_{\tilde{i}}$ with $\tilde{i} < i$ and close to s_i. For a rectangular mesh this is straightforward.

The initial point is s_0, which is mapped to the initial point (u_0, λ_0), and Φ_0 is any orthonormal basis for the null space of the Jacobian at the initial point; see Fig. 3.25.

Each mesh point is computed in sequence. At step i, $(u_{\tilde{i}}, \lambda_{\tilde{i}})$ is the boundary point.

The neighborhood of the boundary point is not explicitly represented, but might be defined as the mesh cells with index less than i that include $s_{\tilde{i}}$ as a vertex. The only point without a mapping to M is the point i. The mapped point (u_i, λ_i) is found by projecting $s_i - s_{\tilde{i}}$ orthogonally onto M by using the

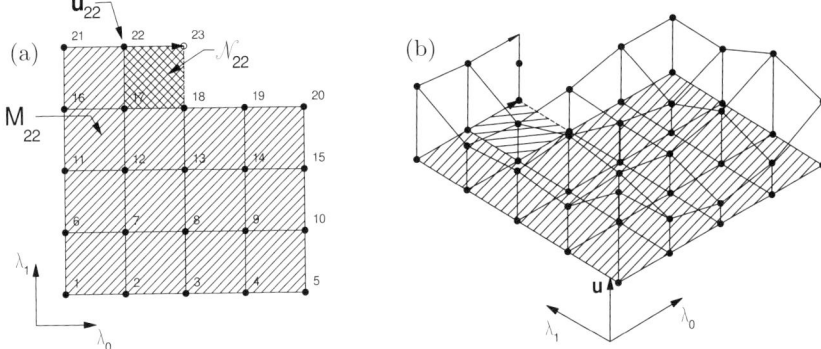

Fig. 3.25. Rheinboldt's moving-frame algorithm when using a rectangular mesh, shown as the natural parameter version (a) and as the solution-space version (b). The natural order (e.g., left to right and bottom to top) is used on the mesh vertices, and the continuation advances in the same order. Moves from scan line to scan line use the point at the beginning of the scan as an initial guess.

tangent space $\Phi_{\tilde{i}}$. The new tangent space is found by first solving the linear system

$$F_{(\mathbf{u}, \lambda)}(\mathbf{u}_i, \lambda_i)\tilde{\Phi}_i = 0\,,$$

$$\Phi_{\tilde{i}}^T \Phi_i = I$$

and then orthonormalizing Φ_i. The merge operation is trivial as it just involves incrementing i.

3.2.6 Brodzik's Tiling Algorithm

Brodzik and Rheinboldt's continuation method for 2-manifolds [10] and Brodzik's extension to arbitrary dimension [9] represent M as a k-dimensional Delaunay triangulation with vertices that lie on M.

   The initial simplicial complex is a reference Delaunay triangulation of a k-dimensional spherical ball. The vertices on M are found by projecting the set of vertices $\mathbf{s}_i$ onto M by using the tangent space at the initial point.

   A boundary point is a vertex on any simplex that has a $(k-1)$-dimensional face without two simplices on opposite sides of the face.

   The neighborhood is the projection of the vertices of the reference decomposition of the spherical ball orthogonally onto M, although not all vertices are projected.

   The difficult step in this algorithm is the merge. The points of the current Delaunay triangulation which are near the boundary point are projected into the tangent space at the new point

$$\tilde{\mathbf{s}}_j = \Phi_i^T \left((\mathbf{u}_j, \lambda_j) - (\mathbf{u}_i, \lambda_i)\right).$$

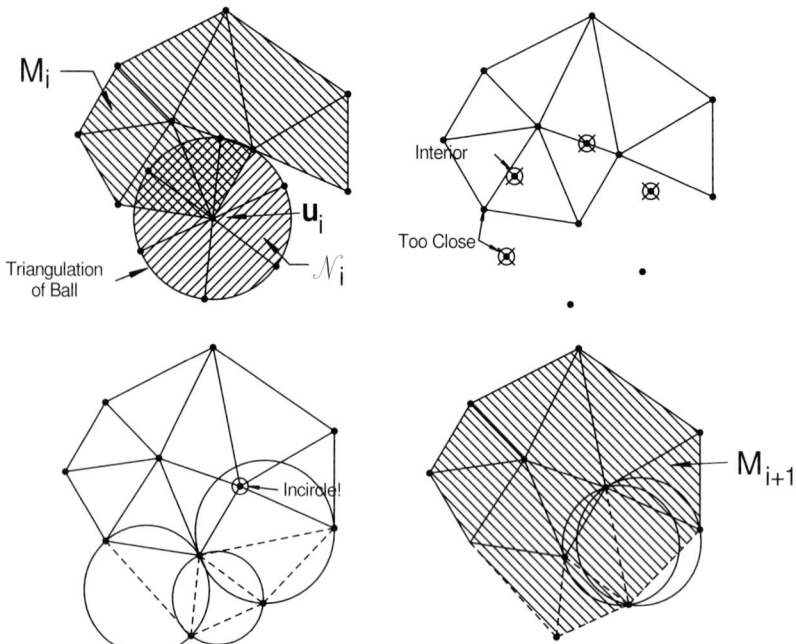

Fig. 3.26. The natural parameter version of Brodzik and Rheinboldt's algorithm [10] and Brodizk's extension to arbitrary dimension [9].

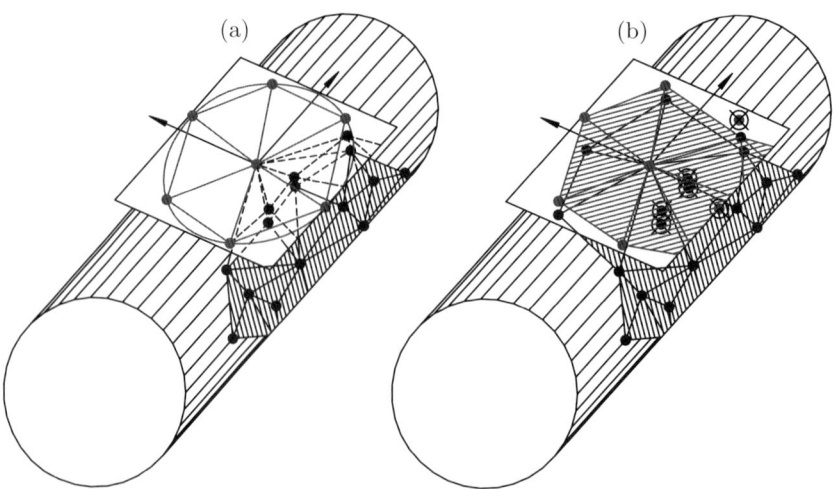

Fig. 3.27. The arclength continuation version of Brodzik and Rheinboldt's algorithm [10] and Brodizk's extension to arbitrary dimension [9].

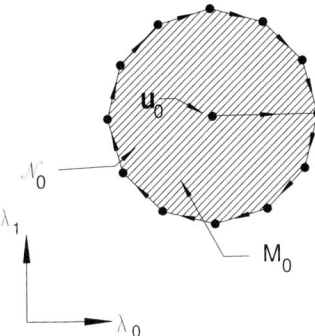

Fig. 3.28. A homotopy in the radius R of the circle about λ_i for obtaining a starting point on the boundary of the neighborhood of λ_i for Melville and Mackey's boundary algorithm.

The previously computed vertices carry with them part of the entire Delaunay triangulation. First, points from the spherical ball that lie inside this projected triangulation are discarded. Next, points from the sphere that lie too close to a vertex of the triangulation are also discarded. Finally, the remaining points are projected orthogonally onto M and the Delaunay triangulation is updated to include the new vertices. This last step is done in the tangent space. It involves removing some of the cells from the triangulation and replacing them with new cells; see Fig. 3.26.

The extension to an arclength continuation is done as follows. The points that are near the boundary point are projected onto the tangent space at the boundary point. Then one proceeds as if the tangent space coordinates were the parameters in the natural parameter continuation; Fig. 3.27.

3.2.7 Melville and Mackey's Tiling Algorithm

Melville and Mackey's algorithm [29] is strictly two-dimensional. M is not explicitly represented, rather the boundary of the triangulation is represented as a polygonal curve with vertices on M.

The initial polygon is found using PSALC, starting at the initial solution, and solving the modified system

$$F(\mathbf{u}, \lambda) = 0 \,,$$
$$|\mathbf{u} - \mathbf{u_0}|^2 + |\lambda - \lambda_0|^2 - R^2 = 0 \,.$$

To obtain an initial solution for this system, a homotopy can be performed in the radius R. To this end, one starts at $R = 0$ when $(\mathbf{u_0}, \lambda_0)$ is a solution and increases R to a final value that is chosen to be roughly the desired resolution of M. With a point on the intersection of M and the sphere, PSALC can be used to trace the intersection of M and the sphere clockwise relative to Φ_0. The polygon formed by the computed points is the initial boundary

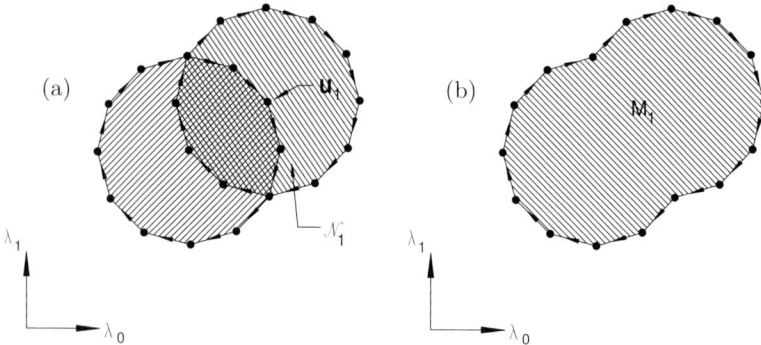

Fig. 3.29. The natural parameter version of Melville and Mackey's boundary representation algorithm for $k = 2$. Starting on the boundary of the region of interest, a neighborhood is constructed by continuing around a circle (a); this neighborhood is then merged with the region of interest (b).

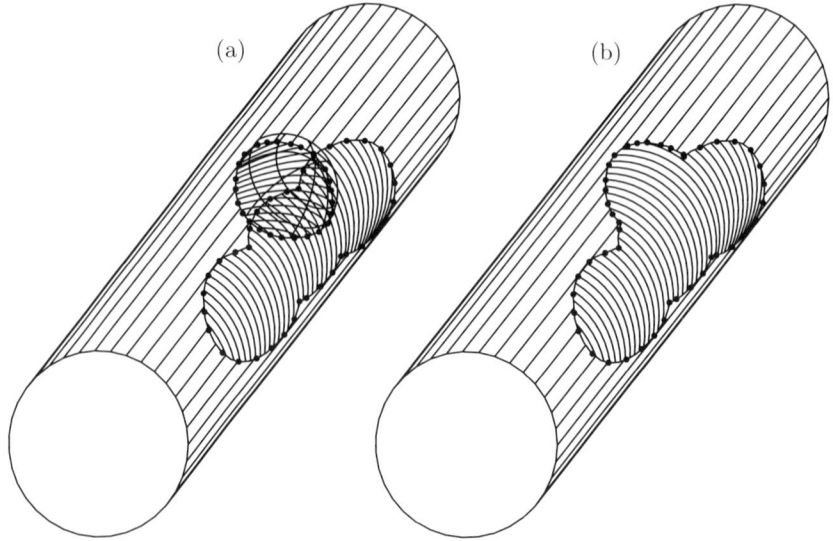

Fig. 3.30. Melville and Mackey's boundary representation algorithm for $k = 2$ and $n = 3$.

polygon. The tangent space Φ_i at each point on the boundary is found as in the moving-frame algorithm, which preserves the orientation of tangent space; see Fig. 3.28.

With an explicit representation of the boundary, any vertex will serve as a boundary point.

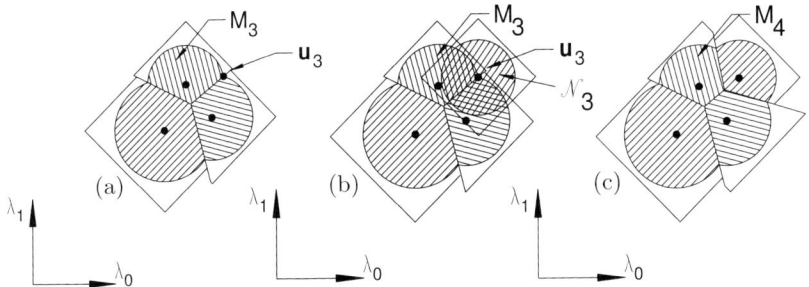

Fig. 3.31. The natural parameter version of the author's covering algorithm.

The neighborhood of the boundary point is a circular disk of radius R, represented by its boundary. Two PSALCs are done in the tangent space of the boundary point. The first to the circle, and the second around the circle.

The merge operation starts by projecting the part of the existing boundary whose vertices are near the boundary point into the tangent space at the boundary point. This leaves a relatively simple two-dimensional problem of clipping the boundary of the neighborhood against the existing boundary; see Figs. 3.29 and 3.30.

3.2.8 The Author's Covering Algorithm

This algorithm [20] represents the solution manifold M as a union of the projection of spherical balls. Each ball lives in the tangent space of a point $\mathbf{u}_i$ on M, and is represented by its radius R_i, an orthonormal basis for the tangent space Φ_i, and a restricted Laguerre-Voronoi polyhedral k-cell P_i.

The initial condition is found by finding Φ_0, estimating the size of the ball (cf. Sect. 3.2.3), and setting P_0 to a cube slightly larger than the spherical ball.

The boundary point is found by selecting a ball with a polyhedral vertex that lies outside its spherical ball. If the ratio of the radii of neighboring balls is within $\sqrt{2}$ then the origin (the center of the ball) is inside the polyhedron. Then a boundary point can be found by finding the intersection of a line between the origin and the exterior vertex and the sphere.

The neighborhood of the boundary point is the projection of a spherical ball in the tangent space at the boundary point onto M. The polyhedron for the boundary point is initialized to a cube centered about the origin.

The merge requires finding neighboring balls and subtracting complementary half-spaces from their polyhedra. The half-spaces are found by projecting each center into the tangent space of the neighbor and using a ball of the same radius, but now in the other tangent space; see Figs. 3.31 and 3.32.

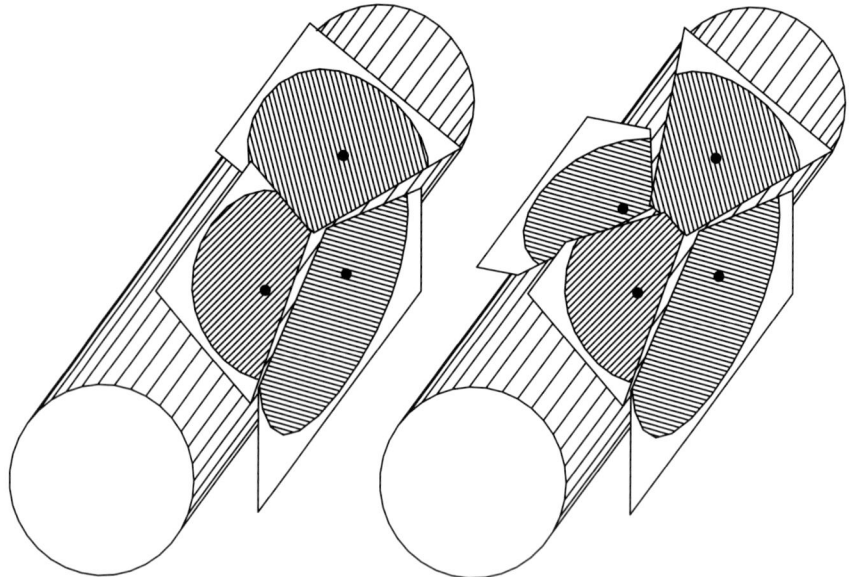

Fig. 3.32. The author's covering algorithm in solution space. To update the polyhedra P_i the centers of neighboring neighborhoods are projected into the tangent space at $\mathbf{u}_i$.

3.3 Performance of the Algorithms

We have tried to cast each of these algorithms as an iteration of the three basic steps. Suppose that M_i is the part of the manifold that has been computed up to step i. Then M_{i+1} is found by selecting either a point or transverse face $\mathbf{u}_i$ that lies on the boundary of M_i, constructing a neighborhood $\mathcal{N}(\mathbf{u}_i)$ of that point, and merging the neighborhood into M_i to get M_{i+1}. The five algorithms differ in how they perform these steps, but there are two common challenges that they must address.

The first challenge is that all algorithms use a simplicial or cellular approximation to the manifold. To perform the merge the simplices in the neighborhood must be compatible with those in M_i, that is, they must either be disjoint or have an intersection which is a face. Allgower and Schmidt's pattern algorithm maintains compatibility by selecting simplices from a decomposition of the entire region of interest, as does Rheinboldt's moving-frame algorithm. Therefore, compatibility is guaranteed for these two algorithms unless the mapping from simplices to M becomes singular. Melville and Mackey do not attempt to store a representation of M_i, but instead keep its boundary as a set of polygons, which is then updated. Their approach is limited to $k = 2$, although, in principle, a simplicial approximation could be maintained for the boundary in higher dimensions. However, in that case the compatibility issue arises again. Brodzik's algorithm maintains a Delaunay triangulation on M

and deals with compatibility by removing points and recomputing the triangulation locally. Finally, the author's algorithm avoids the compatibility issue by representing M_i as a set of overlapping neighborhoods. However, underneath this covering is a type of Voronoi triangulation of M_i, whose dual is a Delaunay triangulation, which aids in finding boundary points. So it is possible to view the algorithm as having a merge — but by way of a method that is not easily described without the Voronoi cells.

The second challenge is to avoid computing part of the manifold more than once. This means that the boundary must be maintained in some form or other. In Allgower and Schmidt's pattern algorithm care is taken not to pivot to a simplex that is already in M_i. This is done by using a clever integer coding of the simplices and keeping a coded list of simplices for M_i. Rheinboldt's moving-frame algorithm does not address the issue, but since there are a finite number of 'boundary' vertices in the reference k-dimensional complex, the algorithm at least terminates and does not cycle. Melville and Mackey explicitly represent the boundary and update the boundary of M_i to find the boundary of M_{i+1}. Brodzik's algorithm detects the overlap of simplices, but simply removes one of the overlapping simplices to leave a gap. The author's algorithm indirectly represents the boundary in terms of polyhedra in the tangent space, which ensures that new points are within a given tolerance of the boundary.

To illustrate how the five algorithms actually perform we implemented them all, except Brodzik's, and applied them to the simple example of computing a sphere given by

$$F(\mathbf{u}) = |\mathbf{u}|^2 - 1$$

for $k = 2$ and $n = 3$.

For these dimensions Allgower and Schmidt's algorithm produces a decomposition of the sphere; see Fig. 3.33(a). Even though the simplices in $\mathbb{R}^3$ are fairly uniform, their intersection with the sphere is not, so that the method can be expected to produce small triangles.

As described above, for a rectangular mesh Rheinboldt's moving-frame algorithm covers parts of the sphere more than once. Although square cells were used in the k-dimensional reference space, the projection to the sphere results in clustering of vertices near the poles (the initial point and tangent space define an equator); see Fig. 3.33(b).

Melville and Mackey's algorithm is able to cover the sphere, and to avoid covering it more than once. Each vertex obtained lies on a curve on the sphere, namely on the intersection of the spherical neighborhood in $\mathbb{R}^3$ that is used to define a curve about the boundary point. In this case these curves are circular arcs; see Fig. 3.33(c).

Brodzik's algorithm not only leaves gaps, but the triangles on the sphere are of a range of sizes; see Fig. 3.34.

The author's algorithm successfully covers the entire sphere, and produces polygonal Voronoi regions on the sphere; see Fig. 3.33(d).

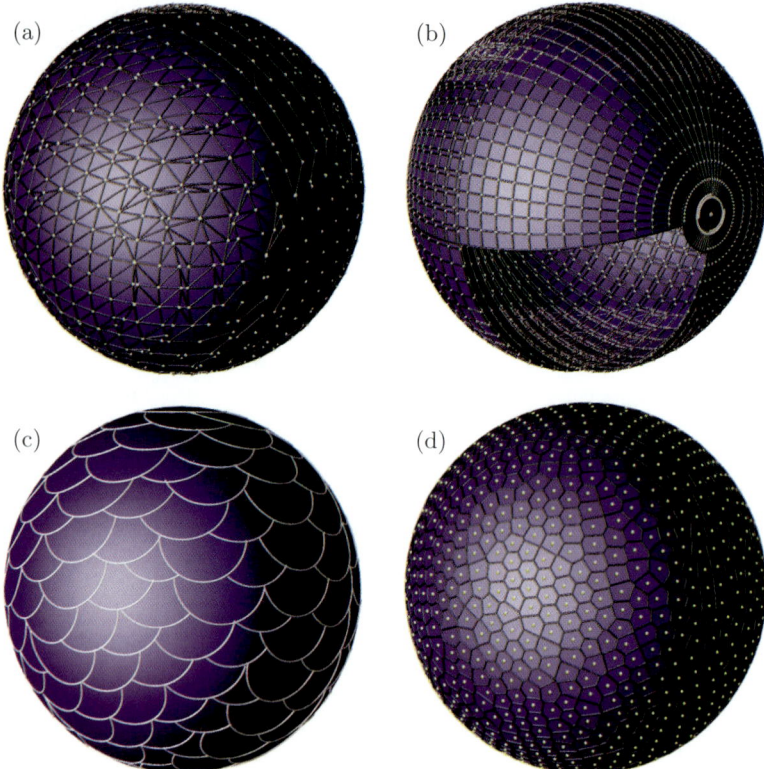

Fig. 3.33. Comparison of different methods when applied to a sphere. Panel (a) is the result of Allgower and Schmidt's pattern algorithm with tetrahedra selected from a decomposition of $\mathbb{R}^3$ with $5 * 20 * 20 * 20 = 40000$ tetrahedra; the five-tetrahedral decomposition of the cube was used, with eight copies reflected so that there are no incompatibilities. Panel (b) is the result of Rheinboldt's moving-frame algorithm with a 50x50 mesh of size 0.1. Panel (c) is the result of Melville and Mackey's algorithm, which produces a set of boundaries (curves) on the sphere, which in this case are circular arcs. Panel (c) is the result of the author's algorithm with the maximum radius of 0.1 and tolerance $\varepsilon = 0.01$.

The result of these computations is an approximation to a branched manifold, but no mention has been made yet of how to compute the singular boundary manifolds or how to move from one branch to another. In [21] there is a discussion of branch switching for manifolds that, although it is cast in terms of the author's algorithm, could be applied to all of the algorithms except Allgower and Schmidt's pattern algorithm. The idea is that moving from a point on M to a point projected orthogonal to the tangent space is one step of PSALC. Therefore, singular points may be detected as usual in PSALC. Branch switching is a little more complicated. For simplicial continuation each

(a) 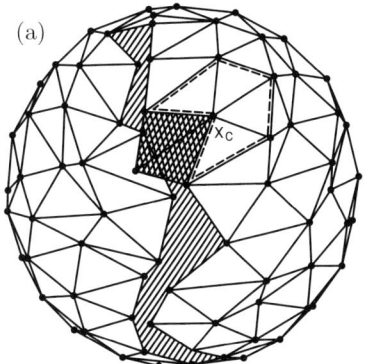   (b)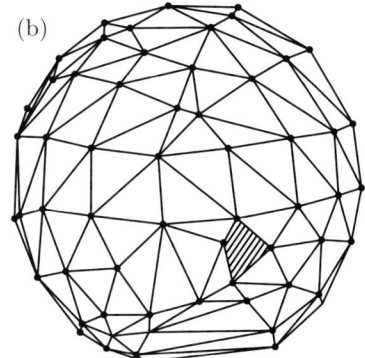

Fig. 3.34. Brodzik's algorithm applied to a sphere (with a larger simplex edge than those above). Panel (a) shows the 'gap' that is left when the neighborhoods of points like $\mathbf{x}_c$ are dropped; these overlap neighborhoods whose centers, as measured around the boundary, are not close to x_c. Panel (b) shows the same computation with a different tolerance, which allows triangles to be more skewed; this additional flexibility results in a smaller gap, but a lower quality triangulation. From M.L. Brodzik, The computation of simplicial approximations of implicity defined p-dimensional manifolds, *Computers Math. Applic.* 36(6) (1998) 93–113 © 1998 by Elsevier Science; reprinted with permission.

of the n-dimensional faces of the simplices are checked for points on M, so unless a singular point lies at a vertex of the complex, there is no difficulty.

3.4 Conclusions

How a mathematical object is represented can affect the complexity of an algorithm. The five algorithms for higher-dimensional continuation of implicitly defined surfaces discussed here all use different representations of either the solution manifold or the neighborhood of a boundary point. They draw on techniques from algebraic topology, computational geometry, linear programming (the algorithm for subtracting a half-space from a polyhedron), and differential geometry. When the results are placed side by side, it can be seen that all five algorithms are a variation of constructing an approximately spherical neighborhood of a point (or face) on the boundary, and merging the neighborhood into the whole.

Lastly, a practical note. Some work has been done on visualizing solution manifolds [27], but it is still not clear how one should visualize a 3-manifold computed with any of these algorithms.

Acknowledgments

The author is grateful to Elsevier Science for permission to reproduce Fig. 3.34.

References

1. E. L. Allgower and K. Georg. Simplicial and continuation methods for approximations, fixed points and solutions to systems of equations. *SIAM Review*, 22:28–85, 1980.
2. E. L. Allgower and S. Gnutzmann. An algorithm for piecewise-linear approximation of implicitly defined two-dimensional surfaces. *SIAM J. Numer. Analysis*, 24(2):452–469, 1987.
3. E. L. Allgower and P. H. Schmidt. An algorithm for piecewise-linear approximation of an implicitly defined manifold. *SIAM J. Numer. Analysis*, 22(2):322–346, 1985.
4. F. Aurenhammer. Power diagrams: properties, algorithms and applications. *SIAM J. Comp.*, 16(1):78–96, 1987.
5. F. Aurenhammer. Improved algorithms for discs and balls using power diagrams. *J. Algor.*, 9:151–161, 1988.
6. F. Aurenhammer and H. Edelsbrunner. An optimal algorithm for constructing the weighted Voronoi diagram in the plane. *Pattern Recognition*, 17(2):251–257, 1984.
7. F. Aurenhammer. Voronoi diagrams – a survey of a fundamental geometric data structure. *ACM Comp. Surveys*, 23(3):345–405, 1991.
8. A. Bowyer. Computing Dirichlet tessellations. *The Computer Journal*, 24(2):162–166, 1981.
9. M. L. Brodzik. The computation of simplicial approximations of implicity defined *p*-dimensional manifolds. *Computers Math. Applic.*, 36(6):93–113, 1998.
10. M. L. Brodzik and W. C. Rheinboldt. On the computation of simplicial approximations of implicitly defined two-dimensional manifolds. *Computers Math. Applic.*, 28(9):9–21, 1994.
11. W. Brostow, J.-P. Dussault, and B. L. Fox. Construction of Voronoi polyhedra. *J. Comp. Phys.*, 29:81–92, 1978.
12. P.-C. Chen, P. Hansen, and B. Jaumard. On-line and off-line vertex enumeration by adjacency lists. *Oper. Res. Lett.*, 10:403–409, October 1991.
13. D. P. Dobkin and M. J. Laszlo. Primitives for the manipulation of three-dimensional subdivisions. *Algorithmica*, 4:3–32, 1989.
14. J. L. Dupont. *Curvature and Charateristic Classes*. Lecture Notes in Mathematics 640. (Springer–Verlag, Berlin, 1978).
15. S. Fortune. Voronoi diagrams and Delaunay triangulations. In D.-Z. Du and F. K. Hwang, editors, *Computing in Euclidean Geometry*, pages 193–233. (World Scientific Publishing, Singapore, 1992).
16. P. W. Gross and P. R. Kotiuga. Data structures for geometric and topological aspects of finite element algorithms. *Progr. Electromagn. Res. PIER 32*, 32:151–169, 2001.
17. B. Grunbaum. *Convex Polytopes*. (Interscience Publishers, London, 1967).

18. L. Guibas and J. Stolfi. Primitives for the manipulation of general subdivisions and the computation of Voronoi diagrams. *ACM Transactions on Graphics*, 4(2):74–123, April 1985.

19. P. M. Hanrahan. Creating volume models from edge–vertex graphs. *ACM SIGGRAPH Comp. Graph.*, 16(3):77–84, July 1982.

20. M. E. Henderson. Multiple parameter continuation: Computing implicitly defined k–manifolds. *Internat. J. Bifur. Chaos Appl. Sci. Engrg.*, 12(3):451–476, 2002.

21. M. E. Henderson. Multiparameter parallel search branch switching. *Internat. J. Bifur. Chaos Appl. Sci. Engrg.*, 15(3):967–974, 2005.

22. H. Imai, M. Iri, and K. Murota. Voronoi diagram in the Laguerre geometry and its applications. *SIAM J. Comp.*, 14(1):93–105, 1985.

23. H. B. Keller. Numerical solutions of bifurcation and nonlinear eigenvalue problems. In Paul Rabinowitz, editor, *Applications of Bifurcation Theory*, pages 359–384. (Academic Press, New York, 1977).

24. R. Klein. *Concrete and Abstract Voronoi Diagrams*, volume 400 of *Lecture Notes in Computer Science*. (Springer-Verlag, Berlin, 1987).

25. H. W. Kuhn. Some combinatorial lemmas in topology. *IBM J. Res. and Develop.*, 45(5):518–524, 1960.

26. S. Lefschetz. *Introduction to Topology*. (Princeton University Press, 1949).

27. J. H. Maddocks, R. S. Manning, R. C. Paffenroth, K. A. Rogers, and J. A. Warner. Interactive computation, parameter continuation, and visualization. *Internat. J. Bifur. Chaos Appl. Sci. Engrg.*, 7:1699–1715, 1997.

28. W. S. Massey. *Algebraic Topology, An Introduction*. (Harcourt, Brace & World, New York, 1967).

29. R. Melville and D. S. Mackey. New algorithm for two-dimensional numerical continuation. *Computers Math. Applic.*, 30(1):31–46, 1995.

30. V. T. Rajan. Optimality of the Delaunay triangulation in $\mathbb{R}^d$. In *Proceedings of the seventh annual symposium on Computational geometry*, pages 357–363. (ACM Press, New York, 1991).

31. W. C. Rheinboldt. On a moving-frame algorithm and the triangulation of equilibrium manifolds. In T. Küpper, R. Seydel, and H. Troger, editors, *ISNM79: Bifurcation: Analysis, Algorithms, Applications*, pages 256–267. (Birkhäuser Verlag, Basel, 1987).

32. W. C. Rheinboldt. On the computation of multi-dimensional solution manifolds of parameterized equations. *Numer. Math.*, 53(1/2):165–181, July 1988.

33. J. R. Shewchuk. A condition guaranteeing the existence of higher-dimensional constrained delaunay triangulations. In *Proceedings of the Fourteenth Annual Symposium on Computational Geometry*, pages 76–85, (ACM Press, New York, 1998).

34. D. F. Watson. Computing the n-dimensional Delaunay tessellation with application to Voronoi polytopes. *The Computer Journal*, 24(2):167–172, 1981.

35. H. Whitney. *Geometric Integration Theory*. (Princeton University Press, 1957).

4

Computing Invariant Manifolds via the Continuation of Orbit Segments

Bernd Krauskopf and Hinke M Osinga

Department of Engineering Mathematics, University of Bristol, United Kingdom

A key feature of packages such as AUTO [12, 15], CONTENT [24] and MATCONT [11] is a collocation solver for two-point boundary value problems (BVPs); see also Chaps. 1 and 2. In conjunction with pseudo-arclength continuation, it is possible to find the solution of a two-point BVP and then continue it in parameters. This basic idea will be known to most readers as the standard technique to compute a one-parameter branch of periodic orbits. However, the continuation of BVPs is a much more versatile tool and the solution need not be a periodic orbit, but may be any specified orbit segment. For example, the continuation of a suitable orbit segment is utilized in the HOMCONT extension for the computation of connecting orbits; see [7] and also [13, 21, 43].

In this chapter we focus on the idea of representing an invariant global manifold of a dynamical system as a family of orbit segments, which can then be computed as a solution family of a suitable BVP. Note that the thus computed object lies entirely in the phase space, rather than the product of phase space and parameter space. More specifically, we consider an n-dimensional autonomous vector field

$$\dot{x} = f(x, \lambda) \tag{4.1}$$

where $x \in \mathbb{R}^n$ and $f : \mathbb{R}^n \times \mathbb{R}^m \to \mathbb{R}^n$ is sufficiently smooth. The multi-dimensional parameter $\lambda \in \mathbb{R}^m$ remains fixed in most methods presented here, in which case we drop the dependence of f on λ for notational convenience. The vector field (4.1) induces a flow φ^t on $\mathbb{R}^n$ that determines the dynamics.

The global dynamics of (4.1) is determined by its equilibria, periodic orbits and invariant tori, together with their global stable and unstable manifolds. Equilibria and their stability properties can often be found analytically. However, since φ^t is rarely known explicitly, the task of finding periodic orbits, invariant tori and global manifolds generally requires the use of numerical techniques. We focus on the computation of an invariant manifold of (4.1) that can be viewed as a one-parameter family of orbit segments. In other words, in terms of the underlying vector field, we consider two-dimensional manifolds. This allows us to make use of the boundary value solver in AUTO [12, 15] to continue the respective orbit segments in order to build up the manifold. Note that the same set-up can be used for higher-dimensional manifolds, but then one needs to employ multi-parameter continuation methods; see [2, 29] and Chap. 3.

Let us assume that (4.1) has an equilibrium x_0, that is, $f(x_0) = 0$. We further suppose that x_0 is a hyperbolic saddle, which means that the Jacobian matrix $Df(x_0)$ at x_0 has eigenvalues in both the open left-half and the open right-half of the complex plane, but not on the imaginary axis. The stable manifold $W^s(x_0)$ associated with x_0 has the property that all orbits contained in $W^s(x_0)$ tend to x_0 in forward time:

$$W^s(x_0) = \left\{ x \in \mathbb{R}^n \mid \varphi^t(x) \to x_0 \text{ as } t \to \infty \right\}.$$

Similarly, the unstable manifold $W^u(x_0)$ is defined as the set of all orbits that tend to x_0 in backward time:

$$W^u(x_0) = \left\{ x \in \mathbb{R}^n \mid \varphi^t(x) \to x_0 \text{ as } t \to -\infty \right\}.$$

The Stable Manifold Theorem [45] guarantees the existence of global stable and unstable (immersed) manifolds that are as smooth as f. Furthermore, the dimension of the (un)stable manifold is equal to the number of (un)stable eigenvalues.

Equivalent notions of stable and unstable manifolds exist if (4.1) has a saddle periodic orbit Γ with period T_Γ. That is, Γ has Floquet multipliers both inside and outside the unit circle of the complex plane, and only the trivial Floquet multiplier (associated with the direction tangent to Γ) on the unit circle. As before, the Stable Manifold Theorem [45] guarantees the existence of global stable and unstable (immersed) manifolds $W^s(\Gamma)$ and $W^u(\Gamma)$ that are as smooth as f. They are defined as the set of orbits that tend to Γ in forward and backward time, respectively:

$$W^s(\Gamma) = \left\{ x \in \mathbb{R}^n \mid \varphi^t(x) \to \Gamma \text{ as } t \to \infty \right\}$$

and

$$W^u(\Gamma) = \left\{ x \in \mathbb{R}^n \mid \varphi^t(x) \to \Gamma \text{ as } t \to -\infty \right\}.$$

Note that the dimensions of $W^s(\Gamma)$ and $W^u(\Gamma)$ are equal to one plus the number of stable and unstable Floquet multipliers, respectively.

This chapter is organized as follows. We first consider in Sect. 4.1 the computation of the one-dimensional intersections of a two-dimensional manifold with a chosen Poincaré section. As we show in Sect. 4.2, our set-up is particularly useful in the context of slow-fast systems and even allows us to compute slow manifolds near a folded node. We then discuss in Sect. 4.3 how one can compute two-dimensional global manifolds as a family of orbit segments. In Sect. 4.3.3 we explain how one can compute two-dimensional invariant tori with quasiperiodic dynamics in this set-up. We end with conclusions and an outlook to future research in Sect. 4.4.

4.1 One-Dimensional Global Manifolds in a Poincaré Section

The Poincaré map of an n-dimensional vector field as given by (4.1) is defined on an $(n-1)$-dimensional section $\Sigma \subset \mathbb{R}^n$. The image $P(x)$ of a point $x \in \Sigma$ under the Poincaré map P is then defined as the next (or kth for some integer $k > 0$) intersection of the orbit through x with Σ. Considering a Poincaré map is a standard tool to study the properties of a periodic orbit Γ of a vector field. The section Σ is chosen transverse to Γ and P is defined such that an (isolated) intersection point γ_0 of Γ with Σ is a fixed point, that is, $P(\gamma_0) = \gamma_0$. Then locally near γ_0 the map P is a diffeomorphism, so that the dynamics of P on Σ near γ_0 describes the local dynamics near Γ; see, for example, [27, 41, 53].

Suppose now that the vector field (4.1) has a saddle periodic orbit Γ. To keep this exposition simple, we assume that the dimension of the phase space is $n = 3$. In this case the periodic orbit Γ has two-dimensional stable and unstable manifolds $W^s(\Gamma)$ and $W^u(\Gamma)$, respectively. Furthermore, we consider the standard situation that the Poincaré section Σ is a suitable two-dimensional plane transverse to Γ, so that there exists a point $\gamma_0 \in \Gamma \cap \Sigma$. When properly defined in a neighborhood of γ_0 as the kth return to Σ, the Poincaré map P is a diffeomorphism and γ_0 is a saddle fixed point of P. As a consequence, the Stable Manifold Theorem guarantees the existence of one-dimensional smooth (un)stable manifolds $W^s(\gamma_0)$ and $W^u(\gamma_0)$ in the region where P is a diffeomorphism. Moreover, globally we can view $W^s(\gamma_0)$ and $W^u(\gamma_0)$ as the intersections of the corresponding stable and unstable manifolds of Γ. That is,

$$W^s(\gamma_0) = W^s(\Gamma) \cap \Sigma$$

and

$$W^u(\gamma_0) = W^u(\Gamma) \cap \Sigma.$$

Note that this definition has the advantage of being very geometrical. It follows from the properties of the manifolds $W^s(\Gamma)$ and $W^u(\Gamma)$ in $\mathbb{R}^3$ that the (generalized) global manifolds $W^s(\gamma_0)$ and $W^u(\gamma_0)$ exist and are a union of manifolds in Σ.

Indeed, $W^s(\gamma_0)$ and $W^u(\gamma_0)$ are not necessarily single connected curves, because P is typically not a global diffeomorphism. In general, P is discontinuous at points where the flow is tangent to Σ and we define the discontinuity locus C as

$$C := \{x \in \Sigma \mid f(x) \cdot \boldsymbol{n}_x = 0\}, \qquad (4.2)$$

where $\boldsymbol{n}_x$ is the normal to Σ at x and the dot denotes the inner product in $\mathbb{R}^3$. (If Σ is a hyperplane then $\boldsymbol{n}_x = \boldsymbol{n}$ is independent of x.) The complement of C consists of open regions of Σ. In the region that contains γ_0, we can define P in a continuous and smooth manner and $W^s(\gamma_0)$ and $W^u(\gamma_0)$ are well defined as the global manifolds of the fixed point γ_0 of the map. The intersections of $W^s(\Gamma)$ and $W^u(\Gamma)$ with Σ become more interesting as soon as $W^s(\gamma_0)$ and $W^u(\gamma_0)$ cross the discontinuity locus C.

Let us consider a simple example, where the three-dimensional vector field has a saddle periodic orbit Γ and we can choose a two-dimensional planar Poincaré section Σ that intersects Γ transversely at exactly two points, say, γ_0 and γ_1. Near both γ_0 and γ_1 one can define the Poincaré map P as a local diffeomorphism (the second return to Σ). In the simplest case, the set C is a single smooth curve that divides the plane Σ into the two regions of local definition of P. In this situation, the stable and unstable manifolds of γ_0 and γ_1 in Σ may cross C. One possibility is that a branch of $W^s(\gamma_0)$ coincides with a branch of $W^s(\gamma_1)$, effectively connecting γ_0 and γ_1; see already Fig. 4.2(c). This corresponds in the three-dimensional phase space to the case that Σ intersects one side of $W^s(\Gamma)$ in a single curve. There are other, more complicated possibilities of how Σ may intersect $W^s(\Gamma)$. For example, the intersection in Σ may consist of a closed curve that corresponds to a piece of $W^s(p_0)$ that is not connected to any saddle; see already Fig. 4.4.

4.1.1 The MANBVP Method

From an algorithmic point of view, it is not a good idea to compute $W^s(\gamma_0)$ and $W^u(\gamma_0)$ by finding $W^s(\Gamma)$ or $W^u(\Gamma)$ first and then determine their intersection with Σ. Rather we will find $W^s(\gamma_0)$ and $W^u(\gamma_0)$ directly as subsets of Σ by using what we call the MANBVP method. The key idea is to consider the manifolds as one-parameter families of solutions to two-point boundary value problems. Our method as reviewed here was introduced in [16], where more details can be found.

The main essence of the MANBVP method is that one should think of the Poincaré map not only as assigning $P(x)$, but as assigning the entire orbit segment

$$\{\varphi^t(x) \mid 0 \le t \le t_x\}.$$

Here t_x is the appropriate return time to the section Σ, which depends continuously on x. In particular, for $\gamma_0 \in \Gamma \cap \Sigma$ we have $P(\gamma_0) = \gamma_0$ and $t_{\gamma_0} = T_\Gamma$ is the period of Γ. The orbit segment that constitutes the action of P on x is the solution $\mathbf{u}$ of the two-point boundary value problem that solves (4.1) subject to the boundary conditions $\mathbf{u}(0) = x \in \Sigma$ and $\mathbf{u}(t_x) \in \Sigma$.

The MANBVP method has the key advantage that the locus C is no longer a discontinuity boundary in the space of boundary value problems. It is explained here for the computation of a one-dimensional *unstable* manifold $W^u(\gamma_0)$ of a saddle fixed point $\gamma_0 \in \Sigma$ that is associated with a saddle periodic orbit Γ of (4.1). A one-dimensional stable manifold can be computed in exactly the same way by reversing time. We assume that the single unstable Floquet multiplier λ^u of Γ is positive, i.e. $\lambda^u > 1$, so that the associated Poincaré map P is orientation preserving on $W^u(\gamma_0)$; if $\lambda^u < -1$, then Γ must be covered twice to define the image of γ_0 under P, that is, $t_{\gamma_0} = 2\,T_\Gamma$.

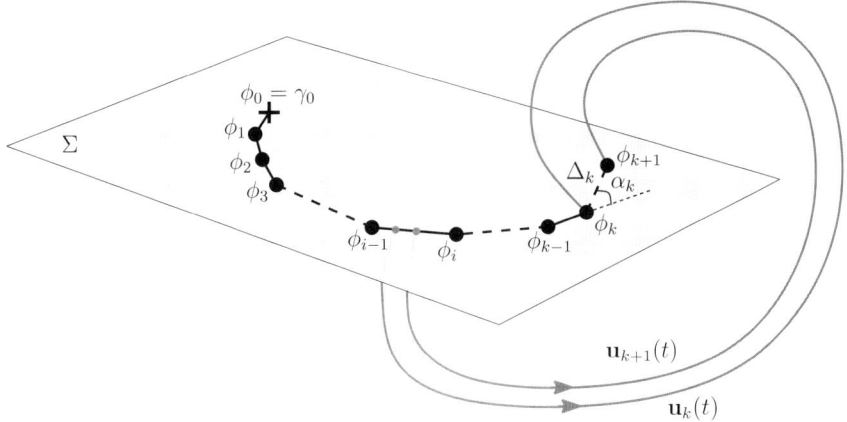

Fig. 4.1. Graphical illustration of the ManBVP algorithm.

The MANBVP method approximates $W^u(\gamma_0)$ as an ordered list of mesh points $M = \{\phi_0, \phi_1, \ldots, \phi_N\}$ up to a prescribed arclength L. The computation starts from the saddle periodic orbit Γ, which we find by continuation with AUTO [12, 15]. The periodic orbit is viewed as an orbit segment that defines the image of γ_0 under the Poincaré map P. We continue this orbit segment in AUTO [12, 15] not as a periodic orbit but as an orbit with both boundary points in Σ. The first boundary point of the orbit segment is initially varied along the unstable eigenvector of γ_0 in Σ. In this way, the other boundary point of the orbit segment begins to trace out an approximation of the initial piece of $W^u(\gamma_0)$. We then allow the first boundary point to vary along this initial piece and subsequently computed parts of the manifold, so that the other boundary point continues to trace out more and more of the unstable manifold.

Figure 4.1 gives an impression of how the MANBVP method works. The variation of the first boundary point along the computed piece is done using a piecewise-linear approximation of $W^u(\gamma_0)$ between mesh points in M. The distance between consecutive points $\phi_k \in M \subset \Sigma$ varies depending on the local curvature of $W^u(\gamma_0)$; in this sense, the method can be seen as an adaptation and refinement of the method in [36] to the specific context of general Poincaré maps.

Set-Up of the Two-Point Boundary Value Problem

To set up a two-point boundary value problem that has solutions with end points on $W^u(\gamma_0)$ we look for solutions of the system

$$\mathbf{u}'(t) = T\, f(\mathbf{u}(t)), \qquad (4.3)$$

with boundary conditions

$$\mathbf{u}(0) \in \Sigma, \tag{4.4}$$

$$\mathbf{u}(1) \in \Sigma. \tag{4.5}$$

Note that (4.3) is in a scaled form so that all orbit segments have total time 1. Hence, the orbit segment $\{\varphi^t(x) \mid 0 \le t \le t_x\}$ associated with the image $P(x)$ of a point $x \in \Sigma$ is rescaled as $\{\mathbf{u}(t/t_x) \mid 0 \le t \le 1\}$ and the 'period' $T = t_x$ is now a parameter in the system.

We obtain a one-parameter family of solutions to (4.3)–(4.5) by letting $\mathbf{u}(0)$ vary along the one-dimensional piecewise-linear approximation of (a first piece of) $W^u(\gamma_0)$ instead of the two-dimensional space Σ. To this end, we let

$$L_i(\tau) = (1 - \tau)\,\phi_{i-1} + \tau\phi_i, \quad 0 \le \tau \le 1,$$

denote the (parametrized) line segment between the already computed mesh points ϕ_{i-1} and ϕ_i. Then boundary condition (4.4) is replaced by

$$\mathbf{u}(0) - L_i(\tau) = \mathbf{0}, \tag{4.6}$$

where the index i varies as the computation progresses. Note that (4.6) automatically ensures that $\mathbf{u}(0) \in \Sigma$ in case Σ is a hyperplane. For non-planar Σ this introduces an error that is of the same order as the mesh error on M; we refer to [16] for details.

In order to detect during a computation that the end of the line segment $L_i(\tau)$ has been reached we introduce the user-defined functions

$$\mathrm{UZ}(0) = \tau,$$
$$\mathrm{UZ}(1) = \tau - 1.$$

When such an event is detected the algorithm switches either to the next or the previous line segment. Indeed, it is necessary to allow switches to the previous line segment, that is, to let $\mathbf{u}(0)$ trace $W^u(\gamma_0)$ backwards toward γ_0; namely, this occurs as soon as $W^u(\gamma_0)$ has crossed the discontinuity locus C.

The selection of the mesh points in M is based on the same accuracy conditions as the method for diffeomorphisms in [36]. To decide when to generate the next point ϕ_{k+1} we calculate the distance

$$\Delta_k := \| \mathbf{u}(1) - \phi_k \|$$

between the end boundary point and the last computed mesh point, and the angle

$$\alpha_k := \angle(\phi_{k-1}, \phi_k, \mathbf{u}(1))$$

between the last two computed mesh points and the end boundary point during the continuation. We then monitor the user-defined accuracy conditions

$$\mathrm{UZ}(2) = \alpha_{\max} - \alpha_k,$$
$$\mathrm{UZ}(3) = (\Delta\alpha)_{\max} - \Delta_k \alpha_k,$$
$$\mathrm{UZ}(4) = \Delta_{\min} - \Delta_k$$

where $\alpha_{\max}$, $(\Delta\alpha)_{\max}$, and $\Delta_{\min}$ are prespecified bounds set by the user; for the examples below and in Sect. 4.2.1 we used $\alpha_{\max} = 0.3$, $(\Delta\alpha)_{\max} = 10^{-5}$, and $\Delta_{\min} = 10^{-4}$. Note that the above conditions minimize the number of mesh points required for achieving the accuracy conditions, so that the mesh selection with respect to the curvature of the manifold is more refined here than that in [36].

The first segment $L_1(\tau)$ is defined by the line through γ_0 in Σ in the eigendirection associated with the unstable Floquet multiplier λ^u; see [16] for details on how to find this direction. We assume that $\tau = 1$ when a prespecified maximum distance δ along this initial segment is reached. Indeed, this initial distance δ has an important influence on the overall error of the approximation to $W^u(\gamma_0)$; see [36]. Starting from the scaled orbit that represents Γ, we continue the one-parameter family of solutions along $L_1(\tau)$ while monitoring the distance

$$\Delta_1 = \| \mathbf{u}(1) - \mathbf{u}(0) \|$$

and the angle

$$\alpha_1 = \angle(\gamma_0, \mathbf{u}(0), \mathbf{u}(1)).$$

When either of the accuracy conditions are met then the continuation stops and we set $\phi_1 = \mathbf{u}(0)$ and $\phi_2 = \mathbf{u}(1)$. We then continue the computation along the line segment $L_2(\tau)$. Hence, the initial line segment between ϕ_0 and ϕ_1 may, in fact, be shorter than the prespecified maximum distance δ.

During the computation we also record the orbit segments $\mathbf{u}_k$ that are used to find the points $\phi_k = \mathbf{u}_k(1)$ of M. These orbit segments give a good impression of the relevant part of the two-dimensional manifold $W^u(\Gamma)$ in the full phase space and provide further insight into the geometry of the manifolds.

Semiconductor Laser with Optical Injection

In this section we illustrate the performance of the MANBVP method by computing the stable and unstable manifolds of a model for an optically injected semiconductor laser. Manifolds in this system were also considered in [9] and we show here how the MANBVP method is able to compute a one-dimensional manifold across the discontinuity boundary C. The laser system is modeled by the so-called rate equations

$$\begin{cases} \dot{E} = K + \left(\frac{1}{2}\left(1 + i\alpha\right)n - i\omega\right)E\,, \\ \dot{n} = -2\gamma n - (1 + 2Bn)\left(|E|^2 - 1\right), \end{cases} \tag{4.7}$$

and is well known to have rich dynamics; see Chap. 6 and [58]. Here, $E = (E_x, E_y)$ is the complex electric field and n is the population inversion (the number of electron-hole pairs). The two main parameters are the injected field strength K, and the detuning ω between the frequency of the free-running laser and the injected frequency. The material properties of the laser are described by the values of α, B and γ.

We use the parameter values $\alpha = 2$, $B = 0.015$, $\gamma = 0.035$, and choose $\omega = 0.270$ and $K = 0.290$, as in [9]. We study the Poincaré map P on the section given by the plane $\Sigma = \{(E, n) \mid n = 0\}$. For this choice of parameters a saddle periodic orbit Γ intersects Σ at four points denoted γ_0, γ_1, γ_2, and γ_3 in consecutive order. The discontinuity boundary C, where the flow is tangent to Σ, is given by $\dot{n} = 0 \Leftrightarrow |E| = 1$ and $\dot{n} > 0$ for points with $|E| < 1$, whereas $\dot{n} < 0$ for points with $|E| > 1$. Figure 4.2 gives an overview of the computations of $W^s(\Gamma) \cap \Sigma$ (black curves). The intersection points of the periodic orbit are indicated by crosses. Panel (a) also shows a period-four sink (triangles) and how $W^u(\Gamma)$ intersects Σ (grey curves). Since the manifold $W^u(\Gamma)$ does not interact with C, it is of less interest here.

Figure 4.2(b) shows the branches of stable manifolds that spiral in toward singular points on Σ. Notice that the two branches starting from γ_1 and γ_3 never interact with C. This means that the associated Poincaré map is a diffeomorphism along the entire branch, and there is no particular difficulty to calculate these branches. On the other hand, the two branches starting from γ_0 and γ_2 cross C several times. A similar observation can be made in Fig. 4.2(c) and we discuss here the behavior of the branch of $W^s(\gamma_0)$ in more detail. Figure 4.2(d) shows two branches of the stable manifold that are disconnected from the four saddle points, which we also discuss in more detail below.

Let us first focus on the left branch of $W^s(\gamma_0)$ as an example of a branch that crosses C. This branch of $W^s(\gamma_0)$ is shown in Fig. 4.2(c) and the crossing through C is more clearly illustrated in Fig. 4.3. Each row in Fig. 4.3 shows the orbit segments used in the computation of an initial piece of $W^s(\gamma_0)$ together with the time profile of the last orbit segment computed for this part. The black dots in the time profiles are the mesh points used by AUTO [12, 15] and the green line is Σ. As can be seen in Fig. 4.3(a1), the boundary points $\mathbf{u}(0)$ of the orbit segments all lie near γ_0. Since time is reversed to compute a stable manifold, the orbit segments first flow downward from Σ. They next intersect Σ near γ_3 and so on until they return to Σ for the fourth time. The boundary points $\mathbf{u}(1)$ are added to the mesh M representing the initial piece of $W^s(\gamma_0)$. The time profile in Fig. 4.3(a2) is representative for all orbit segments in Fig. 4.3(a1) and shows three intersections with Σ (green line) between $\mathbf{u}(0)$ and $\mathbf{u}(1)$; note that the end of the orbit segment is flowing downward into the section. Row (b) in Fig. 4.3 shows the computation exactly up to the point where $W^s(\gamma_0)$ intersects C. The time profile in panel (b2) shows that the last computed orbit segment still has three intersections with the section between $\mathbf{u}(0)$ and $\mathbf{u}(1)$, but now the orbit segment is tangent to Σ at $\mathbf{u}(1)$. As we continue to compute $W^s(\gamma_0)$ past C, the ends of the added orbit segments now flow upward into Σ. Row (c) of Fig. 4.3 shows this change in behavior between orbit segments before and past C. The time profile in panel (c2) shows that there are now four intersections of the last computed orbit segment with Σ between $\mathbf{u}(0)$ and $\mathbf{u}(1)$. This orbit segment is again representative for all orbit segments that end on the other side of C. As we

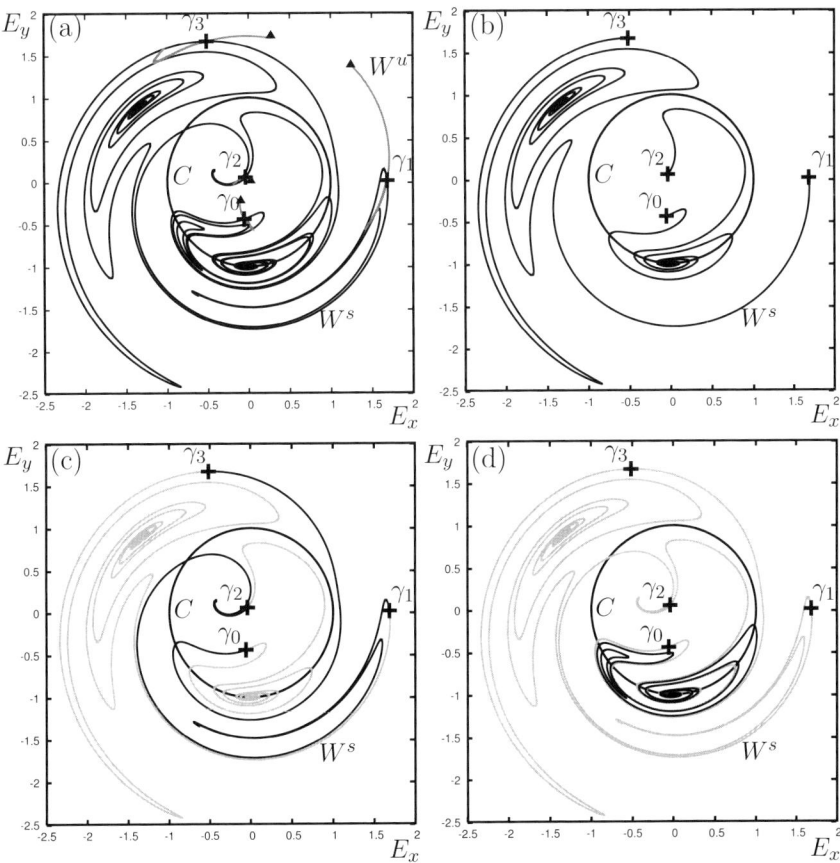

Fig. 4.2. The stable and unstable manifolds, as computed by the MANBVP algorithm, of the four-periodic saddle $\{\gamma_0, \gamma_1, \gamma_2, \gamma_3\}$ (crosses) of (4.7) in $\Sigma = \{n = 0\}$ with $\alpha = 2$, $B = 0.015$, $\gamma = 0.035$, $\omega = 0.270$, and $K = 0.290$. The flow is tangent to Σ along the unit circle labeled C. Panel (a) shows all computed parts of the stable and unstable manifolds of $\{\gamma_0, \gamma_1, \gamma_2, \gamma_3\}$. Panel (b) highlights branches of stable manifolds that spiral into singular points, panel (c) those that join two of the saddle points, and panel (d) two disjoint pieces of manifold. From J.P. England, B. Krauskopf and H.M. Osinga, Computing one-dimensional global manifolds of Poincaré maps by continuation, *SIAM J. Appl. Dyn. Sys.* 4(4) (2005) 1008–1041 © 2005 by the Society for Industrial and Applied Mathematics; reprinted with permission.

already observed in Fig. 4.2(c), $W^s(\gamma_0)$ continues until it reaches γ_3. Indeed, since γ_3 is the next intersection of Γ when following Γ backward in time from γ_0, the extra intersection picked up once $\mathbf{u}(1)$ crosses C makes this possible. Note that the same branch can also be computed as one of the branches of

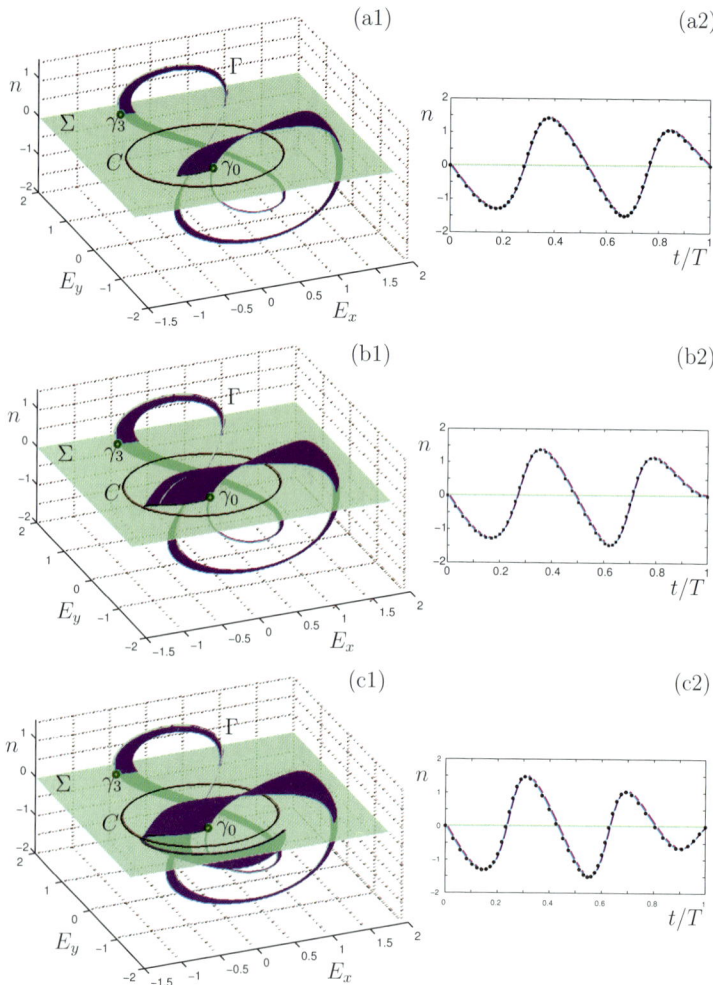

Fig. 4.3. A demonstration of how $W^s(\gamma_0)$ of system (4.7) is computed across C. Rows (a)-(c) show the computations before, at, and after the intersection with C, respectively. Panels in the left column show orbit segments on $W^s(\Gamma)$ used to compute $W^s(\gamma_0)$. The time profiles of the last orbit segment computed in each case are shown in the right column (blue line), where the black dots indicate the mesh points and Σ is the green line. From J.P. England, B. Krauskopf and H.M. Osinga, Computing one-dimensional global manifolds of Poincaré maps by continuation, *SIAM J. Appl. Dyn. Sys.* 4(4) (2005) 1008–1041 © 2005 by the Society for Industrial and Applied Mathematics; reprinted with permission.

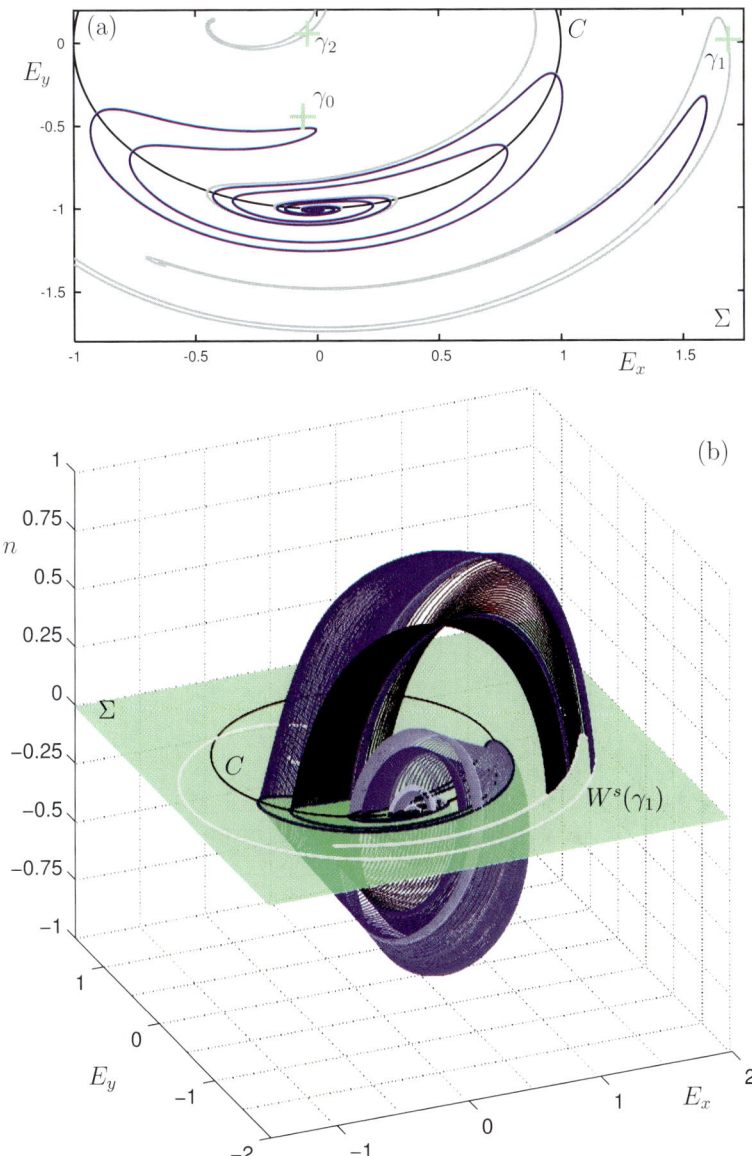

Fig. 4.4. An enlargement of Fig. 4.2(d) in panel (a) shows a disjoint intersection of $W^s(\Gamma)$ with Σ (blue). The colored segment of $W^s(\gamma_1)$ maps to this disjoint piece. The orbit segments used in the computation are shown in panel (b). From J.P. England, B. Krauskopf and H.M. Osinga, Computing one-dimensional global manifolds of Poincaré maps by continuation, *SIAM J. Appl. Dyn. Sys.* 4(4) (2005) 1008–1041 © 2005 by the Society for Industrial and Applied Mathematics; reprinted with permission.

$W^s(\gamma_3)$, where the computation starts out with orbit segments that intersect Σ three times in between $\mathbf{u}(0)$ and $\mathbf{u}(1)$, and which become orbit segments that intersect Σ only two times in between the boundary points as soon as C is crossed.

Figure 4.4(a) shows an enlarged view of Fig. 4.2(d). The blue branch that spirals around C is not connected to any of the intersection points of Γ. However, all points on this disjoint segment lie on orbits that intersect Σ on $W^s(\gamma_1)$. This part of $W^s(\gamma_1)$ is also colored blue in Fig. 4.4(a). Figure 4.4(b) shows the orbits segments used in the computation of this disjoint part. All orbit segments have $\mathbf{u}(0)$ on $W^s(\gamma_1)$ and this part of $W^s(\gamma_1)$ is traversed back and forth several times as $\mathbf{u}(1)$ crosses C. The segment traced out by $\mathbf{u}(1)$ is an isolated submanifold that is due to the way $W^s(\Gamma)$ intersects Σ. We are able to compute this disjoint piece by choosing an already computed point on $W^s(\gamma_1)$ and integrating it backward in time until it intersects Σ again. We then correct the solution such that it satisfies the boundary conditions and use the MANBVP method to trace the entire isolated branch.

4.2 Global Manifolds of Slow-Fast Systems

In its simplest form, a slow-fast system can be written as

$$\begin{cases} \dot{x} = g(x, y, \varepsilon), \\ \varepsilon\dot{y} = f(x, y, \varepsilon), \end{cases} \tag{4.8}$$

where $(x, y) \in \mathbb{R}^n$ (in this section we again consider the case $n = 3$) and ε determines the separation of slow and fast time scales. It is well known that a slow-fast system of the form (4.8) with $\varepsilon \ll 1$ typically displays extreme sensitivity to the initial condition [33]. However, the inherent accuracy of solving two-point boundary problems (for example, with AUTO's collocation routines as in our case) allows one to deal efficiently with the problem of a possible sensitivity of the initial value problem. As a consequence, the MANBVP method is able to compute global manifolds reliably also in the context of systems with multiple time scales. This is demonstrated below in Sect. 4.2.1 where we compute stable and unstable manifolds in a Poincaré section for a Van der Pol-Duffing oscillator modeled by Koper [4, 35]. As is shown in Sect. 4.2.2, a slight variation of the MANBVP method can be used to compute slow manifolds and their intersections, which are known as maximal canard orbits; see also Chap. 8.

4.2.1 The Koper Model

We consider the Van der Pol-Duffing oscillator [4, 18, 35] as modeled by the equations

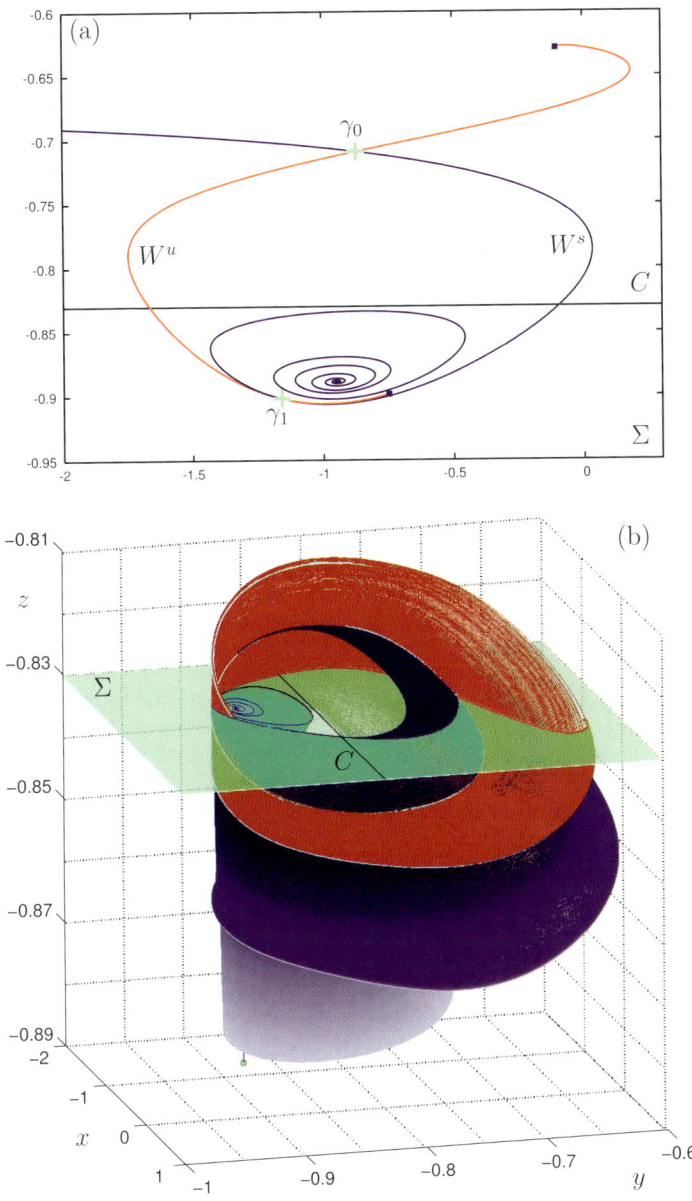

Fig. 4.5. The stable (blue) and unstable (red) manifolds of (4.9) with $\varepsilon = 0.1$, $k = -22.5$, and $\lambda = 18$, computed in the section $z = -0.83$ using the MANBVP method. The points γ_0 and γ_1 are the two intersections of a saddle periodic orbit Γ with Σ. The orbit segments used in the computation are shown in panel (b). From J.P. England, B. Krauskopf and H.M. Osinga, Computing one-dimensional global manifolds of Poincaré maps by continuation, *SIAM J. Appl. Dyn. Sys.* 4(4) (2005) 1008–1041 © 2005 by the Society for Industrial and Applied Mathematics; reprinted with permission.

$$\begin{cases} \dot{x} = ky - x^3 + 3x - \lambda, \\ \varepsilon\dot{y} = x - 2y + z, \\ \varepsilon\dot{z} = y - z, \end{cases} \qquad (4.9)$$

where ε determines the difference between two time scales. System (4.9), which is often referred to as the Koper model, is also one of the examples in Chap. 2; note that, while a slow-fast system can be written in different ways (on the slow or the fast time scale) its implementation in AUTO is always the same due to the introduction of the time variable T. We take $\varepsilon = 0.1$, such that the x-variable evolves on a time scale ten times faster than y and z. A saddle periodic orbit Γ exists for $k = -22.5$ and $\lambda = 18$, together with an attracting orbit of roughly twice the period. We choose the plane $\Sigma = \{(x, y, z) \mid z = -0.83\}$, which Γ intersects at two points γ_0 and γ_1. The flow of (4.9) is tangent to Σ along the line $C = \{y = z = -0.83\}$; $\dot{z} > 0$ for points with $y < -0.83$ and $\dot{z} < 0$ for points with $y > -0.83$.

Figure 4.5(a) shows the stable and unstable manifolds of γ_0 and γ_1 in Σ, along with the computed orbit segments in panel (b). The saddles γ_0 and γ_1 are indicated by green crosses. The period-doubled attractor intersects Σ only twice, at the points indicated by blue squares. One branch of $W^u(\gamma_0)$ and one branch of $W^s(\gamma_0)$ cross C and connect to γ_1. The other branch of $W^u(\gamma_0)$ tends to one of the sinks, as does one branch of $W^u(\gamma_1)$. The other branch of $W^s(\gamma_0)$ tends to $-\infty$. Finally, the branch of $W^s(\gamma_1)$ that does not connect to γ_0 spirals in toward a special point, namely the intersection point of the one-dimensional stable manifold of a saddle equilibrium located below Σ; see the green point below Σ in panel (b).

It is particularly challenging to compute $W^s(\gamma_0)$ and $W^s(\gamma_1)$. Not only are the Floquet multipliers λ^u and λ^s of Γ negative, so that we need to use a double covering, but λ^s is also very strongly contracting. Namely, $\lambda^u \approx -3.03$ and $\lambda^s \approx -9.25 \times 10^{-4}$. Hence, the contraction along the stable manifold is $1/(\lambda^s)^2 = O(10^6)$. It is precisely this difference in scales that makes it impossible to compute even the Poincaré map by solving an initial value problem. However, the MANBVP method is able to compute both manifolds of γ_0 and γ_1 in Σ; indeed, $\mathbf{u}(0)$ varies only on the order of $O(10^{-9})$ during the computation of the entire branch of $W^s(\gamma_0)$ that connects to γ_1.

As Fig. 4.5(b) shows, many orbit segments are needed during the computation. In particular, we are effectively computing the part of the two-dimensional manifold $W^s(\Gamma)$ that below Σ accumulates on the segment of the stable manifold of the (green) saddle equilibrium.

4.2.2 Slow Manifolds and Canards Near a Folded Node

In this section we compute slow manifolds associated with a singularity that is known as a *folded node* of a slow-fast system in $\mathbb{R}^3$. Suppose that the slow variable x in (4.8) is two-dimensional (and, hence, $y \in \mathbb{R}$) and that the critical manifold $S := \{(x, y) \in \mathbb{R}^3; f(x, y, 0) = 0\}$ is a nondegenerate folded surface

in a neighborhood of the origin. Hence, the associated slow subsystem is singular along the fold curve. Such a folded critical manifold is the generic case
where Fenichel theory [20, 33] cannot be applied because the required normal
hyperbolicity breaks down along the fold curve. In particular, this situation
creates the interesting phenomenon (especially for $\varepsilon \neq 0$) that the two sheets
of the associated slow manifold, one of which is locally attracting or stable and
one locally repelling or unstable, may intersect along special orbits known as
canard solutions. A canard solution has the unusual property that it follows
the unstable sheet for a certain amount of time before being repelled; see also
Chap. 8 and, for example, [5, 26, 54].

A folded node is a singular point on the fold line that corresponds to a
node equilibrium of the desingularized slow subsystem (where time has been
rescaled by the factor $\sqrt{\varepsilon}$) [5, 26, 54, 57]. A blow-up procedure, followed by
the restriction to a specific chart on the blown-up locus and desingularization
lead to a normal form near the folded node. Following [57], the normal form
can be written as the vector field

$$\begin{cases} \dot{x} = \frac{1}{2}\mu y - (\mu + 1)z + O(\sqrt{\varepsilon}), \\ \dot{y} = 1, \\ \dot{z} = x + z^2 + O(\sqrt{\varepsilon}). \end{cases} \tag{4.10}$$

Here μ is the ratio of the strong and the weak eigenvalue of the folded singularity of the reduced flow; the critical manifold of the original system is
represented by the set $\{x + z^2 = 0\}$.

We consider here the so-called *maximal canards* near the folded node,
which correspond to intersections of the stable sheet C^- and the unstable
sheet C^+ of the slow manifold for $\varepsilon = 0$ in (4.10). To this end, we compute
the sheet C^- as a family of orbit segments with a slight modification of the
MANBVP method. Note that the sheet C^+ is the image of C^- under the
symmetry $(x, y, z, t) \mapsto (x, -y, -z, -t)$.

We consider orbit segments between the x-section $\Sigma_\xi = \{x = -\xi\}$ and the
y-section $\Sigma_\eta = \{y = \eta\}$ where ξ and η are suitable constants. The manifold C^-
is a smooth perturbation of the stable sheet of the critical manifold (where
z is negative) away from the fold curve (at $z = 0$) where Fenichel theory
fails. Hence, for sufficiently large ξ the intersection curve of the sheet C^-
with the x-section Σ_ξ is approximated well by the one-dimensional line $L =
\{(x, z) = (-\xi, \sqrt{\xi})\}$; in our computations we used $\xi = 100$ throughout. The
idea is now to continue the solution of the family of orbit segments $\mathbf{u}$ with
$\mathbf{u}(0) \in L$ and $\mathbf{u}(1) \in \Sigma_\eta$, while monitoring the accuracy condition in terms
of the curve traced out by $\mathbf{u}(1)$ in Σ_η as is explained in Sect. 4.1.1. In other
words, loosely speaking C^- is the 'unstable manifold of the line L'. As starting
condition we use the explicit solution $\gamma_s(t) = (-\frac{\mu^2}{4}t^2 + \frac{\mu}{2}, t, \frac{\mu}{2}t)$, which is
known as the strong maximal canard; there is a second explicit solution given
by $\gamma_w(t) = (-\frac{1}{4}t^2 + \frac{1}{2}, t, \frac{1}{2}t)$, which is known as the weak maximal canard; see
[57].

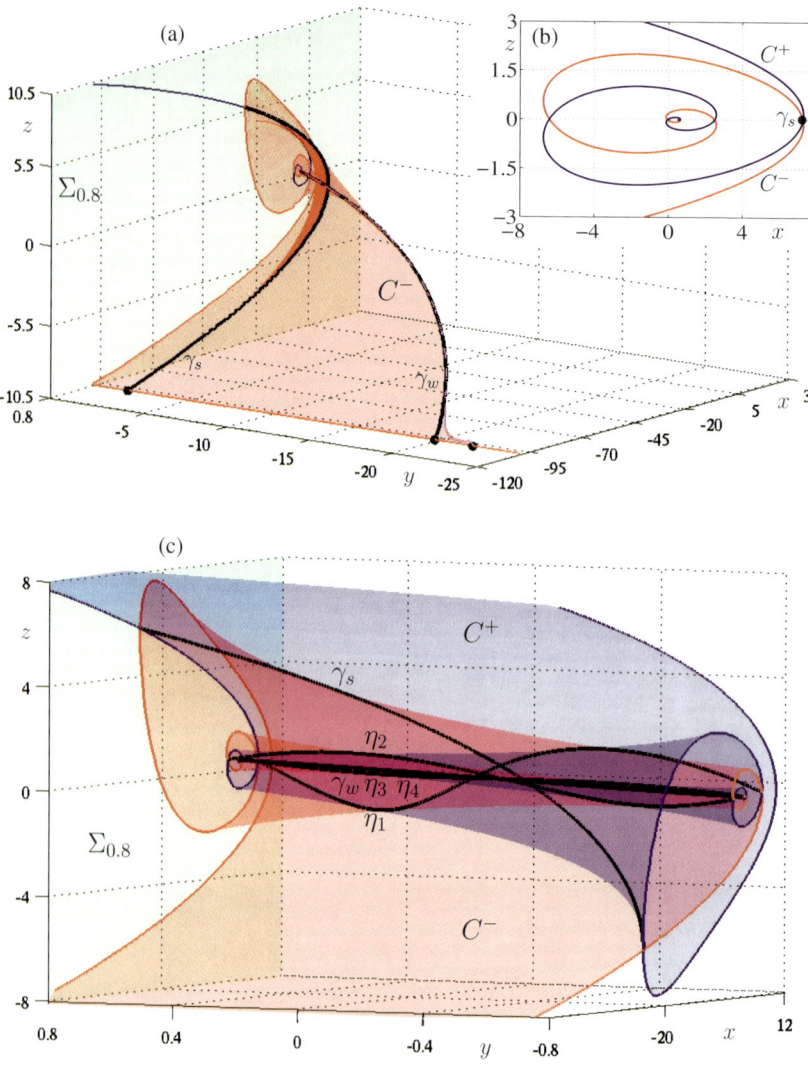

Fig. 4.6. The slow manifolds C^- and C^+ of system (4.10) for $\mu = 14.5$. Panel (a) shows C^- up to the y-section $\Sigma_{0.8}$, the inset panel (b) shows the intersection curves of C^- and C^+ in the y-section $\Sigma_{0.0}$, and panel (c) shows how the surfaces C^- and C^+ between the y-sections $\Sigma_{-0.8}$ and $\Sigma_{0.8}$ intersect in maximal canards γ_s, γ_w and η_1 to η_4.

Figure 4.6 shows the slow manifolds C^- and C^+ of (4.10) for $\mu = 14.5$. Panel (a) gives a global overview of the two-dimensional slow manifold C^- (red surface) computed up to the y-section $\Sigma_{0.8}$ (green). This surface was rendered from the orbit segments that were used in the computation of the curve $C^- \cap \Sigma_{0.8}$. The chosen viewpoint emphasizes that C^- is close to the critical manifold $\{x = -z^2\}$ only for large negative x. Indeed near $x = 0$ the manifold C^- starts to spiral and develops several folds. Also shown are the explicitly known maximal canards γ_s and γ_w. Figure 4.6(b) shows the intersections with the y-section $\Sigma_{0.0}$ of the slow manifolds C^- (red curve) and C^+ (blue curve); compare with [57]. The intersection points between both curves are maximal canards, of which the strong maximal canard γ_s is highlighted. Figure 4.6(c) presents an enlarged view of the geometry of C^- (red surface) and C^+ (blue surface) between the y-sections $\Sigma_{-0.8}$ and $\Sigma_{0.8}$. The manifolds C^- and C^+ intersect along the maximal canards γ_s and γ_w, as well as the additional maximal canards η_1 to η_4 (black curves). Notice how the maximal canards η_1 to η_4 spiral around the weak maximal canard γ_w. For a more detailed study of maximal canards in (4.10) and their dependence on the parameter μ we refer to [10].

4.3 Two-Dimensional Global Manifolds of Vector Fields

In the previous sections we computed one-dimensional manifolds in a Poincaré section. Almost as a by-product, the orbit segments that are used in the computation give an impression of a part of the associated two-dimensional manifold of the underlying vector field.

In this section, we show how one can use the idea of continuing a family of orbit segments to find enough and suitable orbit segments such that an entire first part of a two-dimensional manifold is computed. In contrast to how we selected a mesh, and corresponding orbit segments, in the previous section, we are now interested in obtaining a good representation of the entire two-dimensional object, and not just of a one-dimensional intersection curve. Such a continuation of orbit segments, with a number of choices for the boundary condition as detailed below, can be readily implemented in the package AUTO [12, 15]. While in general the resulting mesh has quite a lot of mesh points, the computations are very accurate and remarkably fast; see [39, Sec. 3] for a general overview of this method. The key property of the continuation procedure is that the step size measures the change of the *entire computed orbit segment* (and various parameters), and not just the change at one of the end points. This means that, generally, the computation produces a reasonable distribution of orbit segments along the manifold.

We first explain and demonstrate the method by computing the unstable manifolds of the secondary saddle equilibria or saddle periodic orbits in the Lorenz system [42]. We then discuss briefly how the basic idea of solving suitable boundary value problems plays an important role in our own methods for growing global (un)stable manifolds as a family of geodesic level sets. Finally, we show how a suitable family of orbit segments can be used in a scheme to find and follow a quasiperiodic invariant torus.

4.3.1 A 2D Global Manifold as a Solution Family of BVPs

Let us now assume that the vector field (4.1) has a saddle equilibrium x_0 with a two-dimensional unstable manifold $W^u(x_0)$, meaning that the Jacobian $Df(x_0)$ has exactly two eigenvalues λ_1^u and λ_2^u with positive real part. Suppose further that $\mathbf{v}_1$ and $\mathbf{v}_2$ are the associated (generalized) eigenvectors. By definition, $W^u(x_0)$ consists of all orbits that converge to x_0 in backward time. The Stable Manifold Theorem [45] implies that $W^u(x_0)$ is tangent at x_0 to the plane spanned by $\mathbf{v}_1$ and $\mathbf{v}_2$. Therefore, we approximate $W^u(x_0)$ by the collection of orbit segments that start on this plane at distance δ from x_0. That is, we are looking for solutions of the system

$$\mathbf{u}'(t) = T\, f(\mathbf{u}(t)), \tag{4.11}$$

$$\mathbf{u}(0) = x_0 + \delta(\cos(\theta)\,\mathbf{v}_1 + \sin(\theta)\,\mathbf{v}_2), \tag{4.12}$$

where θ and T are free parameters. If the eigenvalues λ_1^u and λ_2^u are real, then it is advantageous to choose the initial conditions on the ellipse given by the ratio of the eigenvalues as

$$\mathbf{u}(0) = x_0 + \delta\left(\cos(\theta)\,\frac{\mathbf{v}_1}{|\lambda_1^u|} + \sin(\theta)\,\frac{\mathbf{v}_2}{|\lambda_2^u|}\right). \tag{4.13}$$

In other words, boundary condition (4.12) is replaced by (4.13) during the continuation.

A first orbit segment to start the continuation can be generated using AUTO as well. Namely, for arbitrary fixed $\theta = \theta_0$ ($0 \le \theta_0 < 2\pi$) and $T = 0$, the (constant) orbit $\mathbf{u}(t) = x_0 + \delta(\cos(\theta_0)\,\mathbf{v}_1 + \sin(\theta_0)\,\mathbf{v}_2)$ with $0 \le t \le 1$ is a solution for system (4.11)–(4.12). An actual trajectory for the specific value of θ_0 is then obtained using continuation in the free parameter T while keeping the angle θ_0 fixed. While this may seem like a complicated way of integrating from an specific initial condition, it has the benefit that the output files of this first step in AUTO are then compatible with subsequent continuation steps. In the case of computing a stable manifold, one simply generates an orbit for $T < 0$, that is, integrating backward in time.

During the continuation of this first orbit, one can monitor a user-defined function (a suitable end-point condition) that becomes the second boundary condition in the continuation that generates the two-dimensional manifold. That is, once the first orbit is generated, we continue this solution as a boundary value problem where the initial condition on the small circle (4.12), or ellipse (4.13) varies with the angle θ, which is now a free parameter in the continuation. The resulting one-parameter family of orbit segments forms an approximation of (a part of) $W^u(x_0)$. It is important to note that the angle θ, which defines the initial condition, is not the sole continuation parameter that determines the stepsize for the next orbit segment in the family. Instead each continuation step is taken in the full product space of the (discretized) functions $\mathbf{u}(\cdot)$ and the parameters. That is, the continuation stepsize includes variations along the entire orbit, and θ is not fixed a priori, but is one of the variables solved for in each continuation step.

There are several options for the second boundary condition that defines the one-parameter solution family. We discuss several here, but our list is certainly not exhaustive.

1. **Fixed integration time**
   Probably the simplest choice is to compute a first orbit for fixed angle $\theta = \theta_0$ up to a suitable integration time $T = T_0$, after which (4.11)–(4.12) is continued in the angle θ while the parameter T is kept fixed. An example of a computation with this choice for the second boundary condition can be found in [14], where the computation of the unstable manifold of one of the nontrivial equilibria in the Lorenz system [42] is used to find heteroclinic connections to the origin $\mathbf{0}$. Namely, the orbits that (almost) connect to $\mathbf{0}$ spend a very long time near $\mathbf{0}$. Hence, when T is fixed, these connecting orbits stand out by their short arclength; see [14, Fig. 9].

2. **Fixed arclength**
   Another obvious choice is to fix the total arclength of the orbit segments. To this end, one imposes the integral constraint

$$\int_0^1 T \, \|f(\mathbf{u}(s))\| \; ds - L = 0 \qquad (4.14)$$

along the orbit segment, while solving (4.11)–(4.12). During the continuation, both θ and T are free parameters and L is kept at a desired fixed arclength. A computation with a fixed arclength was used in [39, Sec. 3] for the computation of the stable manifold of the origin in the Lorenz system [42].

3. **Fixed product of arclength and integration time**
   It may be advantageous in certain calculations to fix the product $L \times T$, where L is the total arclength along the orbit segment as defined above. This is particularly useful if $W^u(x_0)$ contains connecting orbits with a finite arclength of less than L. Since a connecting orbit is characterized by the fact that $T \to \infty$, keeping the product $L \times T$ fixed allows one to continue orbit segments on $W^u(x_0)$ in θ past connecting orbits.

4. **Constrained end point**
   Similar to what was done in Sect. 4.1 one can restrict the end point $\mathbf{u}(1)$ to a codimension-one section of the phase space. This is done by adding to system (4.11)–(4.12) the equation

$$g(\mathbf{u}(1), \theta, T) - \alpha = 0.$$

Here g is an appropriate functional, chosen to control the end point in a desirable manner, for example, by requiring one coordinate to have a particular fixed value. The free parameters are again the angle θ and the integration time T, while α is kept fixed. We discuss an example of this condition in more detail in the next section.

Indeed there are many other options to restrict the computations to a specific part of $W^u(x_0)$, which makes this approach rather flexible. For example, the integrand of (4.14) in option **2** can be replaced by any other appropriate functional $h = h(\mathbf{u}(s), \theta, T)$ to control the orbit in a desirable manner. Furthermore, one can choose a combination of end point conditions and integral constraints, but this is beyond the scope of this chapter.

2D Unstable Manifolds in the Lorenz System

We now give a more detailed example of how the above set-up can be used to compute (un)stable manifolds even if they have a very complex geometry. To this end, we consider the well-known Lorenz equations [42], which can be written as the vector field

$$
\begin{cases}
\dot{x} = \sigma(y - x), \\
\dot{y} = \varrho x - y - xz, \\
\dot{z} = xy - \beta z.
\end{cases}
\tag{4.15}
$$

As is well known, the now classic parameters values of $\sigma = 10$, $\beta = 8/3$ and $\varrho = 28$ lead to the Lorenz butterfly attractor, arguably the most famous example of a chaotic attractor.

In [14] a detailed study is presented of how the two-dimensional stable manifold $W^s(\mathbf{0})$ of the origin $\mathbf{0}$ and the two-dimensional unstable manifolds $W^u(p^+)$ and $W^u(p^-)$ of the other two saddle equilibria p^+ and p^- intersect in a combinatorial structure of heteroclinic orbits. Note that the points p^+ and p^- and the manifolds $W^u(p^+)$ and $W^u(p^-)$ are each other's image under rotation by π about the z-axis, which is a symmetry of (4.15). Hence, it is sufficient to compute $W^u(p^+)$.

While AUTO has no problems computing one-parameter families of orbit segments on $W^u(p^+)$, it is a challenge to render this manifold as a nice surface. This is because $W^u(p^+)$ accumulates on the Lorenz attractor, which means that it folds back over itself (and its counterpart $W^u(p^-)$) infinitely often and exponentially closely. The key ingredient here is to identify suitable families of orbit segments. To this end, we take the well-established topological point of view of identifying the different sheets to obtain a branched surface, which is known as the template of the Lorenz system [22, 23].

Specifically, we generate an approximate template for $W^u(p^+)$ by considering the two families of orbit segments shown in Fig. 4.7 (a) and (b). Both start along a fixed vector in the unstable eigenspace of p^+ (at the diamonds) and one family of orbit segments winds only around p^+, while the other winds once around p^-. The end points of both families (indicated by the black horizontal lines) are constrained to lie in the plane $\Sigma_\varrho = \{z = \varrho - 1\}$, which passes

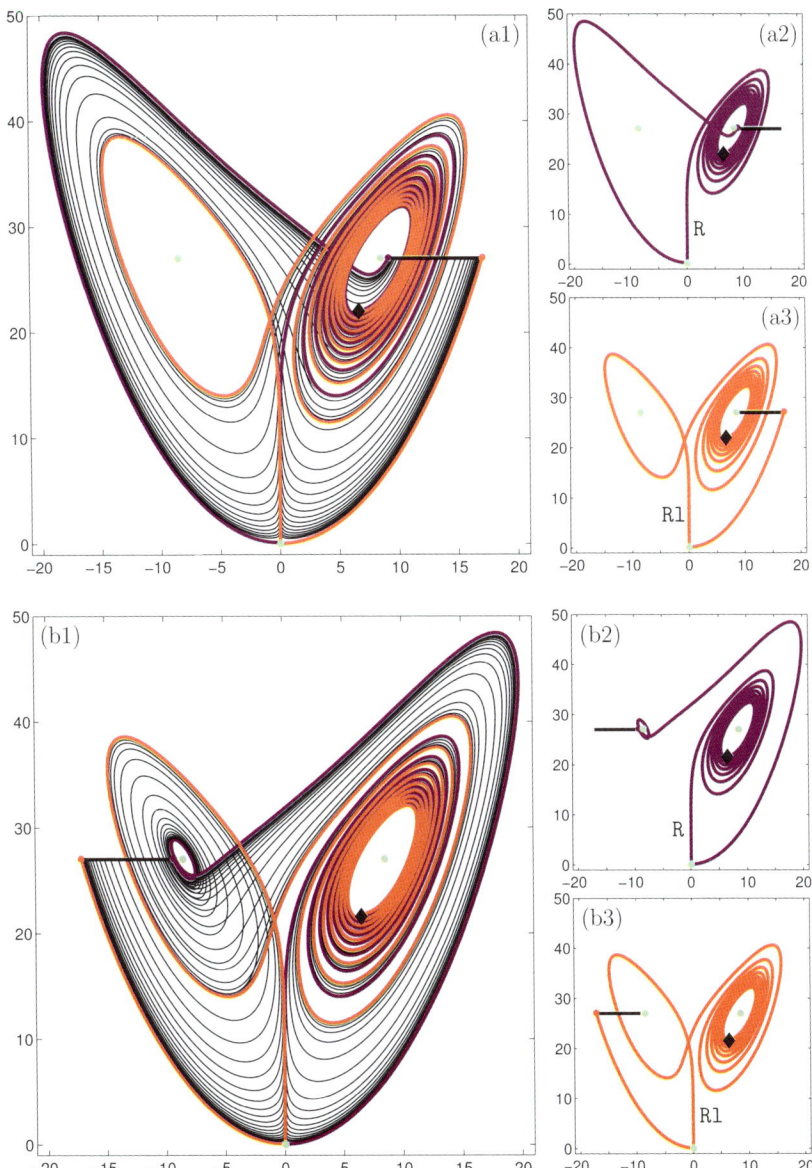

Fig. 4.7. The two families of orbit segments used to compute a template for $W^u(p^+)$ of (4.15). The orbit segments start in the unstable eigenspace (at the $\diamond$) of $W^u(p^+)$ and end in the section $\Sigma_{28} = \{z = 27\}$, either near p^+ (a1) or near p^- (b1). Both families limit on the two singular orbit segments in panels (a2)/(a3) and (b2)/(b3), which are composed of a heteroclinic connection to **0** (of type R or R1) and an initial piece of $W^u(\mathbf{0})$. From E.J. Doedel, B. Krauskopf and H.M. Osinga, Global bifurcations of the Lorenz manifold, *Nonlinearity* 19(12) (2006) 2947–2972 © 2006 by Institute of Physics Publishing; reprinted with permission.

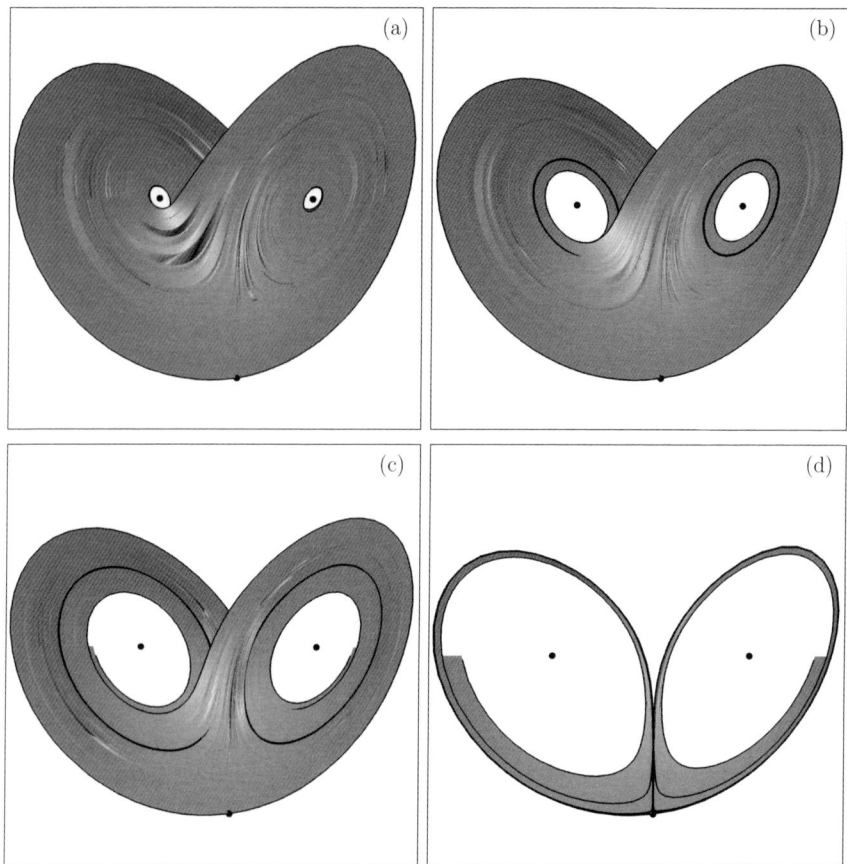

Fig. 4.8. The unstable manifolds of the nontrivial saddles $p^{\pm}$ and of the saddle periodic orbit $\Gamma^{\pm}$, respectively, as represented by the family of orbits in Fig. 4.7, for $\rho = 28.0$ (a), $\rho = 23.0$ (b), $\rho = 19.0$ (c), and $\rho = 14.0$ (d). From E.J. Doedel, B. Krauskopf and H.M. Osinga, Global bifurcations of the Lorenz manifold, *Nonlinearity* 19(12) (2006) 2947–2972 © 2006 by Institute of Physics Publishing; reprinted with permission.

through p^+ and p^-. In other words, we use option **4** of the boundary conditions above by imposing the additional boundary condition that $\mathbf{u}(1) \in \Sigma_{\varrho}$. This choice was made so that the two families 'cover' most of the right and left wings of the Lorenz attractor, respectively.

We now explain this set-up and the different continuation steps in more detail. Since the eigenvalues of p^+ are complex conjugate, which implies that orbits near p^+ spiral around p^+, it is convenient to use the initial condition

$$\mathbf{u}(0) = x_0 + r\,\mathbf{v}_1 \qquad (4.16)$$

instead of (4.12). Here $\mathbf{v}_1$ is the real part of the complex conjugate pair of eigenvectors associated with the eigenvalues of the linearization at p^+, and r is the radial distance from p^+. Note that increasing r from one intersection to the next of the same orbit (a fundamental domain of the return map to this direction) is equivalent to changing the angle θ in (4.12) over 2π.

To generate an initial orbit segment of the family shown in Fig. 4.7(a) we choose $r = r_0 = 5.5$ fixed and continue the solution $\mathbf{u}(t) = x_0 + r_0 \mathbf{v}_1$, $T = 0$ in the free parameter T. Note that $W^u(p^+)$ is extremely flat near p^+ so that a point on the linear approximation at distance $r_0 = 5.5$ from p^+ is still very close to $W^u(p^+)$. This is advantageous, because additional revolutions around p^+, of which there are more and more as $r \to 0$, can only be resolved accurately if a large number of mesh points is used. By using a quite large value for r_0 it suffices to use 800 mesh points. During this first run, we monitor the end-point condition

$$PAR(12) = \mathbf{u}(1)_z - (\varrho - 1) \tag{4.17}$$

where $\mathbf{u}(1)_z$ denotes the z-component of $\mathbf{u}(1)$. Hence, every time $PAR(12)$ is zero, which is detected as a user defined point (UZ in AUTO), the end point $\mathbf{u}(1)$ lies in Σ_ϱ. The user can then select any such orbit segment, which is how we obtained the two families in Fig. 4.7(a) and (b).

The next step of the process then consists of an AUTO run where we continue system (4.11) with (4.16) and (4.17) as boundary conditions, and r and T as free parameters. In the case of the orbit segments in Fig. 4.7(a), in one direction the end point $\mathbf{u}(1)$ moves closer to p^+. The last orbit segment in this run is shown in Fig. 4.7(a2), where we can observe how the orbit passes extremely close to $\mathbf{0}$ before making its loop around p^-. Indeed, the orbit segments accumulate on a heteroclinic connection concatenated with a first piece of the left branch of the one-dimensional unstable manifold $W^u(\mathbf{0})$ of $\mathbf{0}$. The heteroclinic connection is of type R, that is, it only spirals around p^+ and never around p^- before converging to $\mathbf{0}$. When running the continuation in the opposite direction, the end point $\mathbf{u}(1)$ moves further away from p^+ and the orbit segments accumulate on the heteroclinic connection of type R1 concatenated with a first piece of the right branch of $W^u(\mathbf{0})$; the last orbit segment for this run is shown in Fig. 4.7(a3). Note that the initial point $\mathbf{u}(0)$ (indicated by the diamond) virtually does not move during these computations.

The second part of the template for $W^u(p^+)$ is generated in a similar way. Here the starting orbit is generated from an initial condition at distance $r_0 = 5.7$ from p^+ so that it spirals more than once around p^-. The next time when $\mathbf{u}(1)$ passes through Σ_ϱ after the first loop around p^- is where we stop the continuation in T and use the resulting orbit segment to start the second run. The resulting family of orbit segments is shown in Fig. 4.7(b1). The continuation in one direction leads to $\mathbf{u}(1)$ moving closer to p^-, where the last orbit segment is very close to the heteroclinic connection of type R concatenated with a first piece of the right branch of $W^u(\mathbf{0})$; see Fig. 4.7(b2). The continuation in the opposite direction has the effect that the orbit segments accumulate on the heteroclinic connection R1 concatenated with a first piece of the left branch of $W^u(\mathbf{0})$; the last orbit in this run is shown in Fig. 4.7(b3). Again, $\mathbf{u}(0)$ (indicated by the diamond) hardly changes in these runs.

The two families shown in Fig. 4.7(a1) and Fig. 4.7(b1) can be used to represent $W^u(p^+)$. To render $W^u(p^+)$ as a surface we generate a triangulation by connecting corresponding mesh points on neighboring orbit segments. The result as generated by the package GEOMVIEW [46] is shown in Fig. 4.8(a); this image also shows the symmetric counterpart $W^u(p^-)$. The resulting surface is indeed a good representation of the template of the Lorenz system; its boundary is given by pieces of the unstable manifold of the origin.

Topologically the same orbit families as in Fig. 4.7 can be used to compute the unstable manifolds of the saddle equilibria $p^{\pm}$ and the bifurcating saddle periodic orbits $\Gamma^{\pm}$ for different values of the parameter ϱ. Figure 4.8(b)–(c) shows the manifolds $W^u(\Gamma^+)$ and $W^u(\Gamma^-)$ for $\varrho = 23.0$, $\varrho = 19.0$, $\varrho = 14.0$, respectively; the periodic orbits $\Gamma^{\pm}$ are shown as black curves. Notice how for decreasing ϱ the periodic orbits $\Gamma^{\pm}$ (black curves) grow. At the same time, the 'hole' in the respective template around the (now attracting) equilibria grows. Eventually, for $\varrho \approx 13.9265$ the periodic orbits $\Gamma^{\pm}$ disappear in a homoclinic bifurcation with the origin; see [14] for more details of this bifurcation.

One can notice some artifacts in the surface rendering in Fig. 4.8. They occur because the orbit segments that are used to render the surface are not evenly spaced everywhere. In fact, in some regions they are very close together. For example, very many mesh points in Fig. 4.7 are contained in the very first part of $W^u(p_+)$ where the orbit segments spiral around p^+; this part of the surface could be represented with a lot fewer mesh points. Furthermore, the distance between mesh points on a given orbit segment is roughly given by the arclength of the orbit segment divided by the number of mesh points. Therefore, our straightforward way of generating a triangulated surface from the AUTO data results in quite elongated triangles in some parts of $W^u(p^+)$. While the actual error is quite small (since all orbit segments lie on $W^u(p^+)$ in very good approximation), the angles between neighboring triangles may be quite large. Therefore, the artificial lighting scheme of GEOMVIEW generates some color differences on the surface.

More generally, the construction of a 'smoother' triangulation from AUTO data of orbit segments would require some serious post processing. Effectively, one would need to decide where and how to 'thin out' the data to have a uniform bound on the sizes and aspect ratios of the triangles in the overall triangulation.

4.3.2 A BVP Set-Up to Find Geodesic Level Sets

As we have just seen, the generation of a mesh of a prescribed quality from a computed one-parameter family of orbit segments is not a simple matter. While the continuation step size is such that the orbit segments are never far apart, this does not generally guarantee a nice rendering of the two-dimensional surface. This problem occurs because the orbit segments are only required to be good approximations of solutions to the BVP, and there are no restrictions placed on the position of the mesh points with regard to the geometry of the computed surface. In fact, all other methods for computing global manifolds are addressing the mesh generation from a more geometrical point of view; see the recent survey [39] for more details. In particular, Henderson [30, 39] stays closest to the idea of representing a global manifold as a one-parameter family of orbit segments. The difference is that he uses curvature information to cover the manifold with 'fat trajectories', so that the computed mesh points on suitably chosen orbit segments are well distributed over the manifold.

We now explain how the GLOBALIZEBVP implementation [17] of our own method [37, 38] for the computation of two-dimensional global (un)stable manifolds of vector fields makes use of AUTO's collocation and pseudo-arclength continuation routines. The main idea behind this method is to build up the manifold as a collection of geodesic level sets. This is a very natural choice from a geometrical point of view, as the goal is to generate a circular mesh centered around the equilibrium x_0 (or periodic orbit Γ) that grows radially in steps that are dictated by the local curvature. Each circle has the property that the mesh points on it lie at the same geodesic distance from x_0 (or Γ). Hence, the GLOBALIZEBVP method also views the manifold as a one-parameter family of curves — geodesic level circles in the case of a two-dimensional manifold — but these curves are not directly related to the dynamics of the vector field (4.1). In contrast to the method discussed in Sect. 4.3.1, each geodesic circle cannot be expressed as the solution of a family of two-point boundary value problems. Instead, it is generated one point at a time.

Let us consider again the computation of a two-dimensional unstable manifold $W^u(x_0)$ of a saddle point x_0. The GLOBALIZEBVP method starts with N equally-spaced mesh points on the circle (4.12) in the unstable eigenspace centered at x_0 with radius δ. The piecewise-linear closed curve C_δ through these mesh points is an approximation of the first geodesic level set, at distance δ from x_0. Let us denote the last computed geodesic level set by C_r and suppose that we wish to find the next geodesic level set C_b at a distance Δ from C_r.

For each mesh point r_k on C_r we wish to find the point b_k on C_b that lies closest to r_k. To this end, we define a plane $\mathcal{F}_{r_k}$ (approximately) perpendicular to C_r at r_k. The (unknown) intersection $W^u(x_0) \cap \mathcal{F}_{r_k}$ is a one-dimensional curve that is (locally) well defined. Note that any point in $W^u(x_0) \cap \mathcal{F}_{r_k}$ lies on an orbit that passes through C_r since, by definition, orbits on $W^u(x_0)$ come from x_0. The curve $W^u(x_0) \cap \mathcal{F}_{r_k}$ is parametrized locally near r_k by the end points $\mathbf{u}(1) \in \mathcal{F}_{r_k}$ of solutions $\mathbf{u}(t)$, $0 \le t \le 1$ to a one-parameter family of two-point boundary value problems. Namely, we solve system (4.11)

$$\mathbf{u}'(t) = T\, f(\mathbf{u}(t))$$

with the boundary conditions

$$\mathbf{u}(0) \in C_r, \tag{4.18}$$
$$\mathbf{u}(1) \in \mathcal{F}_{r_k}. \tag{4.19}$$

The continuation starts from the trivial solution $\mathbf{u}(t) = r_k$, $0 \leq t \leq 1$ with $T = 0$. We then continue in the parameter T while we monitor the distance

$$\Delta_T = \| \mathbf{u}(1) - r_k \|$$

between the end point $\mathbf{u}(1)$ and the mesh point r_k until the required distance Δ is reached, which defines the point b_k on the new geodesic level set.

Once b_k is found, it is tested against similar accuracy criteria as those explained in Sect. 4.1. Namely, we restrict the angle α between points on three successive geodesic level sets and the product $\Delta\alpha$. This maintains a good resolution of the manifold; if b_k is not acceptable then the geodesic level set currently being computed is discarded and Δ is reduced. If all points on C_b are found and acceptable then the geodesic level set is added to the mesh representation of $W^u(x_0)$. We refer to [17, 37, 38] for further details.

While earlier implementations [37, 38] used a shooting technique for solving the boundary value problem to find the new point b_k, the GLOBALIZEBVP implementation [17] directly calls AUTO's collocation and pseudo-arclength continuation routines. In fact, the set-up of the continuation of the one-dimensional curve $W^u(x_0) \cap \mathcal{F}_{r_k}$ is conceptually just as the computation of a one-dimensional global manifold in a Poincaré section discussed in Sect. 4.1.1. Namely, the start point $\mathbf{u}(0)$ is continued along the piecewise-linear representation of the geodesic circle C_r, while the end point $\mathbf{u}(1)$ is confined to lie in the plane F_r; see [17] for details in terms of the respective user-defined functions.

Stable Manifold in Chua's Circuit

We now illustrate the GLOBALIZEBVP method by computing the stable manifold $W^s(\mathbf{0})$ of the origin in Chua's circuit [8] with a smooth cubic nonlinearity; see also [1, 55] and Chap. 7. The system is given by

$$\begin{cases} \dot{x} = \alpha\,(y - a\,x^3 - c\,x), \\ \dot{y} = x - y + z, \\ \dot{z} = -\beta\,y - \gamma\,z, \end{cases} \tag{4.20}$$

where we take $\alpha = 10.0$, $\beta = 15.0$, $\gamma = 0.01$, $a = 0.1$, and $c = -0.2$ as in [17]. For these parameters the origin has one unstable and two stable eigenvalues, which are a complex conjugate pair. The strong spiralling dynamics on $W^s(\mathbf{0})$ makes it particularly challenging to find a good approximation of this global manifold.

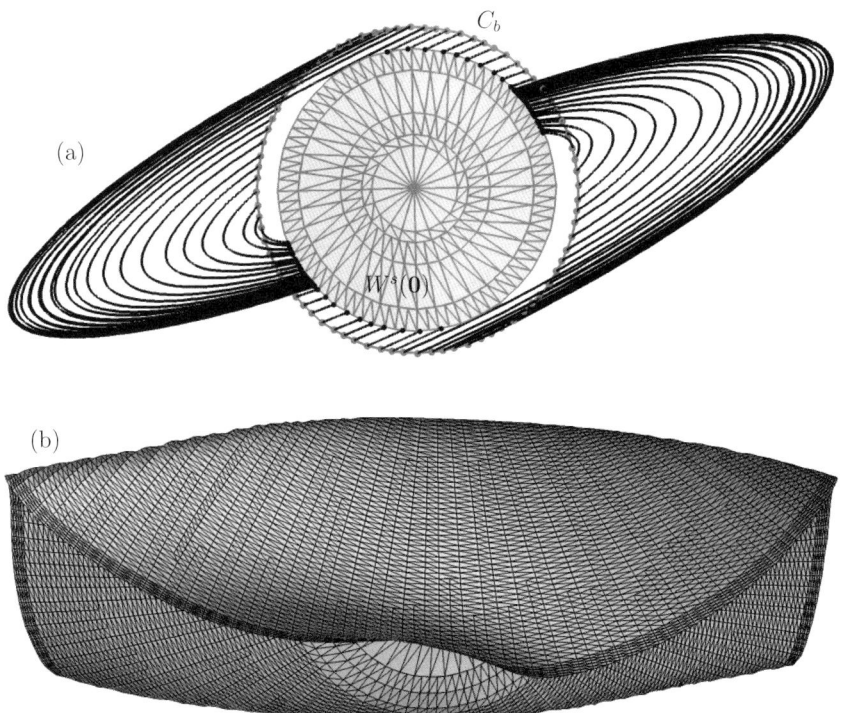

Fig. 4.9. Computation of $W^s(\mathbf{0})$ in Chua's circuit (4.20) with the GLOBALIZEBVP method. Panel (a) shows the orbit segments used in the computation of the geodesic level set C_b at geodesic distance 0.75 from $\mathbf{0}$, and panel (b) the mesh with 64 geodesic level sets when $W^s(\mathbf{0})$ is computed up to geodesic distance 2.1875.

We start the computation with 20 mesh points equally distributed on a circle with radius 0.2 in the stable eigenplane centered around $\mathbf{0}$. The first few geodesic level sets are shown in Fig. 4.9(a). The surface shown is built up from geodesic level sets at distances 0.25, 0.35, 0.55, and 0.65. Figure 4.9(a) shows the orbit segments generated by the GLOBALIZEBVP method to find the next geodesic level set C_b at distance 0.75 by solving system (4.11) with boundary conditions (4.18) and (4.19). Here C_r is the piecewise-linear curve through the mesh points on the geodesic level set at distance 0.65, that is, the boundary of the already computed piece of $W^s(\mathbf{0})$. The grey dots are the new mesh points on C_b and the black dots are the other end points of the orbit segments on C_r.

Notice that about half of the orbit segments move far away from $\mathbf{0}$ before ending on C_b, which is a direct consequence of the nature of the dynamics on $W^s(\mathbf{0})$. We remark that the example of Chua's circuit is quite extreme in this respect. In fact, the required orbit segments become even longer for geodesic level sets at larger distances. This is in contrast to, for example, the two-dimensional stable manifold of the origin of the Lorenz system, which is characterized by two real positive eigenvalues. Generally speaking, the need to find quite long orbit segments by our BVP approach is the price we pay for obtaining the parametrization of the global manifold by geodesic level sets. On the positive side, this parametrization is geometrically optimal in the sense that it is given by the geometry of the manifold and not by the dynamics on it. Furthermore, it allows us to derive rigorous error bounds that go to zero with the user specified accuracy parameters; see [38] for details.

Figure 4.9(b) shows the stable manifold $W^s(\mathbf{0})$ in Chua's circuit (4.20) computed up to geodesic distance 2.1875. Notice how $W^s(\mathbf{0})$ folds quite sharply along the top and bottom edges of the image. The surface is rendered so that the computed mesh is visible. The computed manifold $W^s(\mathbf{0})$ is built up from a total of 64 geodesic level sets, that is, concentric topological circles; the distance between them is determined by the local curvature as was explained above. The radial curves are approximate geodesics. Notice how during the course of the computation many new mesh points are added, which then give rise to new approximate geodesics. Similarly, mesh points may be removed, which can be identified in the mesh as an end point of an approximate geodesic; an example can be seen in the top left of Fig. 4.9(b). For more images of $W^s(\mathbf{0})$ we refer to [17].

Stable Manifold of an Optimal Control Problem

The representation of a global manifold by approximate geodesic level sets can be exploited for visualization purposes. As we demonstrate now with an example of a two-dimensional stable manifold in a four-dimensional phase space, this is a particularly useful feature when the phase space is higher dimensional. Specifically, we consider an inverted planar pendulum that is balanced on a cart subject to a horizontal control force [28, 32, 44]. The system can be written as

$$
\begin{cases}
\dot{x}_1 = x_2, \\
\dot{x}_2 = \dfrac{\frac{g}{l}\sin(x_1) - \frac{1}{2}m_r x_2^2 \sin(2x_1) - \frac{m_r}{m\,l}\cos(x_1)\,u}{\frac{4}{3} - m_r \cos^2(x_1)},
\end{cases}
\tag{4.21}
$$

where x_1 is the angle measured from the upright position (not taken modulo 2π), x_2 is its angular velocity, m_r is the mass fraction of the pendulum with respect to the total mass (of pendulum and cart), l is the length of the pendulum, and g is the Earth's gravitational constant. The function u constitutes a control that is supposed to stabilize the point $(x_1, x_2) = (0, 0)$, which is the unstable equilibrium corresponding to the upright position.

A cost is associated with the stabilization via the instantaneous cost function

$$
Q(x_1, x_2, u) = \mu_1 x_1^2 + \mu_2 x_2^2 + \mu_3 u^2
\tag{4.22}
$$

that penalizes both the state and the control, as long as the origin is not stabilized. Here μ_1, μ_2 and μ_3 are positive parameters. Pontryagin's maximum principle [56] ensures that an optimal control u exists that minimizes the cost function Q over the infinite time interval $[0, \infty)$. The optimal solution is represented by points on the two-dimensional stable manifold $W^s(\mathbf{0})$ of the four-dimensional vector field given by the Hamiltonian

$$H(x_1, x_2, p_1, p_2) = Q(x_1, x_2, u^*(x_1, x_2, p_1, p_2)) + p_1 x_2 +$$
$$p_2 f(x_1, x_2) + p_2 c(x_1, x_2) u^*(x_1, x_2, p_1, p_2) \quad (4.23)$$

where $u^*(x_1, x_2, p_1, p_2) = -\frac{1}{2\mu_3} c(x_1, x_2) p_2$. Namely, for any given initial condition (x_1, x_2, p_1, p_2) on $W^s(\mathbf{0})$, the projection of the corresponding trajectory onto the (x_1, x_2)-plane corresponds to a stabilizing solution via the (implicitly defined) feedback control $u = u^*(x_1, x_2, p_1, p_2)$ that locally minimizes (4.22). Indeed, if in this projection $W^s(\mathbf{0})$ covers a point (x_1, x_2) more than once, then typically only one of these solutions is optimal and the others are only suboptimal; see [28, 44] for more details.

Figure 4.10 shows $W^s(\mathbf{0})$ for the parameters in [32], namely $m_r = 0.2$, $l = 0.5\,\mathrm{m}$, and cost function parameters $\mu_1 = 0.1$, $\mu_2 = 0.05$ and $\mu_3 = 0.01$. The two-dimensional manifold $W^s(\mathbf{0})$ was computed up to a geodesic distance of approximately 26.25. It is rendered transparent in Fig. 4.10 and shown as four projections onto the three-dimensional subspaces that one obtains by setting one of the coordinates to zero. The transparent rendering allows one to see how $W^s(\mathbf{0})$ 'sits' in each of the three-dimensional projections. Note that the computed part of $W^s(\mathbf{0})$ is a topological disk that is parametrized by the geodesic level sets. In particular, the boundary of the computed manifold has the same geodesic distance to the origin, which lies in the center of the manifold.

To help further with the interpretation of the geometry of $W^s(\mathbf{0})$ in $\mathbb{R}^4$, a single geodesic band can be 'moved' over the manifold to observe how its geometry changes simultaneously in all four projections. As an example, Fig. 4.10 shows the geodesic band covering the range 19–20 in a different shade. Note that the band is unknotted and that it divides the manifold $W^s(\mathbf{0})$ into an inner disk and an outer annulus, which is not so obvious in Fig. 4.10(d).

4.3.3 A Two-Point Boundary Value Problem Set-Up for the Computation of Quasiperiodic Invariant Tori

An invariant torus of a given dynamical system of the form (4.1) is a two-dimensional compact invariant manifold that can be viewed as the two-dimensional analogue of an equilibrium or a periodic orbit. Invariant tori are born, for example, in a Neimark-Sacker bifurcation, where a pair of complex conjugate Floquet multipliers of a periodic orbit crosses the unit circle in the complex plane, or they arise from the uncoupled case in a system of coupled oscillators. The bifurcating torus is initially normally hyperbolic [31],

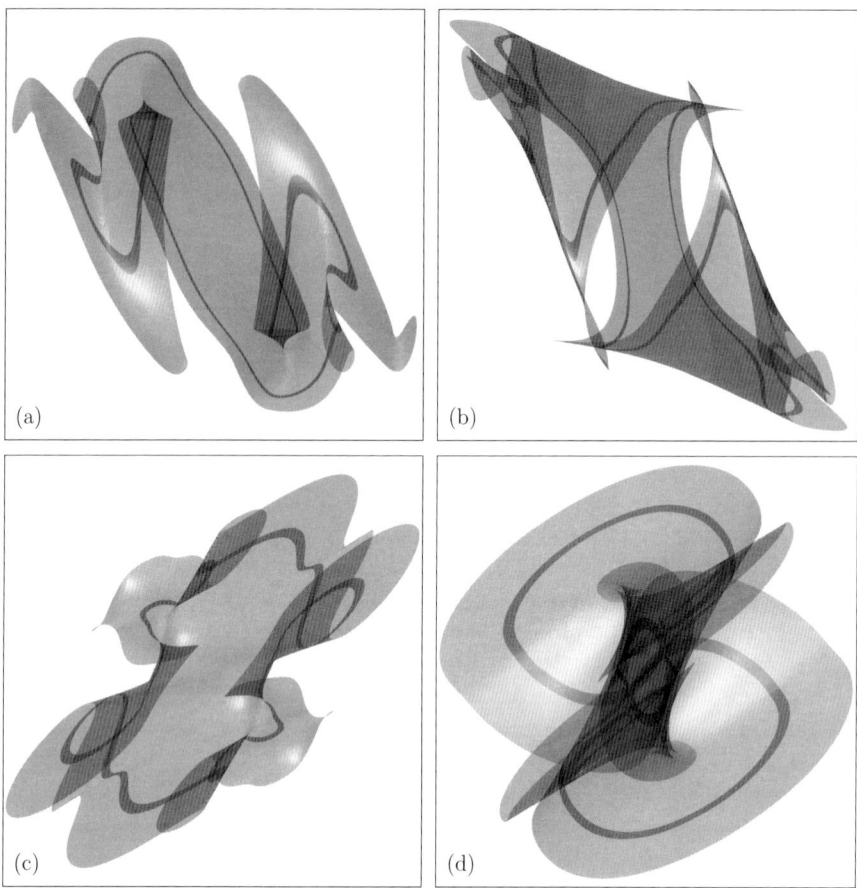

Fig. 4.10. The two-dimensional stable manifold $W^s(\mathbf{0})$ in (x_1, x_2, p_1, p_2)-space of the controlled inverted pendulum (4.23). The surface is rendered transparent and is shown simultaneously in the four projections $p_2 = 0$ (a), $p_1 = 0$ (b), $x_2 = 0$ (c), and $x_1 = 0$ (d); the differently shaded band covers geodesic distances 19.0–20.0.

which means that it is persistent under small perturbations. Depending on the rotation number, the dynamics on the torus is either quasiperiodic (equivalent to a parallel flow on the standard torus) or locked, in which case there is an attracting and a repelling periodic orbit on the torus. Locked dynamics occurs in so-called resonance or Arnol'd tongues, parameter regions that are bounded by saddle-node bifurcations of the locked periodic orbits on the torus. The resonance tongues emerge from points on a Neimark-Sacker curve where the rotation number is rational, while the quasiperiodic tori exist along one dimensional curves that start from points where the rotation number is irrational. The overall picture is truly two-dimensional in parameter space; see, for example, [3, 41, 52] as entry points to the extensive literature. A main difficulty is that invariant tori may lose their normal hyperbolicity in many different ways, some of which are still not fully understood.

It may not be too surprising that the computation of invariant tori is still considered a very challenging task. In fact, only a few specialized algorithms exist; see [49] for a recent overview. We review here the algorithm presented in [50] for the computation of quasiperiodic invariant tori, because it utilizes the two-point boundary value solver of AUTO. Namely, the idea is to view a quasiperiodic invariant torus $\mathbb{T}$ of (4.1) as an invariant circle $\mathbb{T}_\Sigma$ in a suitable Poincaré section Σ that is chosen transverse to $\mathbb{T}$. The associated Poincaré map that leaves $\mathbb{T}_\Sigma$ invariant can be formulated as a two-point boundary value problem in the same way as was done in Sect. 4.1.

To explain the method from [50] in detail we need the dependence of (4.1) on the parameter λ and, for convenience, we assume that it is of the form $\lambda = (\alpha, \beta)$. We restrict our discussion to the case of a two-dimensional torus $\mathbb{T}$, but it is straightforward to generalize the method to higher-dimensional tori. The idea is to construct a two-point BVP in two steps: we start with the invariance condition for invariant circles of the time-T map of (4.1) and then replace the time-T map with a BVP.

An invariant circle $u(\theta)$ of the time-T map φ^T of (4.1), where $\theta \in \mathbb{S}^1$ is parametrized over $[0, 2\pi]$, satisfies the invariance condition

$$u(\theta + 2\pi\varrho) = \varphi^T(u(\theta), \alpha, \beta), \tag{4.24}$$

where ϱ is the rotation number of φ^T restricted to the invariant circle [6, 34]. The invariance condition (4.24) is discretized by approximating u with a Fourier polynomial

$$u(\theta) = c_1 + \sum_{k=1}^{N} c_{2k} \sin\theta + c_{2k+1} \cos\theta,$$

where $c_j \in \mathbb{R}^n$ are real coefficient vectors, and requiring that (4.24) holds at the $Q = 2N + 1$ collocation points $\theta_j = 2\pi j/Q$, $j = 1, \ldots, Q$. Note that the discretized solution is not unique. First of all, T is still unknown and, secondly, the solution $u(\theta)$ is not isolated because for any phase shift $s \in \mathbb{S}^1$ the shifted function $u(\theta + s)$ is also a solution.

By identifying $x_j(0) = u(\theta_j)$ and $x_j(1) = u(\theta_j + 2\pi\varrho)$ and substituting the initial value problem $\{\dot{x}_j = f(x_j, \alpha, \beta),\ x_j(0) = u(\theta_j)\}$ for the time-T map φ^T in (4.24), one obtains the BVP

$$\dot{x}_j = Tf(x_j, \alpha, \beta), \tag{4.25}$$

$$\dot{c}_j = 0, \tag{4.26}$$

$$x_j(0) = u(\theta_j), \tag{4.27}$$

$$x_j(1) = u(\theta_j + 2\pi\varrho), \tag{4.28}$$

$$\sum_{i=1}^{n} \int_0^{2\pi} \tilde{u}_i'(0)u_i(0) \, d\theta = 0, \tag{4.29}$$

$$\sum_{i=1}^{n} \int_0^1 (x_{1,i}(t) - \tilde{x}_{1,i}(t))x_{1,i}'(t) \, dt = 0. \tag{4.30}$$

Here boundary conditions (4.27) and (4.28) represent the invariance condition (4.24), while (4.29) and (4.30) are phase conditions that determine a unique return time T and phase shift s. Note that s does not occur explicitly in (4.29) and that (4.30) cannot be simplified similarly to (4.29), because x_1 is not periodic in t; see also [47, 48, 50]. Condition (4.26) on the Fourier coefficients can be omitted, but is necessary if one wants to implement (4.25)–(4.30) in AUTO.

System (4.25)–(4.30) consists of $2Q$ n-dimensional ODEs and $2Q$ n-dimensional boundary conditions for the functions x_j and c_j, and two scalar phase conditions for T and the parameter α. Continuation of a solution of this system with respect to β is a way to compute a codimension-one family of quasiperiodic invariant tori with the fixed rotation number ϱ.

Tori in an Electronic Circuit Model

To demonstrate the method we consider the model of an electronic circuit that was investigated in detail in [49, 51]. The system can be written in vector-field form as the parametrically forced system

$$\begin{cases} \dot{x} = y, \\ \dot{y} = (\beta/2 - \alpha)y - (\beta - \alpha)y^3 - (1 + \alpha \sin 2t)x, \\ \dot{t} = 1, \end{cases} \tag{4.31}$$

where α is the forcing amplitude and β determines the nonlinear damping. For $\alpha = 0$ system (4.31) is autonomous and its flow $\varphi^t(x, y, t)$ in the phase space $\mathbb{R} \times \mathbb{R} \times \mathbb{S}^1$ is a superposition of the flow in the two-dimensional (x, y)-space with the constant flow given by $\dot{t} = 1$. For $\beta > 0$ there is a limit cycle in (x, y)-space and, hence, system (4.31) has a normally attracting invariant torus, which persists as an invariant manifold for sufficiently small forcing amplitude α [19]. For a specified rotation number ϱ a start solution for $\alpha = 0$ is found by continuing the β-family of periodic solutions in the (x, y)-plane and interpolating it as a function of the rotation number $\varrho = \pi/T(\beta)$. Here $T(\beta)$ is the period of the periodic solutions that for (4.31) increases monotonically as a function of β.

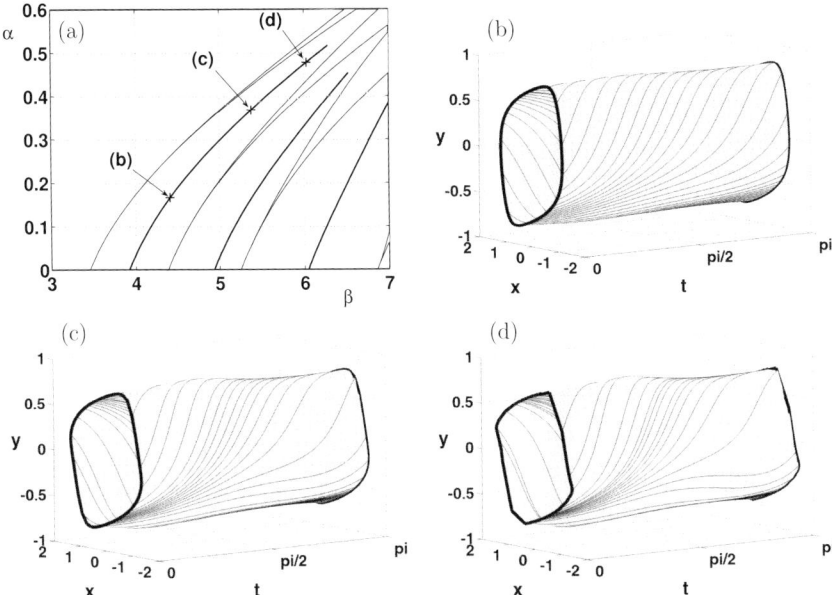

Fig. 4.11. The bifurcation diagram (a) of (4.31) with selected resonance tongues and curves of quasiperiodic tori, together with three quasiperiodic tori (b)–(d) computed as solutions to (4.25)–(4.30). The torus is shown as a gray surface and the closed curves at $t = 0$ and $t = \pi$ represent the same invariant circle. The thin black curves connecting these circles are orbit segments used in the computation; only every fourth orbit is shown for clarity.

Figure 4.11 shows the result of computations of quasiperiodic invariant tori of (4.31) with the above BVP set-up (4.25)–(4.30). The computations were performed with $N = 64$ Fourier modes and a second-order Gauß collocation scheme with 300 mesh points. Panel (a) shows from left to right the Arnol'd tongues for the rotation numbers $\varrho \in \{\frac{3}{7}, \frac{2}{5}, \frac{3}{8}, \frac{1}{3}\}$ as grey wedges and the quasiperiodic curves for the rotation numbers $\varrho = \frac{N}{140}\sqrt{2}$ with $N \in \{41, 38, 35\}$ as bold black curves. Figure 4.11(b)–(d) shows sample solutions of the BVP (4.25)–(4.30), that is, representations of quasiperiodic invariant tori that were obtained at the marked positions in panel (a) during the continuation along the left-most quasiperiodic curve. The tori are shown as gray tubes that are bounded by bold black closed curves representing the invariant circle at $t = 0$ and $t = \pi$. The thin black curves are orbits on the tori that connect the points $x_j(0) = u(\theta_j)$ with $x_j(1) = u(\theta_j + 2\pi\varrho)$, which are part of the solution of (4.25)–(4.30); for clarity, only every fourth orbit is shown. Notice the change to the torus from Fig. 4.11(b) to Fig. 4.11(d) when α is increased towards the top of the locus where the torus loses normal hyperbolicity and the computation stops.

4.4 Summary and Outlook

The basic idea of continuing the solution of a two-point boundary value problem is very versatile. As we have demonstrated in this review, it can also be used to compute different types of global manifolds, including one-dimensional (un)stable manifolds in Poincaré maps, two-dimensional global manifolds of vector fields, slow manifolds in slow-fast systems, as well as quasiperiodic invariant tori. The common theme here is that one needs to consider a suitable family of orbit segments. The continuation of this family can then be achieved by making use of the collocation solver and continuation routines of AUTO.

In this chapter we have given a flavor of how the BVP solver of a package such as AUTO can be used in less obvious ways. We hope that this will stimulate further research into the computation of invariant manifolds, and we mention two directions of ongoing research.

As the wide range of applications indicates, there are other types of invariant objects that can be computed within the presented general framework. In particular, there is an interest in manifolds that are associated with a certain subset of the stable or unstable eigenvalues or Floquet multipliers [31, 59]. Examples are the computations of a weak unstable manifold in a delay equation model of a laser with phase conjugate feedback [25], and of a two-dimensional surface (with the method in Sect. 4.3.1) that separates forward and backward phase resets in a model of a cardiac pacemaker [40].

Another interesting possibility is the computation of higher-dimensional global manifolds in a similar fashion, where the manifold is now represented as a family of orbit segments that is parametrized by more than one continuation parameter. Combining the general BVP set-up presented here with higher-dimensional continuation as discussed in Chap. 3 provides a quite natural approach for computing, for example, two-dimensional manifolds in Poincaré maps and three-dimensional global manifolds in vector fields. However, there remain interesting open questions concerning the representation and visualization of the resulting data.

Acknowledgments

The material reviewed in Sects. 4.1.1, 4.2.1 and 4.3.2 is joint work with James England, who we thank for his contributions during our collaboration. We also thank Mathieu Desroches and Frank Schilder for their help with text and figures in Sects. 4.2.2 and 4.3.3, respectively. The material reviewed in Sect. 4.3.1 is joint work with Sebius Doedel, and we would like to thank Sebius for his continuing influence on our research and his invaluable support throughout our careers. Our visits to Montréal are always productive as well as enjoyable, and we blame Sebius for our addiction to Montréal croissants! Finally, we are grateful to the Society for Industrial and Applied Mathematics and to Institute of Physics Publishing for permission to reproduce previously published material from [16] and from [14].

References

1. A. Algaba, M. Merino, E. Freire, E. Gamero, and A. J. Rodríguez-Luis. Some results on Chua's equation near a triple-zero linear degeneracy. *Internat. J. Bifur. Chaos Appl. Sci. Engrg.*, 13(3):583–608, 2003.
2. E. L. Allgower and K. Georg. Numerical path following. In P. G. Ciarlet and J. L. Lions, editors, *Handbook of Numerical Analysis*, volume 5, pages 3–207. North Holland Publishing, 1996.
3. V. I. Arnol'd. *Geometrical Methods in the Theory of Ordinary Differential Equations.* Springer-Verlag, Berlin, 1983.
4. J. Boissonade and P. DeKepper. Transition from bistability to limit cycle oscillations. *J. Phys. Chem.*, 84:501–506, 1980.
5. M. Brøns, M. Krupa, and M. Wechselberger. Mixed mode oscillations due to the generalized canard phenomenon. *Fields Institute Communications*, 49:39–63, 2006.
6. E. Castellà and À. Jorba. On the vertical families of two-dimensional tori near the triangular points of the bicircular problem. *Celestial Mech. Dynam. Astronom.*, 76(1):35–54, 2000.
7. A. R. Champneys, Yu. A. Kuznetsov, and B. Sandstede. A numerical toolbox for homoclinic bifurcation analysis. *Internat. J. Bifur. Chaos Appl. Sci. Engrg.*, 6(5):867–887, 1996.
8. L. O. Chua, M. Komuro, and T. Matsumoto. The double scroll family. *IEEE Trans. Circuits and Systems*, CAS-33:1073–1118, 1986.
9. P. Collins and B. Krauskopf. Entropy and bifurcation in a chaotic laser. *Phys. Rev. E*, 66:056201, 2002.
10. M. Desroches, B. Krauskopf, and H. M. Osinga. The geometry of slow manifolds near a folded node. Technical report, University of Bristol, 2007, to appear.
11. A. Dhooge, W. Govaerts, and Yu.A. Kuznetsov. MATCONT: A Matlab package for numerical bifurcation analysis of ODEs. *ACM TOMS*, 29(2):141–164, 2003. available via `http://www.matcont.ugent.be/`.
12. E. J. Doedel. AUTO, a program for the automatic bifurcation analysis of autonomous systems. *Congr. Numer.*, 30:265–384, 1981.
13. E. J. Doedel and M. Friedman. Numerical computation of heteroclinic orbits. Continuation techniques and bifurcation problems. *J. Comput. Appl. Math.*, 26(1-2):155–170, 1989.
14. E. J. Doedel, B. Krauskopf, and H. M. Osinga. Global bifurcations of the Lorenz manifold. *Nonlinearity*, 19(12):2947–2972, 2006.
15. E. J. Doedel, R. C. Paffenroth, A. R. Champneys, T. F. Fairgrieve, Yu. A. Kuznetsov, B. E. Oldeman, B. Sandstede, and X. J. Wang. AUTO2000: Continuation and bifurcation software for ordinary differential equations. available via `http://cmvl.cs.concordia.ca/`.
16. J. P. England, B. Krauskopf, and H. M. Osinga. Computing one-dimensional global manifolds of Poincaré maps by continuation. *SIAM J. Appl. Dyn. Sys.*, 4(4):1008–1041, 2005.
17. J. P. England, B. Krauskopf, and H. M. Osinga. Computing two-dimensional global invariant manifolds in slow-fast systems. *Internat. J. Bifur. Chaos Appl. Sci. Engrg.*, 16(3), 2007. at press.
18. I. R. Epstein, K. Kustin, P. DeKepper, and M. Orban. Oscillating chemical reactions. *Sci. Am.*, 3(248):112–123, 1983.

19. N. Fenichel. Persistence and smoothness of invariant manifolds for flows. *Indiana Univ. Math. J.*, 21:193–226, 1971.

20. N. Fenichel. Geometric singular perturbation theory. *J. Diff. Eq.*, 31:53–98, 1979.

21. M. Friedman and E. J. Doedel. Numerical computation and continuation of invariant manifolds connecting fixed points. *SIAM J. Numer. Anal.*, 28(3):789–808, 1991.

22. R. Ghrist, P. J. Holmes, and M. C. Sullivan. *Knots and Links in Three-Dimensional Flows*, volume 1654 of *Lecture Notes in Mathematics*. Springer-Verlag, Berlin, 1997.

23. R. Gilmore and M. Lefranc. *The Topology of Chaos: Alice in Stretch and Squeezeland.* Wiley-Interscience, New York, 2004.

24. W. Govaerts, Yu.A. Kuznetsov, and B. Sijnave. Continuation of codimension-2 equilibrium bifurcations in CONTENT. In E. J. Doedel and L. S. Tuckerman, editors, *Numerical methods for Bifurcation Problems and Large-Scale Dynamical Systems*, pages 163–184. Springer-Verlag, New York, 2000. available via `http://ftp.cwi.nl/CONTENT/`.

25. K. Green, B. Krauskopf, and K. Engelborghs. One-dimensional unstable eigenfunction and manifold computations in delay differential equations. *J. Comput. Phys.*, 197(1):86–98, 2004.

26. J. Guckenheimer, K. Hoffman, and W. Weckesser. The forced Van der Pol equation I: the slow flow and its bifurcations. *SIAM J. Appl. Dyn. Sys.*, 2:1–35, 2003.

27. J. Guckenheimer and P. Holmes. *Nonlinear Oscillations, Dynamical Systems and Bifurcations of Vector Fields.* Springer-Verlag, New York/Berlin, 2nd edition, 1986.

28. J. Hauser and H. M. Osinga. On the geometry of optimal control: the inverted pendulum example. *Proceedings of the American Control Conference, Arlington, VA*, Vol. 2, pages 1721–1726, 2001.

29. M. E. Henderson. Multiple parameter continuation: Computing implicitly defined k-manifolds. *Internat. J. Bifur. Chaos Appl. Sci. Engrg.*, 12(3):451–476, 2002.

30. M. E. Henderson. Computing invariant manifolds by integrating fattened trajectories. *SIAM J. Appl. Dyn. Sys.*, 4(4):832–882, 2005.

31. M. W. Hirsch, C. C. Pugh, and M. Shub. *Invariant Manifolds*, volume 583 of *Lecture Notes in Mathematics*. Springer-Verlag, Berlin, 1977.

32. A. Jadbabaie, J. Yu, and J. Hauser. Unconstrained receding horizon control: stability and region of attraction results. *IEEE Trans. Automat. Control*, 1999.

33. C. K. R. T. Jones. Geometric singular perturbation theory. In R. Johnson, editor, *Dynamical Systems (Montecatini Terme, 1994)*, volume 1609 of *Lecture Notes in Mathematics*, pages 44–120. Springer-Verlag, New York, 1995.

34. À. Jorba. Numerical computation of the normal behaviour of invariant curves of n-dimensional maps. *Nonlinearity*, 14(5):943–976, 2001.

35. M. T. M. Koper. Bifurcations of mixed-mode oscillations in a three-variable autonomous Van der Pol-duffing model with a cross-shaped phase diagram. *Physica D*, 80:72–94, 1995.

36. B. Krauskopf and H. M. Osinga. Growing 1D and quasi-2D unstable manifolds of maps. *J. Comput. Phys.*, 146(1):406–419, 1998.

37. B. Krauskopf and H. M. Osinga. Two-dimensional global manifolds of vector fields. *Chaos*, 9(3):768–774, 1999.

38. B. Krauskopf and H. M. Osinga. Computing geodesic level sets on global (un)stable manifolds of vector fields. *SIAM J. Appl. Dyn. Sys.*, 2(4):546–569, 2003.

39. B. Krauskopf, H. M. Osinga, E. J. Doedel, M. E. Henderson, J. Guckenheimer, A. Vladimirsky, M. Dellnitz, and O. Junge. A survey of methods for computing (un)stable manifolds of vector fields. *Internat. J. Bifur. Chaos Appl. Sci. Engrg.*, 15(3):763–791, 2005.

40. T. Krogh-Madsen, L. Glass, E. J. Doedel, and M. R. Guevara. Apparent discontinuities in the phase-resetting response of cardiac pacemakers. *J. Theor. Biol.*, 230:499–519, 2004.

41. Yu. A. Kuznetsov. *Elements of Applied Bifurcation Theory*. Springer-Verlag, New York/Berlin, 3rd edition, 2004.

42. E. N. Lorenz. Deterministic nonperiodic flows. *J. Atmosph. Sci.*, 20:130–141, 1963.

43. B. E. Oldeman, A. R. Champneys, and B. Krauskopf. Homoclinic branch switching: a numerical implementation of Lin's method. *Internat. J. Bifur. Chaos Appl. Sci. Engrg.*, 13(10):2977–2999, 2003.

44. H. M. Osinga and J. Hauser. The geometry of the solution set of nonlinear optimal control problems. *J. Dynamics and Differential Equations*, 2006.

45. J. Palis and W. de Melo. *Geometric Theory of Dynamical Systems*. Springer-Verlag, New York/Berlin, 1982.

46. M. Phillips, S. Levy, and T. Munzner. Geomview: An interactive geometry viewer. *Not. Am. Math. Soc.*, 40(8):985–988, 1993. available via http://www.geom.uiuc.edu/.

47. F. Schilder. Algorithms for Arnol'd tongues and quasi-periodic tori: a case study. In D. H. van Campen, M. D. Lazurko, and W. P.J . M. van den Oever, editors, *Proceedings of the Fifth EUROMECH Nonlinear Dynamics Conference*, pages ID 16–100, 2005.

48. F. Schilder. Computing Arnol'd tongue scenarios: some recent advances. In *Proceedings of the 2006 International Symposium on Nonlinear Theory and its Applications (NOLTA2006), 11-14 September*, Bologna, Italy, 2006.

49. F. Schilder, H. M. Osinga, and W. Vogt. Continuation of quasi-periodic invariant tori. *SIAM J. Appl. Dyn. Sys.*, 4(3):459–488, 2005.

50. F. Schilder and B. B. Peckham. Computing Arnol'd tongue scenarios. *J. Comput. Phys.*, 220(2):932–951, 2007.

51. F. Schilder, W. Vogt, S. Schreiber, H. M. Osinga. Fourier methods for quasi-periodic oscillations. *Internat. J. Numer. Methods Engrg.*, 67(5):629–671, 2006.

52. A. Shil'nikov, L. P. Shil'nikov, and D. Turaev. On some mathematical topics in classical synchronization: a tutorial. *Internat. J. Bifur. Chaos Appl. Sci. Engrg.*, 14(7):2143–2160, 2004.

53. S. H. Strogatz. *Nonlinear Dynamics and Chaos*. Addison-Wesley Publishing Company, 1996.

54. P. Szmolyan and M. Wechselberger. Canards in $\mathbb{R}^3$. *J. Diff. Eq.*, 177(2):419–453, 2001.

55. A. Tsuneda. A gallery of attractors from smooth Chua's equation. *Internat. J. Bifur. Chaos Appl. Sci. Engrg.*, 15(1):1–49, 2005.

56. A. J. van der Schaft. *L_2-Gain and Passivity Techniques in Nonlinear Control*. Springer-Verlag, Berlin, 1994.

57. M. Wechselberger. Existence and bifurcation of canards in $\mathbb{R}^3$ in the case of a folded node. *SIAM. J. Appl. Dyn. Sys.*, 4(1):101–139, 2005.

58. S. M. Wieczorek, B. Krauskopf, T. B. Simpson, and D. Lenstra. The dynamical complexity of optically injected semiconductor lasers. *Phys. Reports*, 416(1–2):1–128, 2005.
59. S. Wiggins. *Normally Hyperbolic Invariant Manifolds in Dynamical Systems*. Springer-Verlag, New York, 1994.

5

The Dynamics of SQUIDs and Coupled Pendula

Donald G Aronson[1] and Hans G Othmer[2]

[1] School of Mathematics and Institute for Mathematics and Its Applications, University of Minnesota, Minneapolis, USA
[2] School of Mathematics and Digital Technology Center, University of Minnesota, Minneapolis, USA

Josephson [8] predicted in 1962 that a DC tunnel current would flow between two superconductors connected by a thin insulating layer of thickness less than about 20 Å in the absence of a voltage difference, an effect now called the DC Josephson effect. The quantum-mechanical current, called the super-conducting current, arises from the tunneling of Cooper pairs of electrons of opposite spin and momenta and is given by

$$I_s = I_c \sin \phi, \tag{5.1}$$

where I_c is the critical current and ϕ is the difference of the phases of the wave functions of the two superconductors. This gives the ideal current through a junction, but in real circuits there are resistive and capacitive currents as well. One of the standard models of a more realistic circuit is the so-called Stewart-McCumber resistively-shunted-junction (or RSJ) model, which is described by the following equation for the current [6, 9]:

$$\frac{hC}{2e}\frac{d^2\phi}{dt^2} + \frac{h}{2eR}\frac{d\phi}{dt} + I_c \sin \phi = I. \tag{5.2}$$

Here h is Planck's constant, e is the charge on an electron, $h/2e$ is the flux quantum, C is the capacitance, R is the resistance, and I is the imposed bias current. To simplify (5.2) define the frequency $\Omega = \sqrt{2eI_c/hC}$ and the scaled time $\tau = \Omega t$; then (5.2) becomes

$$\ddot{\phi} + \varepsilon\dot{\phi} + \sin \phi = i, \tag{5.3}$$

where $\varepsilon = (\Omega RC)^{-1}$, $i = I/I_c$, and the dot denotes derivation with respect to the rescaled time τ.

A very useful correspondence of this system to a pendulum provides insight into the dynamics studied later. In fact, the pendulum will serve as the basic

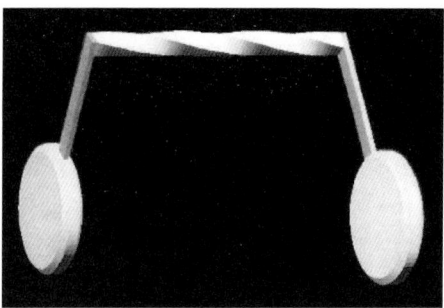

Fig. 5.1. Two pendula coupled via a torsion bar. From D.G. Aronson, E.J. Doedel and H.G. Othmer, The dynamics of coupled current-biased Josephson junctions II, *Internat. J. Bifur. Chaos Appl. Sci. Engrg.* 1(1) (1991) 51–66 ©1991 by World Scientific Publishing; reprinted with permission.

physical model; see also [1]. Suppose that a pendulum consists of a bob of mass m that is attached to a (weightless) rod of length L. Then the equation of motion is

$$\Lambda\frac{d^2\phi}{dt^2} + \eta\frac{d\phi}{dt} + mgL\sin\phi = T, \tag{5.4}$$

where $\Lambda = mL^2$ is the moment of inertia of the pendulum, g is the gravitational acceleration, η is the damping, ϕ is the angle between the bob and vertical measured from the downward position, and T is the applied torque. After non-dimensionalization this leads to (5.3).

When a ring of superconducting material contains two Josephson junctions, the result is a superconducting quantum interference device (SQUID), so called because the wave functions of the Cooper pairs at each junction interfere. SQUIDS are among the most sensitive devices for detecting magnetic fields — a SQUID is capable of detecting magnetic fields of around 2 picotesla, i.e., at the quantum flux level. The coupling between phases across the junctions is proportional to the difference of phases, and therefore, the system of equations governing a SQUID is

$$\begin{cases} \ddot{\phi}_1 + \varepsilon\dot{\phi}_1 + \sin\phi_1 = \gamma(\phi_2 - \phi_1) + I, \\ \ddot{\phi}_2 + \varepsilon\dot{\phi}_2 + \sin\phi_2 = \gamma(\phi_1 - \phi_2) + I. \end{cases} \tag{5.5}$$

Here, γ is the coupling coefficient, and the dimensionless bias current I is assumed to be the same for both junctions.

An identical pair of equations governs the motion of two pendula coupled by a linear torsional spring or bar, and forced with an applied torque I; see Fig. 5.1. We use this system as the paradigm in this chapter and we attempt to synthesize the results of [3, 5] and the unpublished study [2], which are all written in collaboration with Eusebius Doedel. The work involves extensive numerical studies that were carried out using DsTool, MatLab, and primarily, Auto. In the next section we analyze the equilibria of (5.5). Section 5.2

considers the undamped undriven case, which is part of the unpublished results in [2]. We analyze both equilibria and periodic orbits for this case, and also discuss the computation of heteroclinic connections. Finally, Sect. 5.3 shows the existence of so-called rotations, periodic solutions with a period that is an integer multiple of the forcing frequency. We discuss their stability in Sect. 5.4 and draw some conclusions in Sect. 5.5.

5.1 Equilibria and Their Stability

We begin by analyzing the existence and stability of equilibria for the coupled system. Clearly (5.5) is invariant under the transformation $\phi_i \to \phi_i + 2\pi$ and, thus, defines a flow on the product space $\{\mathbb{S}^1 \times \mathbb{R}\}^2$. In addition, (5.5) is invariant under the transformations $\phi_i \to \phi_{i+1}$ (mod 2) and $(\phi_i, I) \to (-\phi_i, -I)$. Therefore, we assume that $I \geq 0$ from now on. In order to analyze (5.5) it is convenient to introduce the variables

$$r = \frac{1}{2}(\phi_1 - \phi_2) \quad \text{and} \quad s = \frac{1}{2}(\phi_1 + \phi_2),$$

where r is (half) the instantaneous phase difference and s is the average phase difference. In these variables, and when written as a first-order system, (5.5) becomes

$$\begin{cases} \dot{r} = u, \\ \dot{s} = v, \\ \dot{u} = -\varepsilon u - \sin r \, \cos s - 2\gamma r, \\ \dot{v} = -\varepsilon v - \cos r \, \sin s + I. \end{cases} \tag{5.6}$$

If $\gamma = 0$ then the pendula are uncoupled, and if $r = 0$ then they are in phase or synchronized. The subspace $r = u = 0$ is invariant under the flow associated with this system and we refer to it as the *in-phase subspace*. The dynamics on this subspace are well characterized, even when the forcing is time dependent, because the fourth-order system reduces to a second-order system [4, 13].

The equilibria of (5.6) are given by $(R, S, 0, 0)$, where R and S are solutions to the system

$$\begin{cases} \sin R \, \cos S = -2\gamma R, \\ \cos R \, \sin S = I. \end{cases} \tag{5.7}$$

Solutions with $|R| > 0$ are called asynchronous equilibria, and those with $R = 0$ are called synchronous equilibria. Clearly the existence of equilibria is independent of the damping, but the forcing must be small enough ($|I| < 1$) to have an equilibrium. In the SQUID context this means that the bias current must be smaller than the superconducting current. In addition, solutions must satisfy $|R| < 1/2\gamma$, and therefore the asynchronous solutions approach synchronous solutions as the coupling strength increases.

With a slight abuse of notation we use the abbreviation (R, S) for equilibria, and in this notation we have the following: for $\gamma = 0$ system (5.7) has two countably infinite families of equilibria $\{(R_m, S_m)\}$ and $\{(R_n, S_n)\}$ that satisfy

$$R_m = \arccos{(-1)}^m I, \quad S_m = \frac{2m+1}{2}\pi \tag{5.8}$$

and

$$R_n = n\pi, \quad S_n = (-1)^n \arcsin I, \tag{5.9}$$

respectively. If we define $\sigma = \arcsin I$ then the complete set of solutions to (5.8) is generated by 2π-translates in S and π-translates along the diagonal of the basic sets

$$(R, S) = \left(\pm\left[\frac{\pi}{2} - \sigma\right], \frac{\pi}{2}\right) \quad \text{and} \quad (R, S) = (0, \{\sigma, \pi - \sigma\}). \tag{5.10}$$

For example, the π-translation along the diagonal of the basic set results in the four equilibria

$$\left(\frac{3\pi}{2} - \sigma, \frac{3\pi}{2}\right), \left(\frac{\pi}{2} + \sigma, \frac{3\pi}{2}\right), (\pi, 2\pi - \sigma), \text{ and } (\pi, 2\pi + \sigma).$$

When the applied torque vanishes (5.7) reduces to

$$\sin R = \pm 2\gamma R. \tag{5.11}$$

This equation has infinitely many solutions at $\gamma = 0$, and the number of solutions decreases to zero by a sequence of saddle-node bifurcations as $|\gamma|$ increases. The last solution disappears at the value for which the line $y = \pm 2\gamma R$ is first tangent to the curve $y = \sin R$.

The local stability of any equilibrium of (5.6) is determined by the eigenvalues of the Jacobian of the right-hand side of (5.6), which is

$$J = \begin{bmatrix} 0 & 0 & 1 & 0 \\ 0 & 0 & 0 & 1 \\ -\cos R \cos S - 2\gamma & \sin R \sin S & -\varepsilon & 0 \\ \sin R \sin S & -\cos R \cos S & 0 & -\varepsilon \end{bmatrix} = \begin{bmatrix} 0 & I_2 \\ L & -\varepsilon I_2 \end{bmatrix}.$$

The eigenvalues of J are solutions to the pair of quadratic equations

$$\lambda^2 + \varepsilon\lambda - \kappa_\pm = 0.$$

Here $\kappa_\pm$ are the eigenvalues of L, that is,

$$\kappa_\pm = \frac{1}{2}\left(\text{trace}(L) \pm \sqrt{(\text{trace}(L))^2 - 4\det(L)}\right),$$

where

$$\text{trace}(L) = -2(\cos R \cos S + \gamma),$$
$$\det(L) = (\cos R \cos S)^2 - (\sin R \sin S)^2 + 2\gamma \cos R \sin S.$$

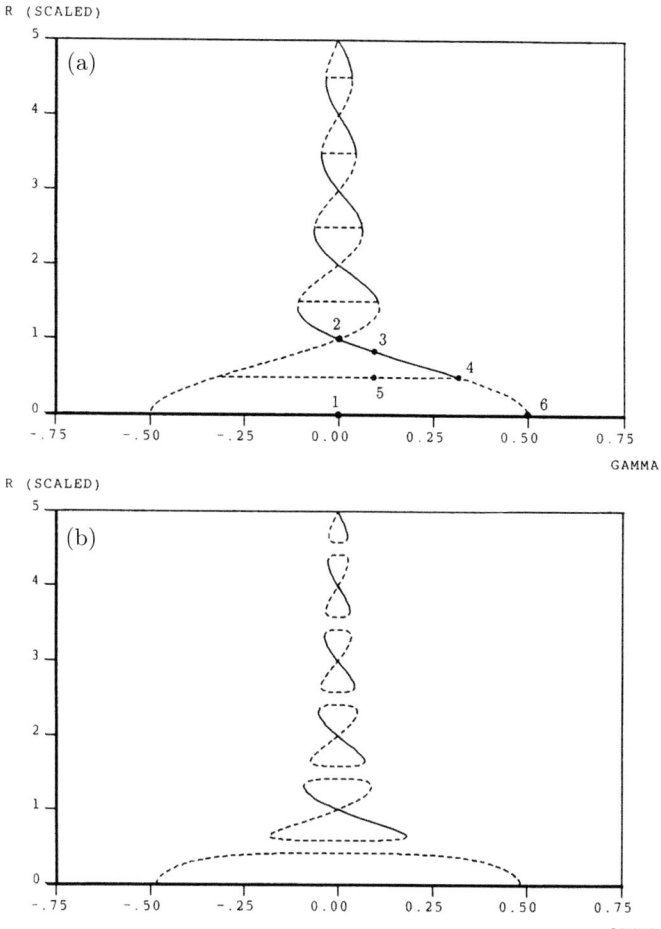

Fig. 5.2. The first five sets of equilibria of (5.6) for $\varepsilon = 0.15$ as a function of the coupling strength γ for $I = 0$ (a) and for $I = 0.25$ (b). Solid and dashed curves denote stable and unstable solutions, respectively; saddle-node bifurcations arise at the values of γ where there is a vertical tangent. From D.G. Aronson, E.J. Doedel and H.G. Othmer, The dynamics of coupled current-biased Josephson junctions II, *Internat. J. Bifur. Chaos Appl. Sci. Engrg.* 1(1) (1991) 51–66 ©1991 by World Scientific Publishing; reprinted with permission.

Equilibria for which $\kappa_\pm$ are both negative are stable, those with $\kappa_+ \kappa_- < 0$ have a one-dimensional stable manifold, and those for which $\kappa_\pm$ are both positive have a two-dimensional unstable manifold.

If $\gamma = 0$, the characteristic equation of L is

$$(\kappa + \cos R \cos S)^2 - (\sin R \sin S)^2 = 0.$$

and, therefore, for (R, S) given by (5.10) we have that

- $\kappa_+ = \kappa_- = -\cos\sigma < 0$ for $(R,S) = (0,\sigma)$,
- $\kappa_+ \kappa_- \quad = -\cos^2\sigma < 0$ for $(R,S) = (\pm(\frac{\pi}{2}-\sigma),\frac{\pi}{2})$,
- $\kappa_+ = \kappa_- = \quad \cos\sigma > 0$ for $(R,S) = (0,\pi-\sigma)$.

Thus, of the four points, only $(R,S) = (0,\sigma)$ is stable, and the same pattern is found for all the translates of these points. In terms of the phase angles of the individual pendula, the existence and stability of equilibria at $\gamma = 0$ can be summarized as follows. The three types of solutions are:

1. solutions with $\phi_1,\phi_2 \in (0,\pi/2)$; these are asymptotically stable for any $\varepsilon > 0$.
2. solutions with $\phi_1 \in (0,\pi/2)$ and $\phi_2 \in (\pi/2,\pi)$; these have a three-dimensional stable manifold and a one-dimensional unstable manifold.
3. solutions with $\phi_1,\phi_2 \in (\pi/2,\pi)$; these have a two-dimensional stable manifold and a two-dimensional unstable manifold.

At $\gamma = 0$ there exists an infinite number of other equilibria for which $R \neq 0$, each of which can be continued for small $|\gamma|$ because none of the equilibria that exist at $\gamma = 0$ is critical in the sense that the Jacobian has one or more eigenvalues on the imaginary axis. These equilibria and their continuations are naturally grouped into families of four equilibria, as determined above, by the various choices of ϕ_1 and ϕ_2 at $\gamma = 0$. By translation in S, each of these families determines an equivalence class of families modulo 2π. For any $\gamma \neq 0$ only finitely many of these exist, the remaining ones having disappeared via saddle-node bifurcations. Each of the families contains four equilibria at $\gamma = 0$, from which the entire family can be generated by continuation. The resulting families are shown in Fig. 5.2(a) for zero forcing and in Fig. 5.2(b) for $I = 0.25$; here we plotted the phase difference R versus γ. For the nth family, $n \neq 0$, the four solutions at $\gamma = 0$ can be denoted as $(\phi_d, \phi_d - 2n\pi)$, $(\phi_u, \phi_d - 2n\pi)$, $(\phi_d, \phi_u - 2n\pi)$, and $(\phi_u, \phi_u - 2n\pi)$, where $\phi_d = \sigma$, and $\phi_u = \pi - \phi_d$; see also [5].

5.2 Hamiltonian Dynamics

In the absence of damping and forcing (5.5) reduces to the Hamiltonian system

$$\ddot{\phi}_1 + \sin\phi_1 = \gamma(\phi_2 - \phi_1),$$
$$\ddot{\phi}_2 + \sin\phi_2 = \gamma(\phi_1 - \phi_2), \tag{5.12}$$

where the energy is given by

$$\mathcal{H} = \frac{1}{2}\left(\dot{\phi}_1^2 + \dot{\phi}_1^2\right) - (2 + \cos\phi_1 + \cos\phi_2) + \frac{\gamma}{2}(\phi_1 - \phi_2)^2.$$

In this section we summarize a portion of the results on the undamped undriven case from the unpublished work [2].

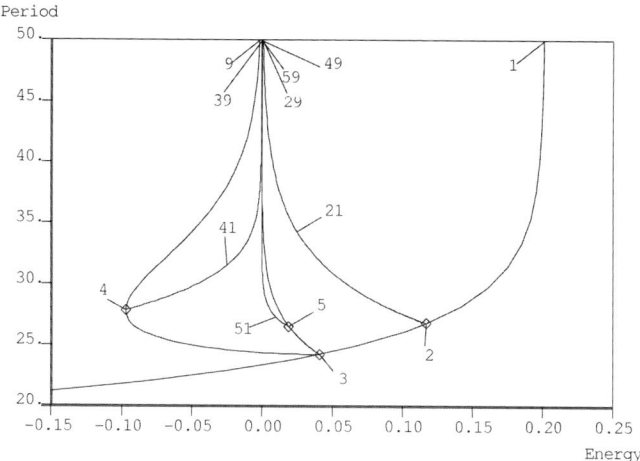

Fig. 5.3. The branch of antidiagonal solutions and some bifurcating branches for $\gamma = 0.01$. Where necessary, periods have been doubled to ensure that branches connect continuously.

The uncoupled system (5.12) with $\gamma = 0$ has equilibria at all points $e_{m,n} = (m\pi, n\pi)$ in the (ϕ_1, ϕ_2) configuration plane. In the basic square $(-\pi, \pi) \times (-\pi, \pi)$ there are heteroclinic orbits joining the diagonal points $e_{-1,-1}$ and $e_{1,1}$, the antidiagonal points $e_{1,-1}$ and $e_{-1,1}$, as well as all four pairs of neighboring corner points. The system is doubly periodic, which generates the entire plane.

For $\gamma \neq 0$ the doubly periodic structure is destroyed since the energy surfaces are bounded by the zero-velocity cylinders

$$- (2 + \cos \phi_1 + \cos \phi_2) + \frac{\gamma}{2}(\phi_1 - \phi_2)^2 = 0.$$

The symmetry that comes from translation by 2π along the diagonal remains.

When (5.12) is written in (r, s) variables, one sees from (5.6) that the diagonal $r \equiv 0$ and the antidiagonal $s \equiv 0$ are invariant. On the diagonal there are equilibria at $(k\pi, 0)$ for all integers k. These equilibria are centers for k even, while they are saddles for k odd. On the antidiagonal the equilibria are $(\rho, 0)$, where ρ is given by the solution of (5.11) with the negative sign. Let us suppose that there is a minimal positive solution $\rho = \rho_1$. To construct an antidiagonal solution to (5.6) (with $I = \varepsilon = 0$) we solve the initial value problem

$$\ddot{r} + \sin r + 2\gamma r = 0, \quad r(0) = 0 \text{ and } \dot{r}(0) = p. \tag{5.13}$$

The first integral for this problem is

$$\frac{1}{2}\dot{r}^2 - \cos r + \gamma r^2 = \frac{1}{2}p^2 - 1.$$

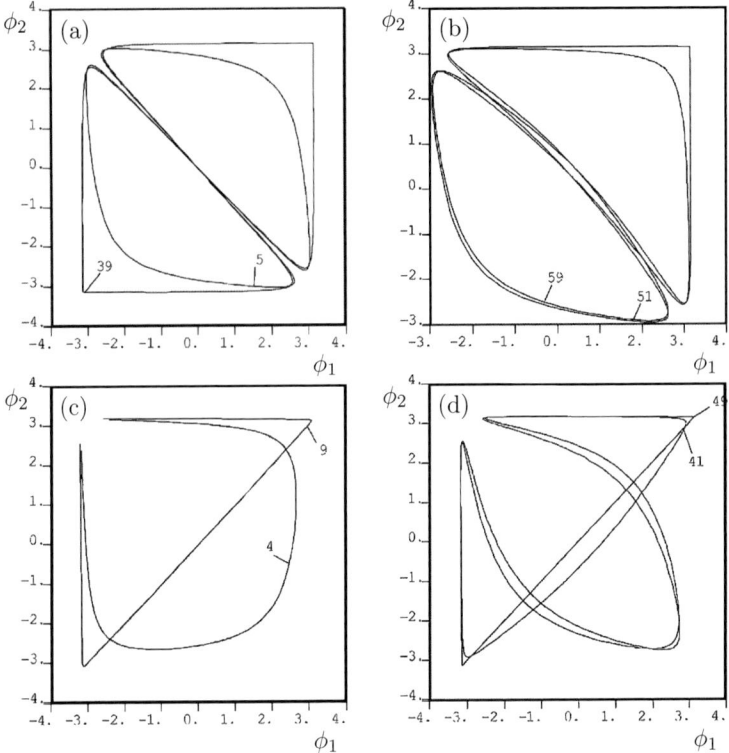

Fig. 5.4. The (ϕ_1, ϕ_2) component of selected solutions in Fig. 5.3.

The solution corresponds to a heteroclinic orbit joining the equilibria $(\pm\rho_1, 0)$ if

$$p(\rho_1) = p_1 = \sqrt{2 - 2\cos(\rho_1) + 2\gamma\rho_1^2}.$$

For $\rho < \rho_1$ the solution $p(\rho)$ corresponds to a periodic orbit about the origin. For these solutions the energy is given by

$$\mathcal{H}(p) = p^2 - 4.$$

For $\mathcal{H} = \mathcal{H}_1 := \mathcal{H}(p_1)$ the solution to (5.13) has infinite period. As $\mathcal{H}$ (and therefore p) is reduced, there is a value $\mathcal{H}_2 \in (0, \mathcal{H}_1)$ such that the periodic orbit is hyperbolic for $\mathcal{H} \in (\mathcal{H}_2, \mathcal{H}_1)$ and elliptic for $\mathcal{H} < \mathcal{H}_2$. Moreover, a new solution branch bifurcates from the antidiagonal solution at $\mathcal{H} = \mathcal{H}_2$. The bifurcation diagram is given in Fig. 5.3 and selected solutions on the bifurcating branch are shown in Fig. 5.4; the bifurcation at $\mathcal{H} = \mathcal{H}_2$ is label 2 in Fig. 5.3. Solutions on the new branch connect the zero-velocity surfaces about $(-\pi, \pi)$ and $(\pi, -\pi)$. Note that the period becomes infinite as $\mathcal{H} \searrow 0$, and at $\mathcal{H} = 0$ this branch seems to be generated by a concatenation of the

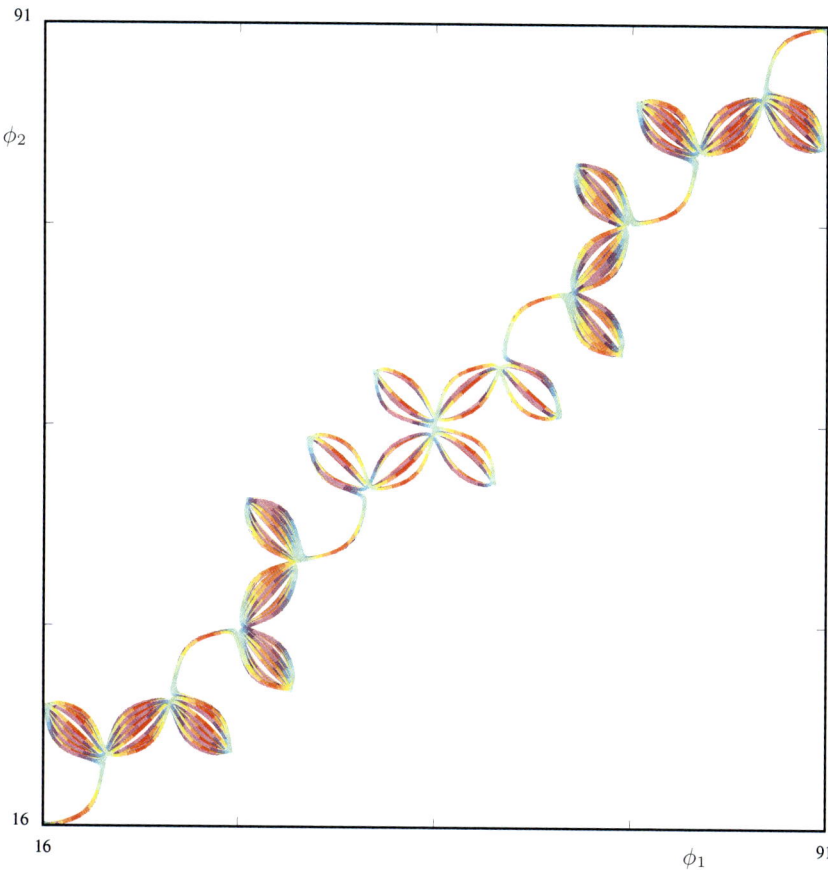

Fig. 5.5. The long-time integration of certain initial conditions on the zero-energy surface results in an intriguing petal structure. Color indicates the velocity of the solution, where blue/green is slow and red is fast. (Courtesy of John Guckenheimer.)

heteroclinic orbits that connect the zero-velocity surfaces about $(\pm\pi, \mp\pi)$ to the equilibrium point at $(-\pi, -\pi)$.

With further reduction of $\mathcal{H}$ a value $\mathcal{H}_3 \in (0, \mathcal{H}_2)$ is reached (label 3 in Fig. 5.3) at which a degenerate period-doubling bifurcation occurs. This bifurcation generates a branch of 'butterfly'-shaped orbits that undergo additional bifurcations and disappear in an infinite-period orbit at $\mathcal{H} = 0$; two of these butterfly-shaped orbits are shown in Fig. 5.4(a), and further bifurcated orbits are shown in Fig. 5.4(b). The bifurcation also generates a branch of 'horseshoe'-shaped orbits that are symmetric about the antidiagonal; see Fig. 5.4(c) and (d). As we will see, the horseshoes play an important role in the overall dynamics of the system. The horseshoe branch persists through

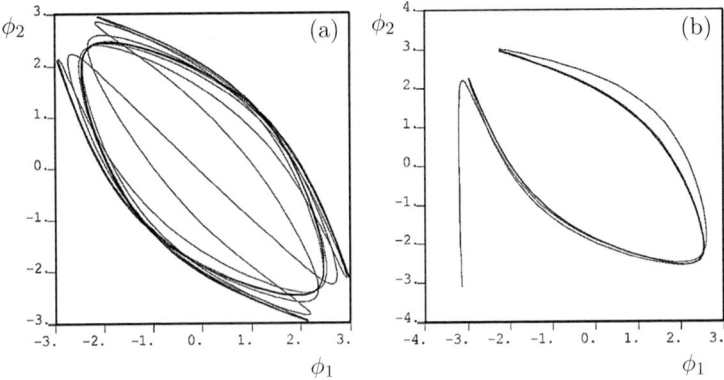

Fig. 5.6. Heteroclinic orbits in the (ϕ_1, ϕ_2)-plane. Panel (a) shows an orbit in the unstable manifold of the horseshoe orbit for $\gamma = 0.036377$. This orbit connects to the stable manifold of the symmetric partner of the horseshoe orbit. Panel (b) connects the unstable equilibrium $(-\pi, 0, -\pi, 0)$ to the stable manifold of the horseshoe orbit for $\gamma = 0.03638$.

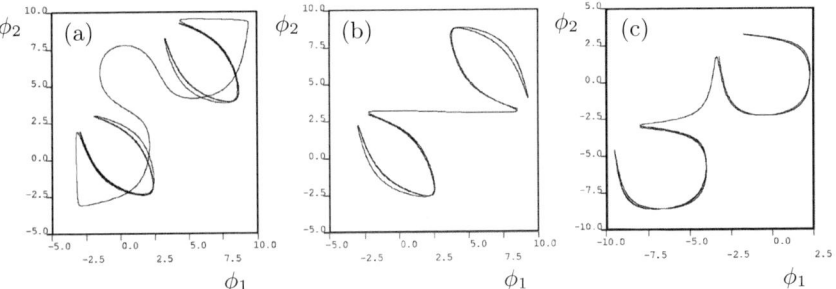

Fig. 5.7. Connecting orbits between horseshoe orbits in neighboring cells. Here, $(\gamma, \mathcal{H}) = (0.05, 0)$ in panel (a), $(\gamma, \mathcal{H}) = (0.0268297698, 0)$ in panel (b), and $(\gamma, \mathcal{H}) = (0.05, -0.22049)$ in panel (c).

$\mathcal{H} = 0$. There is a value $\mathcal{H}_4 < 0$ (label 4 in Fig. 5.3) at which there is a saddle-node bifurcation leading to a second branch of horseshoes. The second branch ends in an infinite-period orbit at $\mathcal{H} = 0$, where it appears to be the concatenation of three hetero- or homoclinic orbits; two examples are shown in Fig. 5.4(c). Note that for $\mathcal{H} \in (\mathcal{H}_4, 0)$ there are two distinct branches of horseshoes.

Figure 5.5 shows the result of a long-time integration in configuration space of system (5.12) with $\gamma = 0.01$. The trajectory lies on the zero-energy surface and has initial condition

$$\phi_1(0) = -\pi, \quad \dot{\phi}_1(0) = 0, \quad \phi_2(0) = -3.141585,$$

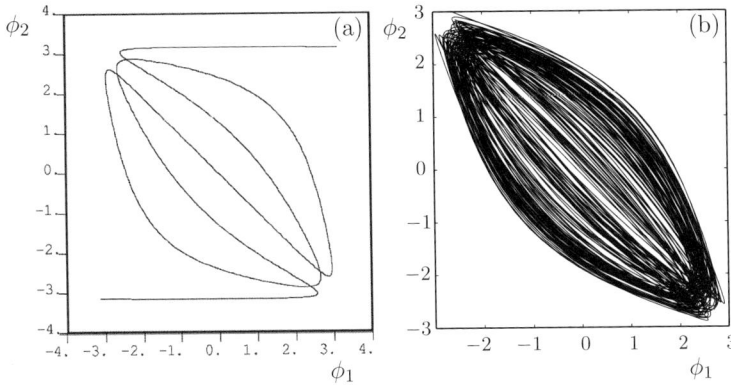

Fig. 5.8. Solutions to (5.12) on the zero-energy surface in the (ϕ_1, ϕ_2)-plane. Panel (a) shows a heteroclinic orbit for $\gamma = 0.03638$ that connects the unstable equilibrium $(-\pi, 0, -\pi, 0)$ to its diagonal translate $(\pi, 0, \pi, 0)$. Panel (b) shows a solution integrated for time $T \in [0, 700]$ with $\gamma = 0.01$; the initial condition is $\phi_1(0) = -\phi_2(0) = 2.690233$, and $\dot{\phi}_1(0)$ and $\dot{\phi}_2(0)$ are chosen so that the energy is zero.

with $\dot{\phi}_2(0)$ determined by the requirement of zero energy. The unbounded trajectory forms an intriguing petal structure. The equilibria where the two pendula are in the upright position lie along the diagonal at the base of the 'petals'. In this motion each pendulum crosses the upright position and changes its direction many times. The petals facing out from the diagonal of the vine-like structure are created when one pendulum crosses the upright position while the other stops and reverses direction. The trajectory moves back and forth along the diagonal in a seemingly erratic fashion. This and similar trajectories are induced by the rich structure of connecting orbits joining the horseshoes and the equilibria, as we now describe briefly.

Figure 5.6(a) shows a heteroclinic connection between the unstable manifold of the horseshoe orbit for $\gamma = 0.036377$ and the stable manifold of its symmetric partner (i.e., its reflection in the main diagonal). Figure 5.6(b) shows a heteroclinic orbit that connects the unstable equilibrium $(-\pi, 0, -\pi, 0)$ to the stable manifold of the horseshoe orbit for $\gamma = 0.03638$. Figure 5.7 shows connections between horseshoes in neighboring cells. Panels (a) and (b) show two solutions in the zero-energy surface, while panel (c) shows a solution for $\mathcal{H} = -0.22049$. Figure 5.8(a) shows a heteroclinic connection between the unstable equilibrium $(-\pi, 0, -\pi, 0)$ and its diagonal translate $(\pi, 0, \pi, 0)$. These connections provide the escape routes from cell to cell, but not all trajectories are ejected from their initial cells; many are simply trapped inside the horseshoes, as shown in Fig. 5.8(b).

5.3 Rotations

In addition to equilibria and periodic solutions, system (5.5) may also have *running solutions* or *k-rotations*. These are are solutions for which there exists a time $T > 0$ such that $\phi_j(t + T) = \phi_j(t) + 2k\pi$ for some integer $k \geq 1$. One can anticipate that rotations exist only for the appropriate relationship between the damping and the applied torque in the dissipative case. If we map the configuration space onto a cylinder then these solutions are periodic with period T. To construct a k-rotation we solve the initial value problem for system (5.6) with

$$s(0) = 0, \quad \dot{s}(0) = p_2, \quad r(0) = p_3, \quad \text{and} \quad \dot{r}(0) = p_4.$$

Solutions depend on the three 'state' parameters $\mathbf{p} = (p_2, p_3, p_4)$ and the three 'system' parameters ε, γ, I. For simplicity we regard the coupling strength $\gamma > 0$ as fixed and only deal with the two system parameters $\mathbf{q} = (\varepsilon, I)$. A solution to the initial value problem, written in the form

$$[s(t; \mathbf{p}, \mathbf{q}), \, r(t; \mathbf{p}, \mathbf{q})],$$

is a k-rotation if there exists a minimal $T > 0$ such that

$$\begin{cases} s(T; \mathbf{p}, \mathbf{q}) - 2\pi k = 0 \\ \dot{s}(T; \mathbf{p}, \mathbf{q}) - p_2 = 0, \\ r(T; \mathbf{p}, \mathbf{q}) - p_3 = 0, \\ \dot{r}(T; \mathbf{p}, \mathbf{q}) - p_4 = 0. \end{cases} \tag{5.14}$$

It is easy to see that in the Hamiltonian case ($\mathbf{q} = \mathbf{0}$) there exist $T_0 > 0$ and state parameters $\mathbf{p}_0$ such that $[s(t; \mathbf{p}_0, \mathbf{0}), r(t; \mathbf{p}_0, \mathbf{0})]$ is a k-rotation with period T_0 for each $k \geq 1$. We define $\mathcal{P}_0 = (T_0, \mathbf{p}_0, \mathbf{0})$ and say that a continuation of the solution $[s(t; \mathbf{p}_0, \mathbf{0}), r(t; \mathbf{p}_0, \mathbf{0})]$ to a neighborhood of $\mathcal{P}_0$ is *regular* if there is a distinguished state parameter and a distinguished system parameter such that the remaining state and system parameters are all smooth functions of the distinguished ones in a full neighborhood of $\mathcal{P}_0$.

It is an easy consequence of the Implicit Function Theorem that in-phase rotations ($k = 1$) always have a regular continuation. Here, we consider the general case, which is more complicated. The differential of (5.14) at $\mathcal{P}_0$ is the matrix

$$\Delta = (\Delta_0 \mid \zeta_\varepsilon \mid \zeta_I),$$

where

$$\Delta_0 = \begin{bmatrix} \dot{s} & s_{p_2} & s_{p_3} & s_{p_4} \\ \ddot{s} & \dot{s}_{p_2} & \dot{s}_{p_3} & \dot{s}_{p_4} \\ \dot{r} & r_{p_2} & r_{p_3} & r_{p_4} \\ \ddot{r} & \dot{r}_{p_2} & \dot{r}_{p_3} & \dot{r}_{p_4} \end{bmatrix}_{\mathcal{P}_0}, \quad \zeta_\varepsilon = \begin{pmatrix} s_\varepsilon \\ \dot{s}_\varepsilon \\ r_\varepsilon \\ \dot{r}_\varepsilon \end{pmatrix}, \quad \text{and} \quad \zeta_I = \begin{pmatrix} s_I \\ \dot{s}_I \\ r_I \\ \dot{r}_I \end{pmatrix}.$$

Here, Δ_0 is the differential of the Hamiltonian continuation problem, that is, the problem of continuing $(s, r)(\mathcal{P}_0)$ to a neighborhood of $\mathcal{P}_0$ in the subspace

$\mathbf{q} = \mathbf{0}$. For the Hamiltonian problem, conservation of energy provides a relationship between s, $\dot{s}$, r, and $\dot{r}$ so that, generically, only three of them are
independent, i.e., generically

$$\operatorname{rank}(\varDelta_0) = 3. \tag{5.15}$$

When (5.15) holds then there is a regular continuation of $(s,r)(\mathcal{P}_0)$ if there
exists

$$\zeta^* \in \operatorname{span}\{\zeta_\varepsilon, \zeta_I\} \quad \text{such that} \quad \zeta^* \notin \operatorname{range}(\varDelta_0), \tag{5.16}$$

that is, ζ^* can be chosen such that

$$\operatorname{rank}(\varDelta_0 \mid \zeta^*) = 4.$$

Let $X(t)$ denote the fundamental matrix solution to the variational system
associated with (5.6) at $(s,r)(\mathcal{P}_0)$. As is shown in [3], the differential $\varDelta_0$ then
becomes

$$\varDelta_0 = (\zeta_T \mid \xi_2 \mid \xi_3 \mid \xi_4),$$

where

$$\zeta_T = \begin{pmatrix} \dot{s} \\ \ddot{s} \\ \dot{r} \\ \ddot{r} \end{pmatrix}\Bigg|_{\mathcal{P}_0} = \begin{pmatrix} p_2^0 \\ 0 \\ p_4^0 \\ -2\gamma p_3^0 - \sin p_3^0 \end{pmatrix}$$

and the ξ_j for $j = 2,3,4$ are the corresponding columns of $X(T_0) - \operatorname{Id}$. In
order to satisfy (5.15) and (5.16) there are two possibilities: either

$$\operatorname{rank}(X(T_0) - \operatorname{Id}) = 3 \quad \text{and} \quad \zeta_T \in \operatorname{range}(X(T_0) - \operatorname{Id}) \tag{5.17}$$

or

$$\operatorname{rank}(X(T_0) - \operatorname{Id}) = 2 \quad \text{and} \quad \zeta_T \notin \operatorname{range}(X(T_0) - \operatorname{Id}). \tag{5.18}$$

If (5.17) holds there is no distinguished state parameter and hence no regular
continuation. We note that this case was never observed in any of the numerical studies reported in [3]. On the other hand, possibility (5.18) is known to
occur. Suppose (5.18) holds and let $\{i_1, i_2, i_3\}$ be a permutation of $\{2, 3, 4\}$
such that

$$\operatorname{range}(X(T_0) - \operatorname{Id}) = \operatorname{span}\{\xi_{i_1}, \xi_{i_2}\}.$$

Then there is a Hamiltonian continuation of $(s,r)(\mathcal{P}_0)$ with T, p_{i_1}, and p_{i_2}
expressed as smooth functions of p_{i_3} in a neighborhood of $p_{i_3}^0$. If, in addition,
(5.16) holds then there is a regular continuation in a neighborhood of $\mathcal{P}_0$.

The choice of distinguished system parameter is arbitrary. Suppose, for in-
stance, that (s,r) is a regular continuation of $(s,r)(t; \mathbf{p}_0, \mathbf{0})$ with distinguished
system parameter ε. Then we have a relation of the form $I = H(\cdot, \varepsilon)$, where
H is a smooth function. However, k-rotations must satisfy the *kinetic energy
relation*

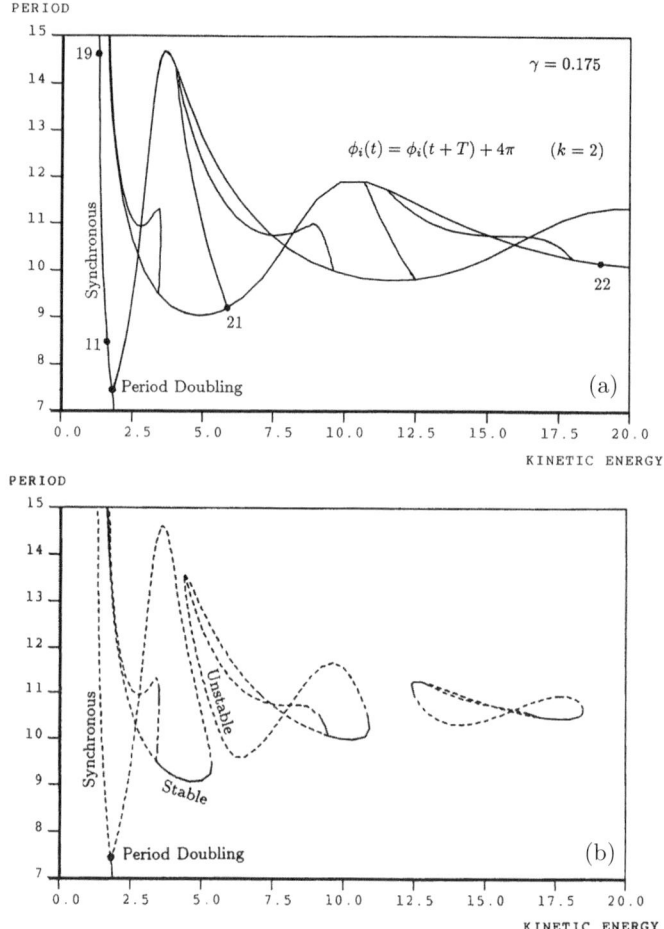

Fig. 5.9. Solutions of (5.12) for $\gamma = 0.175$ as a function of the kinetic energy for $I = 0$ (a) and variable I (b). From D.G. Aronson, E.J. Doedel and H.G. Othmer, The dynamics of coupled current-biased Josephson junctions II, *Internat. J. Bifur. Chaos Appl. Sci. Engrg.* 1(1) (1991) 51–66 ©1991, with permission from World Scientific Publishing; reprinted with permission.

$$I = \frac{\varepsilon}{2k\pi} \int_0^T (\dot{s}^2 + \dot{r}^2)\, dt.$$

It follows that

$$\left.\frac{\partial H}{\partial \varepsilon}\right|_{\mathcal{P}_0} > 0.$$

Therefore, we can invert H to obtain $\varepsilon = h(\cdot, I)$ and we get a regular continuation with I as the distinguished system parameter.

Figure 5.9(a) shows some branches of 2-rotations with $\mathbf{q} = \mathbf{0}$ and $\gamma = 0.175$; the plot shows the period T versus the kinetic energy. Figure 5.9(b) shows the equivalent solutions for $\mathbf{q} = (0.01, I)$ with I variable. The bifurcation at label 21 in Fig. 5.9(a) does not persist as ε is increased since there is a regular continuation at this point. The bifurcations to the right of label 21 do persist, because condition (5.16) fails in this case.

5.4 Stability and Bifurcations of the In-Phase Rotations

Most of the solution branches shown in Fig. 5.9(a) are 2-rotations for the Hamiltonian system. The exception is the left-most branch of synchronous or in-phase rotations. Here we investigate the stability and bifurcation properties of these solutions and their extensions to the damped/driven regime. To simplify the analysis we scale the damping ε and the torque I together by assuming that $I = \varepsilon A$ for some fixed $A > 0$. It is clear that system (5.5) has a one-parameter family of rotations $\Omega_\tau(\varepsilon)$ defined by $\phi_1(t) = \phi(t)$ and $\phi_2(t) = \phi(t + \tau)$ for each $\tau \in \mathbb{R}$ when $\gamma = 0$. It was shown in [5] that the only member of this family that can be continued for $\gamma \neq 0$ is the in-phase rotation $\Omega_0(\varepsilon)$. In this section we discuss the stability of $\Omega_0(\varepsilon)$ as the parameters ε and γ are varied.

It is known that for fixed $A > 4/\pi$ and each $\varepsilon > 0$, the equation

$$\ddot{\phi} + \sin\phi = \varepsilon(A - \dot{\phi}) \tag{5.19}$$

has a unique rotation solution for which $\dot{\phi} > 0$ [12]. If we translate time so that $\phi(0) = 0$ then there is a unique positive $\xi(\varepsilon) > 2$ such that the rotation solution satisfies $\dot{\phi}(0) = \xi(\varepsilon)$. As $\varepsilon \to 0$, we have $\xi(\varepsilon) \to \xi_0$, where $\xi_0 = \xi_0(A) > 2$ is the unique solution of

$$2\pi A = \int_0^{2\pi} \sqrt{\xi^2 - 2 + 2\cos\theta}\, d\theta.$$

Note that ξ_0 can have any value in the interval $(2, \infty)$ depending on the choice of $A > 4/\pi$. We denote the rotation solution by $\phi^*(\varepsilon)$ and its period by $T^*(\varepsilon)$.

In order to determine the stability of the in-phase rotation $\phi^*(\varepsilon)$ we must find the associated Floquet multipliers, which are the eigenvalues at $t = T^*(\varepsilon)$ of the fundamental matrix solution to the variational system associated with (5.6) at $\phi^*(\varepsilon)$. For this purpose it is convenient to order the variables as (r, u, s, v). Then we have to solve the system

$$\dot{V} = \begin{bmatrix} 0 & 1 & 0 & 0 \\ -\cos\phi^*(\varepsilon) - 2\gamma & -\varepsilon & 0 & 0 \\ 0 & 0 & 0 & 1 \\ 0 & 0 & -\cos\phi^*(\varepsilon) & -\varepsilon \end{bmatrix} V \tag{5.20}$$

subject to the initial condition $V(0) = \mathrm{Id}$. System (5.20) decomposes into the two 2×2 subsystems

$$\dot{X} = \begin{bmatrix} 0 & 1 \\ -\cos \phi^*(\varepsilon) - 2\gamma & -\varepsilon \end{bmatrix} X, \qquad X(0) = \mathrm{Id} \qquad (5.21)$$

and

$$\dot{X} = \begin{bmatrix} 0 & 1 \\ -\cos \phi^*(\varepsilon) & -\varepsilon \end{bmatrix} Y, \qquad Y(0) = \mathrm{Id}. \qquad (5.22)$$

Subsystem (5.22) determines stability with respect to the in-phase subspace and subsystem (5.21) determines stability with respect to the orthogonal complement of this subspace. It is easy to see that $Y = (\dot{\phi}, \ddot{\phi})^T$ is a $T^*(\varepsilon)$-periodic solution of (5.22). Therefore, the Floquet multipliers associated with subsystem (5.22) are 1 and $\exp(-\varepsilon T^*(\varepsilon))$, regardless of the value of γ. Thus, to determine the stability of $\phi^*(\varepsilon)$ it suffices to study the 2×2 system (5.21).

Let $\Psi(t, \gamma, \varepsilon)$ denote the fundamental matrix solution to (5.21). Then the Floquet multipliers are the eigenvalues of $\Psi(T^*(\varepsilon), \gamma, \varepsilon)$, that is, the roots of

$$\lambda^2 - \Theta(\gamma, \varepsilon)\lambda + \exp(-\varepsilon T^*(\varepsilon)) = 0,$$

where

$$\Theta(\gamma, \varepsilon) = \mathrm{trace}(\Psi(T^*(\varepsilon), \gamma, \varepsilon).$$

Therefore, the multipliers are

$$\lambda^{\pm} = \frac{1}{2}\left(\Theta \pm \sqrt{\Theta^2 - 4\exp(-\varepsilon T^*(\varepsilon))}\right),$$

and it follows that $\phi^*(\varepsilon)$ is stable if $|\Theta| < 1 + \exp(-\varepsilon T^*(\varepsilon))$ and unstable if $|\Theta| > 1 + \exp(-\varepsilon T^*(\varepsilon))$.

When $\varepsilon \to 0$, the second-order equation associated with system (5.21) reduces to a Hill equation

$$\ddot{x} + (2\gamma - q(t))x = 0, \qquad (5.23)$$

where the potential is given by

$$q(t) = -\cos \phi^*(\varepsilon)(t)|_{\varepsilon=0}. \qquad (5.24)$$

Moreover,

$$\Theta(\gamma, \varepsilon) \to \Theta_0(\gamma) = \left(\psi_1(T_0) + \dot{\psi}_2(T_0)\right),$$

where ψ_1 and ψ_2 are solutions to (5.17) that satisfy $\psi_1(0) = \dot{\psi}_2(0) = 1$ and $\dot{\psi}_1(0) = \psi_2(0) = 0$, and $T_0 = T^*(0)$.

The general theory for Hill's equation [11] shows that there exists a sequence of eigenvalues

$$-\infty < \gamma_0 < \gamma_1 \leq \gamma_2 < \gamma_3 \leq \gamma_4 < \cdots$$

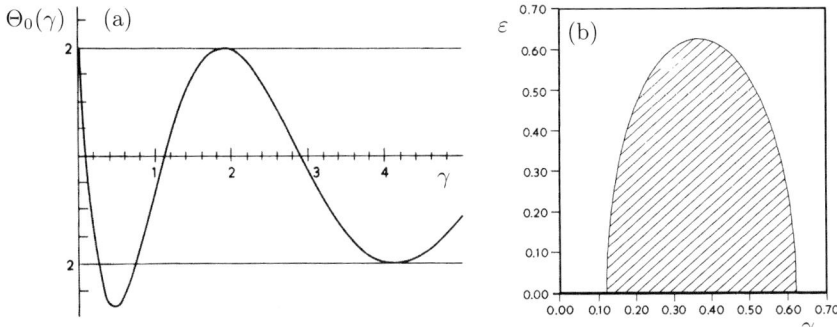

Fig. 5.10. The graph of $\Theta_0(\gamma)$ (a) and the locus of period-doubling bifurcations from $\phi^*(\varepsilon)$ for $A = 5/3$ (b). The in-phase rotation is unstable in the hatched region. From E.J. Doedel, D.G. Aronson and H.G. Othmer, The dynamics of coupled current-biased Josephson junctions I, *IEEE Trans. Circ. Sys.* 35(7) (1988) 810–817 ©1988 by IEEE; reprinted with permission.

with $\gamma_j \to \infty$ as $j \to \infty$, such that

$$
\begin{array}{lll}
|\Theta_0| < 2 \text{ and } \dot{\Theta}_0 < 0 & \text{on} & (\gamma_0, \gamma_1) \cup (\gamma_4, \gamma_5) \cup \cdots, \\
|\Theta_0| < 2 \text{ and } \dot{\Theta}_0 > 0 & \text{on} & (\gamma_2, \gamma_3) \cup (\gamma_6, \gamma_7) \cup \cdots, \\
\Theta_0 > 2 & \text{on} & (-\infty, \gamma_0) \cup (\gamma_3, \gamma_4) \cup (\gamma_7, \gamma_8) \cup \cdots, \\
\Theta_0 < -2 & \text{on} & (\gamma_1, \gamma_2) \cup (\gamma_5, \gamma_6) \cup \cdots.
\end{array}
\tag{5.25}
$$

At $\gamma = 0$ system (5.21) reduces to (5.22), which has a T_0-periodic solution. Thus, 0 is an eigenvalue of (5.23) with T_0-periodic boundary conditions, and we know from previous remarks that the associated eigenfunction is strictly positive. It follows from Sturm-Liouville theory that 0 is the smallest eigenvalue for this problem and, therefore, $\gamma_0 = 0$. Consequently, $\Theta_0(\gamma) > 2$ for all $\gamma < 0$.

According to Goldberg's theorem [7], equation (5.23) has exactly one finite interval of instability if and only if the potential q is periodic and integrable, and satisfies

$$
\ddot{q} = 3q^2 + \alpha q + \beta
\tag{5.26}
$$

for some constants α and β, that is, if and only if q is an elliptic function. In the present case, since $\phi^*(0)$ is a rotation it follows from (5.24) that q is periodic. Moreover, using (5.19) with $\varepsilon = 0$ and its first integral, one can verify that (5.26) is satisfied. Thus, there is precisely one finite interval of instability for $\gamma > 0$, and the numerical computations performed in [5] show that this interval is (γ_1, γ_2). It follows that $\gamma_{2j-1} = \gamma_{2j}$ for all $j > 2$ and that $|\Theta_0(\gamma)| = 2$ for $\gamma = \gamma_{2j}$ with $j > 1$. The graph of $\Theta_0(\gamma)$ is shown in Fig. 5.10(a).

The second-order equation corresponding to (5.21) is

$$
\ddot{x} + \varepsilon \dot{x} + (2\gamma + cos\phi^*(\varepsilon))x = 0.
\tag{5.27}
$$

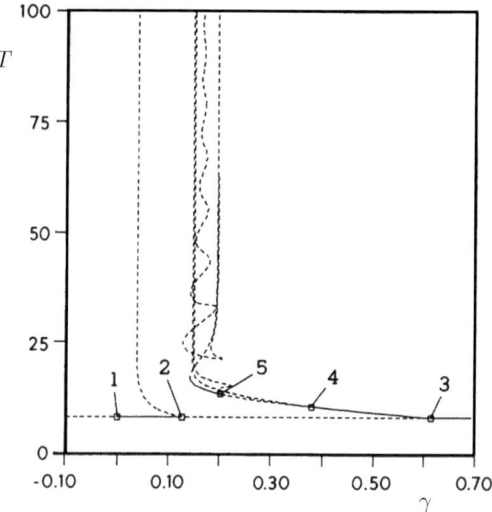

Fig. 5.11. The bifurcation diagram for (5.6), represented as the period T versus γ for $A = 5/3$. From E.J. Doedel, D.G. Aronson and H.G. Othmer, The dynamics of coupled current-biased Josephson junctions I, *IEEE Trans. Circ. Sys.* 35(7) (1988) 810–817 ©1988 by IEEE; reprinted with permission.

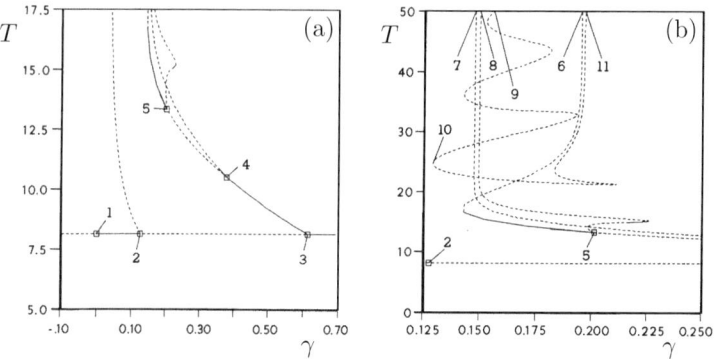

Fig. 5.12. Details of the bifurcation diagram shown in Fig. 5.11. The points in panel (a) labeled 1, 4, and 5 are transcritical bifurcations, while those labeled 2 and 3 are period-doubling bifurcations. There are two regions of stability along the asynchronous branch that bifurcates at label 3. Panel (b) shows a blow-up of the region around label 5 in panel (a). From E.J. Doedel, D.G. Aronson and H.G. Othmer, The dynamics of coupled current-biased Josephson junctions I, *IEEE Trans. Circ. Sys.* 35(7) (1988) 810–817 ©1988 by IEEE; reprinted with permission.

If $\gamma = 0$ then $x_1(t) = \dot{\phi}^*(\varepsilon)(t)$ is a strictly positive $T^*(\varepsilon)$-periodic solution to (5.27) for any $\varepsilon > 0$. Thus, $\gamma = 0$ is an eigenvalue of (5.27) for any $\varepsilon > 0$, and it is a simple eigenvalue because

$$x_2(t) = \dot{\phi}^*(\varepsilon)(t) \int_0^t \frac{e^{-\varepsilon\tau}}{[\dot{\phi}^*(\varepsilon)(\tau)]^2} \, d\tau$$

is a linearly independent non-periodic solution. Using these solutions one can construct the fundamental matrix solution to (5.21) and show that $\Theta_\gamma(0, \varepsilon) < 0$. It follows that the infinite instability interval $(-\infty, 0)$ remains invariant for $\varepsilon > 0$ and that $\phi^*(\varepsilon)$ becomes stable as γ increases through 0. Numerical computations show that there is a 'vertical' bifurcation from $\phi^*(\varepsilon)$ at $\gamma = 0$; the numerical results are described in more detail below.

By continuity, the unique instability interval (γ_1, γ_2) for $\varepsilon = 0$ persists for sufficiently small $\varepsilon > 0$. The numerical computations done in [5] strongly suggest that the remaining eigenvalues γ_{2j} disappear for $\varepsilon > 0$. Consequently, for sufficiently small $\varepsilon > 0$ the rotation solution $\phi^*(\varepsilon)$ is unstable for $\lambda \in (-\infty, 0) \cup (\gamma_1(\varepsilon), \gamma_2(\varepsilon))$ and stable otherwise. Furthermore, there exists $\tilde{\varepsilon} = \tilde{\varepsilon}(A) > 0$ such that $\gamma_1(\tilde{\varepsilon}) = \gamma_2(\tilde{\varepsilon})$, and $\phi^*(\varepsilon)$ is unstable on $\mathbb{R}^-$ and asymptotically stable on $\mathbb{R}^+$ whenever $\varepsilon > \tilde{\varepsilon}$; see Fig. 5.10(b). Note that the bifurcations at $\gamma = \gamma_j(\varepsilon)$, $j = 1, 2$, are period-doubling bifurcations, because the multiplier passes through -1; see Fig. 5.10(a). In the Hamiltonian case these two bifurcations project onto the point labeled 'Period Doubling' in Fig. 5.9(a).

We now discuss the numerically computed bifurcation behavior in the interval (γ_1, γ_2) for the value $\varepsilon = \varepsilon^* = 0.15$ used in [10]. Then the period-doubling bifurcations from $\phi^*(\varepsilon)$ are at $\gamma_1 = 0.1275$ and $\gamma_2 = 0.6132$. Solution branches that bifurcate from $\phi^*(\varepsilon^*)$ at $\gamma = \gamma_1$ and $\gamma = \gamma_2$ are shown in Fig. 5.11 with enlarged views given in Fig. 5.12. Rotations that correspond to some of the labels in Fig. 5.12 are shown in Fig. 5.13.

All rotations in Figs. 5.11 and 5.12 have winding number 2, so that $\phi_i(T) - \phi_i(0) = 4\pi$, where T is the integration time. The bifurcating branches then connect continuously to the horizontal branch $\phi^*(\varepsilon^*)$. The solutions with labels 2 and 3 in Fig. 5.12(a) denote the two period-doubling bifurcations from $\phi^*(\varepsilon^*)$. The branch that emanates from label 2 terminates at an orbit of infinite period. The same holds for the branch that emanates from label 3. The solution with label 6 in Fig. 5.12(b) can be thought of as an approximation to the infinite-period orbit that terminates the branch; it is shown in Fig. 5.13(a). The infinite-period orbit is a 'double homoclinic loop', i.e., an orbit that passes through the same saddle point twice. Note that this branch contains two regions of stable rotations. The solutions with labels 4 and 5 in Fig. 5.12(a) are secondary transcritical bifurcations, not period-doubling bifurcations. The bifurcating tertiary branches from labels 4 and 5 also terminate in infinite-period orbits past the solutions labeled 7 and 8, respectively; compare Figs. 5.13(b) and (c). These branches also contain stable portions and further bifurcations that are not shown in Figs. 5.11 and 5.12.

The 'oscillating' branch of solutions in Fig. 5.11 is shown in a blow-up in Fig. 5.12(b); it contains the solutions labeled 9, 10, and 11, which are plotted in Figs. 5.13(d), (e), and (f), respectively. This branch terminates in an

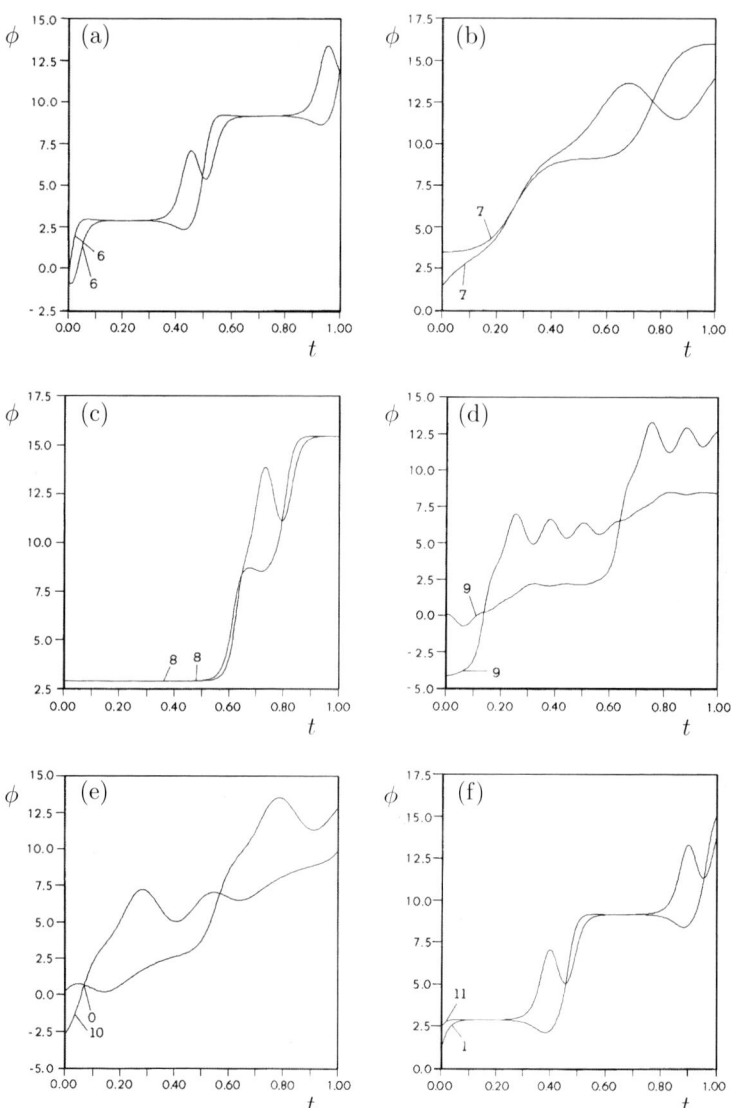

Fig. 5.13. Some rotations corresponding to the labels in Fig. 5.12. From E.J. Doedel, D.G. Aronson and H.G. Othmer, The dynamics of coupled current-biased Josephson junctions I, *IEEE Trans. Circ. Sys.* 35(7) (1988) 810–817 ©1988 by IEEE; reprinted with permission.

infinite-period orbit at both end points of the γ interval in which it exists. The solutions with labels 9 and 11 can be considered as approximations to these orbits. Solution 9 approximates an infinite-period orbit containing two distinct unstable equilibria, each of which has one complex conjugate pair of eigenvalues; see Fig. 5.13(d). Solution 11, at the other end of the branch, is a double homoclinic loop; see Fig. 5.13(f). Along this branch there are small intervals of stable behavior near the limit points. For example, one such stable interval is near the solution with label 10 on the lower part of the branch. These stable regions are so small that they cannot be distinguished in Fig. 5.12(b). They are bounded by bifurcations that lead to more complicated solution types. Many of these more complicated, stable solutions can be observed numerically by careful choice of initial data and accurate integration. For example, a stable rotation such as solution 10 in Fig. 5.13(e) can be obtained by accurately choosing initial data near solution 10 in Fig. 5.12(b), on the small portion of the branch that consists of stable rotations.

5.5 Conclusions

We presented an overview of the dynamics of a model of a ring of superconducting material that contains two Josephson junctions, which is known as a SQUID. The resulting system equations are identical (in non-dimensional form) to the equations modeling two pendula that are coupled by a linear torsional spring or bar and forced with an applied torque. Our analysis involved intensive use of AUTO to construct bifurcation diagrams and rotating solutions that can only be found explicitly for the Hamiltonian case.

A complex bifurcation structure organizes the existence of rotation solutions. We focused particularly on the case of fixed coupling parameter γ and varying damping coefficient ε and forcing I. We found the conditions under which in-phase rotation solutions that exist for $\varepsilon = 0$ persist; the resulting branch leads to a series of bifurcating branches. In particular, there is a region of relatively small values of γ and ε in which the in-phase rotation is unstable.

Maginu [10] was the first to observe the instability for intermediate values of γ and suitable ε and I. His numerical studies indicate the presence of chaos in the unstable range. The results discussed here provide a more detailed, though still incomplete, understanding of the transitions in dynamics suggested in [10].

Acknowledgments

All of the work described in this chapter was performed in close collaboration with Sebius Doedel and we thank him for his contributions. We also thank John Guckenheimer, Àngel Jorba, and Björn Sandstede for their contributions to the material on Hamiltonian Systems in Sect. 5.2. Furthermore, we are grateful to World Scientific Publishing and IEEE for permission to reproduce previously published material from [3] and from [5], respectively.

References

1. E. Altshuler and R. Garcia. Josephson junctions in a magnetic field: Insights from couppled pendula. *Am. J. Phys.*, 71(4): 405–408, 2003.
2. D. G. Aronson, E. J. Doedel, J. Guckenheimer, A. Jorba, and B. Sandstede. On the dynamics of torsion-coupled pendula. Unpublished, 2002.
3. D. G. Aronson, E. J. Doedel, and H. G. Othmer. The dynamics of coupled current-biased Josephson junctions II. *Internat. J. Bifur. Chaos Appl. Sci. Engrg.*, 1(1): 51–66, 1991.
4. V. N. Belykh, N. F. Pedersen, and O. H. Soerensen. Shunted-Josephson-junction model. II. The nonautonomous case. *Physical Review B*, 16(11): 4860–71, 1977.
5. E. J. Doedel, D. G. Aronson, and H. G. Othmer. The dynamics of coupled current-biased Josephson junctions I. *IEEE Trans. Circ. Sys.*, 35(7): 810–817, 1988.
6. L. Finger. The Josephson junction circuit family: Network theory. *Int. J. Circ. Th. Appl.*, 28(4): 371–420, 2000.
7. W. Goldberg. Necessary and sufficient conditions for determining a Hill's equation from its spectrum. *J. Math. Anal. Appl.*, 55: 549–554, 1976.
8. B. D. Josephson. Supercurrents thorough barriers. *Phys Letts*, 1: 201, 1962.
9. Konstantin K. Likharev. *Dynamics of Josephson Junctions and Circuits*. Gordon and Breach Publishers, New York, 1986.
10. K. Maginu. Spatially homogeneous and inhomogeneous oscillations and chaotic motion in the active Josephson junction line. *SIAM J. Appl. Math.*, 43(2): 225–243, 1983.
11. W. Magnus and S. Winkler. *Hill's Equation*. Dover Publ., New York, 1979.
12. N. Minorsky. *Nonlinear Oscillations*. Robert E. Krieger Publishing Co., Malabar, Fla., 1983.
13. F. M. A. Salam and S. S. Sastry. The Complete Dynamics of the Forced Josephson Junction Circuit: The Regions of Chaos. In J. Chandra, editor, *Chaos in Nonlinear Dynamical Systems*, pages 43–55. SIAM, Philadelphia, 1984.

6

Global Bifurcation Analysis in Laser Systems

Sebastian M Wieczorek

Mathematics Research Institute, University of Exeter, United Kingdom

Nonlinear dynamics of lasers has been a lively theoretical and experimental field since the invention of the laser in 1960. Its focus in the last two decades have been instabilities in widely used semiconductor lasers. Nonlinear studies of laser systems contributed to the field of dynamical systems with general phenomena including chaos, (chaotic) synchronization of coupled oscillators, competition, excitability, delay-induced instabilities, unfolding of high-codimension bifurcations, bifurcation cascades, and spatial patterns; see [1, 28, 30, 36, 51, 63] for general reading and further references. These studies also deepened the understanding of nonlinear phenomena that are important for technological applications, e.g. external-modulation response of semiconductor lasers for faster Internet connections [57]. Furthermore, nonlinear analysis of laser systems stimulated and helped to validate the feasibility of novel, chaos-based applications including secure communication schemes [4, 50], chaotic radars [34], and instability-based laser sensors [56].

Much of the recent progress in the field of laser dynamics is owing to the application of numerical continuation techniques. The study of lasers with tools from bifurcation theory started already in 1987 with the work of Maloney and coworkers on the nonlinear dynamics of three-level molecular lasers [37]. By now, there are over one hundred publications where tools from bifurcation theory are used to investigate dynamics of various laser systems. To explain the strong impact that numerical continuation techniques had and are still having on the field of nonlinear laser dynamics we mention here four key properties that we found to be very influential in our research. Namely, numerical continuation techniques:

1. facilitate immensely the systematic search of an extensive and many-dimensional parameter space to identify the important contributions. Interesting phenomena may be missed by the traditional approach of simulation of the governing equations by direct time integration. Indeed, numerical continuation supplements numerical simulation by enabling parameter

studies to be performed to the necessary detail and accuracy, and with relatively modest computational resources;

2. allow the study of global homoclinic and heteroclinic bifurcations, which are often associated with interesting nonlinear effects but cannot be studied otherwise;

3. supplement and expand analytical bifurcation studies, which are themselves invaluable but generally restricted to small neighborhoods of parameter space, for example, near isolated individual bifurcations. Numerical bifurcation analysis is the tool of choice for those problems that cannot be addressed with analytical techniques but are of great importance to physicists, chemists, biologists, and engineers. For example, using the analytical results on bifurcation curves expected near a codimension-two bifurcation point, continuation techniques give answers to questions such as where these bifurcation curves go and to which other codimension-two bifurcation points they connect. Do phenomena occupy regions of the parameter space that can be experimentally detectable or that are of any practical interest? From a more general bifurcation theory point of view, this question concerns a better understanding of the organizing properties of bifurcations;

4. may actually stimulate real laser experiments where bifurcation diagrams are used as 'road maps' to guide experimentalists through the complexity and variety of nonlinear laser dynamics.

The aim of this chapter is to give a taste of how continuation techniques can be used to understand complicated dynamics in laser systems. At the same time, laser systems emerge as natural candidates to study how global bifurcation phenomena manifest themselves in a real system. In this sense, this chapter should also be seen as a contribution to the wider field of dynamical systems. Specifically, we present here the following two concrete examples of a global bifurcation analysis in semiconductor laser systems.

- In Sect. 6.1 we consider structures of global n-homoclinic bifurcations that lead to the phenomenon of multi-pulse excitability in semiconductor lasers with optical injection [58, 62, 64]; and

- in Sect. 6.2 we present the backbone of the bifurcation set for two back-to-back coupled lasers in which we find the counter-intuitive appearance of chaos at practically vanishing coupling [53, 54].

Both examples feature interesting global bifurcation structures that have a physical meaning and actually stimulated real laser experiments. The word 'global' in this context refers to objects in phase space that are due to certain arrangements of stable and unstable invariant manifolds, as well as to associated bifurcation structures in parameter space. In terms of the parameter space we use the physically motivated approach of calculating k-parameter bifurcation sets (usually for $k = 2$) for several fixed values of an additional $(k + 1)$st parameter. These k-dimensional bifurcation sets, which consist of various local and global bifurcations, are slices that are influenced or even determined by unfoldings of certain codimension-$(k + 1)$ bifurcations or singularities. These so-called organizing centers provide links between various types of bifurcations that appear at first glance to be (and are often thought of as) unrelated. While analytical techniques typically force investigators to focus on particular bifurcations, which imposes some sense of isolation of the particular phenomenon under investigation, numerical bifurcation analysis allows one to connect seemingly unrelated pieces. The goal is to get to a deeper understanding of the dynamics by obtaining a consistent and global bifurcation picture of the given (laser) system; see also Chaps. 2 and 7.

6.1 Multi-Pulse Excitability and n-Homoclinic Orbits in an Optically Injected Laser

This section is based on [58] and describes intricate structures of n-homoclinic orbits and their bifurcations in the rate equation model of an injection laser. The analysis reveals how codimension-two and -three homoclinic bifurcations act as organizing centers of the bifurcation diagram. First, we find heteroclinic cycles known as T-point bifurcations; we are dealing here with the case that both saddles involved have a pair of complex conjugate eigenvalues. Such T-point bifurcations were found in systems from applications [2, 19, 20, 22, 26, 32, 47, 68] and their unfolding is known to involve n-homoclinic orbits for any n [9, 10]. Secondly, we find double-homoclinic orbits to a saddle-focus, where there are two different homoclinic connections to a single saddle-focus. (This should not be confused with a 2-homoclinic orbit.) This codimension-two global bifurcation has been studied in an abstract setting in [27, 40, 44]. The bifurcations of 1-homoclinic orbits are known, but the possible unfoldings are not yet fully understood. We present sketches of relevant bifurcation curves associated with these global bifurcations and show with numerical bifurcation diagrams how they manifest themselves in the optically injected laser model. Also, we explain how these n-homoclinic bifurcations give rise to the phenomenon of multi-pulse excitability.

All curves of global bifurcations and the associated homoclinic and heteroclinic orbits were calculated with the HOMCONT [11, 12] part of the continuation package AUTO [16]; the invariant manifolds and time series illustrating multi-pulse excitability were computed with the package DsTOOL [5].

6.1.1 Optically Injected Laser

From the dynamical systems point of view, a free-running class-B laser (the active-medium polarization decays much faster than the population inversion and electric field) is a damped nonlinear oscillator characterized by a stable equilibrium with two complex-conjugate eigenvalues. This situation can be changed drastically when the laser is subjected to an external optical signal,

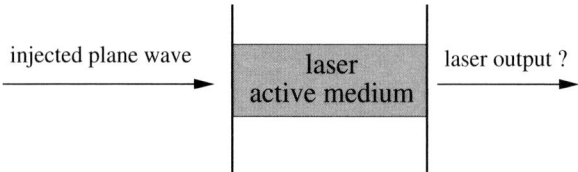

Fig. 6.1. A laser that is being injected with an external optical field.

as discussed here and sketched in Fig. 6.1. With optical injection it becomes a driven nonlinear oscillator: one of the nicest physical systems to show a fascinating array of nonlinear dynamics. Several kinds of complex and chaotic dynamics were discovered; see, for example, [3, 17, 21, 35, 29, 59, 63]. Of particular importance is the fact that this system is very well described by a set of three autonomous ordinary differential equations for the complex electric field $E = E_x + iE_y$ and the population inversion n (the number of electron-hole pairs in the case of a semiconductor laser) [48, 59]. These so-called *single-mode rate equations* for this system can be written in dimensionless form as

$$\begin{cases} \dot{E} = K + \left(\frac{1}{2}(1 + i\alpha)n - i\omega\right)E \\ \dot{n} = -2\Gamma n - (1 + 2Bn)(|E|^2 - 1) \ . \end{cases} \tag{6.1}$$

The two main parameters are the injected field amplitude K and the detuning ω, the frequency difference between the injected light and the frequency of the laser without injection. The explicit time dependence in the drive term proportional to K was eliminated thanks to the S^1 symmetry of the system [63]. While K and ω can easily be changed in an experiment, the parameters B, Γ and α describe material properties of a given laser. Specifically, B is the rescaled lifetime of photons in the laser cavity and Γ is the rescaled damping rate of the so-called relaxation oscillations, which are an exchange of energy between the electric field E and the population n of a characteristic frequency ω_R in a free-running laser. We use the realistic values $B = 0.015$ and $\Gamma = 0.035$ throughout in our study.

The material constant α, called the *linewidth enhancement factor*, can be very different for different lasers, and it is known that changing α has a very large effect on the dynamics of the injected laser [59]. The parameter α describes the coupling between the phase and the amplitude of the electric field E, and it is in the range of $\alpha \in [1, 10]$ for typical semiconductor lasers. On the other hand, (6.1) for $\alpha = 0$ models injected solid-state and CO_2 lasers, which have a negligible phase-amplitude coupling. This is our motivation for studying how the bifurcation set in the (K, ω)-plane depends on α, that is, on the main material property of the particular laser under consideration.

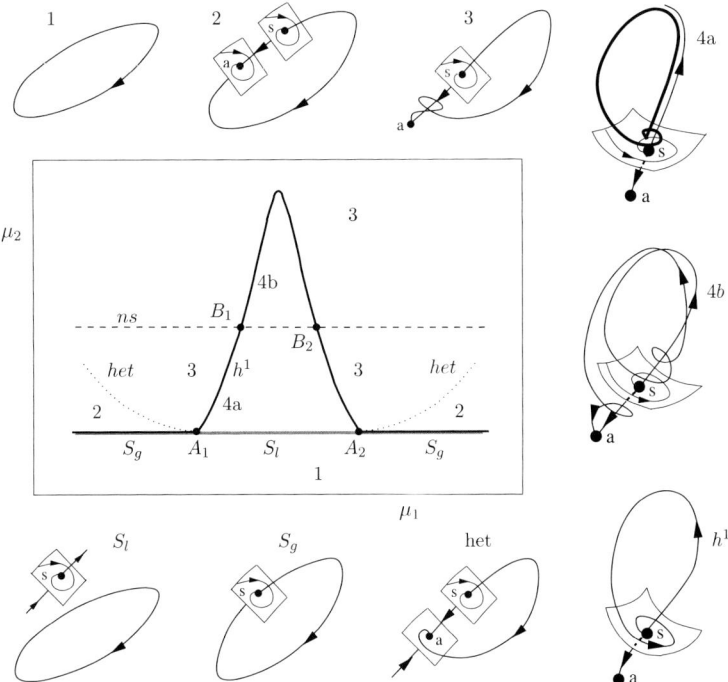

Fig. 6.2. Sketches of phase portraits near the boundary of a homoclinic tooth in two unfolding parameters μ_1 and μ_2. From S. Wieczorek and B. Krauskopf, Bifurcations of n-homoclinic orbits in optically injected lasers, *Nonlinearity* 18(3) (2005) 1095–1120 © 2005 by Institute of Physics Publishing; reprinted with permission.

6.1.2 Homoclinic Teeth

The phenomena we are interested in appear in what we call 'homoclinic teeth'. What we mean by a homoclinic tooth is sketched in Fig. 6.2. It is the region bounded by the curve h^1 of a 1-homoclinic bifurcation and the grey part of the curve S_l of local saddle-node bifurcations. The two curves meet at two points A_1 and A_2 of codimension-two non-central saddle-node homoclinic bifurcations. This codimension-two bifurcation was identified in [31] as an organizing center for multi-pulse excitability (single pulse excitability in this case of a curve of 1-homoclinic bifurcations); see [6, 14, 15] for more details on its unfolding. Figure 6.2 shows sketches of phase portraits for different locations of parameter space near the homoclinic tooth. Notice, in particular, that the saddle-node bifurcation takes place on a periodic orbit along the parts marked S_g (where g stands for global), but this is not the case along S_l (where l stands for local).

In the part of region 3 that is close to the curves h^1 and S_g the laser is 1-excitable: a small perturbation to above the stable manifold of the saddle s will lead to a large excursion, which follows the one-dimensional unstable manifold of s, before the laser relaxes back to the attractor a. By comparison, in region 2 there is a smooth invariant circle, but the laser is still excitable close to the curve S_g: a perturbation beyond the stable manifold of s will lead to a large excursion around the invariant circle and back to a. In fact, phase portraits 2 and 3 are topologically equivalent. However, away from the codimension-two points A, they may relate to different physical phenomena. On the one hand, phase portrait 2 represents phase locking because the smooth invariant circle is centered at the origin of the complex E-plane. Hence, an excitable response associated with phase portrait 2 is mainly of the form of a 2π phase slip with only slight variations in the electric field intensity. On the other hand, the upper branch of the unstable manifold in phase portrait 3 evolves away from the origin of the complex E-plane so that an excitable response leads to a short (30 ps in our case) and distinct intensity pulse. Near A, the difference in the excitable response for the phase portraits 2 and 3 disappears. The curve het is not a bifurcation curve, but when crossing it there is a change of the direction from which the relevant branch of the unstable manifold of the saddle approaches the attractor. Consequently, the closure of the unstable manifold of the saddle is a smooth curve in region 2, while this is not the case in region 3; see [31] for more details.

The homoclinic tooth is shown to intersect with the dashed curve ns where the saddle is neutral, that is, the absolute values of the real parts of the real eigenvalue and of the pair of complex conjugate eigenvalues are equal. What the dynamics looks like inside the tooth crucially depends on whether one is above or below ns. Along the parts of h^1 below ns, often called a simple Shil'nikov case, the homoclinic orbit bifurcates into an attracting periodic orbit [Fig. 6.2(4a)]. On the other hand, along the parts of h^1 above ns, often called a chaotic Shil'nikov case, the bifurcating periodic orbit is no longer stable [Fig. 6.2 (4b)]. Breaking this type of homoclinic orbit leads to the creation of n-homoclinic orbits for any n. While the curve ns is not a bifurcation curve, each of its intersection points B_1 and B_2 with h^1 is a codimension-two homoclinic bifurcation, known as a Belyakov point [7, 24]. Belyakov points mark the transition between the two cases of homoclinic orbits and, hence, give rise to an intricate structure of n-homoclinic orbits.

How homoclinic teeth arise in (6.1) is shown in Fig. 6.3 with panels of the (K, ω)-plane of (6.1) near the locking region for increasing values of α as indicated. It shows the curves S of saddle-node bifurcations and the curves H of Hopf bifurcations (both gray), the supercritical parts of which bound the locking region of the injected laser; see [59]. Also shown is the neutral saddle curve ns. All these curves are given by local conditions at equilibria of (6.1) and can be found analytically. The curve h^1 of 1-homoclinic bifurcations, on the other hand, cannot be found analytically. It was computed with the AUTO/HOMCONT. The computations do not distinguish between a generic (codimension-zero) homoclinic connection along the parts S_g in Fig. 6.2 and the codimension-one homoclinic bifurcation along h^1. In other words, when the bold black curve in Fig. 6.3 coincides with S then the saddle-node bifurcation takes place on a periodic orbit. If it leaves S we find a homoclinic tooth.

For $\alpha = 0$ (the case of a solid-state or CO_2 laser) the (K, ω)-plane is symmetric and there are no homoclinic teeth. As α is increased, homoclinic teeth start to grow along the saddle-node bifurcation curve S that forms the lower boundary of the locking range. (The other boundary is the Hopf bifurcation curve H.) Initially the teeth are quite small [panels (b)–(c)] but then they grow in size with α and the bifurcation diagram changes qualitatively, showing the existence of codimension-three phenomena. At $\alpha = 1.21$ [panel (c)] the first tooth starts to intersect the neutral saddle curve ns. What is more, new teeth start to appear between already present teeth [panel (e)]. All teeth keep growing, and the tooth closest to the saddle-node Hopf point G_1 develops a rather bizarre shape [panels (f)–(i)]. On top of this, when α increases neighboring teeth may merge, meaning that the curve h^1 detaches from the curve S. This occurs at codimension-three points when two neighboring non-central saddle-node homoclinic bifurcation points come together and vanish. Furthermore, one notices the appearance of codimension-two homoclinic bifurcation points (dots along the curve h^1 in panels (h) and (i)). They are created when the section given by fixed α crosses a minimum in the respective codimension-two bifurcation curve, which is discussed in detail in Sect. 6.1.5.

To study how new teeth are born and neighboring teeth merge we continued with HomCont the curve of codimension-two non-central saddle-node homoclinic bifurcations in (K, ω, α)-space [6, 45]. The projection of this curve onto the (α, ω)-plane is shown in Fig. 6.4(a), while Fig. 6.4(b) shows a sketch of a non-central saddle-node homoclinic orbit. The left-hand fold points of the curve in Fig. 6.4(a) are points where teeth are born, while right-hand fold points are points where two neighboring teeth merge. This figure clearly shows that there are no teeth for $\alpha < 0.5$. New teeth are then born one-by-one as α is increased. Secondary teeth appear from about $\alpha = 2$ on. Merging teeth can be observed from about $\alpha = 2.2$ onward when the first two teeth (nearest G_1) merge. Successively teeth for larger negative detuning ω also merge. In fact for $\alpha > 7.5$ there appears to be one giant tooth, if one still wants to call it that. It is already clear that the situation becomes increasingly complicated with α.

6.1.3 Complex Structure of n-Homoclinic Bifurcations

Complex structures of global homoclinic and heteroclinic bifurcations arise inside the homoclinic teeth as a result of interactions of the curves of 1-homoclinic orbits. The fact that the curve ns intersects the first homoclinic tooth, for example, for $\alpha = 2.0$ in Fig. 6.3 (d), giving rise to two Belyakov points, already allows us to conclude from general theory [7, 24] that there must be further curves of n-homoclinic orbits. We remark that the exact combinatorics of these n-homoclinic orbits is still not fully understood [24].

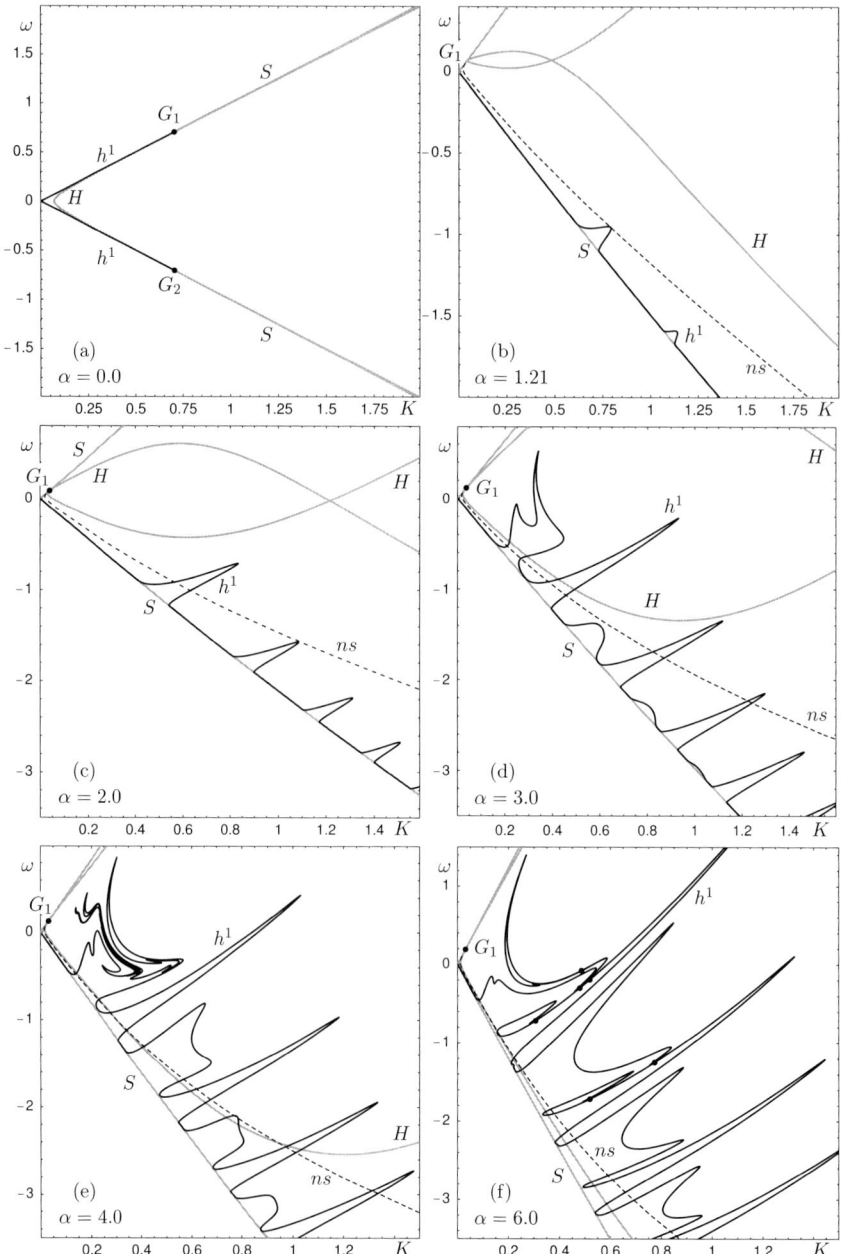

Fig. 6.3. Homoclinic teeth in the locking region as a function of α. From S. Wieczorek and B. Krauskopf, Bifurcations of n-homoclinic orbits in optically injected lasers, *Nonlinearity* 18(3) (2005) 1095–1120 © 2005 by Institute of Physics Publishing; reprinted with permission.

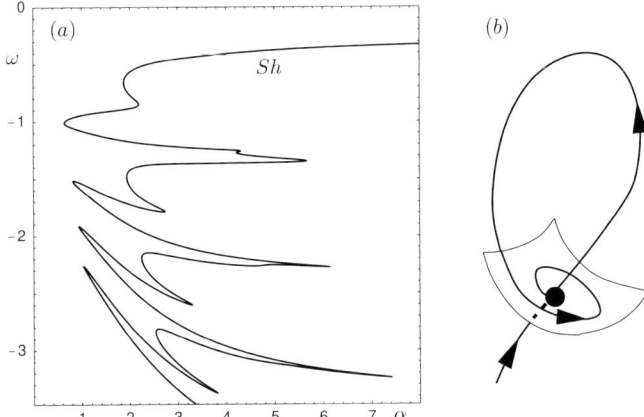

Fig. 6.4. Curve of codimension-two saddle-node homoclinic bifurcations, projected onto the (α, ω)-plane (a), and a sketch of a codimension-two saddle-node homoclinic orbit (b). From S. Wieczorek and B. Krauskopf, Bifurcations of n-homoclinic orbits in optically injected lasers, *Nonlinearity* 18(3) (2005) 1095–1120 © 2005 by Institute of Physics Publishing; reprinted with permission.

The question is how these n-homoclinic orbits are organized inside the homoclinic teeth. At the same time, we obtain an impression of a Belyakov bifurcation in a concrete system. Furthermore, one may ask where the associated n-homoclinic bifurcation curves go and to which other codimension-two points they connect. In short: what is the bifurcation diagram, as far as one can assemble it? These questions cannot be addressed by analytical studies in a neighborhood near codimension-two points but they require the use of continuation techniques. From a bifurcation theory point of view, this is the next step towards the understanding of the organizing properties of global bifurcations. Physically, we reveal structures that stretch over large regions in the parameter plane and become experimentally accessible, that is, potentially relevant for real applications of optically injected lasers.

Figure 6.5 (a1) shows curves h^n of n-homoclinic orbits for $n \leq 4$ inside the first tooth for $\alpha = 2.0$, while Fig. 6.5 (a2) is an enlargement near the saddle-node bifurcation curve S. Many of these curves extend from the region above ns to below ns and in crossing ns have further Belyakov points on them. The picture that emerges is that of a complicated arrangement of nested n-homoclinic bifurcation curves. Most interestingly, several curves extend to very near the curve S, and some even attach to S at points of non-central saddle-node n-homoclinic orbits.

We now focus on what happens to the infinite number of h^n-tongues when the Belyakov points are gone, that is, the homoclinic tooth is entirely below the curve ns. One straightforward scenario would be that all the h^n curves disappear when B_1 and B_2 merge. However, this is not the case here. Fig-

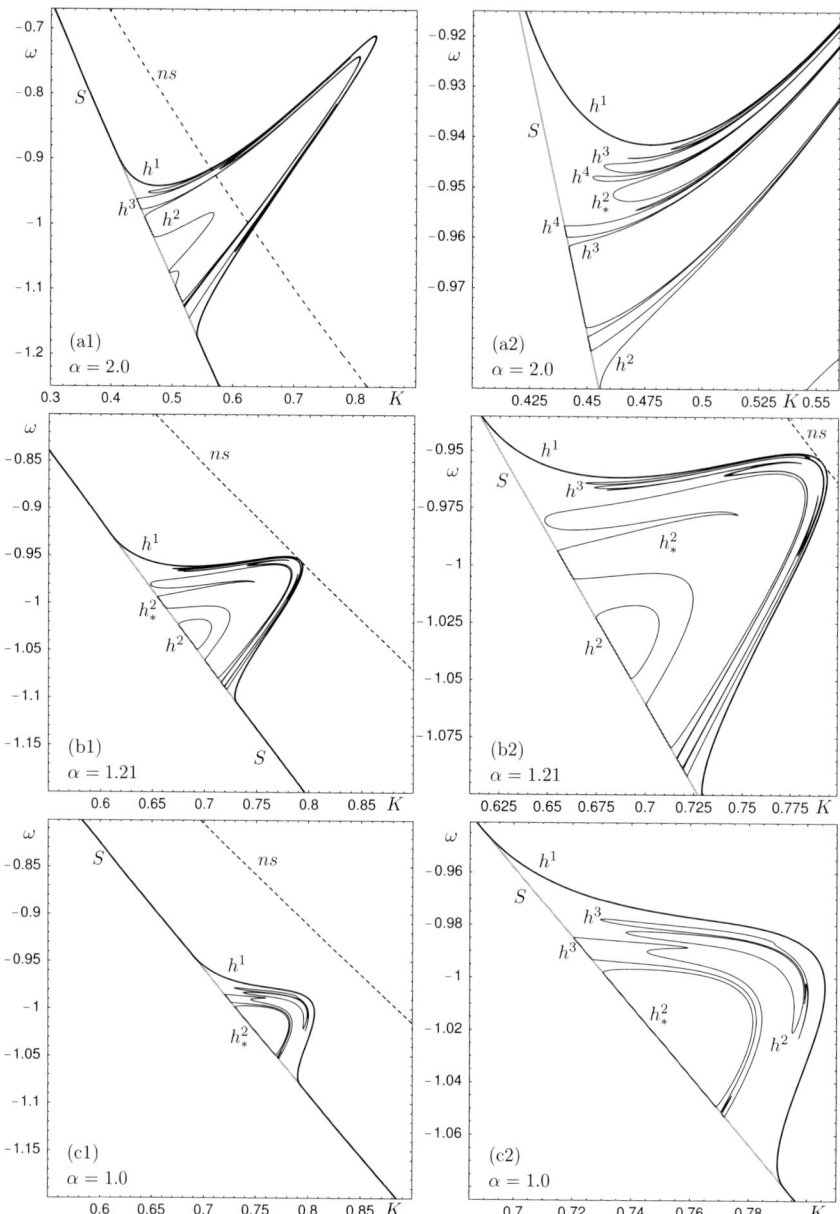

Fig. 6.5. Bifurcation curves of n-homoclinic curves inside a homoclinic tooth for three different values of α as indicated in the panels. From S. Wieczorek and B. Krauskopf, Bifurcations of n-homoclinic orbits in optically injected lasers, *Nonlinearity* 18(3) (2005) 1095–1120 © 2005 by Institute of Physics Publishing; reprinted with permission.

ure 6.5 (b1)–(b2) shows the first tooth for $\alpha = 1.21$, just as it touches the curve ns. This is a codimension-three phenomenon in (K, ω, α)-space where two Belyakov points coincide and then disappear when α is decreased, as is shown in Fig. 6.5 (c1)–(c2). (The curve B of Belyakov points in (K, ω, α)-space has a minimum.) Even though the tooth is well below the curve ns for $\alpha = 1.0$, there are still curves of n-homoclinic orbits inside it. In particular, we find that the curves h^2 and h^3 are attached to S.

Our numerical investigation suggests that there are only finitely many curves of n-homoclinic orbits for $\alpha < 1.21$. To illustrate how subsequent curves h^n appear with increasing α we marked one of them with a star. For $\alpha = 1.0$ [Fig. 6.5 (c2)] h^2_* is the last homoclinic curve that just emerged from the saddle-node bifurcation curve S. As α is increased above $\alpha = 1.0$, the curve h^2_* develops two extra noncentral-homoclinic points on S, forms a sort of bridge, and provides space for the next homoclinic curve to emerge [Fig. 6.5 (b2)]. This process seems to repeat, such that for $\alpha > 1.21$ there exist infinitely many curves h^n.

6.1.4 Multi-Pulse Excitability

The regions bounded by h^2 and h^3 near S appear to be large enough to be experimentally accessible [60]. In such a region the laser exhibits multi-pulse excitability. We remark that our study shows that this phenomenon can be found even for surprisingly low values of α; see also [31, 62]. An example is shown in Fig. 6.6 for $\alpha = 1.0$. The phase portrait in Fig. 6.6 (a1) is as that of region 3 in Fig. 6.2 — the laser is 1-pulse excitable. A small perturbation above the excitability threshold, given by the stable manifold of the saddle point, results in the laser sending out a single pulse; see Fig. 6.6 (a2). In the region bounded by h^2, on the other hand, the phase portrait is close to a 2-homoclinic orbit and the laser produces two pulses in reaction to a single perturbation; see Fig. 6.6 (b1)–(b2). Finally, three pulses result in the region bounded by the curve h^3, as is illustrated in Fig. 6.6 (c1)–(c2). Indeed, it is possible to find n-pulse excitability for any n, but the regions for $n > 4$ become impractically small.

It is important to note a key ingredient for multi-pulse excitability to occur, namely the fact that the respective curve h^n extends all the way below ns. For the parameters above ns the h^n-tongues are so narrow that they become hard to distinguish, even numerically. Furthermore, there exist an infinite number of unstable periodic orbits in the phase space for parameters outside the tongues. As a result, the excitable response is often irregular and unpredictable as the trajectory bounces between the unstable orbits before it decides to return to the stable equilibrium. On the other hand, below ns the tongues are easily distinguishable and the phase portraits are simpler as there are no unstable periodic orbits. Consequently, the system can be prepared to be well within h^n (certainly for $n \leq 3$) where the excitable response is predictable and consist of a certain number of pulses.

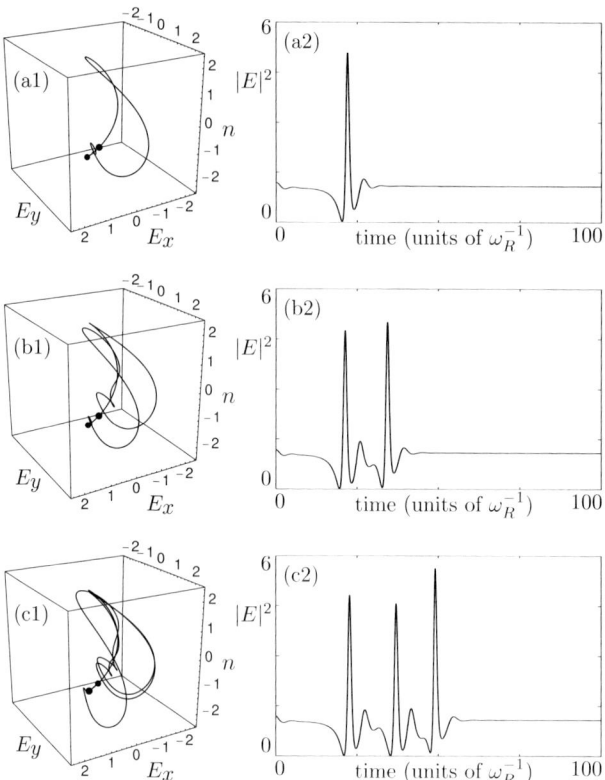

Fig. 6.6. Examples of multi-pulse excitability. The left column shows the phase portrait and the right column the reaction of the laser to a small perturbation above the excitability thresholds. Throughout $\alpha = 1.0$ and from (a) to (c) (K,ω) takes the values $(0.71, -0.95)$, $(0.745. - 1)$ and $(0.735, -0.993)$. From S. Wieczorek and B. Krauskopf, Bifurcations of n-homoclinic orbits in optically injected lasers, *Nonlinearity* 18(3) (2005) 1095–1120 © 2005 by Institute of Physics Publishing; reprinted with permission.

6.1.5 Codimension-Two Homoclinic Bifurcations

We now study in considerable detail the structure and bifurcations associated with the curve h^1 that forms the boundary of the homoclinic teeth. In particular, we show that codimension-two double-homoclinic and T-point bifurcations play a prominent role in organizing the dynamics.

Figure 6.7 shows an enlargement near the first homoclinic tooth (or what is left of it) for $\alpha = 4.5$; compare with Fig. 6.3. Notice the two points D_1 and D_2 where additional homoclinic bifurcation curves emerge. The phase

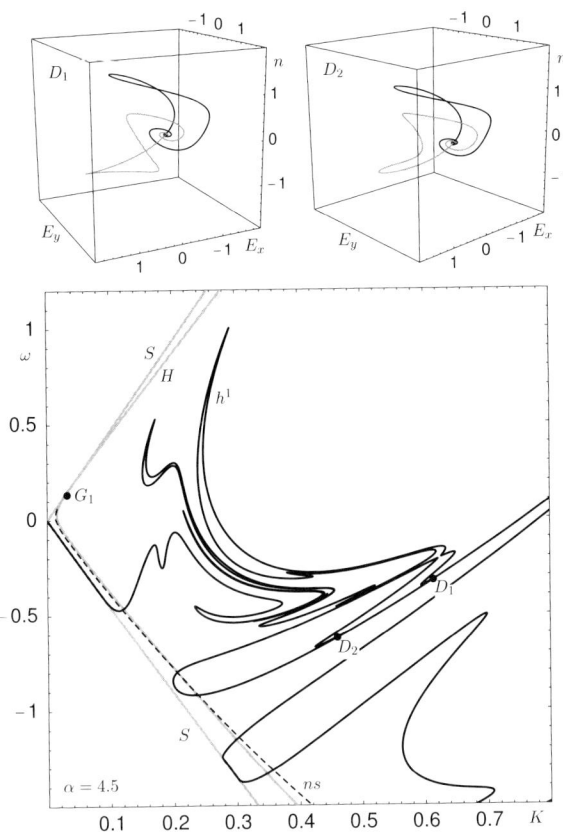

Fig. 6.7. The (K, ω)-plane for $\alpha = 4.5$ near the point G_1 (compare with Fig. 6.3) and the phase portraits at the codimension-two double-homoclinic bifurcation points D_1 and D_2. From S. Wieczorek and B. Krauskopf, Bifurcations of n-homoclinic orbits in optically injected lasers, *Nonlinearity* 18(3) (2005) 1095–1120 © 2005 by Institute of Physics Publishing; reprinted with permission.

portrait at D_1 and D_2 show that we are dealing with a codimension-two *double-homoclinic orbit* [9, 40]: both branches of the unstable manifold spiral back to the saddle point. This means that there are simultaneously two individual homoclinic orbits associated with the same saddle point.

The two phase portraits at D_1 and D_2 are topologically equivalent and both lie on the primary branch of the curve h^1. This can be seen in the further enlargement of the (K, ω)-plane in Fig. 6.8, where panels (a)-(e) show the 1-homoclinic orbit in phase space at the indicated parameter points along h^1. As D_2 is approached the unstable manifold forming a homoclinic orbit comes closer and closer (from below) to the saddle and then leaves a neighborhood of the saddle roughly along the other branch of the unstable manifold. Finally, at D_2 there are two simultaneous homoclinic orbits, one for each branch of the

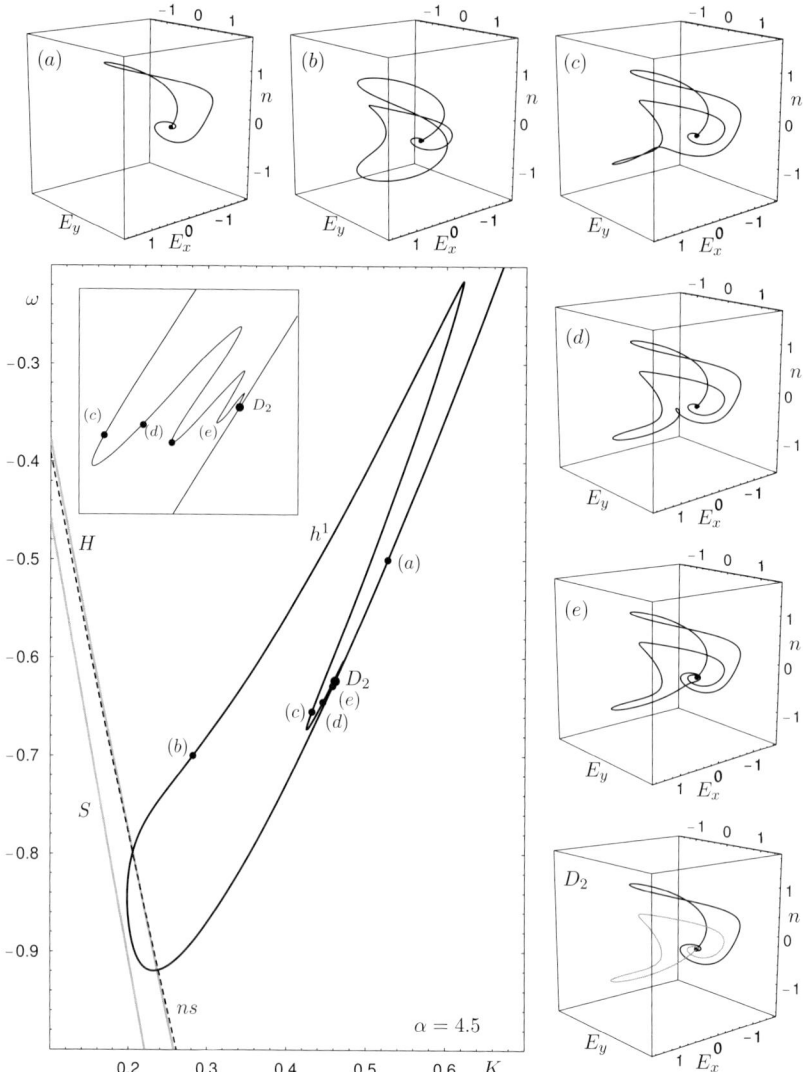

Fig. 6.8. The (K, ω)-plane for $\alpha = 4.5$ near the point D_2 (compare with Fig. 6.7) and the phase portraits along the homoclinic curve h^1 as D_2 is approached. From S. Wieczorek and B. Krauskopf, Bifurcations of n-homoclinic orbits in optically injected lasers, *Nonlinearity* 18(3) (2005) 1095–1120 © 2005 by Institute of Physics Publishing; reprinted with permission.

unstable manifold. Effectively, the original 1-homoclinic orbit along the curves h^1 has split into two homoclinic orbits. Notice that the curve h^1 accumulates back on itself at D_2, as is also sketched in the inset of Fig. 6.8.

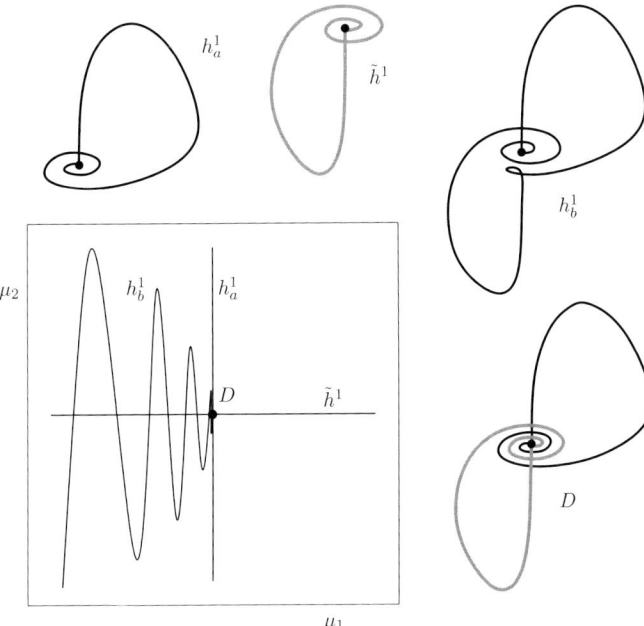

Fig. 6.9. Sketch of phase portraits and the bifurcation diagram (in two unfolding parameters μ_1 and μ_2) of a double-homoclinic point D, as described in [9, 40]. From S. Wieczorek and B. Krauskopf, Bifurcations of n-homoclinic orbits in optically injected lasers, *Nonlinearity* 18(3) (2005) 1095–1120 © 2005 by Institute of Physics Publishing; reprinted with permission.

This scenario agrees with what is known in the literature about the double-homoclinic bifurcation [9, 40]. Again, not all details of this codimension-three global bifurcation are known, but key features are as sketched in Fig. 6.9 (for the case of a saddle focus as we encounter it here). The double-homoclinic orbit D exists at the intersection point of two curves h_a^1 and $\tilde{h}^1$ of two different homoclinic orbits to the same saddle that contain each a different branch of the unstable manifold of the saddle. As sketched, there is a third curve h_b^1 of homoclinic orbits that accumulates on the curve h_a^1. The accumulation is as shown when the saddle quantity is larger than one [9, 40], which is the case we encounter, because all double-homoclinic orbits occur above the curve ns.

Note that the analysis in the literature is in terms of a small tubular neighborhood around the double-homoclinic orbits as sketched in panel D. In this neighborhood the curves h_a^1 and h_b^1 are unrelated. However, as can be seen in Fig. 6.8, they may be one and the same curve accumulating back on itself. In fact, we find this to be the typical situation in system (6.1). We finally stress that the points D_i that we encounter here are of codimension-two because the two simultaneous homoclinic orbits are not related by symmetry. Unlike in the case of a codimension-one symmetric double-homoclinic orbit, it

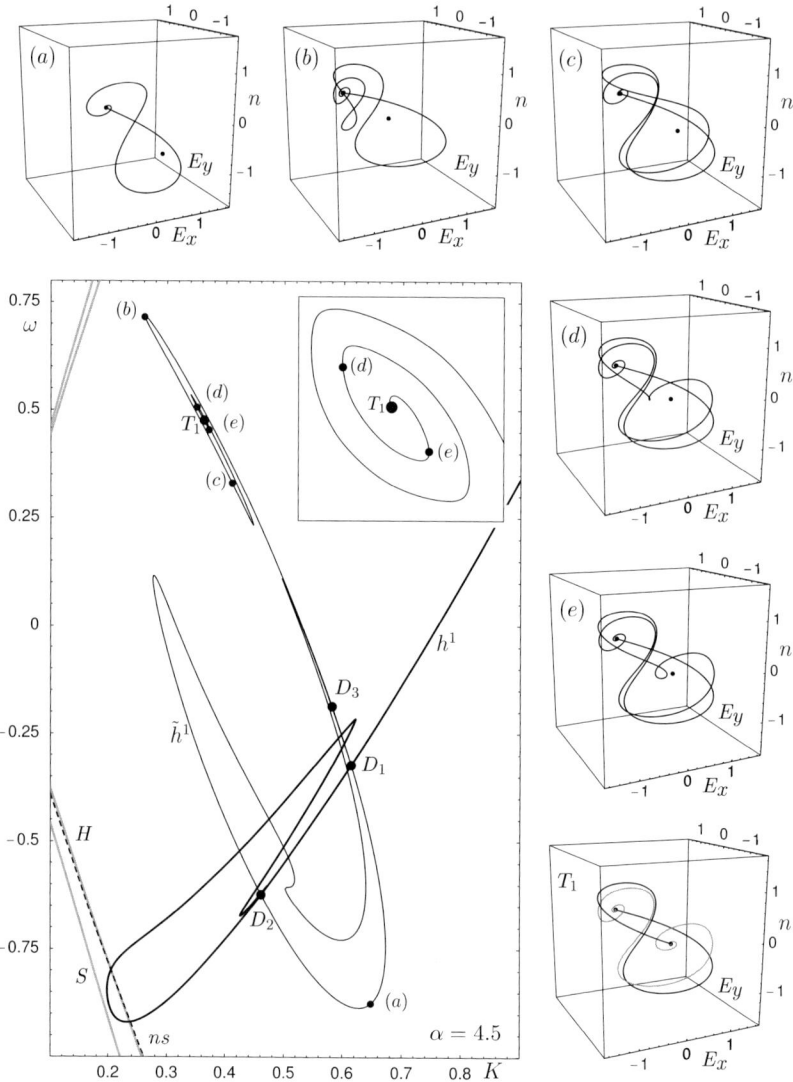

Fig. 6.10. The (K, ω)-plane for $\alpha = 4.5$ near the points D_1, D_2, D_3, and T_1 with phase portraits as T_1 is approached. From S. Wieczorek and B. Krauskopf, *Bifurcations of n-homoclinic orbits in optically injected lasers*, *Nonlinearity* 18(3) (2005) 1095–1120 © 2005 by Institute of Physics Publishing; reprinted with permission.

is possible to perturb parameters such that one of the homoclinic connections is broken and the other persists.

In Fig. 6.7 and Fig. 6.8 we found the double-homoclinic points D_1 and D_2 as the end points of the curve h^1 as it accumulates on itself. However, we know from Fig. 6.9 that there must be a curve $\tilde{h}^1$ of a second homoclinic orbit

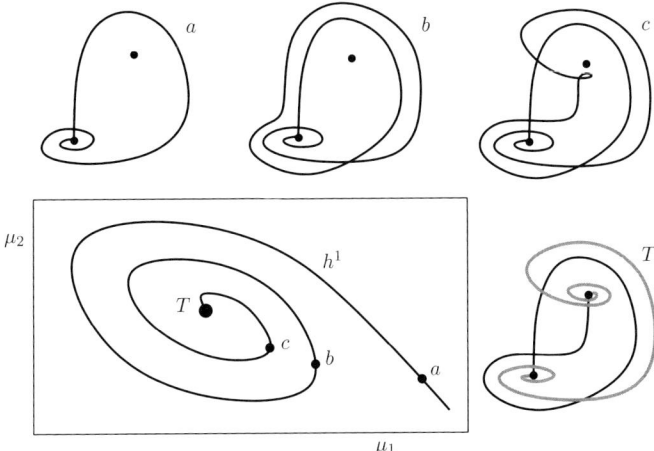

Fig. 6.11. Sketches of the phase portraits along the homoclinic bifurcation curve approaching a T-point in the plane of two unfolding parameters μ_1 and μ_2. From S. Wieczorek and B. Krauskopf, Bifurcations of n-homoclinic orbits in optically injected lasers, *Nonlinearity* 18(3) (2005) 1095–1120 © 2005 by Institute of Physics Publishing; reprinted with permission.

crossing at D_i. In order to find this new homoclinic orbit we split off the new homoclinic orbit from the data of the approximate double-homoclinic orbit at D_i (given as the end point of the curve h^1). We then follow this second codimension-one homoclinic orbit in the (K, ω)-plane.

The result is shown in Fig. 6.10. The point D_2 is indeed the intersection point of two curves of codimension-one homoclinic orbits. The new curve $\tilde{h}^1$ also contains D_1 and has two end points. One end point is a point D_3 of a double-homoclinic orbit, which lies on the curve $\tilde{h}^1$ itself. The other end point is a point denoted by T_1 that is reached in a spiraling fashion, as is also sketched in the inset.

At the point T_1 we encounter a bifurcation that is now generally referred to as a *T-point bifurcation*. This type of codimension-two heteroclinic cycle was studied in a general system, that is, one without any symmetry, in [9, 10] in a tubular neighborhood around the heteroclinic orbits at the T-point. Note that the T-point bifurcation is often associated with vector fields that have the $\mathbb{Z}_2$-symmetry of a rotation by π around an invariant axis. In this case, the heteroclinic cycle involves two saddle-foci, which are each others images under the symmetry, and the origin (more generally, a point in the invariant subspace of the symmetry), which is also a saddle-focus. This $\mathbb{Z}_2$-symmetric T-point bifurcation was initially found and studied in the Lorenz system [22], but also occurs in other systems with rotational symmetry, such as an optically pumped three-level laser [20], an electronic oscillator [19], and a semiconductor laser with phase-conjugate feedback [26]. It was recently also discovered in systems with the $\mathbb{Z}_2$-symmetry of point-reflection [2, 32]; see also Chap. 7.

What we find is a general T-point bifurcation (that is, in a system without symmetry) for the case that both saddles involved are of saddle foci. The approach to the point T is illustrated in Fig. 6.10(a)–(e) with images of the 1-homoclinic orbit in phase space at the indicated parameter points along the curve $\tilde{h}^1$.

The situation (a part of the bifurcation diagram near a T-point) is sketched in Fig. 6.11. As the point T is approached along h^1, the homoclinic orbit approaches a second saddle focus, passing closer and closer by the saddle. At the point T_1 there are two heteroclinic connections: a codimension-two heteroclinic connection (black) where the one-dimensional unstable manifold of the first saddle coincides with the one-dimensional stable manifold of the second saddle, and a generic (codimension-zero) heteroclinic connection (gray), given as the intersection curve of the two-dimensional stable manifold of the first saddle and the two-dimensional stable manifold of the second saddle.

According to general theory [9, 10] there must exist a second spiraling curve of homoclinic connection to the other (lower) saddle, leading to another curve in parameter space that spirals into T_1. Furthermore, it is known that there are many more curves of n-homoclinic bifurcations, which pass close to the saddles an arbitrary number of times. We did not attempt to find all these bifurcation curves, but instead concentrated on the structure of 1-homoclinic bifurcation curves. Nevertheless, the injection laser appears to be a good model in which to study global bifurcations near T-point bifurcations in more detail.

The bifurcation diagram in Fig. 6.10 is still quite incomplete. The curve $\tilde{h}^1$ of homoclinic orbits also accumulates on itself at D_3. So, as we did near the double-homoclinic point D_1, we find and follow the second codimension-one homoclinic that must exist near D_3. This gives the continuation of the curve h^1 shown in Fig. 6.12, which ends at the point D_1. Furthermore, we followed from near T_1 the codimension-one homoclinic orbit of the (upper) saddle point to lower values of α (see already Fig. 6.13), where we discovered a second T-point bifurcation T_2. We then followed this T-point back to $\alpha = 4.5$. As can be seen in Fig. 6.12, the point T_2 is the end point of two spirals. In fact both spirals turn out to belong to one and the same closed curve of codimension-one two-homoclinic orbit $\tilde{h}^2$ as is illustrated by the sketch in the inset.

The bifurcation diagram in Fig. 6.12 is quite intricate: it involves several double-homoclinic and T-point bifurcations. Unraveling it required detailed numerical continuation with HomCont, guided by theoretical knowledge of which homoclinic orbits are possible near the different codimension-two points. We finally remark, that Fig. 6.12 shows a 'skeleton' consisting of curves of 1-homoclinic bifurcations. Indeed the existence of the T-points suggests that there are n-homoclinic orbits for arbitrary n.

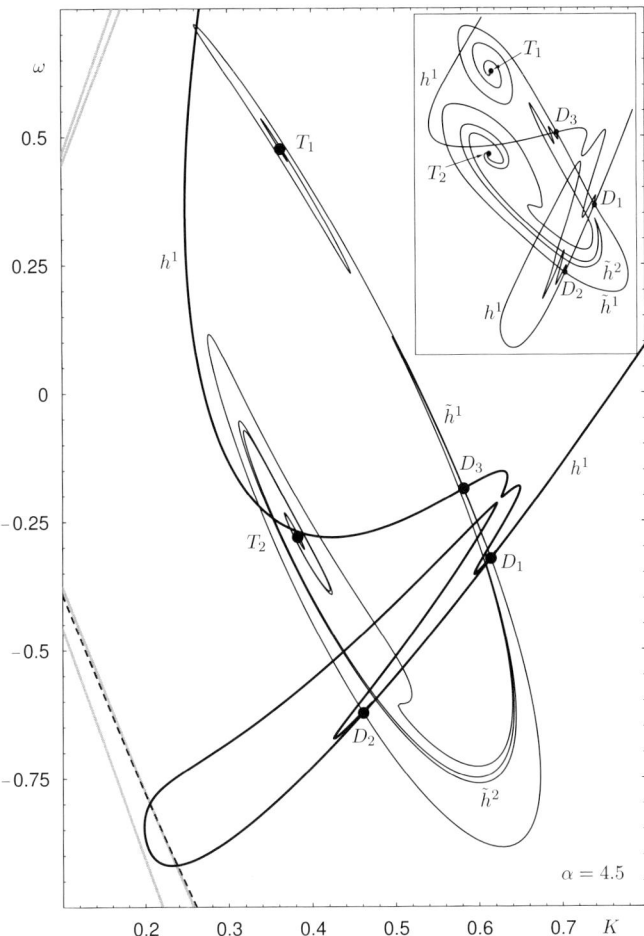

Fig. 6.12. The (K, ω)-plane for $\alpha = 4.5$ near the points T_1 and T_2. From S. Wieczorek and B. Krauskopf, Bifurcations of n-homoclinic orbits in optically injected lasers, *Nonlinearity* 18(3) (2005) 1095–1120 © 2005 by Institute of Physics Publishing; reprinted with permission.

6.1.6 Folds of Codimension-Two Homoclinic Bifurcation Curves

We know from Fig. 6.3 that the complicated structure of codimension-two bifurcations in Fig. 6.12 is not present for smaller values of α. The question arises of how it disappears.

It turns out that an important ingredient in this change of the bifurcation diagram are minima (more generally, a fold) with respect to α of certain curves of codimension-two homoclinic bifurcations in the three-dimensional (K, ω, α)-space. This phenomenon is of codimension three, where one codimension is due to the fold with respect to α. The other two codimensions

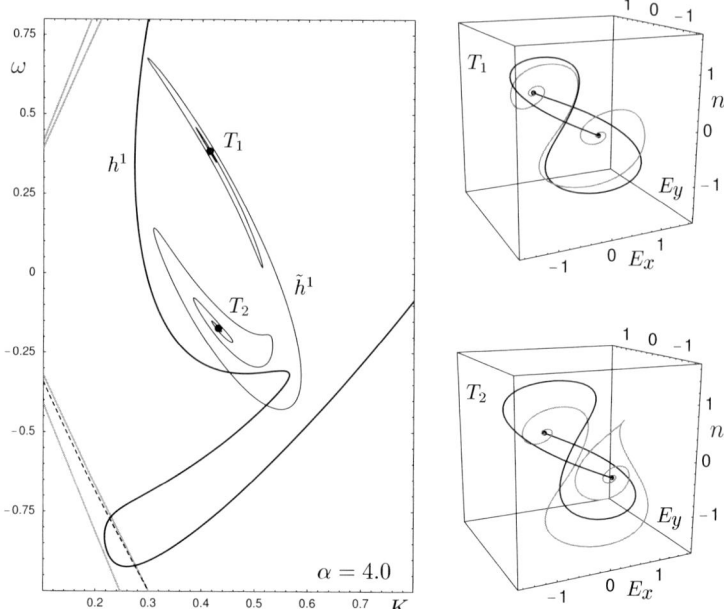

Fig. 6.13. The (K, ω)-plane for $\alpha = 4.0$ near the points T_1 and T_2 and the respective phase portraits at T_1 and T_2. From S. Wieczorek and B. Krauskopf, Bifurcations of n-homoclinic orbits in optically injected lasers, *Nonlinearity* 18(3) (2005) 1095–1120 © 2005 by Institute of Physics Publishing; reprinted with permission.

are due to the special object in phase space, in this case a codimension-two homoclinic bifurcation. One might speak of a codimension-two-plus-one event to distinguish it from codimension-three bifurcations, where all codimensions are due to a codimension-three object in phase space.

We already encountered this phenomenon in the creation and disappearance of points of codimension-two saddle-node homoclinic bifurcation (see the folds with respect to α in Fig. 6.4) and in the creation, with increasing α, of Belyakov points in the tangency between the curves ns and h^1 (see Fig. 6.5). In this section we consider two other examples, namely a fold of a curve of T-point bifurcations and a fold of a curve of double-homoclinic bifurcations. As we will see now, in both these examples the fold of the codimension-two curve is accumulated by singularities in associated surfaces of codimension-one global bifurcations.

We first consider the case of T-point bifurcations. Figure 6.14 shows what happens to the points T_1 and T_2 of T-point bifurcations as α is decreased. After the disappearance of the point D_3, the points T_1 and T_2 move closer and closer to each other. There are a number of codimension-three events where the spiral around T_1 touches that around T_2. Each such event leads to a new closed curve surrounding both T_1 and T_2 and the curve of homoclinic orbits connecting T_1

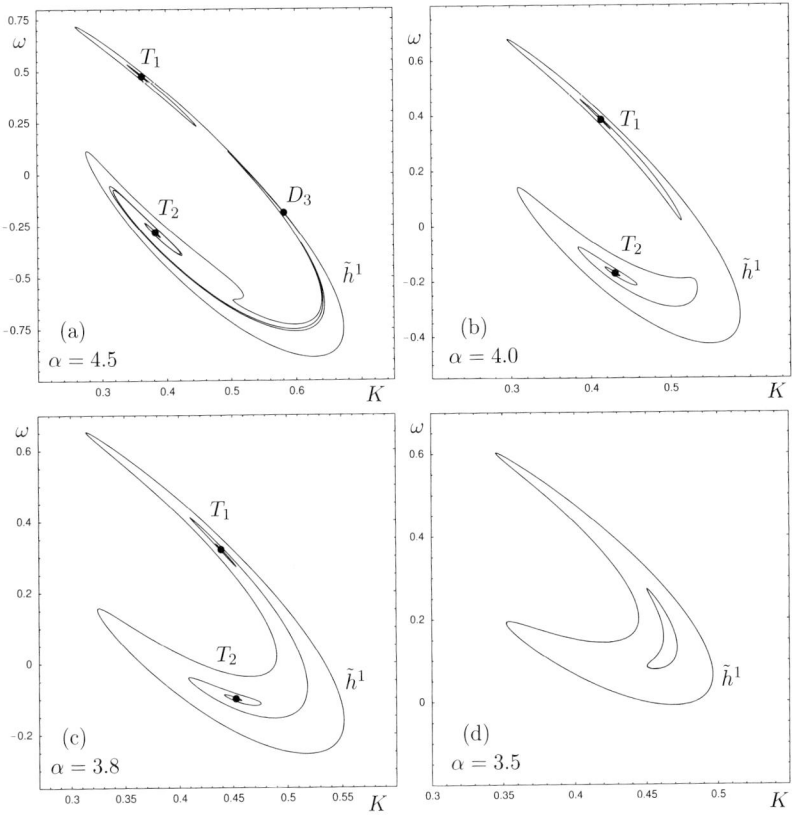

Fig. 6.14. Dependence on α of the (K,ω)-plane near the points T_1 and T_1. From S. Wieczorek and B. Krauskopf, Bifurcations of n-homoclinic orbits in optically injected lasers, *Nonlinearity* 18(3) (2005) 1095–1120 © 2005 by Institute of Physics Publishing; reprinted with permission.

and T_2, as in Fig. 6.14 (c). This process continues until the points T_1 and T_2 finally coincide, leaving behind a number of closed concentric curves of homoclinic orbits, as in Fig. 6.14 (d). These closed curves then disappear one by one as α is decreased further. (We remark that this phenomenon has been found independently in [2] in the $\mathbb{Z}_2$-symmetric Chua's circuit with a cubic nonlinearity; see also Chap. 7.) Finding this transition numerically was quite a challenge because the curves involved are no longer connected. We succeeded by starting from suitable points and continuing the respective homoclinic orbit in α.

The individual changes in the structure of the curve $\tilde{h}^1$ are due to a classical singularity, namely the passage through an α-degenerate point. At such a point, the tangent space to the $\tilde{h}^1$ surface in (K,ω,α)-space does not have an α-component (the derivative with respect to α is zero). There are two

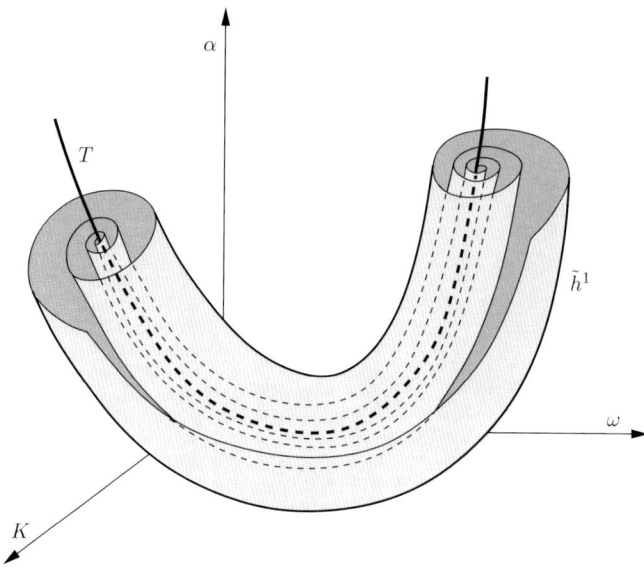

Fig. 6.15. In the (K, ω, α)-space the curve of T-point bifurcation is surrounded by a surface of homoclinic bifurcation $\tilde{h}^1$ that spirals onto the T-curve; compare with Fig. 6.14. From S. Wieczorek and B. Krauskopf, Bifurcations of n-homoclinic orbits in optically injected lasers, *Nonlinearity* 18(3) (2005) 1095–1120 © 2005 by Institute of Physics Publishing; reprinted with permission.

cases depending on the index of the α-degenerate point, namely the transition through a saddle and the transition through an extremum. Note that these singularities are also called the simple bifurcation and the isola bifurcation; see, for example, [23] for details.

This explanation in terms of singularity theory is a consequence of the geometry of bifurcation surfaces and curves in (K, ω, α)-space. In fact, the whole sequence of events of T_1 and T_2 coming together and disappearing can be nicely explained with the sketch in Fig. 6.15 of how the surface $\tilde{h}^1$ of homoclinic bifurcations spirals around the curve T of T-point bifurcations. The curve of T-point bifurcations is a smooth curve with a minimum with respect to α, and it is surrounded by a surface of codimension-one homoclinic bifurcations that spirals towards this curve. The panels in Fig. 6.14 are two-dimensional cross sections for fixed α through this surface. If α is large enough, the curve T is intersected in two points T_1 and T_2 and the spiraling near these two points must be clockwise and counter-clockwise, respectively. The intersection of the surface with the section is a single curve for sufficiently large α. However, nearer the minimum of the curve T the surface has α-degenerate points where its tangent space does not have an α-component. Passing through each such point constitutes a basic codimension-one singularity of the surface of homoclinic bifurcations. More precisely, above the minimum of the curve T

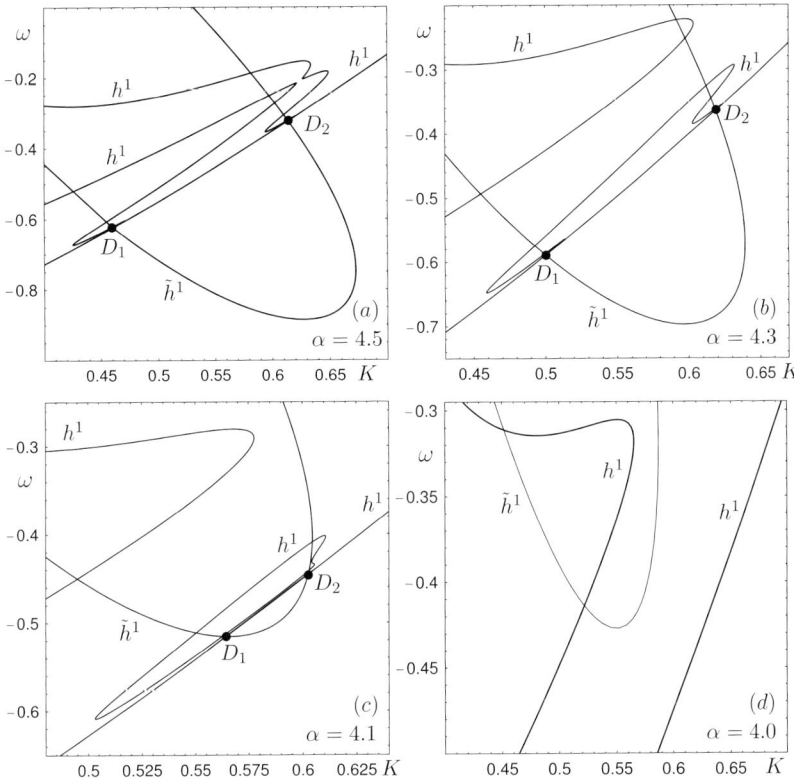

Fig. 6.16. The (K, ω)-plane near D_1 and D_2 for varying α. Points D_1 and D_2 disappear via codimension-three resonant double-homoclinic bifurcation as α is decreased. From S. Wieczorek and B. Krauskopf, Bifurcations of n-homoclinic orbits in optically injected lasers, *Nonlinearity* 18(3) (2005) 1095–1120 © 2005 by Institute of Physics Publishing; reprinted with permission.

there are infinitely many passages through saddles, which accumulate on the minimum of the curve T. Globally, this creates the closed concentric curves by connecting the respective homoclinic curves in a different way. Below the minimum of T, on the other hand, each concentric circle disappears by contracting to a single point, which is the passage through an extremum (with respect to a parameter, in this case α) in a two-dimensional surface $\tilde{h}^1$. We finally remark that it would be quite a challenge to produce a numerical picture of the surface sketched in Fig. 6.15.

Our second example is the merging and disappearance of the points D_1 and D_2 as α is decreased from $\alpha = 4.5$ to $\alpha = 4.0$. Figure 6.16 shows four numerical bifurcation diagrams in this transition. As the points D_1 and D_2 are moving closer together we again encounter a passage through a saddle point. This happens between panels (a) and (b) of Fig. 6.16 and it leads to a

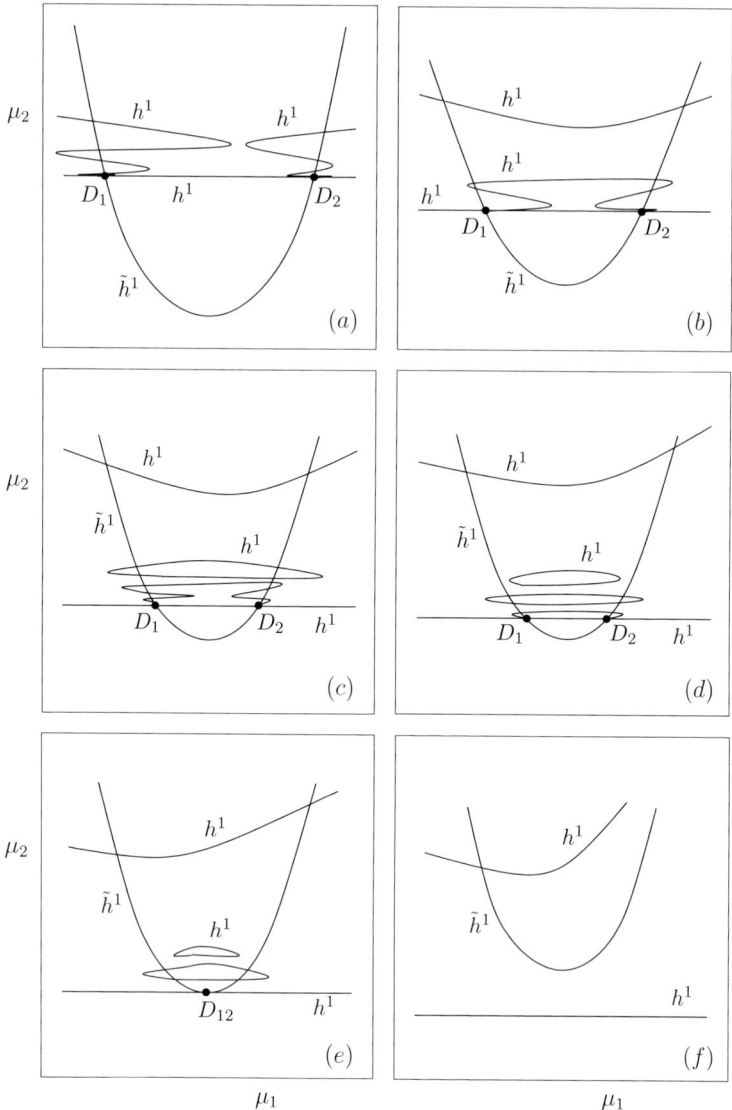

Fig. 6.17. The sketch of how D_1 and D_2 disappear via codimension-three resonant double-homoclinic bifurcation in the two unfolding parameters μ_1 and μ_2. Compare with Fig. 6.16. From S. Wieczorek and B. Krauskopf, Bifurcations of n-homoclinic orbits in optically injected lasers, *Nonlinearity* 18(3) (2005) 1095–1120 © 2005 by Institute of Physics Publishing; reprinted with permission.

change in how the curves h^1 in the cross section in the (K, ω)-plane connect. After this event, the curve h^1 in Fig. 6.16(b) connects the two points D_1 and D_2. In a further passage through a saddle point the curve h^1 pinches off to create an isola, which is the situation shown in Fig. 6.16(c). In fact, the isola is very close to the new connection between D_1 and D_2. Numerical continuation suggests that more and more isolas are formed as D_1 and D_2 come closer together. These isolas then disappear in passages through minima. Furthermore, α passes through the minimum of the D curve in the (K, ω, α)-space. As a result, the curves $\tilde{h}^1$ and h^1 in Fig. 6.16(d) no longer intersect and the points D_1 and D_2 have disappeared.

To clarify the situation, we sketch the transition leading to the disappearance of D_1 and D_2 in Fig. 6.17. It can again be understood by the geometry of bifurcation surfaces in (K, ω, α)-space, which in this case are organized around a minimum (with respect to α) of the curve D of double-homoclinic bifurcations. Figure 6.17(a)–(c) and (f) are topologically as the numerical bifurcation diagrams in Fig. 6.16(a)–(d), respectively. We remark that it becomes more and more difficult to resolve numerically the different, small and disjoint intersection curves of the surface h^1 in (K, ω, α)-space. The sketches in Fig. 6.17(d) and (e) are based on our numerical investigations, and indicate how the transition appears to take place. However, the exact details, in particular, the order in which isolas are created and shrink to points and disappear is yet unknown. Our continuation study suggests the basic ingredients of this transition and can reveal some of the first steps in the specific transition at hand. This scenario agrees with what is known about the (local) codimension-two bifurcation diagrams near a double-homoclinic bifurcation as sketched in Fig. 6.9, but a complete study of this codimension-two-plus-one phenomenon remains a challenge.

The fact that we encounter minima in curves T and D confirms the experience from simulations and experiments that the dynamics and the bifurcation diagram of the injected laser become more complicated as the line-width enhancement factor α is increased [59]. Indeed, when α is increased past these minima extra organizing centers, T-points or double homoclinic bifurcation points, are born. These events are associated with infinitely many transitions through saddles and extrema in surfaces of global bifurcations. Furthermore, general theory shows that the emerging T-points or double homoclinic bifurcation points are organizing centers that give rise to n-homoclinic orbits for any n.

6.1.7 A Self-Similar Cascade Phenomenon

As a final example of the increase in the complexity with α we show in Fig. 6.18 the bifurcation diagram in the (K, ω)-plane for $\alpha = 6.0$ near the point G_1. Notice that we only show the different parts of the curve h^1 of one-homoclinic orbits, which form what is left from the left most homoclinic tooth near G_1; compare with Fig. 6.3(f). Near the points D_1 and D_2, that were already found for $\alpha = 4.5$, we find two new points D_4 and D_5; compare with Fig. 6.7. The different bifurcation curves are very close together, and the inset shows a topological sketch of the bifurcation diagram. Notice further that two extra

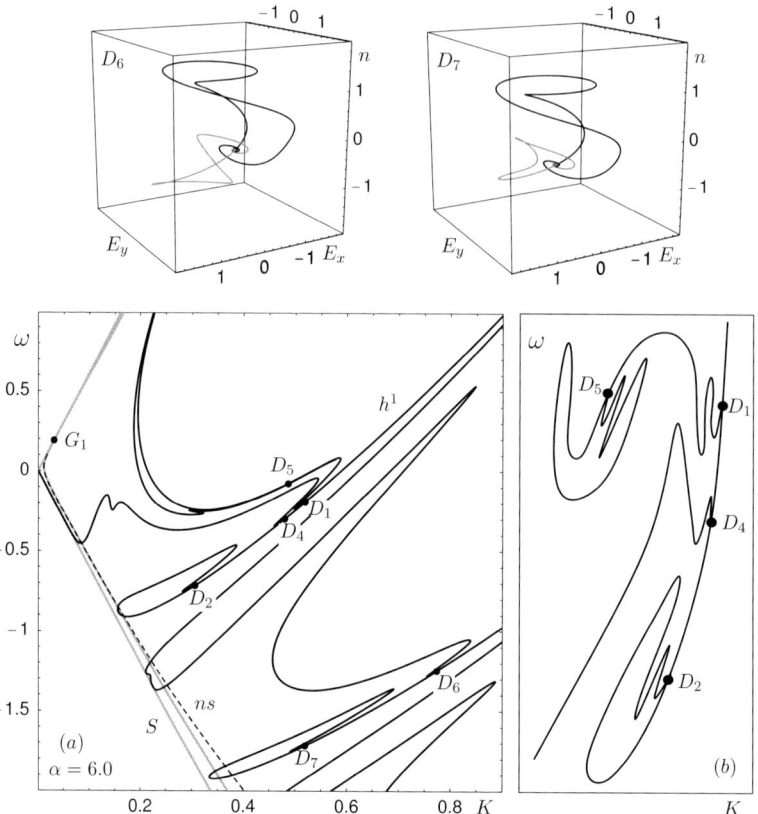

Fig. 6.18. The (K,ω)-plane for $\alpha = 6.0$ near the point G_1 (compare with Fig. 6.3) and the phase portraits at the codimension-two points D_6 and D_6. From S. Wieczorek and B. Krauskopf, Bifurcations of n-homoclinic orbits in optically injected lasers, *Nonlinearity* 18(3) (2005) 1095–1120 © 2005 by Institute of Physics Publishing; reprinted with permission.

double-homoclinic bifurcation points D_6 and D_7 have just been created in the same way as D_1 and D_2 previously. This is another example of the passage through a minimum of a curve D of double-homoclinic orbits; compare with Fig. 6.7.

Figure 6.18 shows that in the injected laser we are dealing with a type of cascade phenomenon: complicated bifurcation scenarios found for one tooth also occur for all the other teeth when α is increased.

6.1.8 Concluding Remarks on Injected Lasers

This section presented a detailed study of the bifurcations of n-homoclinic orbits in the rate equations describing a semiconductor laser with optical injection. The corresponding curves of n-homoclinic bifurcations are organized in what we call homoclinic teeth, experimentally accessible regions inside the locking region of the laser. The analysis of the bifurcation diagram from a global viewpoint proved to provide new insight into the nature of global bifurcations and allowed us to identify a cascade phenomenon where complicated bifurcation scenarios repeat for subsequent homoclinic teeth.

The injection laser rate equations emerged as a concrete vector field in which complicated global bifurcations can be found and studied. Specifically, we found in this three-dimensional vector field (without any additional symmetries) T-point bifurcations and double-homoclinic orbits. By making extensive use of continuation techniques for homoclinic and heteroclinic orbits, it is possible to study these codimension-two global bifurcations themselves, and also to find out how they organize the corresponding bifurcation diagrams.

When changing a third parameter, we found a new phenomenon, namely complicated transitions in two-parameter bifurcation diagrams that are due to folds (in this case, minima) in codimension-two curves of global bifurcations. These 'codimension-two-plus-one events' come with accumulations of singularity transitions through saddles and extrema, which can be explained by the geometry of surfaces of global bifurcations in a three-dimensional parameters space. Our results raise a number of questions of bifurcation theory. In particular, a detailed study of the unfoldings of the 'codimension-two-plus-one events' remains a challenging task.

From the physical point of view, we presented here how the regions in which one may find multi-pulse excitability depend on the linewidth-enhancement factor α. Our results confirm that the dynamics and bifurcations of an injected laser are more complex the larger the linewidth enhancement factor α. Most importantly, they stimulated new laser experiments. The already demonstrated good agreement between theory and experiment on the level of local bifurcations in the injection laser [65, 66] has now been extended to global bifurcations such as the ones described here. In fact, the predicted effect of multi-pulse excitability was recently measured independently by two different groups [8, 25].

6.2 Phase-Locking Anomaly in Two Back-to-Back Coupled Lasers

This section is based on [53, 54] and concerned with the dynamics and bifurcations of two coupled lasers. We model the system with spatial composite-cavity modes that describe the entire coupled-laser structure [13, 43]. This is in contrast to the more usual approach of modeling the individual uncoupled lasers and then introducing the coupling via ad-hoc terms in the equations of motion.

It is generally believed that the transition from weakly-coupled to totally isolated lasers occurs smoothly. This means that, as the coupling approaches

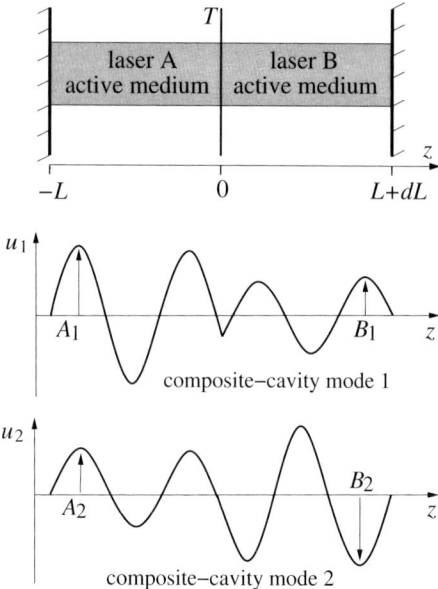

Fig. 6.19. Two back-to-back coupled lasers (top), and a pair of composite-cavity modes (bottom).

zero, there is an uninterrupted vanishing of the detuning range over which phase-locking is achieved. We show here that there is an anomaly in the transition from coupled to totally isolated lasers. Namely, this transition does not occur smoothly but is interrupted with the occurrence of instabilities and even chaotic dynamics. Importantly, there exists an open interval of the coupling strength where phase locking is impossible at any detuning.

To understand these counter-intuitive phenomena we study in detail two-dimensional bifurcation diagrams in the plane of laser detuning and coupling strength for different values of the linewidth enhancement factor α; see Sect. 6.1.1. The analysis reveals various codimension-two bifurcations including saddle-node-Hopf and generalized Hopf bifurcations, 1:1 and 1:2 resonances, and homoclinic-doubling points. In particular, we identify codimension-three bifurcations leading to the appearance of a gap in the phase-locking region with increasing α. Furthermore, we perform a detailed study of bifurcations of periodic orbits, and specifically their origins and mutual connections, to show how the gap is gradually occupied by instabilities and chaos with increasing α.

6.2.1 Composite-Cavity Description of Two Back-to-Back Coupled Lasers

We consider two laser cavities with instantaneous (no time delay) coupling: cavity A of length L is coupled via a common mirror of transmission T to cavity B of length $L + dL$; see Fig. 6.19. For the consistent description of coupling and optical nonlinearities, we expand the spatiotemporal laser field $\mathcal{E}(z,t)$ in terms of standing waves $u_n(z)$ for the *entire* double-cavity cavity structure [13, 43, 53]

$$\mathcal{E}(z,t) = \frac{1}{2} \sum_j \left[E_j(t) e^{-i\Psi_j(t)} u_j(z) + c.c. \right]. \tag{6.2}$$

Such standing waves are called composite-cavity modes. This is in contrast to the usual approach that neglects spatial effects in the coupling. Notice, that in the composite-cavity-mode picture, the resulting equations describe interaction of composite-cavity modes rather than individual lasers [52].

Mathematically, the situation is described by a system of two globally coupled oscillators that can be described by the set of five autonomous ordinary differential equations

$$\dot{E}_j = -\gamma E_j + C_{jj}\gamma \times \sum_{k=1,2} \left\{ [C_{kj}^A(1 + \beta N_A) + C_{kj}^B(1 + \beta N_B)] \cos(\psi_{kj}) \right.$$
$$\left. -\alpha\beta[C_{kj}^A(1 + N_A) + C_{kj}^B(1 + N_B)] \sin(\psi_{kj}) \right\} E_k, \tag{6.3}$$

$$\dot{\Psi}_j = \Omega_j + C_{jj}\gamma \times \sum_{k=1,2} \left\{ \alpha\beta[C_{kj}^A(1 + N_A) + C_{kj}^B(1 + N_B)] \cos(\psi_{kj}) \right.$$
$$\left. +[C_{kj}^A(1 + \beta N_A) + C_{kj}^B(1 + \beta N_B)] \sin(\psi_{kj}) \right\} \frac{E_k}{E_j}, \tag{6.4}$$

$$\dot{N}_{A/B} = \Lambda - (N_{A/B} + 1) - \sum_{k,j=1,2} C_{kj}^{A/B}(1 + \beta N^{A/B}) \cos(\psi_{kj}) E_k E_j, \tag{6.5}$$

where $j = 1, 2$ and $\psi_{kj} = \Psi_k - \Psi_j$ is the phase difference between mode k and mode j. (Note that it is sufficient to consider the equation for the phase difference ψ_{12} between the two electric fields.) Equations (6.4)–(6.5) are coupled to the algebraic constraints

$$\sin\left[\frac{\Omega_j}{c} n_b(2L + dL)\right] = 2\frac{\sqrt{1-T}}{T} \sin\left[\frac{\Omega_j}{c} n_b L\right] \sin\left[\frac{\Omega_j}{c} n_b(L + dL)\right], \tag{6.6}$$

$$A_j^2 L \left[\frac{1}{2} - \frac{\sin\left(2\frac{\Omega_j}{c} n_b L\right)}{4\frac{\Omega_j}{c} n_b L}\right] + B_j^2(L + dL) \left[\frac{1}{2} - \frac{\sin\left[2\frac{\Omega_j}{c} n_b(L + dL)\right]}{4\frac{\Omega_j}{c} n_b(L + dL)}\right]$$
$$+ A_j^2 \frac{2c}{\Omega_j n_b} \frac{\sqrt{1-T}}{T} \sin^2\left(\frac{\Omega_j}{c} n_b L\right) = L, \tag{6.7}$$

$$A_j \sin\left(\frac{\Omega_j}{c} n_b L\right) = -B_j \sin\left[\frac{\Omega_j}{c} n_b (L + dL)\right], \tag{6.8}$$

$$C_{jk}^A(T, dL) = \frac{1}{L} \int_{-L}^{0} dz \, u_j(z) u_k(z), \tag{6.9}$$

$$C_{jk}^B(T, dL) = \frac{1}{L} \int_{0}^{L+dL} dz \, u_j(z) u_k(z). \tag{6.10}$$

Differential equations (6.3)–(6.5) describe the time evolution of the real field amplitudes E_1 and E_2 of the composite-cavity modes, their phase difference ψ_{12}, and the population inversion in lasers A (N_A) and B (N_B). The modal frequencies Ω_1 and Ω_2 are determined from the transcendental equation (6.6), the modal amplitudes $A_{1/2}$ and $B_{1/2}$ are determined from (6.7)–(6.8), and the coupling coefficients C_{jk}^A and C_{jk}^B are determined from the spatial overlap of the composite-cavity modes; see (6.9)–(6.10). More details on the derivation of the model, algebraic constraints, and dimensionless parameters can be found in [13, 52, 53].

The aim is to calculate two-dimensional bifurcation diagrams of system (6.3)–(6.5) in the (T, dL)-plane for different fixed values of the linewidth enhancement factor α. For the other parameters we chose the realistic values, namely for the refractive index $n_b = 3.4$, for the dimensionless gain coefficient $\beta = 9.82$, for the dimensionless excitation rate $\Lambda = 2$ in cavity A and B, and for the ratio of the composite-cavity and population decay rates $\gamma = 10$. Because of the nature of the model, the bifurcation analysis of the coupled-laser system is not as straightforward as in the case of the optically injected laser in Sect. 6.1. The two main issues are:

1. the coupling parameters (which are the main bifurcation parameters) namely, the coupling-mirror transmission T and the cavity-length mismatch dL, appear in (6.3)–(6.5) implicitly through the modal frequencies Ω_1 and Ω_2 and integrals C_{jk}^A and C_{jk}^B as described by (6.6)–(6.10);
2. the system has two types of periodic solutions: those where the phase difference ψ_{12} is bounded within a 2π interval, and those where ψ_{12} is unbounded, that is, $\psi_{12}(t)$ is periodic modulo 2π; the latter oscillations are also called rotations [16].

The first issue can be overcome by appending the algebraic constraints (6.6)–(6.10) to the system of ODEs (6.3)–(6.5) and solving the extended system, that is, by performing continuation of solutions to (6.3)–(6.5) and (6.6)–(6.10) simultaneously. The second issue becomes problematic only near transitions between periodic solutions with bounded and unbounded phase. (Each individual type of periodic solution is readily continued with AUTO.) Such transitions are common near interesting phenomena (e.g., codimension-two saddle-node-Hopf points with re-injection) and may cause technical inconvenience. This issue can be overcome in the case of two laser modes by appropriate change of variables [46]. However, it remains an interesting issue for the bifurcation analysis of multi-mode lasers where phase relations between more than two individual modes need to be taken into account.

6.2.2 Symmetry Properties

It is interesting to discuss symmetries in the presence of two composite-cavity modes. Because each composite mode has different spatial overlap with the active media there is no perfect symmetry in the system of coupled-cavity lasers. However, for long (compared to the wavelength $\lambda = 1\mu$m) cavities this difference is small enough so that the system appears to have some symmetries. If $L \gg \lambda$ (this works well already when $L \sim 10\lambda$), we have that $C_{jj}^A \simeq C_{kk}^B$, and $C_{jk}^A \simeq -C_{jk}^B$ for $j \neq k$. Furthermore, if $L \gg \lambda$ and $dL \sim \lambda$ we have that the symmetry $(C_{jj}^A, dL) \rightarrow (C_{jj}^B, -dL)$. One consequence of the above relations is the (approximate) reflection symmetry $(\psi_{kj}, N^A, N^B, \alpha, \Lambda, dL) \rightarrow (\psi_{kj} \pm \pi, N^B, N^A, \alpha, \Lambda, -dL)$. Hence, the bifurcation diagram in the (T, dL) plane can be symmetric with respect to the change $dL \rightarrow -dL$, provided that both lasers have equal excitation rates Λ and equal linewidth enhancement factors α.

Another consequence is the symmetry in the phase space. Under the assumption of equally pumped lasers, equal losses for both composite-modes, and zero linewidth enhancement factor $\alpha = 0$, if $\{E_1^0, E_2^0, \psi_{12}^0, N_A^0, N_B^0\}$ is an equilibrium, then we notice that $\{E_1^0, E_2^0, \psi_{12}^0 \pm \pi, N_B^0, N_A^0\}$ is an equilibrium too. Each of the two points may sometimes be associated with lasing at a single composite-cavity mode. Whether both of them are stable at the same time depends on the competition between the composite-cavity modes. Strong competition results in bistability between these two equilibria [42].

6.2.3 Chaos in Practically Isolated Microcavity Lasers

In recent studies focusing on coupled lasers little attention has been devoted to dynamical properties of practically uncoupled lasers, although such lasers are encountered in a wide range of applications. By practically uncoupled or isolated lasers, we mean two or more lasers where the desire is for the lasers to operate totally independent of one another, while in practice, only partial isolation is possible. Practically isolated lasers are encountered in the modern technology of micro-optical circuits, where one faces the problem of reducing cross-talk between laser diodes that are densely integrated onto a single chip.

Figure 6.20 depicts the bifurcation diagram in the (T, dL)-plane for two microcavities described by (6.3)–(6.5) with $L = 2.8\mu$m and $\alpha = 2$, where supercritical bifurcations are plotted as solid curves, and subcritical bifurcations as dashed curves. Phase-locking of lasers corresponds to the situation where both lasers emit light of constant intensities and the same frequency. This can be achieved in two ways: through phase locking of the composite-cavity

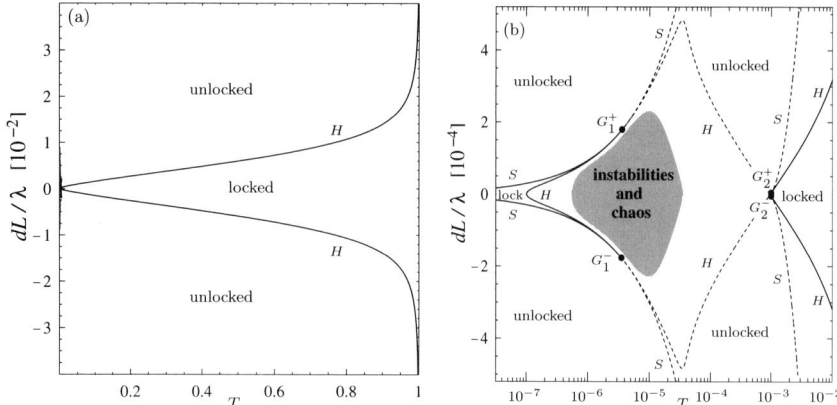

Fig. 6.20. Bifurcation diagram of (6.3)–(6.5) in the $(T, dL/\lambda)$-plane of coupled microcavity lasers (a), and an enlargement for very small T (b). From S. Wieczorek and W.W. Chow, Chaos in practically-isolated microcavity lasers, *Phys. Rev. Lett.* **92** (2004) 213901 © 2004 by the American Physical Society; reprinted with permission.

modes or when both lasers operate at a single composite-cavity mode. In the $\{E_1, E_2, \psi_{12}, N_A, N_B\}$ phase space, phase locking is represented by an equilibrium. The phase-locking region in Fig. 6.20(a), where the lasers operate with a single composite-cavity mode, is indicated by the region between the two supercritical branches of Hopf bifurcation curve H. The locked state is lost when H is crossed towards higher values of $|dL|$. No other transition is visible at this scale and the general features at moderate coupling are similar to what is generally expected. The transition from weakly-coupled to totally isolated lasers appears to occur smoothly, i.e., with an uninterrupted vanishing of the phase locking region, as the coupling approaches zero.

However, the coupled-laser behavior contains an anomaly, whose presence is only noticeable with significant magnification of the (T, dL) parameter space in the vicinity of the origin as shown in Fig. 6.20(b). There, we find the curves of saddle-node and Hopf bifurcation, S and H, respectively. These curves are tangent at four codimension-two saddle-node-Hopf points where they change from sub- to super-critical. These saddle-node-Hopf points are often origins of complex bifurcation structures that give rise to chaos [33]. Starting from the right, the phase-locking region closes near G_2^+ and G_2^- where the two branches of S merge. Phase-locking reappears at G_1^+ and G_1^- and ends at the origin of the (T, dL)-plane. In the notation of [33] G_1 and G_2 belong to different types of saddle-node-Hopf points. Both types are associated with a complex web of bifurcations roughly indicated by the shaded region in Fig. 6.20(b) and studied in more detail in the next section. As the coupling approaches zero, one expects the oscillators to be more independent. Instead, they start interacting in a most complicated way and exhibit mutually induced chaotic oscillations. An interesting question arises as to the origin of this intriguing and counter-intuitive example of coupled nonlinear-oscillator behavior.

6.2.4 Origin of the Interrupted Phase-Locking Region

To explore the dependence of the coupled-laser instabilities on the resonator length L, and to avoid technical difficulties associated with continuation of bifurcation curves across eight orders of magnitude in T (see Fig. 6.20) we now consider longer laser cavities with $L = 280\mu$m.

Bifurcations of Equilibria for $\alpha = 0$

The case of $\alpha = 0$ is highly degenerate and has an uninterrupted phase-locking range as is shown in Fig. 6.21. The two composite-cavity modes are in strong competition leading to bistable locking range. Saddle-node and Hopf bifurcations associated with both stationary points are tangent at $T \approx 0.027$, at four codimension-two saddle-node-Hopf points. Starting inside the locking range and increasing $|dL|$, the locking is lost either via saddle-node bifurcation ($T < 0.027$) or via Hopf bifurcation ($T > 0.027$). Although there are two different bifurcations responsible for the locking-unlocking transition, one can distinguish three different locking-unlocking mechanisms.

For $T < 0.027$ each of the two bifurcating stable equilibria has contributions from both composite modes; see Fig. 6.21(a). Here, frequency separation of the two composite-cavity modes is small, and the laser locking-unlocking transition is a transition between two composite-cavity modes which are phase-locked and two composite-cavity modes which are phase-unlocked. Locking of the lasers arises from phase-locking of the composite-cavity modes. In the $\{E_1, E_2, \psi_{12}, N_A, N_B\}$ space this is represented by two (there is bistability) saddle-node bifurcations of equilibria that take place on a single periodic orbit. Unlocked operation is represented by a single stable periodic orbit.

For $T > 0.15$, each of the two bifurcating stable equilibria has a large contribution from one composite mode, and a vanishing contribution from the other composite mode. The laser locking-unlocking transition is a transition between a single composite-cavity mode (see Fig. 6.21(a)) and two phase-unlocked composite-cavity modes. Locking of the lasers arises from strong competition (owing to strong cross-saturation) between the two composite modes [13, 52]. In the $\{E_1, E_2, \psi_{12}, N_A, N_B\}$ space, this is represented by two supercritical Hopf bifurcations leading to two stable periodic orbits. It also involves saddle-node bifurcation of periodic orbits in which one of these stable orbits disappears. Consequently, unlocked operation is represented by a single stable periodic orbit [53] and no instabilities appear with increasing $|dL|$.

The most interesting region lies in between the two, near the points $G_j^{\pm}$, where neither the composite-mode phase-locking nor the competing

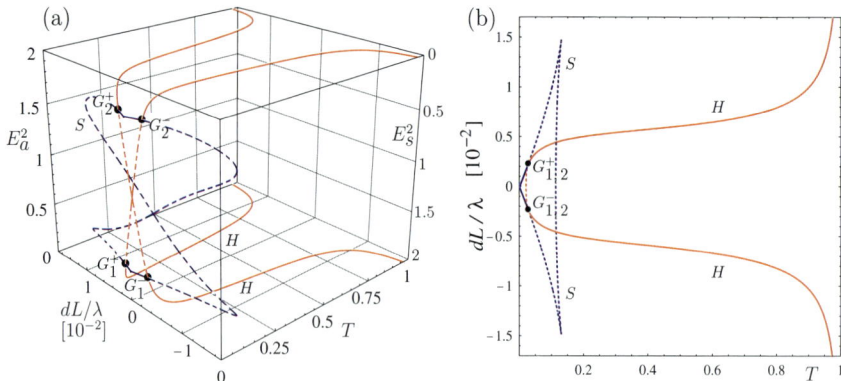

Fig. 6.21. Bifurcations of equilibria of (6.3)–(6.5), shown as a three-dimensional bifurcation diagram in the $(T, dL/\lambda, E_{1/2}^2)$-space (a), and as a projection onto the $(T, dL/\lambda)$-plane (b). The supercritical parts (darker solid curves) of the saddle-node bifurcation curve S (blue) and the Hopf bifurcation curve H (red) bound the phase-locking region. S and H are tangent and change from super- to subcritical at codimension-two saddle-node-Hopf points $G_j^\pm$. From S. Wieczorek and W.W. Chow, Global view of nonlinear dynamics in coupled-cavity lasers-a bifurcation study, *Opt. Comm.*, 246(4–6) (2005) 471–493 © 2004 by Elsevier Science; reprinted with permission.

composite-cavity-mode description is valid. This is where the beat note frequency $\dot{\psi}_{12}$ comes close to the laser's characteristic relaxation oscillation frequency, which is known to give rise to nonlinear resonances and chaos. For $0.027 < T < 0.1$, starting within the phase-locking region, locking of the lasers is lost via undamping of the relaxation oscillation at a Hopf bifurcation. The two stable stationary points become unstable and each of them gives rise to one stable periodic orbit. Outside the phase-locking region and near $G_j^\pm$, these two periodic orbits (bistability in unlocked operation) encounter instabilities leading to complicated dynamics and chaos. It is interesting to note that, as T increases, the transition between the first and the second locking mechanism is clear cut, indicated by $G_j^\pm$. On the other hand, the transition between the second and the third locking mechanism happens continuously and involves saddle-node bifurcation of periodic-orbits.

As a result of the degeneracy of the case $\alpha = 0$, in the projection of the bifurcation diagram onto the $(T, dL/\lambda)$ plane shown in Fig. 6.21(b) bifurcations of different stationary states appear as a single curve or point. How can this degeneracy be removed?

Influence of α on Bifurcations of Equilibria

To explore the dynamics of different types of lasers and to understand how qualitative differences in the behavior of different lasers come about, we focus our attention on the evolution of the phase-locking region with increasing α. The two phase-locking regions in Fig. 6.22, associated with the two stable equilibria, are distinguished by left-inclined and right-inclined patterning, respectively. Let us recall that bifurcation theory predicts four different types of saddle-node Hopf points. All four points $G_j^\pm$ from Fig. 6.22(a) are of type IV in the notation from [33].

Increasing α above zero unfolds the otherwise degenerate bifurcation diagram in Fig. 6.22(a) so that for nonzero α neither bifurcation curves nor $G_1^\pm$ and $G_2^\pm$ fully overlap any longer [Fig. 6.22(b)]. One phase-locking region, associated with G_1^+ and G_1^-, expands along the dL/λ axis and moves in the direction of lower values of T. The other phase-locking region, associated with G_2^+ and G_2^-, moves together with G_2^+ and G_2^- in the direction of higher values of T. Furthermore, the phase-locking region associated with G_1^+ and G_1^- is no longer bounded by the entirely supercritical Hopf bifurcation curve. Codimension-two generalized-Hopf bifurcation points H_g appear where the Hopf curve changes from supercritical to subcritical. Throughout the range of α under consideration, the type of G_2^+ and G_2^- remains unchanged. On the other hand, G_1^+ and G_1^- change from type IV to type III (in the notation from [33]) at $\alpha \simeq 0.5$. The bifurcation diagram in Fig. 6.22(c) shows that the curve H has a cusp at G_1^+ and G_1^- which makes this a very special point, namely a bifurcation of codimension at least three. During the change in the type of G_1^+ and G_1^- the two associated branches of H, one supercritical and the other subcritical, locally exchange their order. This has important consequences to where the chaotic dynamics associated with $G_1^\pm$ appear; see the next subsection for explanation. Increasing α further results in less overlap between the two phase-locking regions to the point where they no longer coalesce [Fig. 6.22(d)]. At $\alpha = 1$ the two phase-locking regions are well separated and a gap appears where the coupled-cavity lasers never lock. This gap increases with further increase of α [Fig. 6.22(e)] so that for $\alpha = 3$ there are two distinct phase-locking regions [Fig. 6.22(f)], one at low coupling-mirror transmissions $0 < T < 0.01$ and the other at relatively high coupling-mirror transmissions $0.45 < T < 1$. Furthermore, the generalized Hopf points H_g are gone and both phase-locking regions are again bounded by the entirely supercritical parts of S and H.

Bifurcations of Periodic Orbits

The next question concerns nonlinear oscillations for parameter settings within the gap between the two phase-locking regions. In a coupled-cavity laser periodic orbits emerge along Hopf bifurcation curves H, along curves S of global saddle-node homoclinic bifurcation, and along homoclinic bifurcation curves h. In particular, we already identified two types of codimension-two bifurcations, namely saddle-node-Hopf points $G_j^\pm$ and generalized Hopf points H_g. Both are sources of bifurcations of periodic orbits and, hence, starting

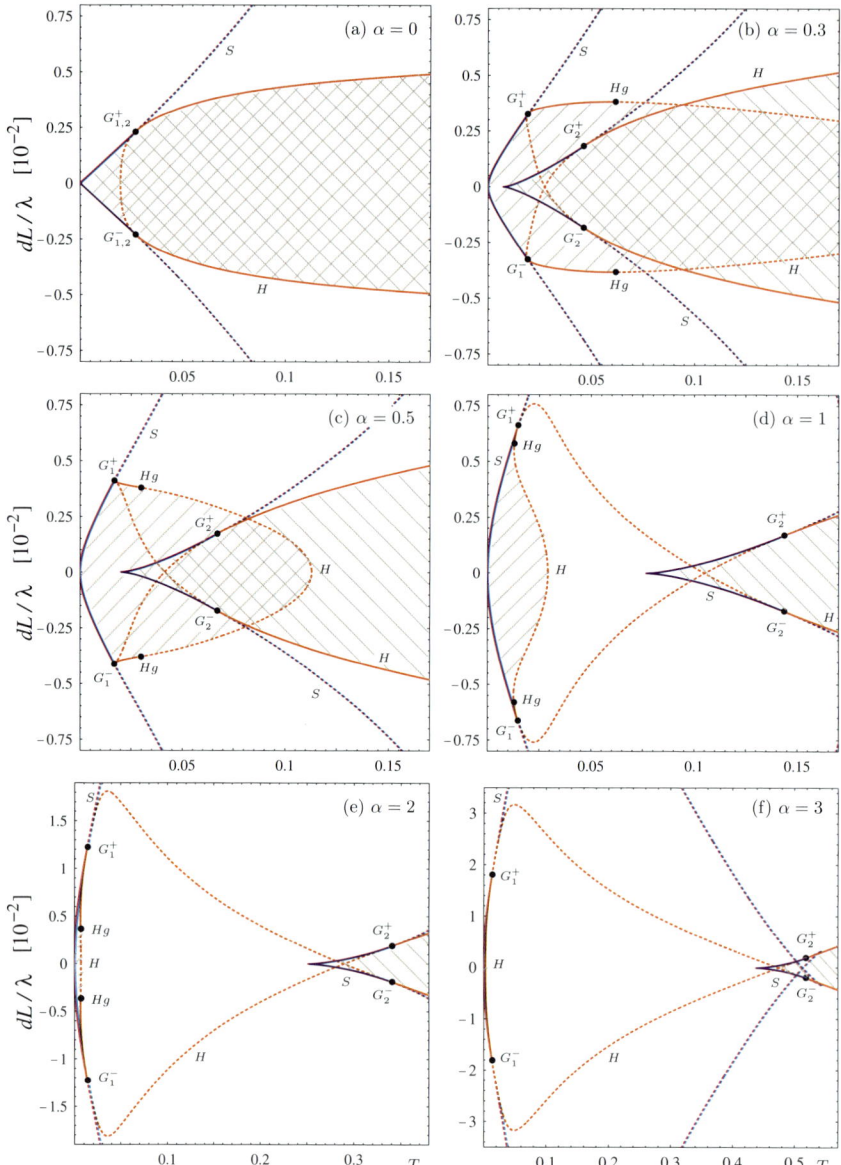

Fig. 6.22. Phase-locking region of (6.3)–(6.5) in the $(T, dL/\lambda)$-plane for different values of the linewidth enhancement factor α. S and H are tangent and change from supercritical to subcritical at codimension-two saddle-node-Hopf points $G_j^{\pm}$. H also changes from supercritical to subcritical at generalized Hopf points H_g. The color coding is as in Fig. 6.21. From S. Wieczorek and W.W. Chow, Global view of nonlinear dynamics in coupled-cavity lasers-a bifurcation study, *Opt. Comm.*, 246(4–6) (2005) 471–493 © 2004 by Elsevier Science; reprinted with permission.

points for further analysis. Note that in the plots of curves of bifurcations of periodic orbits we do not distinguish between super- and subcritical parts.

As expected from general theory [33], there is a torus bifurcation curve T emerging from each point $G_j^\pm$ [Fig. 6.23(a)]. Only two of four T curves are visible (since the case $\alpha = 0$ is degenerate). These torus curves involve two frequencies, the relaxation oscillation frequency and the inter-mode frequency. They are associated with a resonance tongue structure (not shown here) and denote the onset of either quasiperiodic (parameter settings between the tongues) or periodic oscillations (parameter settings within a resonance tongue) when the solid black curves T are crossed from the right to the left [38]. Also, they signal the appearance of chaos via the break-up of a 2-torus when the resonance tongues start to overlap. The curves T terminate at 1:2 resonance points [33, Sec. 9.5.3] where they connect to period-doubling curves PD^1. The PD^1 curves are the first steps in an infinite period-doubling cascade to chaos [18]. The secondary period-doubling curves $PD^{n>1}$ may be arranged in nested or unnested islands of period-doublings [61]. In either case, period-doubling islands are associated with chaotic dynamics. One of the period-doubling curves from Fig. 6.23(a) does not form a closed loop but terminates at two homoclinic-doubling bifurcation points B^1 [39]. Furthermore, there is a non-degenerate saddle-node-of-periodic-orbit curve SL where one of the two stable periodic orbits, born along the degenerate H curve, disappears. The overall dynamical picture for $\alpha = 0$ consists of the uninterrupted bistable phase-locking region and complicated, sometimes chaotic, dynamics found outside of the phase-locking region and near the points $G_j^\pm$.

When α is increased from zero [Fig. 6.23(b)], the degenerate bifurcation diagram unfolds and one clearly sees four torus curves connecting to period-doubling curves at four 1:2 resonance points. Interestingly, regions of complicated dynamics associated with $G_2^\pm$ start to overlap with the phase-locking region associated with $G_1^\pm$. There, depending on the initial condition, the coupled lasers may either be phase-locked or exhibit complicated unlocked oscillations. At the generalized Hopf points H_g, the curve SL attaches to the supercritical branches of H emerging from G_1^+ and G_1^-, causing the curves H to change from supercritical to subcritical. Stable periodic orbits born along these supercritical branches of H disappear at SL.

Increasing α further results in no qualitative changes to saddle-node Hopf points $G_2^\pm$ nor associated torus and period-doubling bifurcations. As the gap between the two phase-locking regions appears [Fig. 6.23(c)–(f)], the points $G_2^\pm$ move in the direction of higher values of T. Concurrently, the two torus curves T emerging from G_2^+ and G_2^- are 'dragged along', causing the two attached period-doubling cascades to shift into the gap between the two phase-locking regions. On the other hand, a number of qualitative changes takes place near $G_1^\pm$. For $\alpha < 0.5$, bifurcations of periodic orbits emerging from $G_1^\pm$ evolve in the direction of increasing $|dL/\lambda|$. Near $\alpha = 0.5$, the type of G_1^+ and G_1^- changes.

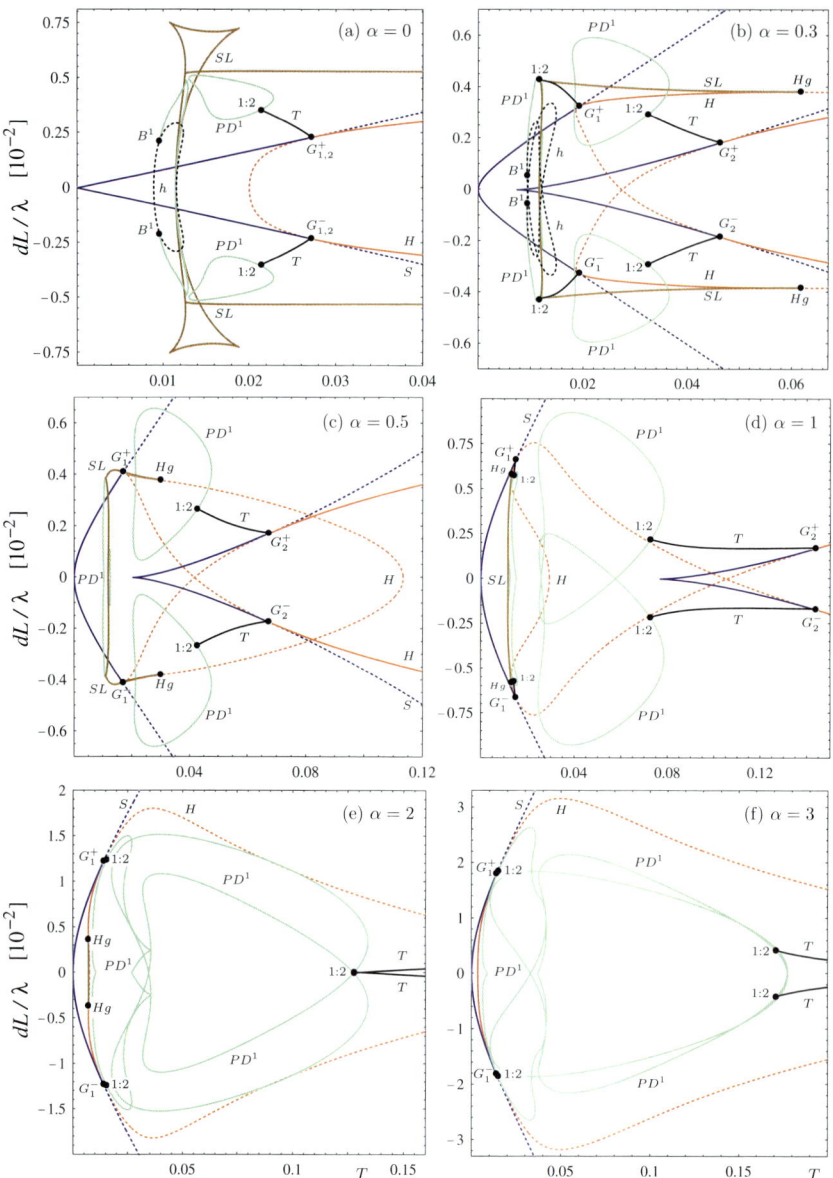

Fig. 6.23. Bifurcation diagram of (6.3)–(6.5) in the $(T, dL/\lambda)$ plane for different values of the linewidth enhancement factor α. Period-doubling bifurcation curves PD are in green, saddle-node bifurcation of periodic orbit curves SL are in brown, torus bifurcation curves T are in solid black, and homoclinic bifurcation curves h are in dotted black. From S. Wieczorek and W.W. Chow, Global view of nonlinear dynamics in coupled-cavity lasers-a bifurcation study, *Opt. Comm.*, 246(4–6) (2005) 471–493 © 2004 by Elsevier Science; reprinted with permission.

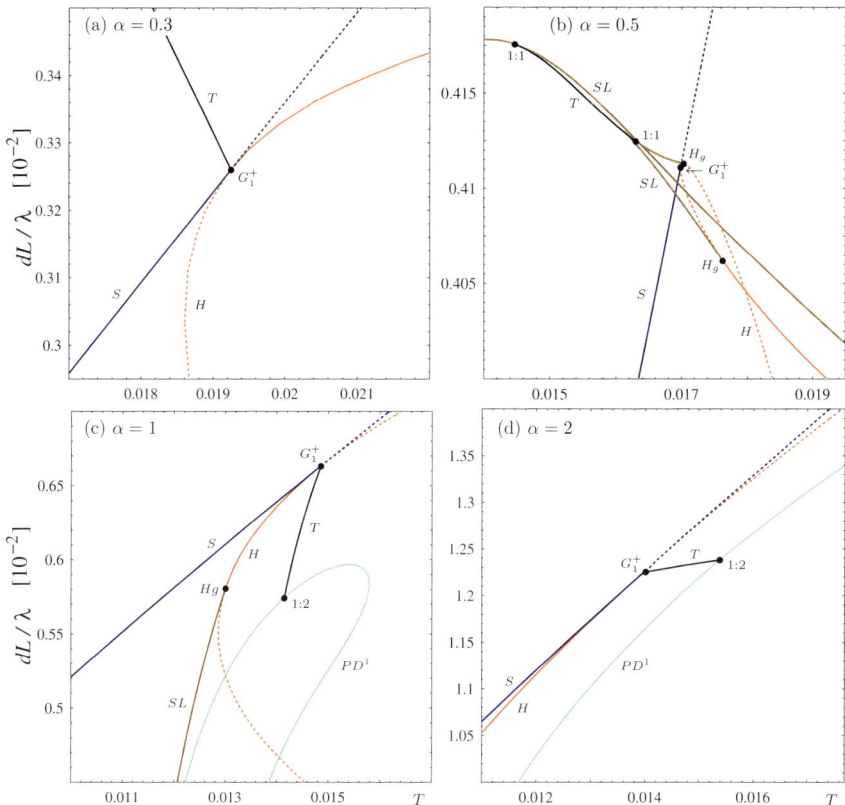

Fig. 6.24. Enlarged bifurcation diagram of (6.3)–(6.5) near saddle-node-Hopf point G_1^+ for different values of the linewidth enhancement factor α. From S. Wieczorek and W.W. Chow, Global view of nonlinear dynamics in coupled-cavity lasers-a bifurcation study, *Opt. Comm.*, 246(4–6) (2005) 471–493 © 2004 by Elsevier Science; reprinted with permission.

The details near G_1^+ are shown in Fig. 6.24. Close to $\alpha \simeq 0.5$ a new curve SL appears that connects to H at the two H_g points, and has a cusp near $(T, dL/\lambda) = (0.016, 0.4125)$ [Fig. 6.24(b)]. The torus curve T detaches from G_1^+ and attaches near the cusp of this new SL curve at 1:1 resonance [33, Sec. 9.5.2]. At $\alpha = 1$ [Fig. 6.24(b)] the extra SL curve is gone. The torus curve T attaches again to G_1^+ but 'flips' from above to below the curve S [Fig. 6.24(c)–(d)]. As a consequence, starting at $\alpha \simeq 1$ [Fig. 6.23(d)], the two torus curves T emerging from G_1^+ and G_1^-, and the attached period-doubling curves PD^1, start filling the gap between the two phase-locking regions [Fig. 6.23(d)–(f)]. For clarity, only parts of the PD^1 curves associated with G_1^+ and G_1^- are plotted in Fig. 6.23(b)–(f).

6.2.5 Concluding Remarks on Coupled Lasers

The interesting dynamics of coupled-cavity lasers arises from two types of nonlinearities: those imparted in the composite-mode properties by the optical coupling (algebraic constraints), and those imparted in the active medium by the population dynamics (differential equations). Their interplay results in a rich display of chaotic oscillations when two conditions are met simultaneously.

(i) The composite-cavity-mode beat note must be close to resonant with the characteristic (relaxation oscillation) frequency of the active medium, thus strongly coupling laser fields and active media;

(ii) an appreciable spatial overlap between composite-cavity modes must be present for a strong coupling of the lasing modes.

Continuation techniques allowed us to uncover a counter-intuitive example of chaos in ultra-weakly coupled nonlinear-oscillators and to explain how this unexpected dynamical picture arises with increasing linewidth enhancement factor α. In particular, an uninterrupted and degenerate (bistable) phase-locking region at $\alpha = 0$ unfolds and develops a gap, which is gradually occupied with instabilities and chaos for $\alpha \neq 0$. The underlying mechanism is a change in the competition between composite-cavity modes that causes a change in the type of codimension-two saddle-node Hopf points via a codimension-three cusp singularity on the Hopf bifurcation curve. Furthermore, several other codimension-two bifurcations, including strong resonances, are identified as sources of instabilities and chaos in coupled-cavity lasers. Many of the phenomena mentioned are interesting from a bifurcation theory point of view and should be studied in more detail.

The dependence of the dynamics on the cavity length reveals effects due to nonlinear optical coupling. With decreasing cavity length, the two conditions that are necessary for the complicated dynamics to occur shift towards the origin of the $(T, dL/\lambda)$ parameter space. Consequently, for short cavities these two conditions may be satisfied at ultra-low optical coupling (e.g. $T < 10^{-5}$ for $L \sim 3\lambda$), where lasers are generally expected to act independently. This bifurcation analysis provided new insight into an overall understanding of coupled-laser behavior.

6.3 Outlook

The field of nonlinear optical/laser systems is expanding in many new directions. Examples of new types of optical systems include nanoscale photonic-crystal lasers [41], optical resonators with quantum coherence [55, 67], and multimode quantum-dot lasers [49]. Owing to their nanoscale and quantum coherence, these systems are expected to have strong optical nonlinearities that are different from those found in conventional optical/laser systems. New nonlinear phenomena are waiting to be uncovered and, based on our experience so far, we believe that continuation techniques are the tool of choice.

The nonlinear analysis of these newly emerging optical systems faces mathematical challenges such as handling high-dimensional multimode systems, ODEs with algebraic constraints, and multiple time scales. It is, therefore, an easy prediction that the bifurcation analysis of newly emerging problems in laser physics and photonics will continue to stimulate and contribute to the further development of continuation techniques themselves.

Acknowledgements

Individual results in this chapter have appeared in previous publications, and I thank my co-workers Bernd Krauskopf, Weng. W. Chow, and Daan Lenstra for their contributions. Text passages and figures have been reproduced with permission, and I thank the American Physical Society, Elsevier Science and Institute of Physics Publishing for permission to use material from [53], [54] and [58], respectively.

References

1. N. B. Abraham, L. A. Lugiato, and L. M. Narducci (Eds.), Special issue on Instabilities in Active Optical Media. *J. Opt. Soc. Am. B*, **2**, 1985.
2. A. Algaba, M. Merino, F. Fernández-Sánchez, and A. Rodriguez-Luis. Closed curves of global bifurcations in Chua's equations: a mechanism for their formation. *Internat. J. Bifurc. Chaos Appl. Sci. Engrg.*, **13**:609–616, 2002.
3. V. Annovazzi-Lodi, S. Donati, and M. Manna. Chaos and locking in a semiconductor laser due to external injection. *IEEE J. Quantum Electron.*, **30**(7):1537–1541, 1994.
4. A. Argyris, D. Syvridis, L. Larger, V. Annovazzi-Lodi, P. Colet, I. Fischer, J. Garcia-Ojalvo, C. R. Mirasso, L. Pesquera, and K. A. Shore. Chaos-based communications at high bit rates using commercial fiber-optic links. *Nature*, **438**:343-346, 2005.
5. A. Back, J. Guckenheimer, M. R. Myers, F. J. Wicklin, and P. A. Worfolk. DsTool: Computer assisted exploration of dynamical systems. *Not. Amer. Math. Soc.*, **39**:303–309, 1992.
6. F. Bai and A. R. Champneys. Numerical computation of saddle-node homoclinic orbits of co-dimension one and two. *J. Dyn. Stab. Syst.*, **11**:325–346, 1996.
7. L. Belyakov. Bifurcation of systems with homoclinic curve of a saddle-focus with saddle quantity zero. *Mat. Zam.*, **36**:681–689, 1984.
8. P. Besnard, Private communication, 2006.
9. V. V. Bykov. The bifurcations of separatrix contours and chaos. *Physica D*, **62**(1-4): 290–299, 1993.
10. V. V. Bykov. Orbit structure in a neighborhood of a separatrix cycle containing two saddle foci in *Methods of qualitative theory if differential equations and related topics. American Math. Soc. Transl. Ser 2*, **200**:87–97, 2000.

11. A. R. Champneys and Yu. A. Kuznetsov. Numerical detection and continuation of codimension-two homoclinic bifurcations. *Internat. J. Bifurc. Chaos Appl. Sci. Engrg.*, **4**:785–822, 1994.

12. A. R. Champneys, Yu. A. Kuznetsov, and B. Sandstede. A numerical toolbox for homoclinic bifurcation analysis. *Internat. J. Bifurc. Chaos Appl. Sci. Engrg.*, **6**:867–887, 1996.

13. W. W. Chow. A composite-resonator mode description of coupled lasers. *IEEE J. Quant. Electron.*, **QE-22**(8):1174–1183, 1986.

14. S-N. Chow and X-B Lin. Bifurcation of a homoclinic orbit with a saddle-node equilibrium. *Diff. Int. Eqns.*, **3**:435–466, 1990.

15. B. Deng. Homoclinic bifurcations with nonhyperbolic equilibria. *SIAM J. Math. Anal.*, **21**:693–720, 1990.

16. E. Doedel, A. R. Champneys, T. Fairgrieve, Yu. A. Kuznetsov, B. Sandstede, and X. Wang. AUTO2000: Continuation and bifurcation software for ordinary differential equations. Available via http://indy.cs.concordia.ca/auto/main.html.

17. T. Erneux, V. Kovanis, A. Gavrielides, and P. M. Alsing. Mechanism for period-doubling bifurcation in a semiconductor laser subject to optical injection. *Phys. Rev. A*, **53**:4372–4380, 1996.

18. M. Feigenbaum. Quantitative universality for a class of nonlinear transformations. *J. Stat. Phys.*, **19**(1):25–52, 1978.

19. F. Fernández-Sánchez, E. Freire, and A. Rodriguez-Luis. T-points in a $\mathbb{Z}_2$-symmetric electronic oscillator:(I) analysis. *Nonlin. Dyn.*, **28**:53–69, 2002.

20. W. Forysiak, J. V. Moloney, and R.G . Harrison. Bifurcations of an optically pumped three-level laser model. *Physica D*, **53**(1):162–186, 1991.

21. A. Gavrielides, V. Kovanis, P. M. Varangis, T. Erneux, and G. Lythe. Coexisting periodic attractors in injection-locked diode lasers. *Quant. Semiclass. Opt.*, **9**:785–786, 1997.

22. P. Glendinning and C. Sparrow. T-points: A codimension two heteroclinic bifurcation. *J. Statist. Phys.*, **43**:479–488, 1986.

23. M. Golubitsky and D. G. Schaeffer, *Singularities and Groups in Bifurcation Theory, Vol. 1* Applied Mathematical Sciences **51** (New York:Springer).

24. S. V. Gonchenko, D. V. Turaev, P. Gaspard, and G. Nicolis. Complexity in the bifurcation structure of homoclinic loops to a saddle-focus. *Nonlinearity*, **10**:409–423, 1997.

25. D. Goulding and G. Huyet, Private communication, 2006.

26. K. Green, B. Krauskopf, and G. Samaey. A two-parameter study of the locking region of a semiconductor laser subject to phase-conjugate feedback. *SIAM J. Appl. Dyn. Sys.*, **2**(2):254–276, 2003.

27. A. J. Homburg. Periodic attractors strange attractors and hyperbolic dynamics near homoclinic orbits to saddle-focus equilibria. *Nonlinearity*, **15**:1029–1050, 2002.

28. D. M. Kane and K. A. Shore (Eds.), *Unlocking Dynamical Diversity: Optical Feedback Effects on Semiconductor Lasers*, (John Wiley & Sons, 2005).

29. V. Kovanis, A. Gavrielides, T. B. Simpson, and J.-M. Liu. Instabilities and chaos in optically injected semiconductor lasers. *Appl. Phys. Lett.*, **67**:2780–2782, 1985.

30. B. Krauskopf and D. Lenstra (Eds.), *Fundamental Issues of Nonlinear Laser Dynamics*, (AIP Conference Proceedings **548**, 2000).

31. B. Krauskopf, K. R. Schneider, J. Sieber, S. M. Wieczorek, and M. Wolfrum. Excitability and self-pulsations near homoclinic bifurcations in semiconductor laser systems. *Opt. Commun.*, **215**(4-6):367–379, 2003.

32. B. Krauskopf and J. Sieber. Bifurcation analysis of an inverted pendulum with delayed feedback control near a triple-zero eigenvalue singularity. *Nonlinearity*, **17**:85–103, 2004.

33. Yu. A. Kuznetsov, *Elements of Applied Bifurcation Theory*, (Springer, New York, 1995).

34. F.-Y. Lin and J.-M. Liu. Chaotic radar using nonlinear laser dynamics. *IEEE J. Quantum Electron.*, **40**(6):815–820, 2004.

35. L. A. Lugiato, L. M. Narducci, D. K. Bandy, and C. A. Pennise. Breathing spiking and chaos in a laser with injected signal. *Opt. Commun.*, **46**:64-68, 1983.

36. P. Mandel, *Theoretical Problems in Cavity Nonlinear Optics*, (Cambridge studies in modern optics, Cambridge University Press, 1997).

37. J. V. Moloney, J. S. Uppal, and R. G. Harrison. Origin of Chaotic Relaxation Oscillations in an Optically Pumped Molecular Laser. *Phys. Rev. Lett.*, **59**:2868–271, 1987.

38. E. Ott, *Chaos in Dynamical Systems* (Cambridge University Press, Cambridge, 1993).

39. B. Oldeman, B. Krauskopf, and A. R. Champneys. Death of period-doublings: locating the homoclinic-doubling cascade. *Physica D*, **146**:100–120, 2000.

40. I. M. Ovsyannikov and L. P. Shilnikov. Systems with a saddle-focus homoclinic curve. *Math. USSR-Sb.*, **58**(2):557–574, 1997.

41. B. Pasenow, M. Reichelt, T. Stroucken, T. Meier, S. W. Koch, A. R. Zakharian, and J. V. Moloney. Enhanced light-matter interaction in semiconductor heterostructures embedded in one-dimensional photonic crystals. *J. Opt. Soc. Am. B*, **22**(9):2039–2048, 2005.

42. M. Sargent III, M. O. Scully, and W. E. Lamb Jr., *Laser Physics*. (Addison-Wesley, 1974).

43. S. A. Shakir and W. W. Chow. Semiclassical theory of coupled lasers. *Opt. Lett.*, **9**:202–204, 1984.

44. M. V. Shashkov and D. V. Turaev. On the complex bifurcation set for a system with simple dynamics. *Internat. J. Bif. Chaos Appl. Sci. Engrg.*, **6**:949–968, 1996.

45. S. Schecter. Numerical computation of saddle-node homoclinic bifurcation points. *SIAM J. Numer. Anal.*, **30**(4):1155–1178, 1993.

46. J. Sieber. Numerical bifurcation analysis for multi-section semiconductor lasers. *SIAM J. Appl. Dyn. Syst.* **1**(2):248–270, 2002.

47. J. Sneyd, A. LeBeau, and D. Yule. Traveling waves of calcium in pancreatic acinar cells: model construction and bifurcation analysis. *Physica D*, **145**(1-2):158–179, 2000.

48. M. B. Spencer and W. E. Lamb Jr. Laser with a transmitting window. *Phys. Rev. A*, **5**(2):884–892, 1972.

49. Y. Tanguy, J. Houlihan, G. Huyet, E. A. Viktorov, and P. Mandel. Synchronization and clustering in a multimode quantum dot laser. *Phys. Rev. Lett.*, **96**(5):053902, 2006.

50. G. D. VanWiggeren and R. Roy. Communication with chaotic lasers. *Science*, **279**(5354):1198–1200, 1997.

51. C. O. Weiss and R. Vilaseca, *Dynamics of Lasers* (VCH Verlagsgesellschaft, Weinheim, Germany, 1991).

52. S. M. Wieczorek and W. W. Chow. Bifurcations and interacting modes in coupled lasers: A strong-coupling theory. *Phys. Rev. A*, **69**:033811, 2004.

53. S. M. Wieczorek and W. W. Chow. Chaos in practically-isolated microcavity lasers. *Phys. Rev. Lett.* **92**:213901, 2004.

54. S. M. Wieczorek and W. W. Chow. Global view of nonlinear dynamics in coupled-cavity lasers-a bifurcation study. *Opt. Commun.*, **246**(4-6):471–493, 2005.

55. S. M. Wieczorek and W. W. Chow. Self-induced chaos in a single-mode inversionless laser. *Phys. Rev. Lett.*, **97**:113903, 2006.

56. S. M. Wieczorek and W. W. Chow. On dynamical sensitivity of lasers and instability-based coherent-signal detection. submitted, 2007.

57. S. M. Wieczorek, W. W. Chow, L. Chrostowski, and C. J. Chang-Hasnain. Improved semiconductor-laser dynamics from induced population pulsation. *IEEE J. Quant. Electron.*, **QE-42**(6):552–562, 2006.

58. S. Wieczorek and B. Krauskopf. Bifurcations of n-homoclinic orbits in optically injected lasers. *Nonlinearity*, **18**:1095–1120, 2005.

59. S. M. Wieczorek, B. Krauskopf, and D. Lenstra. A unifying view of bifurcations in a semiconductor laser subject to optical injection. *Opt. Commun.*, **172**(1):279–295, 1999.

60. S. M. Wieczorek, B. Krauskopf, and D. Lenstra. Mechanisms for multistability in a semiconductor laser with optical injection. *Opt. Commun.*, **183**(1-4):215–226, 2000.

61. S. M. Wieczorek, B. Krauskopf, and D. Lenstra. Unnested islands of period-dublings in an injected semiconductor laser. *Phys. Rev. E*, **64**:056204, 2001.

62. S. M. Wieczorek, B. Krauskopf, and D. Lenstra. Multipulse excitability in a semiconductor laser with optical injection. *Phys. Rev. Lett.*, **88**:063901, 2002.

63. S. M. Wieczorek, B. Krauskopf, T. B. Simpson, and D. Lenstra. The dynamical complexity of optically injected semiconductor lasers. *Phys. Rep.*, **416**(1-2):1–128, 2005.

64. S. M. Wieczorek and D. Lenstra. Spontaneously excited pulses in an optically driven semiconductor laser. *Phys. Rev. E*, **69**:016218, 2004.

65. S. M. Wieczorek, T.B . Simpson, B. Krauskopf, and D. Lenstra. Global quantitative predictions of complex laser dynamics. *Phys. Rev. E*, **65**:045207R, 2001.

66. S. M. Wieczorek, T. B. Simpson, B. Krauskopf, and D. Lenstra. Bifurcation transitions in an optically injected diode laser: theory and experiment. *Opt. Commun.*, **215**(1-3):125–134, 2003.

67. W. Yang, A. Joshi, and M. Xiao. Chaos in an electromagnetically induced transparent medium inside an optical cavity. *Phys. Rev. Lett.*, **95**:093902, 2005.

68. M. G. Zimmermann, S. Firle, M. A. Natiello, M. Hildebrand, M. Eiswirth, M. Bar, A. K. Bangia, and I. G. Kevrekidis. Pulse bifurcation and transition to spatiotemporal chaos in an excitable reaction-diffusion model. *Physica D*, **110**(1-2):92–104, 1997.

Numerical Bifurcation Analysis of Electronic Circuits

Emilio Freire and Alejandro J Rodríguez-Luis

Departamento de Matemática Aplicada II, Escuela Superior de Ingenieros de Sevilla, Spain

It is well known that the creation of the modern geometrical theory of dynamical systems by Poincaré at the end of the 19th century was motivated by problems arising in celestial mechanics [41]. Perhaps it is not so widely known that the dynamics of electronic circuits played an important role at the early stages of the development of this theory. In the 1920s Van der Pol [47] described the periodic oscillations of self-sustained circuits in terms of the limit cycles of Poincaré. He performed experiments with periodically excited circuits and measured, for the first time, complex behavior in a nonlinear system. In the 1930s there was pioneering work of Andronov's Russian school on the theory of oscillations in electronic, mechanical and control systems [11].

It is important to realize that these first applications of Poincaré's qualitative theory to electronic circuits led to new concepts and theoretical results. Examples of this include Liénard's theorems [40, Chap. 3], as motivated by the works of Van der Pol, the development of bifurcation theory for planar systems by Andronov and co-workers [9], and the introduction of the concept of structural stability by Andronov and Pontriaguin [10].

The relationship between the mathematical theory of ordinary differential equations (ODEs) and the dynamics of electronic circuits was initially very close, but this did not continue for very long. In fact, one might speak of a divorce between the two fields in the subsequent development, where electronic devices and systems became ever more complex and of greater dimension. Starting with the invention of the transistor in the 1950s, there was a true explosion in the size of electronic circuits, culminating in the ascent of the microchip — a complex circuit with thousands or even millions of components. In theoretical investigations of such electronic circuits one can hardly find a trace of the geometrical theory of dynamical systems. One of the few exceptions is the almost forgotten work of Hayashi's school in Japan on the global dynamics in experiments with analog computers [46].

On the other hand, in the 1960s and 1970s the mathematical theory of dynamical systems experienced much development, with the introduction of new ideas by Smale, Arnol'd, Lorenz, Yorke, and Feigenbaum, to name just a few of the contributors. The theory as we know it today (see, for example, the textbooks [32, 36, 49]) sheds light on the structure of complex dynamic behavior (by which we mean recurrent, aperiodic and chaotic) that is present in numerous nonlinear models arising in applications; see, for example, the recent survey [35] and references therein.

With these theoretical developments there came a renewed interest in the dynamics of electronic circuits in the early 1980s, when new ideas and methods were introduced to the study of (periodic or non-periodic) oscillations generated by nonlinear electronic circuits of low dimension. However, the theory usually only provides a framework for different phenomena that one may find in a given circuit. To perform an effective study of the actual dynamics it is necessary to resort to numerical methods. In order to obtain a global view of the dynamics in phase space and of the bifurcations in parameter space, one needs to employ numerical methods that go beyond mere numerical simulation. Indeed, what is needed is the technique of numerical continuation that has been developed since about the 1980s and is now available in the form of several software packages, most notably the package AUTO [19]; see also Chaps. 1 and 2.

In this chapter we demonstrate how complicated dynamical behavior and bifurcations can be found and identified in ODE models of electronic circuits. The combination of theoretical methods and numerical techniques allows one to obtain a deep understanding of a wide range of dynamical phenomena. In particular, electronic circuits provide concrete examples of unfoldings of singularities that act as organizing centers of the dynamics.

Specifically, we consider in Sect. 7.1 a three-dimensional modified Van der Pol oscillator as studied in [20]. We show that there are co-existing canard periodic orbits, which we find and continue with AUTO. The core of the chapter is the bifurcation analysis in Sect. 7.2 of a three-dimensional ODE model of a modified Van der Pol-Duffing electronic circuit; see [28] and references therein. This system exhibits very complex dynamics and associated bifurcation structures. We concentrate here on an extensive study of (global) dynamics associated with Arnol'd (or resonance) tongues and on a global bifurcation known as a T-point bifurcation. In the process we identify a number of global bifurcations, including homoclinic bifurcations, Shil'nikov-Hopf bifurcations, T-point bifurcations and T-point-Hopf bifurcations.

7.1 Canards in a Modified Van der Pol Circuit

We consider here the electronic circuit shown in Fig. 7.1. It is obtained from the well-known Van der Pol circuit with a battery by adding a linear RC parallel branch. This circuit was chosen as a convenient system to study *canard periodic orbits* in a three-dimensional phase space.

Canard orbits arise due to the slow-fast nature of the system and have first been found in the Van der Pol equation with a battery (that is, without

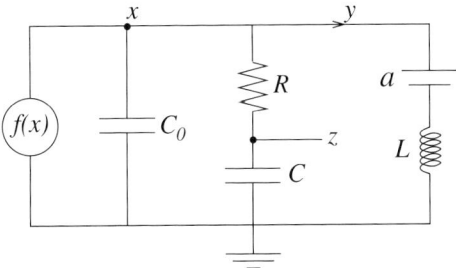

Fig. 7.1. Circuit scheme of a three-dimensional modified Van der Pol circuit with a parallel linear RC branch.

the RC branch in Fig. 7.1). Canards are composed of slow segments that closely follow parts of the S-shaped slow manifold of the Van der Pol system, and they exist in exponentially small parameter regions. Their existence was first shown with techniques from non-standard analysis [15]. The name canard (French for duck) was adopted because the shape of these periodic orbits; see also [18] and Chap. 8.

By applying Kirchhoff's laws and a suitable rescaling of the state and time variables [20], the electronic circuit in Fig. 7.1 can be represented by the vector field

$$\varepsilon \dot{x} = -\alpha \left(\frac{x^3}{3} - x \right) + \frac{z - x}{R} - y,$$
$$\dot{y} = x - a, \tag{7.1}$$
$$\dot{z} = -\frac{z - x}{R}.$$

System (7.1) has exactly one equilibrium, which may undergo a degenerate Hopf bifurcation. A numerical study in [20] with AUTO and DsTOOL [33] corroborates the analytical results and provides evidence of new global bifurcation phenomena, including cusp bifurcations of periodic orbits, period-doubling bifurcations, and the presence of chaotic attractors.

We concentrate here on canard periodic orbits of (7.1). Figure 7.2 shows the situation for $R = 3$, $\alpha = 0.7$ and $\varepsilon = 0.001$. Panel (a) is a plot of the period of a periodic orbit as a function of the parameter a. The branch of periodic orbits emerges from a Hopf bifurcation point H and then is almost vertical in a very narrow interval of the parameter a. Indeed the periodic orbit grows very fast in size in this interval until it takes the typical shape of relaxation oscillations, which exist along the horizontal part of the branch. It is known for the Van der Pol equation that the parameter interval of canard solutions is exponentially small in ε. In other words: the life of *canards* is very short. The sudden growth of the periodic orbit is also referred to as a *canard explosion*.

The enlargement in Fig. 7.2(b) of the narrow a-interval where canard orbits exist shows that the branch of periodic orbits actually has four saddle-node

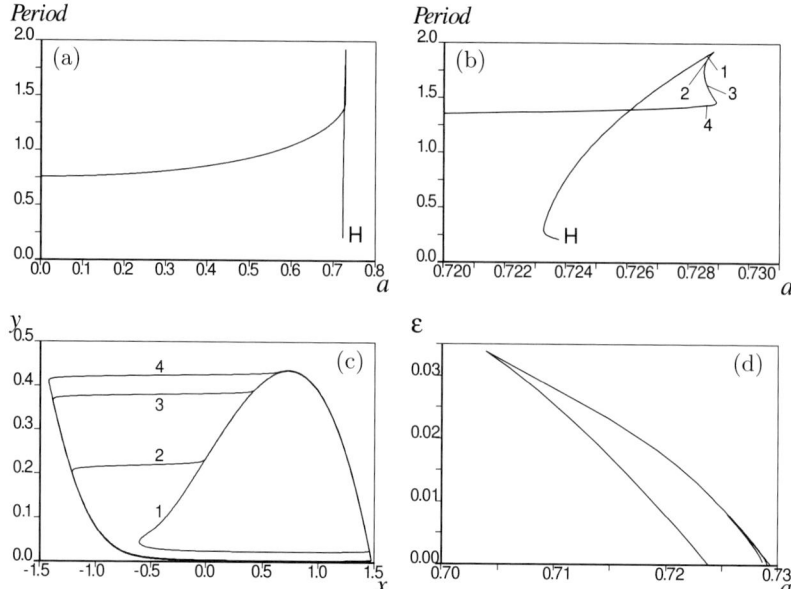

Fig. 7.2. Canard orbits of (7.1) for $R = 3$ and $\alpha = 0.7$. Panel (a) shows the bifurcation diagram for $\varepsilon = 0.001$, and panel (b) is an enlargement. Panel (c) shows four co-existing canard periodic orbits for $a = 0.7286$ and $\varepsilon = 0.001$ in projection onto the (x, y)-plane, and panel (d) the loci of folds in (a, ε)-plane. From E.J. Doedel, E. Freire, E. Gamero and A.J. Rodríguez-Luis, An analytical and numerical study of a modified Van der Pol oscillator, *J. Sound Vibr.* 256 (2002) 755–771 © 2002 by Elsevier Science; reprinted with permission.

bifurcations of periodic orbits (folds with respect to a). As a consequence, there are up to four co-existing canard orbits, for example, those for $a = 0.7286$ (labeled 1 to 4) that are shown in Fig. 7.2(c). Canard orbit 1 does not have a 'head' while orbits 2–4 are canards 'with a head'. Note that the shape of orbit 4 is very close to a relaxation oscillation. The four folds can been continued in the (a, ε)-plane, which results in the curves in Fig. 7.2(d); observe the presence of two cusp points (one of them in the bottom right corner) on the fold curve.

We remark that the bifurcation diagram of Fig. 7.2(a) is very similar to the one for a canard explosion in the Van der Pol equations; see [27] and Chap. 8. According to [50, Fig. 4.7], the period of the periodic orbit does not simply decrease from the value it has when it is a relaxation oscillation. Instead, it first increases up to a maximum (corresponding to the largest canard without a head) and only then decreases very rapidly until the Hopf bifurcation is reached. This agrees with what we find for (7.1) in Fig. 7.2(a) and (b), but note that the situation is richer because of the presence of saddle-node bifurcations.

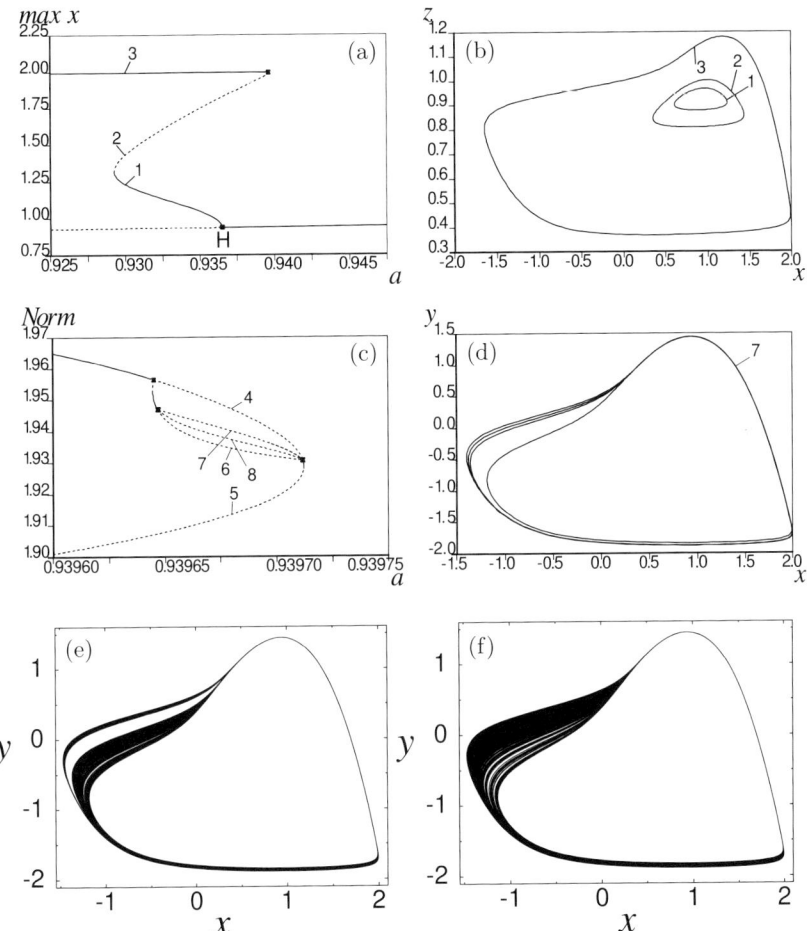

Fig. 7.3. Periodic orbits of (7.1) and their bifurcations for $R = 4$, $\alpha = 2$ and $\varepsilon = 0.25$. Panel (a) shows the bifurcation diagram; solid curves indicate stable objects and dashed curves unstable ones. Panel (b) shows three co-existing periodic orbits for $a = 0.93$. Panel (c) is an enlargement of the bifurcation diagram near the upper fold, where we now show the L_2-norm of the periodic orbits, and panel (d) shows periodic orbit 7 in projection onto the (x, y)-plane. Panels (e) and (f) show chaotic attractors for $a = 0.939648$ and for $a = 0.9396485$, respectively. From E.J. Doedel, E. Freire, E. Gamero and A.J. Rodríguez-Luis, An analytical and numerical study of a modified Van der Pol oscillator, *J. Sound Vibr.* 256 (2002) 755–771 © 2002 by Elsevier Science; reprinted with permission.

Figure 7.3 shows the situation for $R = 4$, $\alpha = 2$ and $\varepsilon = 0.25$, that is, for a larger value of ε. This allows us to study the bifurcation diagram in more detail with AUTO. The bifurcation diagram is shown in panel (a). When decreasing a,

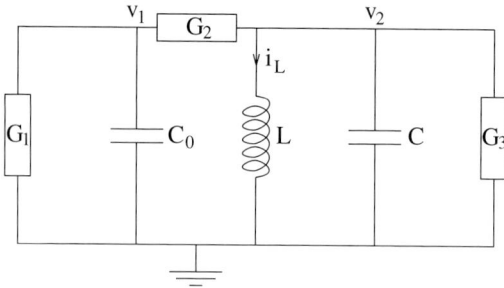

Fig. 7.4. Scheme of the modified Van der Pol-Duffing electronic oscillator.

the equilibrium becomes unstable at the supercritical Hopf bifurcation point
H, where a branch of stable periodic orbits emerges. This branch has two
folds, so that there are up to three co-existing periodic orbits (labels 1–3),
of which two are stable; see Fig. 7.3(b). To study the bifurcation diagram
near the upper saddle-node of periodic orbits, panel (c) shows an enlargement
that also shows period-doubling bifurcations and branches of period-doubled
orbits. Note that the branch of period-two orbits has a fold on the left. Indeed
we find a cascade of period-doubling bifurcations; orbits 4 and 5 are on the
principal branch, orbit 6 on the branch of period-two orbits, orbit 7 on the
branch of period-four orbits, and orbit 8 on the branch of period-eight orbits.
Figure 7.3(d) shows period-four orbit 7 in projection onto the (x, y)-plane.

Because the periodic orbits undergoing successive period-doubling bifurcations are stable, one expects that there exist chaotic attractors in a narrow
a-interval. That this is indeed the case is shown in Fig. 7.3(e) and (f); where
the latter chaotic attractor is 'fully developed'.

7.2 Bifurcations in a Modified Van der Pol-Duffing Circuit

Several different ways have been proposed to modify the classical Van der
Pol circuit to obtain an electronic circuit with three states. Examples of such
circuits can be found in [26, 31, 34, 37, 38, 43]; see also the review of different
configurations in [28].

In this section we consider the modified Van der Pol-Duffing electronic
circuit sketched in Fig. 7.4. This system has been suggested as a random wave-
form generator [43], and it consists of a parallel RCL-circuit and an RC-circuit
that are coupled by a nonlinear conductance; see [28] and references therein.
To arrive at an ODE model, we take the voltages at the capacitors and the
current across the inductance as state variables, and model the current-voltage
characteristics of conductances by means of odd third-order polynomials. In
dimensionless variables the circuit is described by the three-dimensional vector

field

$$\dot{x} = -\left(\frac{\nu + \beta}{r}\right)x + \frac{\beta}{r}y - \frac{A_3}{r}x^3 + \frac{B_3}{r}(y - x)^3,$$

$$\dot{y} = \beta x - (\beta + \gamma)y - z - B_3(y - x)^3 - C_3 y^3, \qquad (7.2)$$

$$\dot{z} = y,$$

where $r > 0$ represents the ratio between the two capacitances. Note that system (7.2) has a $\mathbb{Z}_2$-symmetry due to its invariance under the transformation $(x, y, z) \mapsto (-x, -y, -z)$.

We now summarize the local bifurcations of (7.2) in dependence on the parameters ν, β and γ. The origin is always an equilibrium. A pitchfork bifurcation of equilibria occurs on the plane $\{\nu + \beta = 0\}$, which creates two symmetry-related equilibria that exist for $\nu + \beta < 0$. The origin as well as the nontrivial equilibria undergo Hopf bifurcations. System (7.2) exhibits two different kinds of Takens-Bogdanov bifurcations (double-zero eigenvalue). The first is of homoclinic type and occurs on the straight line $\{(\nu, \beta) = (-\sqrt{r}, \sqrt{r})\}$ where $\gamma \neq -\sqrt{r}$; the second is of heteroclinic type and occurs on the straight line $\{(\nu, \beta) = (\sqrt{r}, -\sqrt{r})\}$ where $\gamma \neq \sqrt{r}$. A detailed analysis of the Hopf and Takens-Bogdanov bifurcations in system (7.2) and their degeneracies can be found in [4].

There are also Hopf-pitchfork bifurcations of (7.2), which occur on the line segment $\{(\nu, \beta) = (\gamma, -\gamma)\}$ where $\gamma^2 < r$. Furthermore, there is a codimension-three singularity corresponding to a degeneracy of the Hopf-pitchfork bifurcation, whose normal form and unfolding is the topic of [3]. How this bifurcation, which occurs for $\gamma \approx -0.4519$, organizes the dynamics of (7.2) is discussed in [5, 6, 7]. Moreover, system (7.2) exhibits a triple-zero bifurcation at the critical values $\nu_c = -\beta_c = \gamma_c = \pm\sqrt{r}$; see [29] for more information.

In this section we focus on two topics, namely the intriguing bifurcation structure inside weak resonance tongues and global bifurcation phenomena, especially those associated with T-point bifurcations.

7.2.1 Analysis of Arnol'd Tongues

We now present some results on the numerical analysis of Arnol'd tongues in the system; see [8] and further references therein. Specifically, we consider (7.2) for fixed $r = 0.6$, $A_3 = 0.3286$, $B_3 = 0.9336$ and $C_3 = 0$.

Figure 7.5(a) shows the Floquet multipliers of the periodic orbit that undergoes a torus (or Neimark-Sacker) bifurcation HH for $\gamma = -0.6$; shown is the the argument (angle between the horizontal axis and the Floquet multiplier of the complex conjugate pair with positive imaginary part) versus parameter β. When the argument fails to vary monotonically we say that the torus bifurcation has an angular degeneracy [39]; this occurs in Fig. 7.5(a) at the

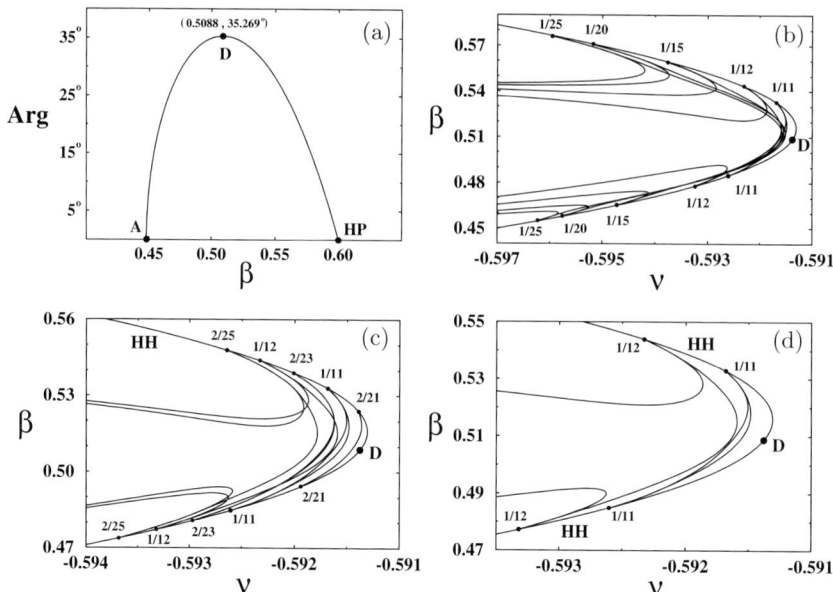

Fig. 7.5. Arnol'd tongues of (7.2) for $\gamma = -0.6$. Panel (a) shows the arguments of the characteristic multipliers versus β along the torus bifurcation curve; D is an angular degeneration point. Panel (b) depicts Arnol'd tongues close to the torus bifurcation curve HH of 1:p resonances for $p = 11$, 12, 15, 20 and 25, close to the HH curve, panel (c) the Arnol'd tongues for the 2:21, 1:11, 2:23, 1:12 and 2:25 resonances, and panel (d) details of the Arnol'd tongues for the 1:11 (closed resonance zone) and 1:12 (first open resonance zone) resonances. From A. Algaba, M. Merino and A.J. Rodríguez-Luis, Takens-Bogdanov bifurcations of periodic orbits and Arnold's tongues in a three-dimensional electronic model, *Internat. J. Bifur. Chaos Appl. Sci. Engrg.* 11 (2001) 513–531 © 2001 by World Scientific Publishing; reprinted with permission.

maximum D of the parabola-like curve (where the Floquet multipliers of the periodic orbit reverse their direction of movement on the unit circle). The appearance of this angular degeneracy is a consequence of the existence of the point A, associated to a double +1 characteristic multiplier of a periodic orbit (since the argument is zero at both endpoints of HH).

The value of the argument at the angular degeneracy point D (of ≈ 35.3 degrees) indicates that 1:p and 2:q resonances will appear on the curve HH only for $p \geq 11$ and $q \geq 21$. Several of these 1:p Arnol'd tongues are shown in Fig. 7.5(b) in a neighborhood of HH. In Fig. 7.5(c) we show, for $\gamma = -0.6$, the angular degeneracy point D on the torus curve HH as well as the first 1:p and 2:q resonance zones close to it. The numerical computations show that the first three (2:21, 1:11, 2:23) resonance zones are closed, which suggests that the point D is of 'banana-type' [39].

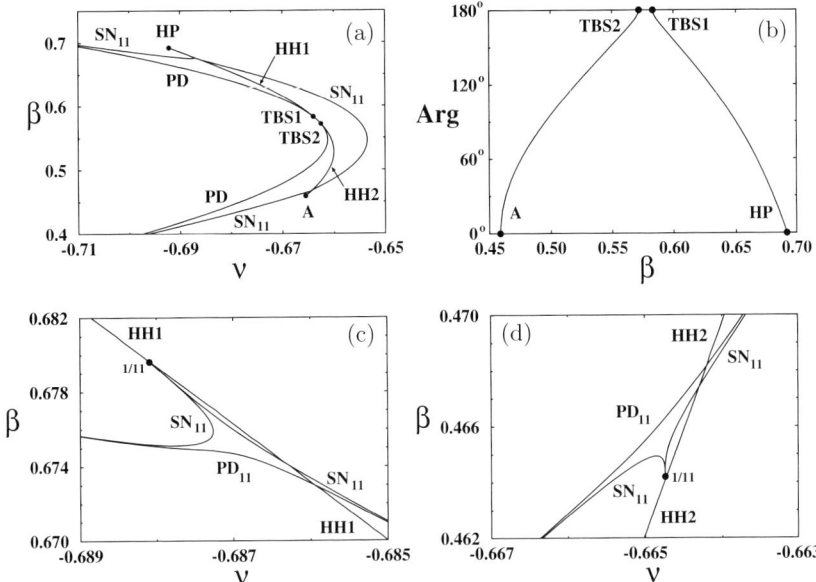

Fig. 7.6. Arnol'd tongues of (7.2) for $\gamma = -0.69217$. Panel (a) shows the partial bifurcation set where the torus curve HH now has two parts, while panel (b) depicts the arguments of the respective characteristic multipliers with the points TBS1 and TBS2. Panels (c) and (d) show details of Arnol'd tongues on HH1 and on HH2, respectively. From A. Algaba, M. Merino and A.J. Rodríguez-Luis, Takens-Bogdanov bifurcations of periodic orbits and Arnold's tongues in a three-dimensional electronic model, *Internat. J. Bifur. Chaos Appl. Sci. Engrg.* 11 (2001) 513–531 © 2001 by World Scientific Publishing; reprinted with permission.

In the remainder of this section we will concentrate on the first two lower 1:p resonances, which are the 1:11 and 1:12 resonances. Note that, for $\gamma = -0.6$, they correspond respectively to a closed resonance zone (1:11) and to the first open resonance zone (1:12) in the (ν, β)-plane; see Fig. 7.5(d).

We now change γ to observe the evolution of the 1:11 Arnol'd tongue. When γ is decreased, we detect that for $\gamma_c \approx -0.69205$ the torus curve collides at a point TBS with a period-doubling curve of the asymmetric periodic orbit that emerges in a Hopf bifurcation of the nontrivial equilibria.

For $\gamma < \gamma_c$ the torus curve appears to be split into two parts, as is shown in Fig. 7.6(a) for $\gamma = -0.69217$. The first part of the torus curve, HH1, joins the Hopf-pitchfork point HP with the point TBS1 (where the periodic orbit has a non-diagonalizable double Floquet multiplier -1). The second part of the torus curve, HH2, connects the points TBS2 and A. At these codimension-two points the periodic orbit has a non-diagonalizable double Floquet multiplier -1 and a diagonalizable double Floquet multiplier $+1$, respectively. In fact, TBS1 and TBS2 correspond to cubic homoclinic-type Takens-Bogdanov bifur-

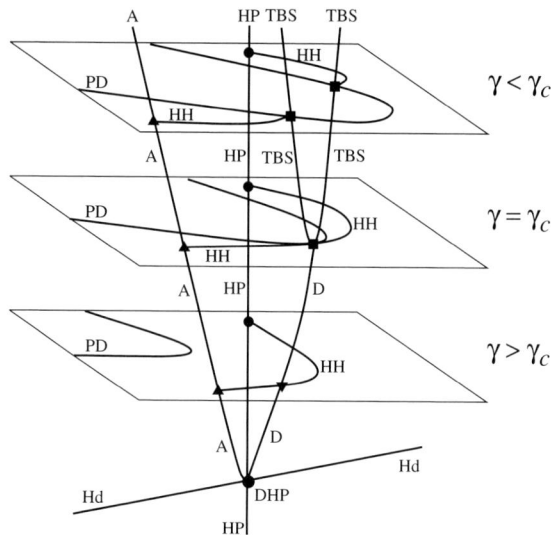

Fig. 7.7. Qualitative partial bifurcation set in (ν, β, γ)-space that explains why the torus curve splits; filled circles are points on the Hopf-pitchfork curve HP, filled squares correspond to points on the Takens-Bogdanov curve TBS, filled triangles stand for points on the Hopf-saddle-node of periodic orbits curve A, and inverted filled triangles indicate points on the angular degeneracy curve D. From A. Algaba, M. Merino and A.J. Rodríguez-Luis, Takens-Bogdanov bifurcations of periodic orbits and Arnold's tongues in a three-dimensional electronic model, *Internat. J. Bifur. Chaos Appl. Sci. Engrg.* 11 (2001) 513–531 © 2001 by World Scientific Publishing; reprinted with permission.

cations of periodic orbits [36]. This splitting of the torus curve induces the disappearance of the angular degeneration point D, as can be seen in Fig. 7.6(b).

Figure 7.7 is a qualitative partial bifurcation set in (ν, β, γ)-space that explains why the torus curve splits. A degenerate Hopf-pitchfork point DHP appears on the curve HP when it intersects with the curve Hd of degenerate Hopf bifurcation of the origin. On one side of DHP a curve of points A appears. The torus surface HH is bounded initially by the curves HP and A. As the Floquet multipliers have argument zero on both curves, an angular degeneracy curve D exists on the torus surface. Since the maximum value of the argument increases when separating from DHP (decreasing γ), there is a point where the curve D ends (when the maximum is 180 degrees). This situation occurs exactly when the surface of period-doubling bifurcations PD reaches the torus surface. From this moment on, the torus surface is also bounded by a parabola-shaped curve of Takens-Bogdanov bifurcations of periodic orbits TBS. This means that the torus surface has a 'parabolic hole': it does not exist between the two branches of TBS. Therefore, the points TBS1 and TBS2 appear in a slice of constant γ.

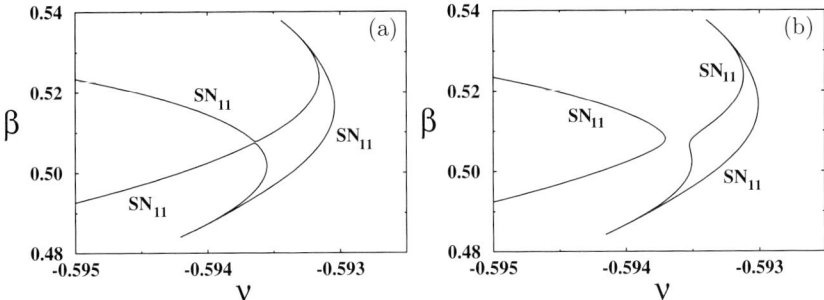

Fig. 7.8. Arnol'd tongue for the 1:11 resonance for $\gamma = -0.60184$ (a), and for $\gamma = -0.6018$ (b). From A. Algaba, M. Merino and A.J. Rodríguez-Luis, Takens-Bogdanov bifurcations of periodic orbits and Arnold's tongues in a three-dimensional electronic model, *Internat. J. Bifur. Chaos Appl. Sci. Engrg.* 11 (2001) 513–531 © 2001 by World Scientific Publishing; reprinted with permission.

Now that we know how the torus loci change due to the presence of Takens-Bogdanov bifurcations of periodic orbits, we focus on Arnol'd tongues and their evolution. We will see that two different types of Takens-Bogdanov bifurcations of periodic orbits will be also present, including a cascade of one of them.

Figure 7.6(c)-(d) shows how the 1:11 Arnol'd tongues emerge from HH1 and HH2, respectively. Note that the right branch of folds that emerges from the 1:11 resonance on HH1 crosses this curve when it moves away from its starting point on HH1; see Fig. 7.6(c). The same happens for the right curve of saddle-node bifurcations starting at the 1:11 resonance on HH2; see Fig. 7.7(d). Such a crossing implies that periodic orbits (of approximately eleven times the period of the principal periodic orbit) exist on both sides of the curves HH1 and HH2. This phenomenon is not a consequence of the splitting of the torus curve (since it also occurs, for example, for $\gamma = -0.65$ when there is only one torus curve). Rather it is due to the evolution of the curves HH and the boundaries (saddle-node bifurcations of periodic orbits) of the 1:11 resonance zones with γ. (In the (ν, β, γ)-space the surfaces HH and SN_{11} intersect independently of the collision of the surface PD with HH.)

The continuation of these saddle-node curves of periodic orbits shows that both right branches are connected, whereas the left branches are disconnected; see Fig. 7.6(a). (Recall that for $\gamma = -0.6$ the 1:11 resonance zone is a closed region.) Moreover, in the present situation the repelling periodic orbit undergoes a period-doubling bifurcation (which is again different from the behavior for $\gamma = -0.6$); see the curve PD_{11} in Fig. 7.6(c) and (d), which has not been included in panel (a) as it would be indistinguishable from the saddle-node curves that limit the 1:11 resonance zone.

The question arises how the resonance zones evolve from being open to being closed when γ changes between $\gamma = -0.69217$ and $\gamma = -0.6$. To see how

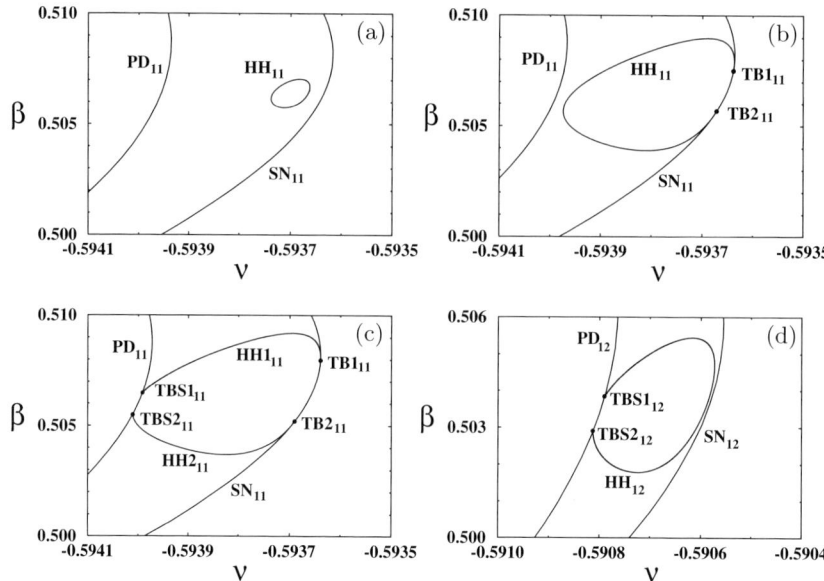

Fig. 7.9. Partial bifurcation set inside the Arnol'd tongues of the open 1:11 resonance zones for $\gamma = -0.601285$ (a), $\gamma = -0.60134$ (b), and $\gamma = -0.60135$ (c). Panel (d) shows the partial bifurcation set inside the 1:12 Arnol'd tongues of the open 1:12 resonance zone for $\gamma = -0.5973$. From A. Algaba, M. Merino and A.J. Rodríguez-Luis, Takens-Bogdanov bifurcations of periodic orbits and Arnold's tongues in a three-dimensional electronic model, *Internat. J. Bifur. Chaos Appl. Sci. Engrg.* 11 (2001) 513–531 © 2001 by World Scientific Publishing; reprinted with permission.

this happens we increase γ from $\gamma = -0.69217$. At $\gamma \approx -0.60184$ the upper and lower left branches of the resonance zone touch, so that, for smaller γ, two separate resonance regions are created; see Fig. 7.8(a) and (b). Note that the saddle-node curves that bound the closed resonance zone emerge from the curve HH but the curves of the open zone are not related to HH.

We conclude that, for the initial value of $\gamma = -0.6$, there exists not only the closed 1:11 resonance zone shown in Fig. 7.5(d) but also the open zone in Fig. 7.8(a). In fact, this open resonance zone exists even before the closed resonance zone (which appears for $\gamma \approx -0.5957$).

Note that inside the closed 1:11 resonance zone the periodic orbits do not exhibit any bifurcation. Both are hyperbolic, one of saddle type and the other repelling. However, the study of the periodic orbits inside the open 1:11 resonance zone for $\gamma = -0.6$ shows that the stability of one of them changes relative to the periodic orbits inside the closed resonance zone for the same value of γ. Now one periodic orbit is a saddle and the other is attracting. The latter undergoes a period-doubling bifurcation when crossing the curve PD_{11}.

Obviously, the change in the stability of one of the periodic orbits in the 1:11 resonance zones indicates that, as γ varies, some additional bifurcation has to be present on the saddle-node curve that bounds the open resonance zone. This bifurcation is necessary to make possible the contact between the open and the closed 1:11 resonance zones. To understand this situation we investigate the partial bifurcation set inside this region for values of γ close to the value where the two regions join.

Moving again the control parameter γ we see in Fig. 7.9(a) that a closed torus curve HH_{11} of the 11-period orbit appears in the parameter plane (this curve does not exist for $\gamma = -0.60128$). As γ decreases, this closed curve approaches the saddle-node curve that bounds the open region of the 1:11 resonance. In this way, two quadratic homoclinic-type Takens-Bogdanov points of periodic orbits $TB1_{11}$ and $TB2_{11}$ are created; see Fig. 7.9(b). For even lower γ, when the torus curve HH_{11} interacts with the curve of period-doubling PD_{11}, it splits into two curves $HH1_{11}$ and $HH2_{11}$; see Fig. 7.9(c). Two new Takens-Bogdanov points of periodic orbits (symmetric and of homoclinic type), $TBS1_{11}$ and $TBS2_{11}$, appear. Note that both curves $HH1_{11}$ and $HH2_{11}$ connect two Takens-Bogdanov points that are on the saddle-node curve SN_{11} and on the period-doubling curve PD_{11}, respectively.

We have checked whether the presence of closed torus curves also occurs for other resonances. The answer is affirmative, but a difference may appear. In the above case of a 1:11 resonance, the torus curve first collides with SN_{11} and later with PD_{11}, whereas in other resonances the torus curve first collides with the period-doubling curve and later with the saddle-node curve. This is illustrated for the 1:12 resonance in Fig. 7.9(d). The curve HH_{12} starts and ends at two symmetric homoclinic-type Takens-Bogdanov points of periodic orbits, $TBS1_{12}$ and $TBS2_{12}$.

Several possible scenarios are proposed in [39] where angular degeneracy points are present, all involve Hopf-Hopf bifurcations and most involve Takens-Bogdanov points. These authors wonder whether there is some model that presents such a behavior in relation to the torus curve and its resonance tongues; see [39, Fig. 5(d)-(e)]. The three-dimensional autonomous model (7.2) considered here exhibits four of the five possible situations for the global continuation of a Hopf-Hopf curve in a two-parameter family, namely:

1. continuation in each direction may terminate at a quadratic Takens-Bogdanov point (non-diagonalizable double Floquet multiplier $+1$); see Fig. 7.9(b);
2. continuation in each direction may terminate at a cubic homoclinic Takens-Bogdanov point (nondiagonalizable double Floquet multiplier -1); this was found for the 1:12 resonance in Fig. 7.9(d);
3. continuation in one direction may terminate at a quadratic Takens-Bogdanov point, while continuation in the other direction terminates at a cubic homoclinic Takens-Bogdanov point; see Fig. 7.9(c); and
4. continuation may provide a closed curve; see Fig. 7.9(a).

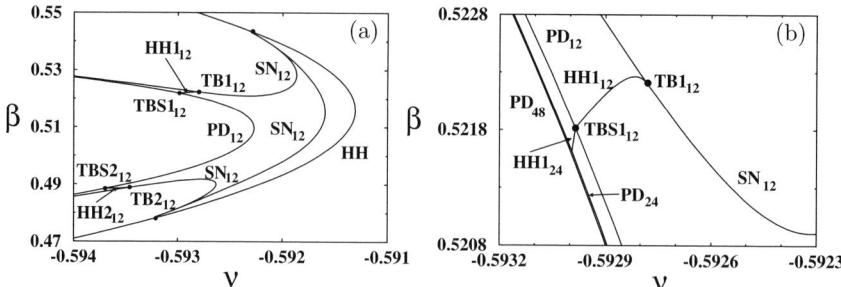

Fig. 7.10. Arnol'd tongues for the 1:12 resonance for $\gamma = -0.6$ (a), and detail inside the 1:12 resonance zone of the upper region (b). From A. Algaba, M. Merino and A.J. Rodríguez-Luis, Takens-Bogdanov bifurcations of periodic orbits and Arnold's tongues in a three-dimensional electronic model, *Internat. J. Bifur. Chaos Appl. Sci. Engrg.* 11 (2001) 513–531 © 2001 by World Scientific Publishing; reprinted with permission.

After we discussed the evolution of the Arnol'd tongues from closed to open (when γ is changed) and how Takens-Bogdanov points of periodic orbits appear, we now describe the bifurcation set in a neighborhood of these Takens-Bogdanov points. To illustrate that the respective dynamical behavior occurs generically for all the weak 1:p resonances (not only for the 1:11 resonance) we consider the 1:12 resonance for $\gamma = -0.6$. As mentioned before, it is the first resonance that appears in an open region.

Figure 7.10(a) shows the principal torus curve HH and the Arnol'd tongues of the 1:12 resonance SN_{12}. Notice how the torus curves $HH1_{12}$ and $HH2_{12}$ connect the saddle-node curves SN_{12} with the period-doubling curve PD_{12}. To see clearly what bifurcations are present we show in Fig. 7.10(b) an enlargement of the region where the curve $HH1_{12}$ exists. We observe a cascade of torus bifurcations $HH1_{12}$, $HH1_{24}$, and so on, that connects SN_{12} with PD_{12}, PD_{12} with PD_{24}, and so on. Therefore, we have found a cascade of Takens-Bogdanov bifurcations of periodic orbits. The first point TB_{12} is of quadratic type whereas the other Takens-Bogdanov points ($TBS1_{12}$, $TBS1_{24}$, and so on) are of cubic homoclinic type.

7.2.2 Isolas, Cusps and Global Bifurcations

We start with a regime that is organized by homoclinic connections of Shil'nikov type. Specifically, we report on a detailed numerical study of one type of periodic orbit that exhibits an isola structure in the bifurcation diagram and cusps of saddle-node bifurcations in a two-parameter bifurcation set; see [23] for details. The mechanism of isola-creation and its relationship with global connections is also described. To this end we consider (7.2) for fixed $\gamma = 0$, $A_3 = 0.3286$, $B_3 = 0.9336$ and $C_3 = 0$.

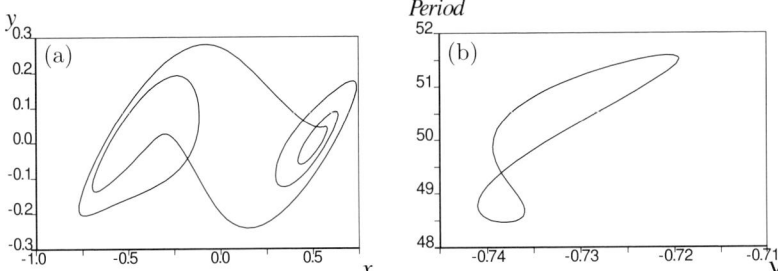

Fig. 7.11. Projection onto the (x, y)-plane of a periodic orbit for $\beta = 0.4$, $r = 0.6$ (a) that gives rise to a figure-8 isola (b) when the period is plotted against ν. From F. Fernández-Sánchez, E. Freire and A.J. Rodríguez-Luis, Isolas, cusps and global bifurcations in an electronic oscillator, *Dynam. Stab. Sys.* 12 (1997) 319–336 © 1997 by Taylor & Francis; reprinted with permission.

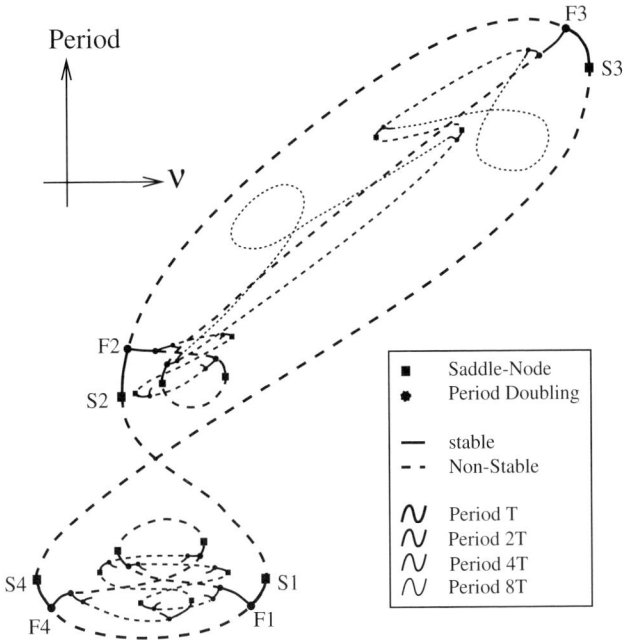

Fig. 7.12. Schematic partial bifurcation diagram for $\beta = 0.4$, $r = 0.6$ of the periodic orbits with periods 1, 2, 4 and 8; solid curves indicate stable orbits, while dashed curves correspond to orbits of saddle-type. For the sake of clarity, only two branches of the 8-periodic orbits are drawn. From F. Fernández-Sánchez, E. Freire and A.J. Rodríguez-Luis, Isolas, cusps and global bifurcations in an electronic oscillator, *Dynam. Stab. Sys.* 12 (1997) 319–336 © 1997 by Taylor & Francis; reprinted with permission.

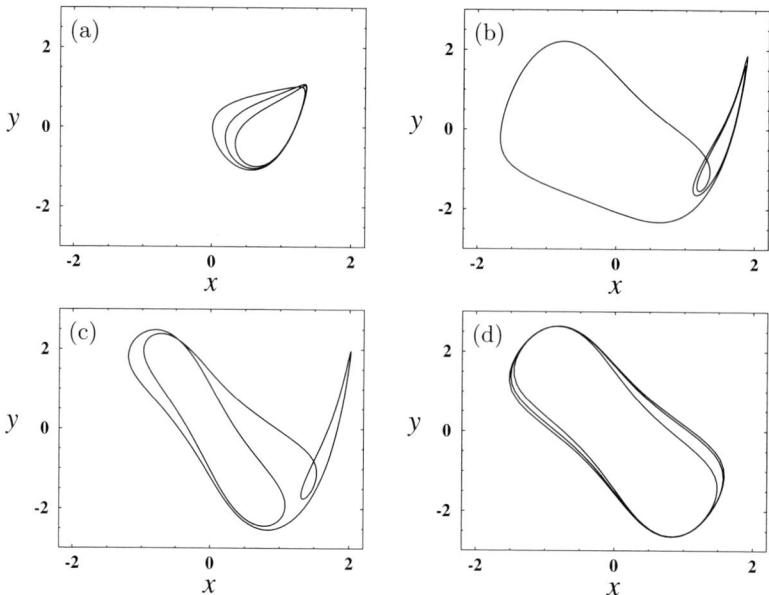

Fig. 7.13. Evolution of the oscillation-sliding phenomenon. Projection of the periodic orbits onto the (x, y)-plane for $\nu = -1.041162$ (a), $\nu = -1.793874$ (b), $\nu = -2.045187$ (c), and $\nu = -2.119680$ (d). From A. Algaba, F. Fernández-Sánchez, E. Freire, E. Gamero and A.J. Rodríguez-Luis, Oscillation-sliding in a modified van der Pol-Duffing electronic oscillator, *J. Sound Vibr.* 249 (2002) 899–907 © 2002 by Elsevier Science; reprinted with permission.

Numerical continuation of the periodic orbit shown in Fig. 7.11(a) shows that its branch in the bifurcation diagram is a closed curve in the shape of a figure-of-eight; see Fig. 7.11(b). The stability of the orbits in this isola is shown in Fig. 7.12. Four saddle-node bifurcations, S1 to S4, and four period-doubling bifurcations, F1 to F4, appear on the figure-8 isola. Also sketched in Fig. 7.12 is the intricate arrangement of the branches of the first sub-harmonic periodic orbits, some of which form closed loops and others are S-shaped.

7.2.3 Oscillation-Sliding Between Two Periodic Regimes

We now consider a type of periodic behavior exhibited by (7.2) near a degenerate Hopf-pitchfork bifurcation [1], where we fix parameters to $r = 0.6$, $A_3 = 0.5$, $B_3 = 0.01$, and $C_3 = -0.1$. A degenerate case of the Hopf-pitchfork bifurcation takes place at $\gamma \approx -0.2473$, and we fix γ close to this special value, namely at $\gamma = -0.24$.

In this context, the principal periodic orbit undergoes a secondary Hopf bifurcation that gives rise to the appearance of an invariant torus. Its break-up results in the presence of resonance phenomena with subharmonic periodic

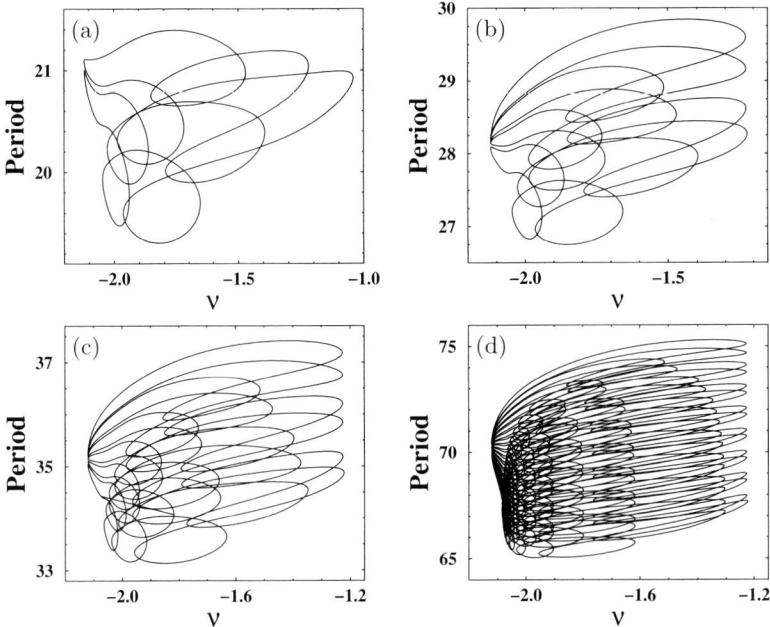

Fig. 7.14. Bifurcation diagram of the period versus ν for 3T-periodic orbits (a), 4T-periodic orbits (b), 5T-periodic orbits (c), and 10T-periodic orbits (d).

orbits. The numerical continuation of a 3T-periodic orbit in Fig. 7.13 shows that the periodic orbit 'slides' between two different oscillation regimes: a small- and a large-amplitude periodic orbit.

The bifurcation diagram of this 3T-periodic orbit corresponds to an isola, shown in Fig. 7.14(a), with sixteen saddle-node bifurcations. But this oscillation-sliding behavior seems to be present as well in all other subharmonic periodic orbits. Figure 7.14(b)–(d) shows the corresponding bifurcation diagrams for the 4T-, 5T- and 10T-periodic orbits, respectively. Observe the nice isolas obtained and some rules that they seems to follow. For instance, the saddle-nodes on the right are organized by pairs that occur approximately at the same parameter value; there are exactly $n-2$ pairs in the nT-periodic orbit branch. A more detailed understanding of this intricate and aesthetic dynamical behavior remains a challenge for future research.

7.2.4 T-Point Bifurcation

In a three-dimensional system with at least two equilibria a T-point occurs when the one-dimensional unstable manifold of one equilibrium and the one-dimensional stable manifold of the other equilibrium coincide. At the same time, the two-dimensional manifolds of the two equilibria have a transversal

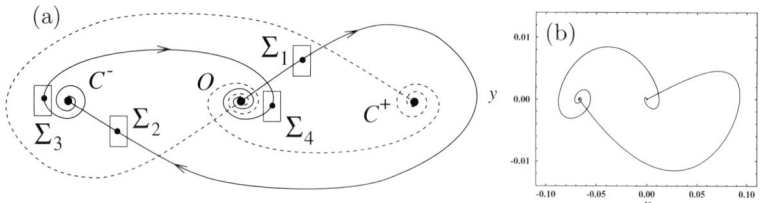

Fig. 7.15. Panel (a) is a sketch of a heteroclinic T-point cycle in phase space; also drawn (dashed line) is its symmetric counterpart. Panel (b) shows the primary T-point heteroclinic orbit in projection onto the (x, y)-plane that exists for $\gamma = -0.6$ and $(\nu, \beta) \approx (-0.7605, 0.7548)$. From F. Fernández-Sánchez, E. Freire and A.J. Rodríguez-Luis, T-points in a $\mathbb{Z}_2$-symmetric electronic oscillator, *Nonlin. Dynam.* 28 (2002) 53–69 © 2002 by Springer; reprinted with permission.

intersection that forms a heteroclinic loop between them. This codimension-two heteroclinic loop is usually referred to as a T-point bifurcation.

The unfolding in the vicinity of a T-point in a system with one real saddle (with three real eigenvalues) and a saddle-focus has been analyzed in the literature. Glendinning and Sparrow [30] consider T-point bifurcations of this sort in the Lorenz systems (which has an extra symmetry); Bykov considers in [12] also the case of two real saddles and in [13, 14] the case where both equilibria are saddle-foci. T-point bifurcations in $\mathbb{Z}_2$-symmetric systems are considered in [24], where it is shown by means of a Shil'nikov-type analysis that three spiral curves of codimension-one global bifurcations emerge from the T-point. The first corresponds to homoclinic orbits to the origin, the second to homoclinic connections of the nontrivial equilibria, and the third to heteroclinic orbits between the nontrivial equilibria. Figure 7.15(a) shows a sketch of the heteroclinic T-point cycle between the equilibrium at the origin and the two (symmetry-related) nontrivial equilibria, $C^{\pm}$. The four planes Σ_i are used in the construction of a Poincaré return map; the flow is divided into four parts; the four corresponding maps are composed [24].

To study T-point bifurcations in (7.2) we fix parameters at $r = 0.6$, $A_3 = 0.3286$, $B_3 = 0.9336$ and $C_3 = 0$ in this section. Then for $\gamma = -0.6$ a primary T-point exists for $(\nu, \beta) \approx (-0.7605, 0.7548)$; it is shown in Fig. 7.15(b). Figure 7.16 shows the three curves HO, HNT and Het of global bifurcations that were introduced above. When plotted in the (ν, β)-plane, as in panel (a), the three spirals are so close that it is almost impossible to distinguish them. To 'open up' these curves (one by one) one may perform successive changes in the parameters (one translation, one rotation and one rescaling). Figure 7.16(b) shows the result of such a rescaling to new parameters $\nu_{\star}$ and $\beta_{\star}$ for the homoclinic orbit of the origin HO. Observe how this curve spirals around the T-point. To open up the other two curves HNT and Het different changes of parameters are needed.

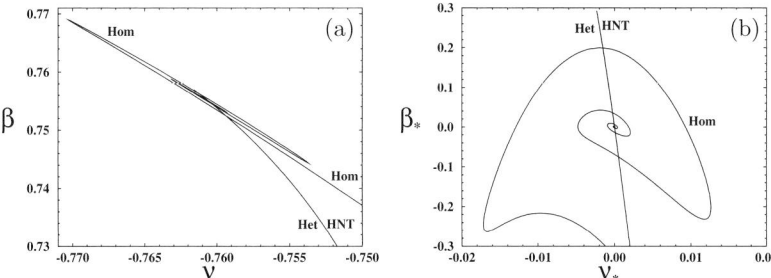

Fig. 7.16. Curves spiraling around the T-point for $\gamma = -0.6$ of homoclinic orbits of the origin, HO, homoclinic orbits of the nontrivial equilibria, HNT, and heteroclinic orbits between the nontrivial equilibria, Het, shown in the (ν, β)-plane (a), and in the $(\nu_\star, \beta_\star)$-plane (b). From F. Fernández-Sánchez, E. Freire and A.J. Rodríguez-Luis, T-points in a $\mathbb{Z}_2$-symmetric electronic oscillator, *Nonlin. Dynam.* 28 (2002) 53–69 © 2002 by Springer; reprinted with permission.

Finally we illustrate in Fig. 7.17 how the homoclinic connection of the origin approaches the nontrivial equilibrium on its way towards the T-point. Panel (a) shows the curve of homoclinic connections of the origin for $\gamma = -0.2$ as it spirals into the T-point at $(\nu, \beta) \approx (-0.7098, 0.4796)$. We focus our attention on three points on this curve (marked by dots) that lie to the left of the T-point and have the same value of β. The corresponding homoclinic connections are shown in Fig. 7.17(b)–(d). Notice how the homoclinic orbit makes, roughly speaking, one more turn around the (left) nontrivial equilibrium when the homoclinic curve 'turns' one more time around the T-point in the parameter plane. This is in accordance with the theoretical predictions [24, Fig. 6].

7.2.5 T-Point-Hopf Bifurcation

It is possible to continue a curve of T-points in a three-dimensional parameter space with the package AUTO. To perform this calculation we consider the linear approximations of the one-dimensional manifolds of the equilibria involved in this heteroclinic loop. Further, we assume that a transversal intersection between the respective two-dimensional manifolds of the equilibria (the orbit that closes the loop) indeed exist. In our calculation we consider system (7.2) for fixed $A_3 = 0.3286$, $B_3 = 0.9336$ and $C_3 = 0$.

The continuation in (ν, β, γ)-space of the primary T-points leads to the curve shown in Fig. 7.18(a). It starts at a triple-zero degeneracy of the origin, TZ, and ends at the point H when it intersects the surface where the origin exhibits a Hopf bifurcation. Two more degenerate points are detected along this curve, both when the Shil'nikov quotient δ (also called the saddle-quantity) of the origin takes the special values $\delta = 1$ at D1 (neutral saddle-focus) and

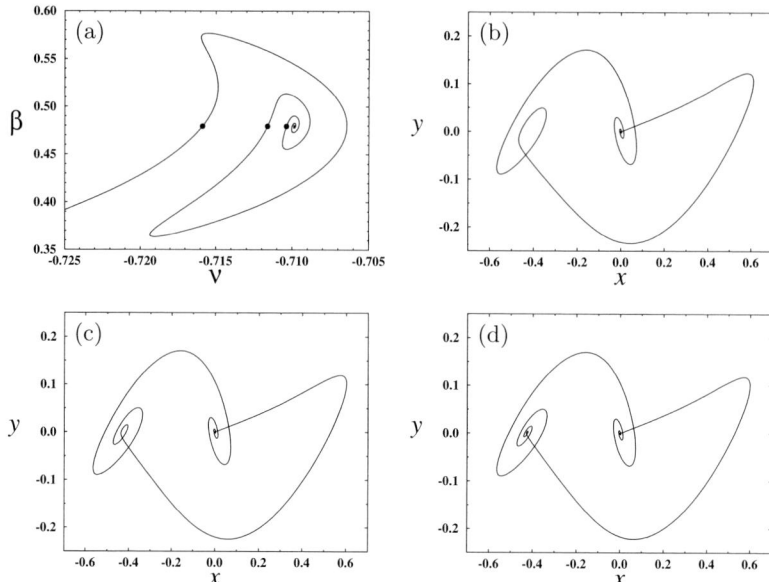

Fig. 7.17. Curve of homoclinic connections of the origin spiraling around the T-point for $\gamma = -0.2$ (a). Panel (b)–(d) are projections onto the (x, y)-plane of three homoclinic orbits corresponding to the three points (from left to right) that are marked by dots in panel (a). From F. Fernández-Sánchez, E. Freire and A.J. Rodríguez-Luis, T-points in a $\mathbb{Z}_2$-symmetric electronic oscillator, *Nonlin. Dynam.* 28 (2002) 53–69 © 2002 by Springer; reprinted with permission.

$\delta = 1/2$ at D05 (neutrally-divergent saddle-focus). Some features of the complex dynamics originating in these codimension-two homoclinic bifurcation can be found in [16] and references therein.

As in each plane of constant γ, the homoclinic curve of the origin spirals around the T-point. When γ is added as the third bifurcation parameter, a surface of homoclinic connections is expected (at least locally) to spiral around the T-point bifurcation curve TP. On this surface, there will appear curves of codimension-two homoclinic connections that end at the corresponding point on the T-point curve. In this way, the curve of degenerate homoclinic bifurcations of the origin with $\delta = 1$ appears in Fig. 7.18(b) in the vicinity of the point D1 where it ends on the curve TP. Similarly, in Fig. 7.18(c) the curve of degenerate homoclinic bifurcations of the origin with $\delta = 1/2$ is shown spiraling around TP. Finally, the curve of Shil'nikov-Hopf homoclinic connections appears in Fig. 7.18(d); it ends at the T-point-Hopf point H in panel (a).

We now investigate the evolution of the primary homoclinic connection in the (ν, β)-parameter plane when the curve of T-points approaches its endpoint, where a T-point-Hopf bifurcation occurs. In fact, the T-point curve TP emerges from a triple-zero degeneracy that occurs for $\gamma \approx -0.7746$. When increasing

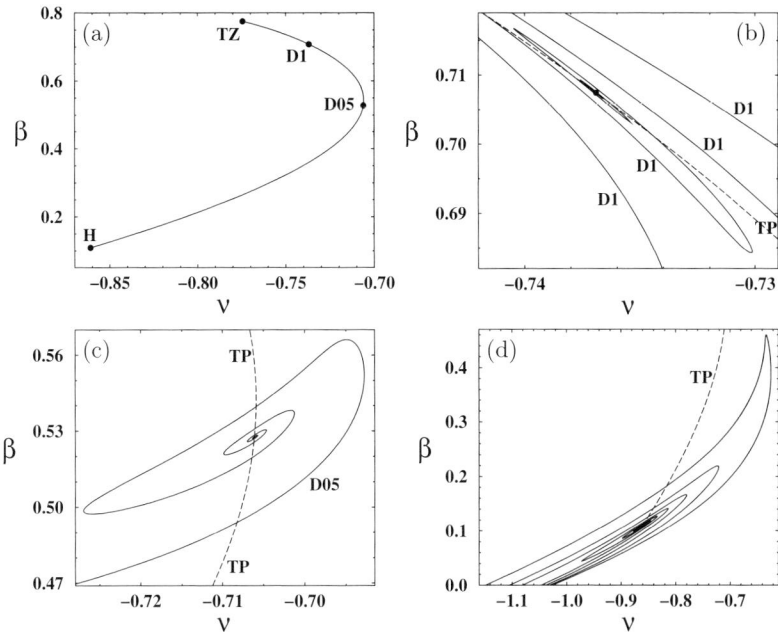

Fig. 7.18. Projections onto the (ν, β)-plane of global bifurcation curves that have been continued in (ν, β, γ)-space. Panel (a) shows the curve of primary T-points; the following points are marked on this curve: triple-zero point TZ at $(\nu, \beta, \gamma) \approx (-0.77457, 0.77457, -0.77457)$; $\delta = 1$ point D1 at $(\nu, \beta, \gamma) \approx (-0.73691, 0.70745, -0.45225)$; $\delta = 1/2$ point D05 at $(\nu, \beta, \gamma) \approx (-0.70614, 0.52746, -0.22966)$; a T-point-Hopf H at $(\nu, \beta, \gamma) \approx (-0.86103, 0.10899, -0.11867)$. Panel (b) shows the curve of degenerate homoclinic bifurcations of the origin with $\delta = 1$, panel (c) the curve of degenerate homoclinic bifurcations of the origin with $\delta = 1/2$, and panel (d) the curve of Shil'nikov-Hopf homoclinic bifurcations; the dashed curve TP is the curve of primary T-points.

γ, for instance to $\gamma = -0.6$, the curve Hom of primary homoclinic bifurcation of the origin emerges from a Takens-Bogdanov bifurcation point and ends spiraling into a T-point; see Fig. 7.16(a). For $\gamma \approx -0.5921$ there is a tangency (at $(\nu, \beta) \approx (-0.6331, 0.4598)$) between the curve Hom and the curve H of Hopf bifurcation of the origin, which gives rise to a so-called non-transverse Shil'nikov-Hopf bifurcation point [17]. (See Sect. 7.2.6 below for some more comments about this bifurcation.) Note that this non-transverse bifurcation appears as a consequence of the upper limit point (with respect to γ) that the Shil'nikov-Hopf curve exhibits; see the upper-right corner of Fig. 7.18(d) (although it shows the (ν, β)-plane).

For γ above the non-transverse critical value, two Shil'nikov-Hopf bifurcations appear. The situation for $\gamma = -0.3$ is drawn in Fig. 7.19(a). One branch of the curve joins the Takens-Bogdanov point TB with a Shil'nikov-Hopf point

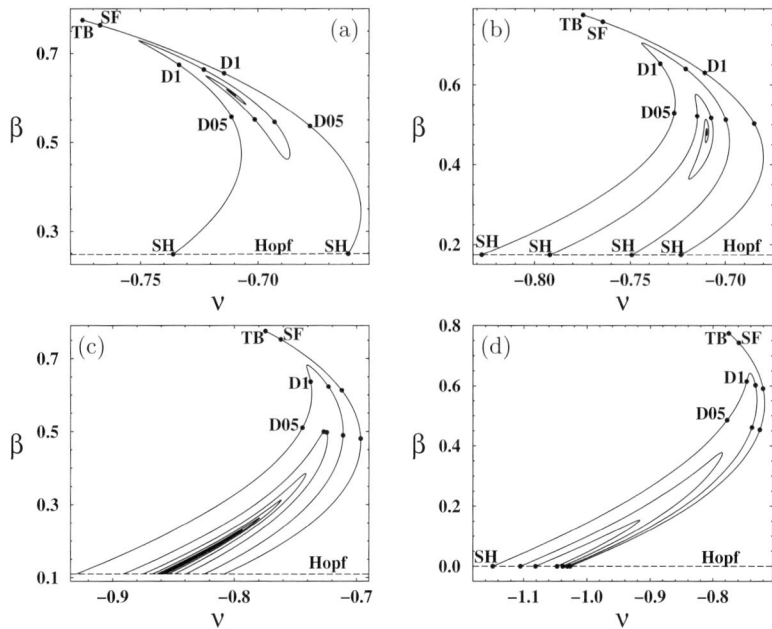

Fig. 7.19. Curve of homoclinic connections to the origin in the (ν, β)-parameter plane for $\gamma = -0.3$ (a), $\gamma = -0.2$ (b), $\gamma = -0.12$ (c), and $\gamma = 0$ (d). The dashed curve `Hopf` corresponds to a Hopf bifurcation of the origin; also marked on the homoclinic curve are points `TB` of Takens-Bogdanov bifurcation, transition points `SF` of the equilibrium from saddle to saddle-focus, points `D1` where $\delta = 1$, and `D05` where $\delta = 0.5$, and Shil'nikov-Hopf points `SH`. For sake of clarity, we have not marked all `SH` points in panels (c) and (d).

`SH`, whereas the second branch emerges from another `SH` point and ends spiraling around the T-point. The codimension-two homoclinic bifurcations that occur when the origin changes from saddle to saddle-focus (marked as `SF`), when $\delta = 1$ (marked as `D1`), and when $\delta = 1/2$ (marked as `D05`) are also indicated in the picture.

When γ is increased to $\gamma = -0.2$, as in Fig. 7.19(b), a new pair of Shil'nikov-Hopf points appear. This is due to the existence of another non-transverse Shil'nikov-Hopf bifurcation for $\gamma \approx -0.2594$. Now three different branches of the homoclinic curve exist.

As the T-point-Hopf for $\gamma \approx -0.11867$ is approached, new non-transverse Shil'nikov-Hopf bifurcations appear and the curve of homoclinic connections splits into more branches. This phenomenon is a direct consequence of the limit points on the Shil'nikov-Hopf curve as it spirals around the T-point-Hopf bifurcation point. For example, for $\gamma = -0.12$ eighteen Shil'nikov-Hopf bifurcations occur; see Fig. 7.19(c). Note that infinitely many Shil'nikov-Hopf points are predicted by the time the T-point-Hopf occurs [25].

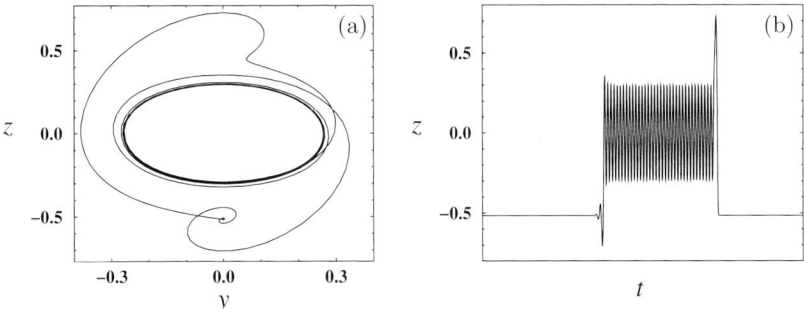

Fig. 7.20. One of the two homoclinic orbits to a nontrivial equilibrium for $\nu = -0.9$. Shown are its projection on to the (x, y)-plane (a), and its time profile of variable z (b).

Past the T-point-Hopf bifurcation, branches of the homoclinic curve disappear as a consequence of non-transverse Shil'nikov-Hopf points (now in the downward-pointing case [17]). For example, for $\gamma = 0$, as in Fig. 7.19(d), the homoclinic curve has four branches and only seven Shil'nikov-Hopf points remain. Furthermore, as Fig. 7.20(a) shows, the homoclinic orbit of the non-trivial equilibria winds closely around the periodic orbit that emerged in the Hopf bifurcation before returning to the equilibrium. In this way, a heteroclinic loop is formed between an equilibrium and the periodic orbit [25]. The time profile in Fig. 7.20(b) shows that the homoclinic orbit indeed spends a lot of time near the saddle periodic orbit. Indeed the orbit in Fig. 7.20 is a nice numerical approximation of the heteroclinic loop. The loop itself consists of an intersection between the two-dimensional unstable manifold of the non-trivial equilibrium and the stable manifold of the saddle periodic orbit, and an intersection between the unstable manifold of the periodic orbit and the one-dimensional stable manifold of the nontrivial equilibrium.

7.2.6 Non-Transverse Shil'nikov-Hopf Bifurcation

The theoretical analysis of the non-transverse Shil'nikov-Hopf bifurcation [17, 25] shows that it contains codimension-two non-transversal homoclinic orbits to equilibria and non-transversal homoclinic tangencies to periodic orbits in its unfolding. Two cases are classified: the downward-pointing and the upward-pointing case, depending on whether the variation of a third parameter causes either the annihilation of a locus of saddle-focus homoclinic orbits to equilibria, or the uncoupling of this locus from the locus of Hopf bifurcations. The downward-pointing case of a non-transverse Shil'nikov-Hopf bifurcation is shown to cause two wiggly curves to coalesce and leave behind finitely many isolas of periodic orbits. The upward-pointing case, on the other hand, causes two wiggly curves to coalesce first into infinitely many and then into finitely many isolas.

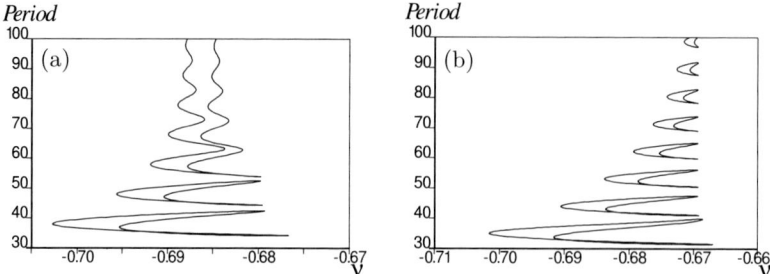

Fig. 7.21. Bifurcation diagrams for $\gamma = -0.65$ for asymmetric periodic orbits in the upward-pointing case for $\beta = 0.62$ (a) and $\beta = 0.58$ (b). From A.R. Champneys and A.J. Rodríguez-Luis, The non-transverse Shil'nikov-Hopf bifurcation: uncoupling of homoclinic orbits and homoclinic tangencies, *Physica D* 128 (1999) 130–158 © 1999 by Elsevier Science; reprinted with permission.

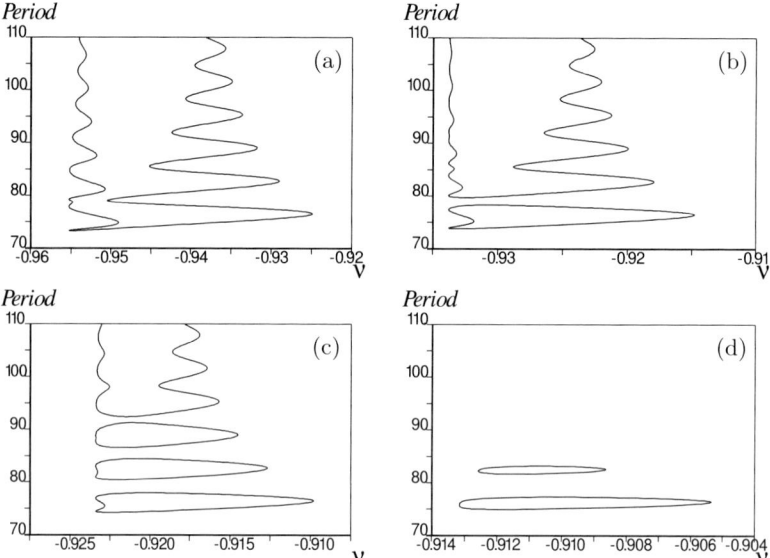

Fig. 7.22. Bifurcation diagrams for $\gamma = 0$ for asymmetric periodic orbits in the downward-pointing case for $\beta = 0.12$ (a), $\beta = 0.14$ (b), $\beta = 0.15$ (c), and $\beta = 0.16$ (d). From A.R. Champneys and A.J. Rodríguez-Luis, The non-transverse Shil'nikov-Hopf bifurcation: uncoupling of homoclinic orbits and homoclinic tangencies, *Physica D* 128 (1999) 130–158 © 1999 by Elsevier Science; reprinted with permission.

Numerical evidence of this bifurcation was found in the electronic circuit (7.2) for fixed $r = 0.6$, $A_3 = 0.3286$, $B_3 = 0.9336$ and $C_3 = 0$.

Figure 7.21 shows two β-slices for the upward-pointing case for $\gamma = -0.65$. Panel (a) for $\beta = 0.62$ is close to the turning point of the homoclinic locus, and the two homoclinic orbits on the primary locus are connected by a single

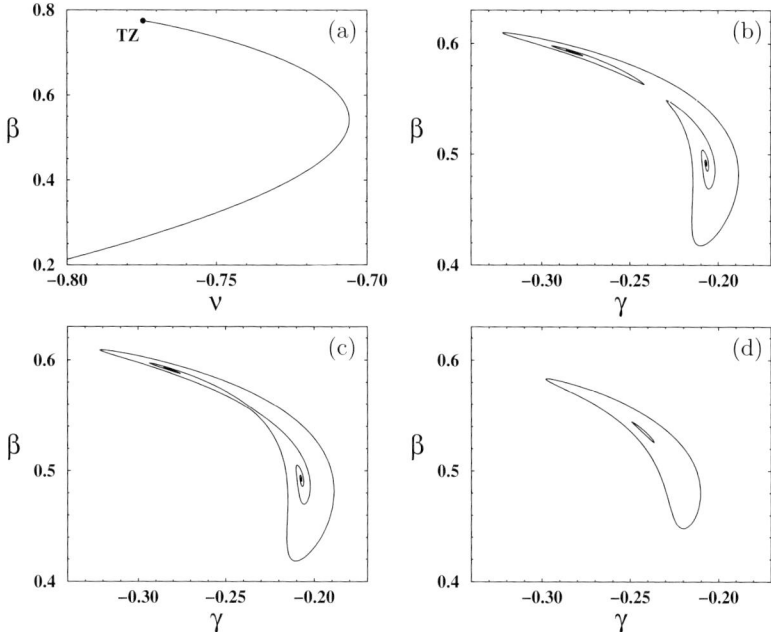

Fig. 7.23. Panel (a) shows the projection of the T-point bifurcation curve onto the (ν, β)-plane. It emerges from a triple-zero degeneracy, TZ, and has a fold for $\nu \approx -0.7059$. Panels (b)–(d) show partial bifurcation sets consisting of curves of homoclinic orbits to the origin for $\nu = -0.7085$, $\nu = -0.7084$ and $\nu = -0.7047$, respectively. From A. Algaba, F. Fernández-Sánchez, E. Freire, M. Merino and A.J. Rodríguez-Luis, Nontransversal curves of T-points: a source of closed curves of global bifurcations, *Phys. Lett. A* 303 (2002) 204–211 © 2002 by Elsevier Science; reprinted with permission.

branch of asymmetric periodic orbits. There are two isolas for small period, and the two wiggly curves approaching the two homoclinic orbits are connected. As β is decreased, the two wiggly curves can be seen to annihilate each other by forming more and more isolas. Figure 7.21(b) shows for $\beta = 0.58$ the first eight isolas in an evident destruction process.

To find the downward-pointing case it is necessary to increase γ to beyond $\gamma \approx -0.1187$, that is, to the other side of the T-point-Hopf bifurcation. Figure 7.22 shows bifurcation diagrams in four β-slices for $\gamma = 0$ that illustrate the process in which isolas are created and destroyed.

7.2.7 Non-Transversal T-Point Bifurcation

A model to explain the existence of closed bifurcation curves of homoclinic and heteroclinic connections in autonomous three-dimensional systems is derived in [2]. This scenario is related to the failure of transversality in a curve of

T-points. The predictions deduced from this model strongly agree with the numerical results obtained for system (7.2) for fixed $r = 0.6$, $A_3 = 0.3286$, $B_3 = 0.9336$ and $C_3 = 0$. This phenomenon was also found in an ODE model of a laser with optical injection; see [48] and Chap. 6.

The presence of a fold in the curve of T-point bifurcations in the (ν, β, γ)-space is observed in Fig. 7.23(a). This curve emerges from a triple-zero degeneracy of the origin, marked TZ [29]. The evolution of the curves of homoclinic orbits to the origin in the vicinity of the non-transversal T-point are shown in the (γ, β)-plane for several values of ν in Fig. 7.23(b)–(d). The first two slices, each with two T-points, show how the first closed curve of homoclinic connections is formed. On the other side of the critical value of ν, where the fold occurs, only closed curves appear; Fig. 7.23(d) shows the last two of them.

We remark that other bifurcation curves in the bifurcation set (for example, saddle-node and period-doubling bifurcations) must be expected to be influenced by these changes to the curves of homoclinic connections. Furthermore, some additional degeneracies may be exhibited by the global connections, which would imply an even richer bifurcation scenario; cf. [16] and Chap. 6.

7.2.8 Bi-Spiraling Curves of Homoclinic Orbits Around a T-Point

In [22] a model was proposed that considers a non-transversal intersection between the two-dimensional manifolds of the saddle-focus equilibria involved in a T-point. The study of this model shows the presence of bi-spiraling curves of homoclinic connections in the parameter plane: the spiral curve that emerges from a T-point between two saddle-focus equilibria ends at the same T-point, which it enters via a different spiral. The predictions deduced from this model strongly agree with the numerical results obtained for (7.2) where we fix $\gamma = 0.6$, $r = 0.6$, $A_3 = 0.3286$, $B_3 = 0.9336$ and $C_3 = 0$.

Due to the $\mathbb{Z}_2$-symmetry, three curves of global connections emerge from every T-point. However, in the case we consider here, only the curve of homoclinic connections to the origin bi-spirals around the T-point TP, as is shown in Fig. 7.24(a). The other two curves (of homoclinic and heteroclinic connections to the nontrivial equilibria, respectively) emerge from TP but do not return to it. This is why they are not shown in the figure. As the bi-spiraling homoclinic curve is very close to itself when shown in the plane (ν, β), we proceed to 'open it up' with a combination of translations, rotations and scalings as we did in Fig. 7.16. The result in the rescaled (ν_*, β_*)-plane is shown in Fig. 7.24(b), where the bi-spiraling is now clearly visible.

Finally, Fig. 7.24(c) and (d) show the two heteroclinic cycles that co-exist at the T-point TP: they both have the same codimension-two connection between their one-dimensional manifolds (solid lines), but differ in the transversal intersection between the two-dimensional manifolds (dashed lines).

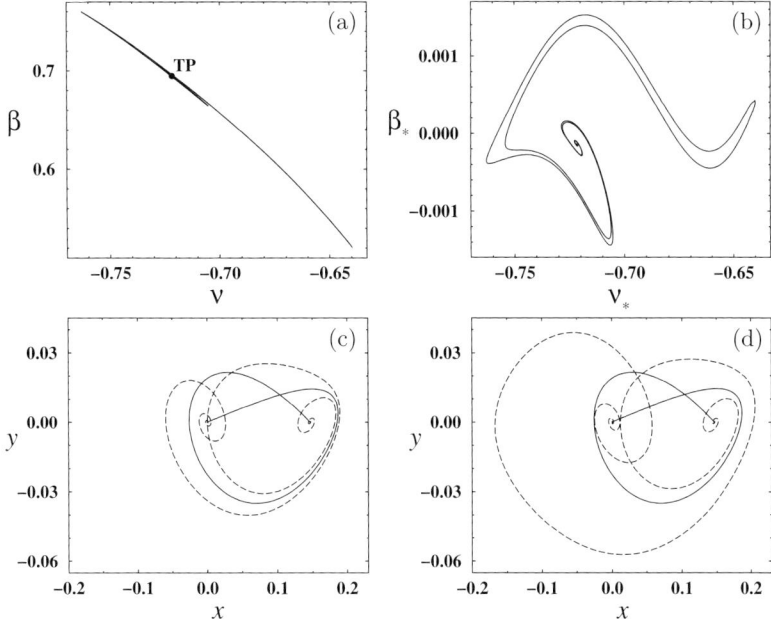

Fig. 7.24. A bi-spiraling curve of homoclinic connections to the origin around the T-point TP, shown in the (ν, β)-parameter plane of (7.2) (a) and in the rescaled (ν_*, β_*)-plane where the bi-spiraling effect is clearly visible (b). Panels (c) and (d) are two heteroclinic cycles that co-exist at the T-point TP; both cycles have the same connection (solid lines) between the one-dimensional manifolds but different intersections between the two-dimensional manifolds (dashed lines). From F. Fernández-Sánchez, E. Freire, L. Pizarro and A.J. Rodríguez-Luis, A model for the analysis of the dynamical consequences of a nontransversal intersections of the two-dimensional manifolds involved in a T-point, *Phys. Lett. A* 320 (2003) 169–179 © 2003 by Elsevier Science; reprinted with permission.

7.3 Conclusions and Outlook

A local analysis of degeneracies of equilibria provides interesting information about a great variety of behaviors, including equilibria, periodic orbits, global connections, and their bifurcations. The validity of such an analysis is restricted, in principle, to a neighborhood in the product of phase and parameter space of the degenerate equilibrium in question. The crux of the matter is that this neighborhood is generally unknown a priori, but it may actually be of relevant size.

   In this survey we have shown with practical examples of electronic circuits how continuation methods can be used to extend local results and to find new phenomena. In an interesting feedback process, new bifurcation phenomena lead to further analytical and numerical study. By contrast, brute-force simulation usually provides poor results at large computational expense.

Analogue electronic circuits are of obvious importance in many applications. What is more, they are particularly useful devices from the dynamical systems point of view, because they allow one, in certain situations, to reproduce the behavior of a given dynamical system [21, 42, 44]. Indeed, many interesting dynamical phenomena may be studied experimentally by assembling inexpensive electronic components that are readily available.

The study of electronic circuits presented here also provides evidence that new theoretical and numerical tools need to be developed for a better understanding of possible dynamical behavior. Some of the numerical challenges are:

1. the detection and continuation of bifurcations of periodic orbits of higher codimension (in the way it is done, for example, in CONTENT and MAT-CONT);
2. the continuation of bifurcations in piecewise linear systems;
3. the continuation of quasiperiodic orbits; and
4. the inclusion of some continuation procedures into the circuit simulator SPICE [45].

Acknowledgements

We would like to thank A. Algaba, F. Fernández-Sánchez, E. Gamero and M. Merino for helpful suggestions, and the editors of this volume for their effort and comments, which have been very much appreciated. The material presented here is the result of a long-term collaborative effort. Apart from the colleagues above, we thank A.R. Champneys, E.J. Doedel, L. Pizarro and E. Ponce for their contributions. We especially thank Sebius Doedel for the friendship he showed during his visits to our research group in Seville. This work has been partially supported by the *Ministerio de Ciencia y Tecnología*, Plan Nacional I+D+I, in the frame of the projects BFM2001–2608, BFM2003–00336 and MTM2004-04066, and by the *Consejería de Educación y Ciencia de la Junta de Andalucía* (TIC-0130).

References

1. A. Algaba, F. Fernández-Sánchez, E. Freire, E. Gamero, and A. J. Rodríguez-Luis. Oscillation-sliding in a modified van der Pol-Duffing electronic oscillator. *J. Sound Vibr.*, 249:899–907, 2002.
2. A. Algaba, F. Fernández-Sánchez, E. Freire, M. Merino, and A. J. Rodríguez-Luis. Nontransversal curves of T-points: a source of closed curves of global bifurcations. *Phys. Lett. A*, 303:204–211, 2002.

3. A. Algaba, E. Freire, and E. Gamero. Hypernormal form for the Hopf-zero bifurcation. *Internat. J. Bifur. Chaos Appl. Sci. Engrg.*, 8:1857–1887, 1998.

4. A. Algaba, E. Freire, E. Gamero, and A. J. Rodríguez-Luis. Analysis of Hopf and Takens-Bogdanov bifurcations in a modified van der Pol-Duffing oscillator. *Nonlin. Dynam.*, 16:369–404, 1998.

5. A. Algaba, E. Freire, E. Gamero, and A. J. Rodríguez-Luis. On a codimension-three unfolding of the interaction of degenerate Hopf and pitchfork bifurcations. *Internat. J. Bifur. Chaos Appl. Sci. Engrg.*, 9:1333–1362, 1999.

6. A. Algaba, E. Freire, E. Gamero, and A. J. Rodríguez-Luis. A three-parameter study of a degenerate case of the Hopf-pitchfork bifurcation. *Nonlinearity*, 12:1177–1206, 1999.

7. A. Algaba, E. Freire, E. Gamero, and A. J. Rodríguez-Luis. A tame degenerate Hopf-pitchfork bifurcation in a modified van der Pol-Duffing oscillator. *Nonlin. Dynam.*, 22:249–269, 2000.

8. A. Algaba, M. Merino, and A. J. Rodríguez-Luis. Takens–Bogdanov bifurcations of periodic orbits and Arnold's tongues in a three-dimensional electronic model. *Internat. J. Bifur. Chaos Appl. Sci. Engrg.*, 11:513–531, 2001.

9. A. A. Andronov, E. A. Leontovich, I. I. Gordon, and A. G. Maier. *Theory of Bifurcations of Dynamic Systems on a Plane*. (Israel Program for Scientific Translations, Jerusalem, 1971).

10. A. A. Andronov and L. Pontriaguin. Systèmes grossiers. *Dokl. Akad. Nauk. SSSR*, 14:247–251, 1937.

11. A. A. Andronov, A. A. Vitt, and S. E. Khaikin. *Theory of Oscillators (transl. by F. Immerzi, first edition in Russian, Moscow, 1937)*. (Pergamon, London, 1966).

12. V. V. Bykov. The bifurcations of separatrix contours and chaos. *Physica D*, 62:290–299, 1993.

13. V. V. Bykov. On systems with separatrix contour containing two saddle-foci. *J. Math. Sci.*, 95:2513–2522, 1999.

14. V. V. Bykov. Orbit structure in a neighborhood of a separatrix cycle containig two saddle-foci. *AMS Translations, Series 2*, 200:87–97, 2000.

15. J. L. Callot, F. Diener, and M. Diener. Le problème de la "chasse au canard". *C. R. Acad. Sci. Paris (Ser. I)*, 23:1059–1061, 1978.

16. A. R. Champneys and Yu. A. Kuznetsov. Numerical detection and continuation of codimension-two homoclinic bifurcations. *Internat. J. Bifur. Chaos Appl. Sci. Engrg.*, 4:785–822, 1994.

17. A. R. Champneys and A. J. Rodríguez-Luis. The non-transverse Shil'nikov–Hopf bifurcation: uncoupling of homoclinic orbits and homoclinic tangencies. *Physica D*, 128:130–158, 1999.

18. F. Diener and M. Diener. Chasse au canard. I. Les canards. *Collect. Math.*, 32:37–74, 1981.

19. E. J. Doedel, A. R. Champneys, T. F. Fairgrieve, Yu. A. Kuznetsov, B. Sandstede, and X. Wang. AUTO97 Continuation and bifurcation software for ordinary differential equations, 1997. Available by anonymous ftp from `ftp.cs.concordia.ca` directory `pub/doedel/auto`.

20. E. J. Doedel, E. Freire, E. Gamero, and A. J. Rodríguez-Luis. An analytical and numerical study of a modified van der Pol oscillator. *J. Sound Vibr.*, 256:755–771, 2002.

21. F. Fernández-Sánchez, E. Freire, L. Pizarro, and A. J. Rodríguez-Luis. Analytical and numerical study of a van der Pol–Duffing oscillator. In J. L. Huertas and A. Rodríguez-Vázquez, editors, *Fourth International Workshop on Nonlinear Dynamics of Electronic Systems, NDES'96, Sevilla*, pages 321–326, 1996.

22. F. Fernández-Sánchez, E. Freire, L. Pizarro, and A. J. Rodríguez-Luis. A model for the analysis of the dynamical consequences of a nontransversal intersections of the two-dimensional manifolds involved in a T–point. *Phys. Lett. A*, 320:169–179, 2003.

23. F. Fernández-Sánchez, E. Freire, and A. J. Rodríguez-Luis. Isolas, cusps and global bifurcations in an electronic oscillator. *Dynam. Stab. Syst.*, 12:319–336, 1997.

24. F. Fernández-Sánchez, E. Freire, and A. J. Rodríguez-Luis. T-points in a $\mathbb{Z}_2$-symmetric electronic oscillator. (I) Analysis. *Nonlin. Dynam.*, 28:53–69, 2002.

25. F. Fernández-Sánchez, E. Freire, and A. J. Rodríguez-Luis. Analysis of the T-point–Hopf bifurcation. Preprint, 2005.

26. E. Freire, L. G. Franquelo, and J. Aracil. Periodicity and chaos in an autonomous electronic system. *IEEE Trans. Circ. Syst.*, 31:237–247, 1984.

27. E. Freire, E. Gamero, and A. J. Rodríguez-Luis. First-order approximation for canard periodic orbits in a van der Pol electronic oscillator. *Appl Math. Lett.*, 12:73–78, 1999.

28. E. Freire, A. J. Rodríguez-Luis, E. Gamero, and E. Ponce. A case study for homoclinic chaos in an autonomous electronic oscillator. A trip from Takens–Bogdanov to Hopf–Shil'nikov. *Physica D*, 62:230–253, 1993.

29. E. Gamero, E. Freire, A. J. Rodríguez-Luis, E. Ponce, and A. Algaba. Hypernormal form calculation for triple–zero degeneracies. *Bull. Belg. Math. Soc.*, 6:357–368, 1999.

30. P. Glendinning, and C. T. Sparrow. T-points: A codimension two heteroclinic bifurcation. *J. Stat. Phys.*, 43:479–488, 1986.

31. M. G. M. Gomes and G. P. King. Bistable chaos II. Bifurcation analysis. *Phys. Rev. A*, 46:3100–3110, 1992.

32. J. Guckenheimer and P. J. Holmes. *Nonlinear Oscillations, Dynamical Systems, and Bifurcations of Vector Fields*, volume 42 of *Applied Mathematical Sciences*. (Springer-Verlag, New York, 1986).

33. J. Guckenheimer and S. Kim. Dstool: A dynamical system toolkit with an interactive graphical interface. Applied mathematics report, Center for Applied Mathematics, Cornell University, Ithaca, New York, 1992.

34. J. J. Healey, D. S. Broomhead, K. A. Cliffe, R. Jones, and T. Mullin. The origins of chaos in a modified van der Pol oscillator. *Physica D*, 48:322–339, 1991.

35. P. Holmes. Ninety plus thirty years of nonlinear dynamics: less is more and more is different. *Internat. J. Bifur. Chaos Appl. Sci. Engrg.*, 15:2703–2716, 2005.

36. Yu. A. Kuznetsov. *Elements of Applied Bifurcation Theory*, volume 112 of *Applied Mathematical Sciences*. (Springer-Verlag, New York, 2004).

37. R. Madan, editor. *Chua's Circuit: a Paradigm for Chaos*, volume 1 of *World Scientific Series on Nonlinear Science, Series B*. (World Scientific Publishing, Singapore, 1993).

38. T. Matsumoto, L. O. Chua, and M. Komuro. The double scroll. *IEEE Trans. Circ. Syst.*, 32:797–818, 1985.

39. B. B. Peckman, C. E. Frouzakis, and I. Kevrekidis. Bananas and banana splits: A parametric degeneracy in the Hopf bifurcation for maps. *SIAM J. Math. Anal.*, 26:190–217, 1995.

40. L. Perko. *Differential Equations and Dynamical Systems*, volume 7 of *Texts in Applied Mathematics*. (Springer-Verlag, New York, 2001).

41. H. J. Poincaré. *Les Méthodes Nouvelles de la Mécanique Céleste*, volumes 1-3. (Gauthiers-Villars, Paris, 1892-1893-1899). (English translation edited by D. Goroff, published by the American Institute of Physics, New York, 1993.)

42. R. Rocha and R. F. Machado L. S. Martins-Filho. A methodology for the teaching of dynamical systems using analogous electronic circuits. *Internat. J. Electr. Eng. Educ.*, 43:334–345, 2006.

43. R. Shinriki, M. Yamamoto, and S. Mori. Multimode oscillations in a modified van der Pol oscillator containing a positive nonlinear conductance. *IEEE Proceedings*, 69:394–395, 1981.

44. M. Suneel. Electronic circuit realization of the logistic map. *Sādhanā*, 31:69–78, 2006.

45. P. Tuinenga. *SPICE: A guide to circuit simulation and analysis using PSpice*. (Prentice-Hall, Englewood Cliffs, NJ, 1995).

46. Y. Ueda. *The Road to Chaos, II*. (Aerial Press Inc., Santa Cruz, CA, 2001).

47. B. van der Pol and B. van der Mark. Frequency demultiplication. *Nature*, 120:363–364, 1927.

48. S. Wieczorek and B. Krauskopf. Bifurcations of n-homoclinic orbits in optically injected lasers. *Nonlinearity*, 18:1095–1120, 2005.

49. S. Wiggins. *Introduction to Applied Nonlinear Dynamical Systems and Chaos*, volume 2 of *Texts in Applied Mathematics*. (Springer-Verlag, New York, 2003).

50. A. K. Zvonkin and M. A. Shubin. Non-standard analysis and singular perturbations of ordinary differential equations. *Russian Math. Surveys*, 39:69–131, 1984.

8

Periodic Orbit Continuation in Multiple Time Scale Systems

John Guckenheimer and M Drew LaMar

Mathematics Department, Cornell University, USA

Continuation methods utilizing boundary value solvers are an effective tool for computing unstable periodic orbits of dynamical systems. AUTO [7] is *the* standard implementation of these procedures. Unfortunately, the collocation methods used in AUTO often require very fine meshes for convergence on problems with multiple time scales. This inconvenience prompts the search for alternative methods for computing such periodic orbits; we introduce here new multiple-shooting algorithms based on geometric singular perturbation theory.

8.1 Mathematical Setting

The systems that we study are *slow-fast* systems of the form

$$\begin{cases} \varepsilon \dot{x} = f(x,y), \\ \dot{y} = g(x,y), \end{cases} \tag{8.1}$$

where $f : \mathbb{R}^m \to \mathbb{R}^m$ and $g : \mathbb{R}^n \to \mathbb{R}^n$ are at least C^1 and $\varepsilon > 0$ is a small parameter determining the ratio of time scales between the fast variable $x \in \mathbb{R}^m$ and the slow variable $y \in \mathbb{R}^n$. The limit $\varepsilon = 0$ is a system of differential algebraic equations in which motion is constrained to the *critical manifold C* defined by $f = 0$. Rescaling time to the 'slow time' $\tau = \varepsilon t$ yields the system

$$\begin{cases} x' = f(x,y), \\ y' = \varepsilon g(x,y). \end{cases} \tag{8.2}$$

Here, the limit $\varepsilon = 0$ is the family of *layer* equations in y, also called the fast subsystems of (8.1). We make two standing assumptions about (8.1) that further constrain the context for our analysis:

1. The critical manifold C of (8.1) is indeed a manifold and its projection Π onto the space of slow variables is generic in the sense of singularity theory;

2. The limit sets of all trajectories for the layer equations are equilibria, i.e., points of C.

At regular points of the projection Π, the manifold C can be represented locally as the graph of a function $x = h(y)$. Substitution of this expression into g yields the *slow flow* on the regular points of C. We shall denote the set of singular points of Π by S.

Trajectories of (8.1) are typically approximated by *candidates*, concatenations of trajectories of the slow flow and the layer equations that form continuous curves. Periodic orbits that contain both segments close to trajectories of the slow flow and segments close to trajectories of the layer equations are called *relaxation oscillations*. Trajectory segments close to an unstable sheet of the critical manifold are called *canards*. Numerical computation of canards by forward solution of an initial value problem is not feasible when ε is sufficiently small due to the instability on the fast time scale [14]. Thus, even stable periodic orbits containing canards cannot be computed by forward numerical integration from initial points in the basin of attraction of these orbits. As mentioned earlier, tracking such orbits with collocation methods also seems to require very fine meshes. Our goal in this paper is to re-examine the computation of relaxation oscillations, including those with canards. We propose a multiple-shooting approach, in which different segments of a periodic orbit are computed differently and then matched with suitably chosen cross-sections.

The next two sections lay out the general framework that we investigate. Section 8.4 presents two numerical examples, comparing the methods introduced here with AUTO computations of the same orbits. Finally, Sect. 8.5 comments on further extension and improvement of these methods.

8.2 Simple Relaxation Oscillations

We consider first the simplest relaxation oscillations, namely, those with a fast-slow decomposition that makes them readily amenable to analysis. We define a relaxation oscillation Γ^ε to be simple if it is approximated by a candidate Γ^0 that satisfies the following conditions:

S1: Γ^0 consists of slow segments α_i and fast segments β_i, $i = 1 \ldots k$, in the order $\alpha_1, \beta_1, \ldots, \alpha_k, \beta_k$. The initial and terminal points of α_i are p_i and q_i. The initial and terminal points of β_i are q_i and $p_{(i+1) \bmod k}$.
S2: Each α_i lies on a stable sheet of the critical manifold.
S3: The points q_i are saddle-node bifurcations of the layer equations and none of their eigenvalues have positive real parts. This assumption implies that there are unique solutions of the layer equations emanating from the points q_i and all nearby fold points of C.
S4: The slow flow satisfies the normal crossing conditions [23] at q_i.
S5: Denote by $\omega(S)$ the forward limits of trajectories of the layer equations emanating from the fold points of S satisfying condition **S3**. We require that $\omega(S)$ is transverse to the slow flow on C at the points p_i.

Note that we have not required that a simple relaxation oscillation be stable or even hyperbolic, although eigenvalues of a return map in the fast directions are stable and, indeed, approach zero as $\varepsilon \to 0$.

We want to establish that well-conditioned multiple-shooting methods can be formulated for the computation of simple relaxation oscillations. Our strategy is to create cross-sections Σ_i to each of the fast segments β_i of a simple relaxation oscillation Γ^ε, compute the flow maps Φ_i^ε from Σ_i to $\Sigma_{(i+1) \bmod k}$ and then solve the multiple-shooting equations

$$z_{(i+1) \bmod k} = \Phi_i^\varepsilon(z_i)$$

for points $z_i \in \Sigma_i$.

Theorem 1. *Let Γ^ε be a hyperbolic simple relaxation oscillation, and Σ_i and Φ_i^ε as described above. For $\varepsilon \geq 0$ sufficiently small, the system of equations*

$$z_{(i+1) \bmod k} = \Phi_i^\varepsilon(z_i), \qquad z_i \in \Sigma_i,$$

is regular and has an isolated solution with $z_i = \Sigma_i \cap \Gamma^\varepsilon$.

Proof. Fenichel theory [10] and a theorem of Levinson [20] imply that the flow maps Φ_i^ε from Σ_i to $\Sigma_{(i+1) \bmod k}$ are smooth maps that converge to smooth maps Φ_i^0 of rank $n-1$ as $\varepsilon \to 0$. We remark that the convergence is continuous but not smooth in ε due to several phenomena; for example, asymptotic expansions of trajectories near the folds involve fractional powers of ε [25]. The point $\Phi_i^0(z_i)$ is obtained by a three-step process:

(1) follow the trajectory of the layer equations with initial condition (z_i) to its limit on the critical manifold;
(2) follow this limit point to its first intersection with a fold of the critical manifold; and
(3) follow the unstable separatrix of the layer equations from this fold point to its intersection with $\Sigma_{(i+1) \bmod k}$.

Denote the image of $\Phi_{(i-1) \bmod k}^0$ by W_i. The dimension of W_i is $n-1$, the dimension of folds of the critical manifold. Condition **S5** implies that the restriction of Φ_i^0 to W_i has full rank $n-1$. The restriction of the equations $z_{(i+1) \bmod k} = \Phi_i^\varepsilon(z_i)$ to W_i has the same structure as the set of equations for a multiple-shooting method based upon cross-sections to the flow. The Jacobian of this system has the block-cyclic structure

$$\begin{pmatrix} -D\Phi_1^0|_{W_1} & I & & & \\ & -D\Phi_2^0|_{W_2} & I & & \\ & & \ddots & \ddots & \\ & & & -D\Phi_{k-1}^0|_{W_{k-1}} & I \\ I & & & & -D\Phi_k^0|_{W_k} \end{pmatrix}.$$

As shown by Guckenheimer and Meloon [15], this matrix has maximal rank if and only if 1 is not an eigenvalue of $\mathrm{diag}(D\Phi_k^0|_{W_k},\ldots,D\Phi_1^0|_{W_1})$, that is, Γ^0 is a hyperbolic fixed point of the composition $\Phi_k^0|_{W_k} \circ \cdots \circ \Phi_1^0|_{W_1}$. On a complementary set of coordinates to the W_i the equations $z_{(i+1)\,\mathrm{mod}\,k} = \Phi_i^\varepsilon(z_i)$ reduce to $z_{(i+1)\,\mathrm{mod}\,k} = 0$ because $\Phi_i^\varepsilon(z_i)$ vanishes in these directions by definition. Thus, the full system of equations on the product of the Σ_i is regular if and only if Γ^0 is hyperbolic. The equations change continuously in the C^1 topology as $\varepsilon \to 0$ [16], so hyperbolicity of Γ^0 implies that the equations are regular and that Γ^ε is hyperbolic for $\varepsilon > 0$ sufficiently small. This proves the theorem. $\square$

For a hyperbolic simple relaxation oscillation a multiple-shooting algorithm based upon the cross-sections described above yields a regular system of equations. In many cases, these equations will be well conditioned uniformly for small ε. If they are not, additional cross-sections can be inserted. The effectiveness of the multiple shooting algorithm will be largely determined by the numerical integration method used to compute the Φ_i^ε.

8.3 Degenerate Slow-Fast Decompositions

The multiple-shooting algorithm described above for locating simple relaxation oscillations can be implemented within a standard continuation framework. The procedure may break down when a family of periodic orbits encounters degenerate slow-fast decompositions resulting from the failure of one of the requirements for the orbit to be simple. Here, we examine modifications of the multiple-shooting algorithm that cope with the instability of canards in the context of two specific examples of degenerate decompositions [11].

8.3.1 Hopf Bifurcation and Canard Explosions

The lowest-dimensional example of a degenerate slow-fast decomposition occurs at Hopf bifurcations of a systems with one slow and one fast direction ($n = m = 1$). The *canard explosion* of the resulting orbits has been studied extensively, especially in the system

$$\begin{cases} \varepsilon\dot{x} = y - \frac{1}{3}x^3 + x, \\ \dot{y} = a - x \end{cases} \tag{8.3}$$

near $a = 1$; see, for example, [2, 8, 9, 13]. (Note that in several studies coordinates have been used that place the point $(1, -2/3)$ of (8.3) at the origin [8].) It has been proven that the periodic orbits of this system grow monotonically as a decreases from 1, 'exploding' in size from $O(\varepsilon)$ to $O(1)$ over a range of a that is $O(\exp(-c/\varepsilon))$ for a suitable $c > 0$. The trajectories in the middle of this explosion contain canards that follow the unstable branch of the critical manifold given by $y = \frac{1}{3}x^3 - x$ for an $O(1)$ distance before jumping right or left to a stable branch of the critical manifold. Trajectories along the canard segments of these trajectories diverge from one another at a rate $\exp(-t(x^2 - 1)/\varepsilon)$. For small values of ε, this divergence effectively prevents accurate computation of a trajectory for times that are larger than $O(\varepsilon)$. However, backward integration along these canards is highly stable.

To compute the periodic orbits with canards in this family, we adopt a shooting strategy that shoots forward and backward from one cross-section of the flow to another cross-section. The cross-sections are chosen so that forward trajectories do not contain canard segments and backward trajectories do not contain segments that track the stable part of the slow manifold. The initial cross-section depends upon where we are in the family, in particular on the direction of the jump away from the canard segment in a periodic orbit. Over part of the family, the jump is to the right and there is a single stable slow segment in the periodic orbit. Over another part of the family, the jump is to the left and there are two stable slow segments and two fast segments in the slow-fast decomposition of the trajectory. The behavior that occurs between these two possibilities is that there is a *maximal canard* that extends over the entire length of the unstable branch of the critical manifold. When the jumps from the canards are to the right, we choose the initial cross-section to be the line $\{x = a\}$ where the vector field is horizontal. For jumps to the left, we choose the initial cross-section $\{x = -1\}$, which is crossed by all trajectories that flow left from the unstable branch of the critical manifold to the stable branch of the critical manifold. In both cases, we take the final target cross-section to be $\{x = a\}$.

The shooting problem that we seek to solve is $\Phi_a^+(y) = \Phi_a^-(y)$ where $\Phi_a^+(y)$ is the flow map from the initial to the final cross-section in the forward time direction and $\Phi_a^-(y)$ is the flow map from the initial to the final cross-section in the backward time direction. There are three remarks that we make about this problem:

1. The derivatives of both $\Phi_a^+(y)$ and $\Phi_a^-(y)$ are small for members of the canard family, but the derivative of $\Phi_a^+(y)$ is much smaller [8, 12], so the periodic orbits are stable;
2. The derivative of $\Phi_a^+(y) - \Phi_a^-(y)$ with respect to a is $O(1)$, so the small derivative of $\Phi_a^+(y) - \Phi_a^-(y)$ with respect to y yields an extreme sensitivity of the solution y to the shooting problem as a function of the parameter a. Therefore, in the middle of the canard family, we fix the initial point on the first cross-section and vary a to locate a periodic orbit passing through the initial point instead of fixing a and trying to locate the solution y of the shooting equation. Alternatively, we regard the shooting equation as a continuation problem in the variables (y, a) and use pseudo-arclength or other continuation strategies to find the curve of solutions to the equation;
3. If we move beyond the range of parameter values for which there are canards, we are likely to find trajectories that do not reach the target cross-section. This happens when there is a stable equilibrium point for

parameters $a > 1$. When a is large enough so that the vector field has a stable node, trajectories that follow the right-hand branch of the critical manifold accumulate at the stable equilibrium point without reaching the line $\{x = a\}$. Implementations of the shooting algorithms need to test for this possibility and take appropriate action if the shooting equation is not defined.

8.3.2 Folded Saddles

The forced Van der Pol system

$$\begin{cases} \varepsilon \dot{x} = y + x - \frac{x^3}{3}, \\ \dot{y} = -x + a \sin(2\pi\theta), \\ \dot{\theta} = \omega \end{cases} \tag{8.4}$$

is a slow-fast system with two slow variables and one fast variable ($n = 2$, $m = 1$). Cartwright and Littlewood studied this system in their seminal work on chaos in dynamical systems [4, 21, 22]. Recently, Haiduc [17] has extended the classical results of Cartwright and Littlewood by using geometric singular perturbation theory; these methods have also been used to investigate the dynamics of this system numerically [3, 14]. Throughout these studies, *folded saddles* play a prominent role in the analysis. Folded saddles are points where the rescaled slow flow equations

$$\begin{cases} \theta' = \omega(x^2 - 1), \\ x' = -x + a \sin(2\pi\theta) \end{cases} \tag{8.5}$$

have a saddle point. In the original system they are points on the fold curves where the normal crossing conditions fail. At the folded saddles the slow flow changes direction from pointing toward the fold to pointing away from the fold. Canards emanate from the folded saddles along stable manifolds of the saddles of (8.5) (lifted back to the unstable sheet of the critical manifold of (8.4)).

Bold et al. [3] used AUTO to track families of periodic orbits in the forced van der Pol system that contain canards emanating from folded saddles. These computations required fine meshes when applied to the system with $\varepsilon = 10^{-3}$ and even more so with $\varepsilon = 10^{-4}$. We develop here modifications similar to the ones described above for Hopf bifurcations to compute these trajectories with multiple-shooting methods that use a small number of cross-sections. We place cross-sections to the flow at the beginning of canard segments and in the middle of jumps that leave the canard segments. Backward integration between these cross-sections is stable since there is a single fast variable and the unstable sheet of the critical manifold is stable for the reversed time flow. At folded saddles of a system with two slow and one fast variables, the flow is parallel to the fold curve. For the forced Van der Pol system this suggests that we choose the cross-section defined by $\theta = \theta_{fs}$, where θ_{fs} is the value of

θ at the folded saddle. All trajectories that jump from the unstable sheet of the critical manifold to a stable sheet intersect one of the planes $\{x = \pm 1\}$, so we choose these planes as the cross-sections for the trajectories that jump from canards. Periodic orbits of different periods yield different sequences of intersections with the cross-sections and different defining equations. Unlike the canard explosions discussed in Sect. 8.3.1, many of the periodic orbits are embedded in chaotic invariant sets and are unstable.

8.4 Numerical Results

We computed families of periodic orbits containing canards in the Van der Pol and forced Van der Pol systems with AUTO [7] and with our shooting methods. Our calculations follow the approach in [13] to address the complexity of periodic orbits in systems with multiple time scales. Thus, we use as many as 1000 mesh points in AUTO and error tolerances for both AUTO and shooting on the order of 10^{-10}.

For the multiple-shooting algorithms, we used PYCONT, a continuation package developed for PYDSTOOL [5]. The shooting algorithms were embedded in a Moore-Penrose continuation framework [1], with the differential equations numerically integrated using Radau [18], an implicit Runge-Kutta method. We set the absolute error tolerances in the calculations with Radau to 10^{-12} and the relative error tolerance to 10^{-9}. These tolerances produced trajectories with sufficient precision so that the truncation errors did not appear to be significant for the Newton iterations within the shooting method.

In the next two sections we show results for each of the two examples mentioned above. After presenting the results, we do some comparisons between AUTO and multiple-shooting with the forced Van der Pol system. Since algorithmic performance is highly dependent on implementation details, e.g., programming language used, we assess algorithmic complexity based on four properties of Newton's method: the size of Jacobians, condition numbers of Jacobians, the number of Newton iterations, and domains of convergence.

8.4.1 Hopf Bifurcation and Canard Explosions

Figure 8.1 displays the results of our computations with multiple shooting of the canard explosion in the Van der Pol oscillator (8.3). Panel (a) shows the bifurcation diagram and panel (b) is a plot of representative limit cycles. Three regions are indicated on the bifurcation diagram in Fig. 8.1(a), where open circles denote the boundaries between them. Region A corresponds to simple relaxation oscillations, where we use the shooting methods described in Sect. 8.2. Regions B and C correspond to jump-left and jump-right canards, respectively; in these regions we use the shooting methods described in Sect. 8.3.1. The open circle that separates regions B and C corresponds to a maximal canard. Finally, the solid black circle denotes the Hopf point.

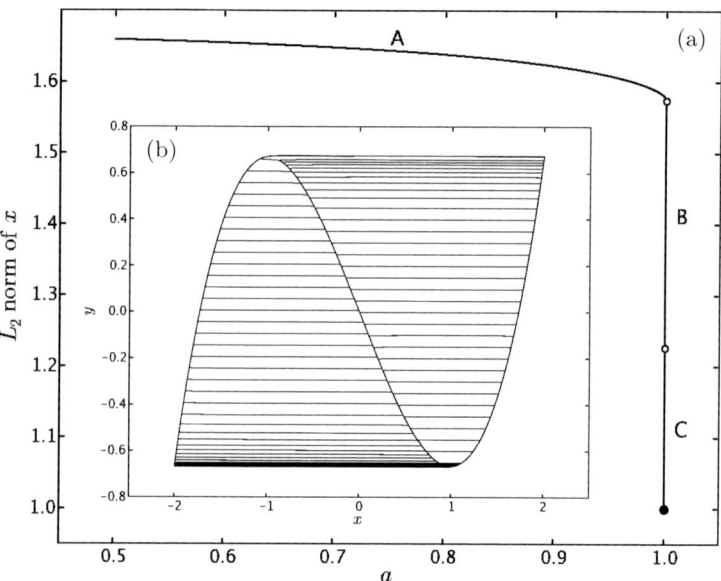

Fig. 8.1. The bifurcation diagram of the Van der Pol oscillator (8.3) near the canard explosion (a) and the corresponding limit cycles with canard segments (b) as computed with our shooting method.

In order to start the continuation of canards in the Van der Pol system with the multiple-shooting algorithm, we proceed as follows. We first find a simple stable relaxation oscillation numerically for the parameter values $(a, \varepsilon) = (0.5, 1.0)$ using Radau integration. We then continue this orbit in AUTO with ε as the free parameter. The continuation terminates at the periodic orbit with $\varepsilon = 10^{-4}$. Starting from this periodic orbit, we subsequently continue the family of orbits in AUTO with a as the free parameter. This continuation terminates at a limit cycle in the middle of region B. This limit cycle is our initial limit cycle for continuation with our shooting methods. We perform continuation in the forward and backward direction with a as the free parameter. When moving in the backward direction, we pass from region B to region A, where canards cease and simple relaxation oscillations exist. At the transition from B to A, backward integration fails, indicating that we should change shooting methods. During the continuation in the forward direction we pass from region B to region C by encountering a maximal canard. In this situation backward integration also fails, and we must switch cross-sections from $\{x = -1\}$ to $\{x = 1\}$. Note that the tangent vector to the bifurcation curve switches: in region B both a and y are increasing, while in region C a is increasing and y is decreasing. Automated detection of periodic orbits that separate these regions (open circles) is an important topic for future work.

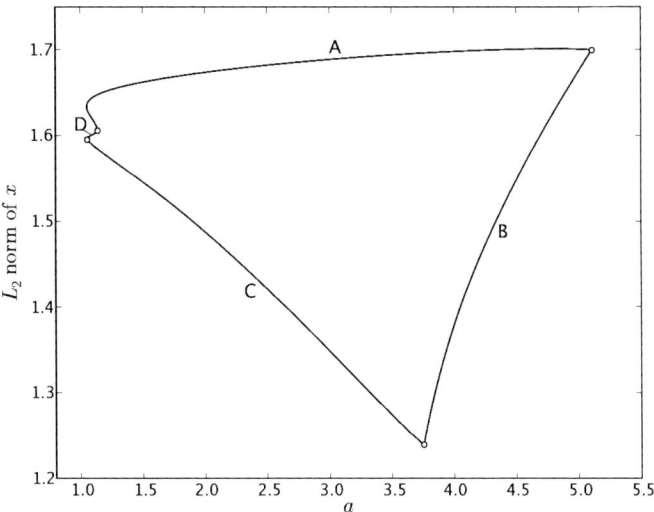

Fig. 8.2. Bifurcation diagram of period-three orbits in the forced Van der Pol system (8.4) computed with shooting. In this example, $\varepsilon = 10^{-4}$ and $\omega = 1.55$.

8.4.2 Folded Saddles

For the continuation of periodic orbits with canards initiated at folded saddles in the forced Van der Pol system, we start with a period-three simple stable relaxation oscillation for the parameter values $(\varepsilon, \omega, a) = (10^{-4}, 1.55, 2.5)$. A bifurcation diagram, calculated with shooting and a as the free parameter, is shown in Fig. 8.2. During the computation of the periodic orbits we take advantage of the symmetry $(\theta, x, y) \mapsto (\theta + 1.5, -x, -y)$, which means that we need only integrate over half the period. We apply the symmetry transformation to the endpoint at the section $\{\theta = \theta_{fs} + 1.5\}$, and compare this point to the endpoint at $\{\theta = \theta_{fs}\}$, giving our matching condition. Our initial cross-sections are $\{x = \pm 1\}$, and thus the matching conditions for shooting occur in a hyperplane with coordinates (y, θ, a). The sign of x on the initial cross-section depends on the location of the periodic orbit in the bifurcation diagram. There are four regions. Region A corresponds to no canards, where we start from the cross-section $\{x = 1\}$ and integrate forward to the cross-section $\{x = -1^{+}\}$ (the superscript denotes crossing in the increasing x-direction). Regions B and D correspond to jump-right canards (right relative to the fast variable x), where we start from the cross-section $\{x = 1\}$, shooting forward to the cross-section $\{\theta = \theta_{fs} + 1.5\}$ and backward to the cross-section $\{\theta = \theta_{fs}\}$. Finally, region C corresponds to jump-left canards, where we start from the cross-section $\{x = -1\}$, shooting forward to the cross-section $\{\theta = \theta_{fs} + 1.5\}$ and backward to the cross-section $\{\theta = \theta_{fs}\}$. Transitions between regions are again denoted by open circles.

8.4.3 Comparisons

Let us now discuss the convergence of the Newton iterations for each method. In AUTO, a collocation boundary value method is used to find periodic orbits. With this method, the Jacobian is of dimension $mnN+b+q+1$, where N is the number of subintervals, m the number of collocation points per subinterval, n the dimension of the vector field, b the number of boundary conditions and q the number of integral conditions. We have already mentioned that, due to the slow-fast structure of the systems, the number of subintervals must be large for accuracy and convergence (we use $N = 1000$). The number of collocation points used per subinterval is set at $m = 4$ in our calculations. Thus, we expect the size of the Jacobian to be approximately $8,000 \times 8,000$ in the Van der Pol system and approximately $12,000 \times 12,000$ in the forced Van der Pol system. In the multiple-shooting methods, the Jacobian is of dimension $(n-1)s+1$, where again n denotes the dimension of the vector field and s is the number of cross-sections that are used. In the forced Van der Pol system the Jacobian is 3×3. Note that we will require more cross-sections when tracking periodic orbits with multiple canard segments.

Computationally, most of the effort in the shooting methods is in the integration, while in collocation most of the computational effort is in solving a large, sparse matrix. The construction of the Jacobian in the multiple-shooting method involves numerical integration of the vector field, which is an efficient and speedy process. The Jacobian for collocation is much larger because it is also used to determine a suitable orbit segment (while in shooting this is done with an integrator whose accuracy must be controlled separately). It should be noted that, although the Jacobian is considerably larger with AUTO, efficient numerical techniques are used to invert the Jacobian in two stages by taking advantage of the sparsity structure of the matrix. The first stage uses a method known as condensation of parameters to perform independent eliminations in N blocks of size $nm \times n(m+1)$. The second stage produces a *reduced* Jacobian of size $(n+b+q+1) \times (n+b+q+1)$; see also Chap. 1. This inversion still requires much more computation than Gaussian elimination on the matrix of size $((n-1)s+1) \times ((n-1)s+1)$ used in our shooting method.

For both AUTO and shooting, we performed a full Newton's method with a maximum of eight iterations and error tolerances on the order of 10^{-10}. During the calculations, we kept track of the number of iterations in the convergence of each step of Newton's method, as well as the condition number of the Jacobian. In the forced Van der Pol system the Jacobians for the shooting methods had $O(10)$ condition numbers, while for AUTO the condition numbers for the reduced Jacobians were $O(10^6)$.

Domains of convergence for the family $\mathcal{F}_\Gamma$ of limit cycles Γ along the branch B of Fig. 8.2 are displayed in Fig. 8.3; panel (a) shows the domain of convergence for the multiple-shooting method and panel (b) that for an AUTO computation (with $N = 1000$ mesh intervals). The thick (approximately) horizontal black line through the origin represents the a-dependent family of limit

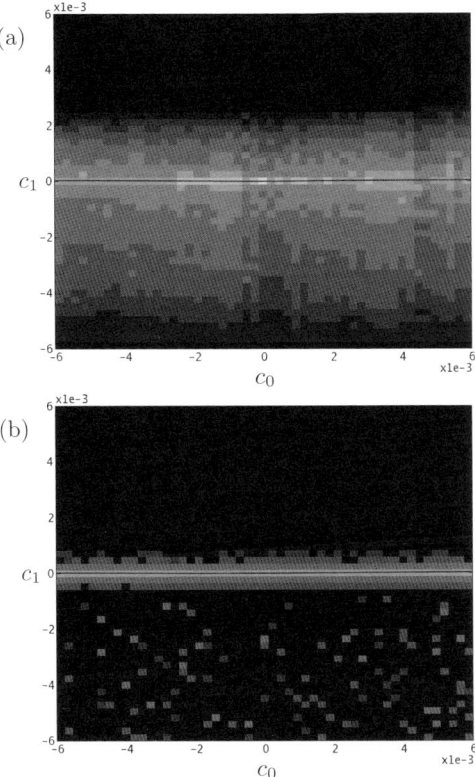

Fig. 8.3. Domains of convergence for the computation of limit cycles Γ along branch B of Fig. 8.2 with the shooting method (a) and with AUTO for $N = 1000$ mesh intervals.

cycles Γ. The figures are shaded according to the number of iterates needed for convergence, where shades white to black represent convergence after one to eight iterations, respectively. In fact, black denotes no convergence of the method using our error tolerances and choice of maximally eight iterates. Figure 8.2(a) contains an additional darkest shade of gray that is used for those points that do not converge in eight iterations but show signs of converging. Specifically, such points are marked as converging if the function values and differences between variable values for the last three iterates are decreasing.

The plots were obtained by starting with a specific limit cycle Γ_0 and its intersection γ_0 with the cross-section Σ defined by $x = 1$. The section Σ is three dimensional with coordinates (y, θ, a). We compute orthonormal vectors ($\boldsymbol{v}_0$ and $\boldsymbol{v}_1$) in Σ at γ_0 so that $\boldsymbol{v}_0$ is tangent to $\mathcal{F}_\Gamma \cap \Sigma$ and $\boldsymbol{v}_1$ lies in the plane spanned by $\boldsymbol{v}_0$ and $(0, 0, 1)$. The coordinates $\boldsymbol{c} = (c_0, c_1)$ in the figure correspond to the points $p_c = \gamma_0 + c_0 \boldsymbol{v}_0 + c_1 \boldsymbol{v}_1 \in \Sigma$ and, thus, the origin represents γ_0. The horizontal black line in the figure is the projection

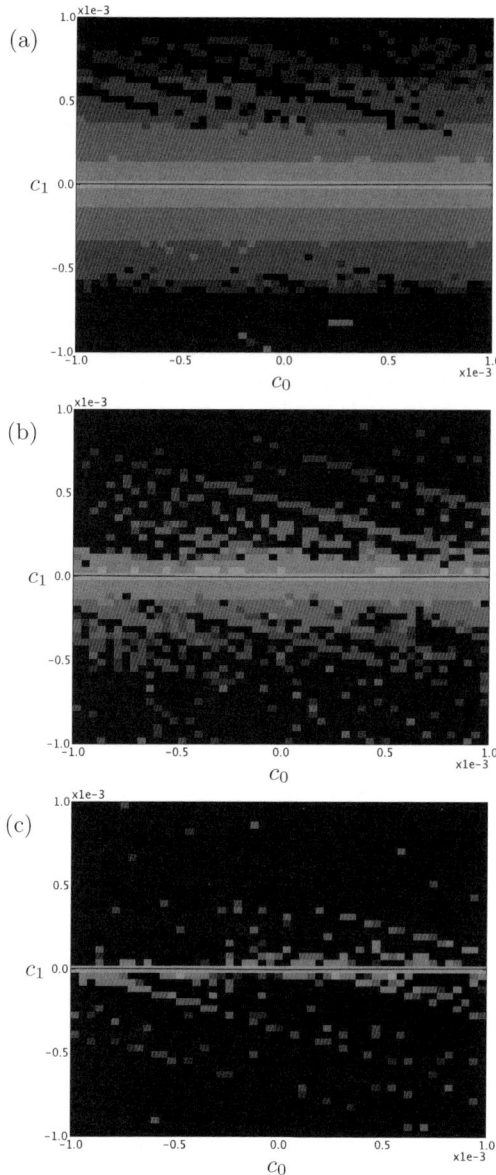

Fig. 8.4. The domains of convergence for the computation of limit cycles Γ along branch B of Fig. 8.2 with AUTO depend on the number of mesh intervals; panel (a) to (c) are for $N = 1000$, $N = 500$ and $N = 250$ mesh intervals, respectively.

of $\mathcal{F}_\Gamma \cap \Sigma$ onto the plane spanned by $(\boldsymbol{v}_0, \boldsymbol{v}_1)$. Note that $\mathcal{F}_\Gamma \cap \Sigma$ appears horizontal because $\boldsymbol{v}_0$ is tangent to $\mathcal{F}_\Gamma \cap \Sigma$ and the domain represented in the figure is small.

For the shooting algorithm, we use the point p_c for each grid point $c = (c_0, c_1)$ in the figure as the initial point for the algorithm and shade the pixel according to the number of iterates required for convergence as described above. Unfortunately, initialization of AUTO requires an entire curve. To obtain a curve from each of the points p_c, we computed trajectories forward and backward as in the shooting algorithm and used the concatenation of these two trajectory segments to initialize AUTO. (AUTO computes a mesh of specified size from the trajectories provided.) Except on $\mathcal{F}_\Gamma$, these curves are not closed: the final points of the forward and backward trajectory segments do not match. Since AUTO is based upon solving the differential equations within the space of closed curves, it might be preferable to initialize AUTO with closed curves. Lacking a natural way to produce closed curves at different distances from $\mathcal{F}_\Gamma$, we did not pursue such a comparison here.

We also tested the dependence of AUTO's domain of convergence on the number of mesh intervals N. Figure 8.4 displays the results of three computations. Panel (a) shows an enlargement of Fig. 8.3(b) (with $N = 1000$). Panels (b) and (c) are for computations with $N = 500$ and 250, respectively. The domain of convergence for AUTO decreases with N, and values of N smaller than 1000 have very small domains of convergence.

8.5 Towards a General Theory

The examples presented above demonstrate that multiple-shooting algorithms adapted to the slow-fast decomposition of trajectories in multiple time scale dynamical systems can be effective for computing periodic orbits. These algorithms are able to exploit the advantages of numerical integration methods for stiff systems to compute canards in vector fields with a single fast variable, where we use integration backward in time. The strategy presented here relies upon two ingredients. First, one must choose cross-sections for shooting that isolate the trajectory segments to be computed in forward time and those to be computed in backward time and, second, the computation must be placed in a continuation setting in which the periodic orbits vary at a moderate rate with respect to the continuation parameter. The rapid change of periodic orbits containing canards with respect to system parameters requires that the root finding procedure performed by the continuation algorithm is able to vary a system parameter as well as phase space variables.

Development of algorithms that make suitable choices of cross-sections and continuation strategy is likely to require good methods for automatically computing aspects of the slow-fast decomposition of trajectories in order to base locating suitable cross-sections upon this information. In particular, such methods should determine where degenerate decompositions are encountered and use information about the types of canards that are associated with these decompositions.

Canards in slow-fast systems with more than one fast variable typically lie along sheets of the critical manifold that consist of saddle points for the layer equations. Accurate computation of these canards cannot be done with either forward or backward numerical integration. Instead, two-point boundary value solvers, methods that are designed for computing normally hyperbolic manifolds (see Chap. 4) or methods for shadowing of trajectories of vector fields [6, 24] will need to be incorporated into shooting methods when one wants to compute relaxation oscillations that contain these canards. These algorithms will require more computation than numerical integration, but they still are likely to provide a good alternative to collocation methods for these problems.

Acknowledgments

This research was partially supported by grants from the National Institutes of Health, the Department of Energy and the National Science Foundation.

References

1. E. Allgower and K. Georg. Continuation and path following. *Acta Numerica 1993*, pages 1–64 (Cambridge University Press, Cambridge, 1993)
2. E. Benoit, J. L. Callot, F. Diener, and M. Diener. Chasse au canard. *Collect. Math.* 32:37–119, 1981.
3. K. Bold, C. Edwards, J. Guckenheimer, S. Guharay, K. Hoffman, J. Hubbard, R. Oliva, and W. Weckesser. The forced van der Pol equation II: Canards in the reduced system. *SIAM J. Appl. Dyn. Syst.* 2:570–608, 2003.
4. M. Cartwright and J. Littlewood On nonlinear differential equations of the second order: II the equation $\ddot{y}-kf(y,\dot{y})\dot{y}+g(y,k) = p(t) = p_1(t)+kp_2(t)$, $k > 0$, $f(y) \geq 1$. *Ann. Math.* 48:472–94, 1947. [Addendum 50:504–505, 1949]
5. R. Clewley, M. D. LaMar, and E. Sherwood. "PyDSTool" available via http://sourceforge.net/projects/pydstool
6. B. Coomes, H. Koçak, and K. Palmer. Rigorous computational shadowing of orbits of ordinary differential equations. *Numer. Math.* 69:401–421 1995.
7. E. J. Doedel, R. C. Paffenroth, A. R. Champneys, T. F. Fairgrieve, Yu. A. Kuznetsov, B. E. Oldeman, B. Sandstede, and X. J. Wang. Auto2000: Continuation and bifurcation software for ordinary differential equations. available via http://cmvl.cs.concordia.ca/.
8. F. Dumortier and R. Roussarie. Canard cycles and center manifolds. With an appendix by Cheng Zhi Li. *Mem. Amer. Math. Soc.* 121, no. 577, x+100 pp, 1996.
9. W. Eckhaus. Relaxation oscillations, including a standard chase on French ducks. *Lecture Notes in Mathematics*, Vol. 985, pages 449–494 (Springer-Verlag, 1983).
10. N. Fenichel, Geometric singular perturbation theory. *J. Diff Eq.* 31:53–98, 1979

11. J. Guckenheimer. Bifurcation and degenerate decomposition in multiple time scale dynamical systems. In J. Hogan, A. Champneys, B. Krauskopf, M. di Bernardo, E. Wilson, H. Osinga, and M. Homer, editors, *Nonlinear Dynamics and Chaos: where do we go from here?*, pages 1–21 (Institute of Physics Publishing, Bristol, 2002).

12. J. Guckenheimer. Bifurcations of relaxation oscillations. In *Normal Forms, Bifurcations and Finiteness Problems in Differential Equations*, pages 295–316, NATO Sci. Ser. II Math. Phys. Chem., Vol. 137 (Kluwer Acad. Publ., Dordrecht, 2004).

13. J. Guckenheimer, K. Hoffman, and W. Weckesser. Numerical computation of canards. *Internat. J. Bifurc. Chaos Appl. Sci. Engrg.*, 10:2669–2687, 2000.

14. J. Guckenheimer, K. Hoffman, and W. Weckesser. The Forced van der Pol Equation I: The Slow Flow and its Bifurcations. *SIAM J. App. Dyn. Sys.* 2:1–35, 2003.

15. J. Guckenheimer and B. Meloon. Computing periodic orbits and their bifurcations with automatic differentiation. *SIAM J. Sci. Comp.*, 22:951–985, 2000.

16. J. Guckenheimer, M. Wechselberger, and L.-S. Young. Chaotic attractors of relaxation oscillators. *Nonlinearity*, 19:701–720, 2006.

17. R. Haiduc. Horseshoes in the forced van der Pol equation, PhD dissertation, Cornell University, 2005.

18. E. Hairer, S. P. Norsett and G. Wanner. *Solving Ordinary Differential Equations I, 2nd. ed.* (Springer-Verlag, 1993).

19. C. K. R. T. Jones. Geometric singular perturbation theory. In *Dynamical Systems, Lecture Notes in Mathathematics*, Vol 1609, pages 44–120 (Springer-Verlag, 1995).

20. N. Levinson. Perturbations of discontinuous solutions of non-linear systems of differential equations. *Acta Math.*, 82:71–106, 1950.

21. J. Littlewood. On nonlinear differential equations of the second order. III. The equation $\ddot{y} - k(1 - y^2)\dot{y} + y = bk\cos(\lambda t + \alpha)$ for large k, and its generalizations. *Acta Math.*, 97:267–308, 1957. [Errata at *Acta Math.*, 98:110, 1957]

22. J. Littlewood. On nonlinear differential equations of the second order. IV. The general equation $\ddot{y} - kf(y)\dot{y} + g(y) = bkp(\phi)$, $\phi = t + \alpha$. *Acta Math.*, 98:1–110, 1957.

23. E. F. Mishchenko, Yu. S. Kolesov, A. Yu. Kolesov and N. Kh. Rhozov. *Asymptotic Methods in Singularly Perturbed Systems. Monographs in Contemporary Mathematics*, (Consultants Bureau, New York, A Division of Plenum Publishing Corporation, 1994).

24. E. Van Vleck. Numerical shadowing near hyperbolic trajectories. *SIAM J. Sci. Comput.* 16:1177–1189, 1995.

25. A. B. Vasil'eva. Asymptotic behaviour of solutions of certain problems for ordinary non-linear differential equations with a small parameter multiplying the highest derivatives. *Russian Mathematical Surveys*, 18:13–84, 1963. (Russian) *Uspehi Mat. Nauk*, 18:15–86, 1963.

Continuation of Periodic Orbits in Symmetric Hamiltonian Systems

Jorge Galán-Vioque[1] and André Vanderbauwhede[2]

[1] Departamento de Matemática Aplicada II, Escuela Superior de Ingenieros de Sevilla, Spain
[2] Department of Pure Mathematics and Computer Algebra, Ghent   University Belgium

The idea of everything returning eventually to its point of departure has a strong hold on humanity, with many historical, philosophical and religious implications. Classical examples are the need to construct a calendar and the subsequent search for orbits in the solar system in which the planets follow a closed track and repeat their history over and over again.

Nature, at its most basic level, has decided to be Hamiltonian; non-Hamiltonian systems come up in physics only as phenomenological models for the more complicated underlying processes. However, Hamiltonian systems are nongeneric dynamical systems with remarkable properties, in particular with respect to periodic orbits. The role of periodic solutions in Hamiltonian systems and their importance in modern physics was first recognized by Poincaré [26]. Today periodic orbits are at the basis of both classical and quantum mechanics [13]. Poincaré conjectured that periodic orbits, that is, solutions that return to their initial conditions after some finite time, are densely distributed among all possible bounded classical trajectories; and he suggested that the study of periodic orbits would provide the clue to the overall behavior of any mechanical system. Quoting the original work [26]:

> "It seems at first that the existence of periodic solutions could not be of any practical interest whatsoever. Indeed, the probability is zero for the initial condition to correspond precisely to those of a periodic solution. But it may happen that they differ by very little. [...] Here is a fact which I have not been able to demonstrate rigorously, but which nevertheless seems very plausible to me. Given equations of the Hamiltonian form and any particular solution of these equations, we can always find a periodic solution (whose period may admittedly be very long) such that the difference between the two solutions is as small as we wish during as long a time as we wish. Besides this, what renders these periodic solutions so precious is that they are, so to speak, the

only opening through which we may try to penetrate into the fortress which has the reputation of being impregnable."

We know from Arnol′d [2] that a literal interpretation of Poincaré's statement is not possible: the completely integrable two-degrees-of-freedom Hamiltonian system corresponding to the Hamiltonian

$$H(I_1, I_2, \theta_1, \theta_2) = I_1 + \sqrt{2}\, I_2$$

(in action-angle variables) admits no periodic solutions whatsoever since the ratio of its frequencies is irrational. However, replacing $\sqrt{2}$ by an arbitrarily close rational number results in a system that has only periodic orbits. This insight leads to a reformulation that says that Poincaré's statement is true for generic Hamiltonian systems; a more precise statement is given by the Closing Lemma due to Pugh and Robinson [27], which states that, given a Hamiltonian and a point in phase space, there exists an arbitrarily close Hamiltonian (in the C^2-sense) for which the given point generates a periodic orbit.

In this chapter we show how two-point boundary value problem continuation can help us to enter the 'fortress' Poincaré was talking about. Specifically, this approach provides us with an efficient tool to compute families of periodic orbits in Hamiltonian systems and allows us to discover numerically how these families bifurcate and connect. Particular attention will be paid to symmetric Hamiltonian systems, where the symmetries and the associated first integrals typically increase the dimensions of these families. Continuation in this context is introduced with the simple model example of the mathematical pendulum, justified theoretically, and then applied to a nontrivial case of computing periodic orbits that are associated with the recently discovered figure-8 solution of the three-body problem. Numerical aspects of the scheme, including comparison with a shooting method, are highlighted with the computation of a periodic solution of the restricted three-body problem.

The chapter is organized as follows. In Sect. 9.1 we discuss several aspects of periodic orbits in Hamiltonian systems, in particular, how they are organized in families and how one can approach the numerical calculation of these families. The discussion is illustrated with the mathematical pendulum and the restricted three-body problem. In Sect. 9.2 we give an outline of theoretical continuation results for periodic orbits and relative equilibria; these results include a set-up for the equations that can be used for numerical computations. In Sect. 9.3 we present continuation results for the three-body problem, where we start from the figure-8 and the Lagrange solutions, respectively. These periodic solutions exist for equal masses, and we use one of the masses as (an external) continuation parameter. It appears that the starting solutions can be continuously connected to a periodic solution of the restricted three-body problem. In Sect. 9.4 we draw some general conclusions and give a brief discussion of open problems.

9.1 Periodic Solutions in Hamiltonian Systems

It is well known that Hamiltonian (or, more generally, conservative) systems are quite different from dissipative systems in terms of periodic orbits, their continuation and their bifurcations. In dissipative systems periodic orbits are generically isolated and, therefore, an external parameter is required in order to continue such periodic orbits. For Hamiltonian systems there is the celebrated Cylinder Theorem [15], which says that periodic orbits appear in one- or more-parameter families and that, under appropriate nondegeneracy conditions, these families are persistent under small Hamiltonian perturbations.

Computationally the problem of finding a periodic orbit can be formulated as a boundary value problem with the period as an additional parameter. In order to avoid phase shifts and to ensure uniqueness one has to introduce an appropriate phase condition, which can be either a boundary condition (Poincaré-type condition) or an integral version of it (see also Chaps. 1, 10 and 11). In dissipative systems this problem is generically well determined: the periodicity condition together with the phase condition give $n + 1$ equations for the n components of the initial point, the period and the external parameter. Generically these equations can be solved by the Implicit Function Theorem, giving a one-dimensional solution curve that can be parametrized by the external parameter.

This scheme does not work for Hamiltonian systems or, more generally, for systems with a first integral. (Another exceptional case are time-reversible systems, but we do not consider them here since they require different arguments.) In conservative systems periodic orbits typically belong to one-parameter families, parametrized by the value of the first integral (the energy in the Hamiltonian case). This 'internal' or 'natural' parameter is not explicitly available in the equations, at least not directly, and this is why the standard continuation scheme fails. Additional complications arise for Hamiltonian systems with several independent constants of motion (symmetries): here periodic orbits belong to families whose dimension is the number of independent integrals. In this case, further 'phase conditions' are required in order to identify a single member of such a family uniquely.

9.1.1 Continuation of Periodic Orbits in Conservative Systems

A straightforward approach to continue periodic orbits in conservative systems is to use the conserved quantity (the energy, in the Hamiltonian case), to eliminate one of the variables. Then one chooses a suitable Poincaré section for the flow and looks for fixed points of the corresponding Poincaré map. This scheme can be extended to the case of several constants of motion and has been extensively used in the literature; see, for example, [30]. This approach requires numerical integration of the differential equations, which can give errors in the case of very stiff equations or very unstable orbits, and the section must be adapted at each step in the continuation process to ensure transversality. Furthermore, it is difficult to use integral phase constraints, which, as explained in Chap. 10 of this book, often have significant computational advantages over the classical Poincaré phase condition. The underlying idea of this scheme is that of the *reduction* of the dimension of the problem by making use of the conserved quantities and/or the symmetries; it is a direct translation of the standard theoretical treatment of the problem. A classical example is the N-body problem (see, e.g., [35]) that will be discussed later in this chapter.

The approach that we propose here extends the dimension rather than reducing it, to get the problem into a form where boundary value continuation methods can be applied directly; a similar approach is discussed in Chap. 10. Our formulation not only allows one to prove and extend some basic continuation results for periodic orbits in Hamiltonian systems — such as the Cylinder Theorem of [17] — but it can also be implemented directly for the numerical calculation of branches of periodic orbits.

Our starting point is a generalization of some continuation results of Sepulchre and MacKay for periodic orbits of systems having a first integral. In [28] the concept of a *normal periodic orbit* is introduced and it is shown that such normal periodic orbits belong to one-parameter families. The key idea of their approach is to embed the conservative equation in a one-parameter family of dissipative systems by adding a small gradient perturbation term to the vector field in such a way that a periodic orbit of the perturbed system can only exist when the perturbation is zero. Under the normality condition one can invoke the Implicit Function Theorem to obtain a continuation result for periodic orbits of the extended system. Because the periodic orbits only exist when the perturbation is zero, one has in fact a continuation result for the unperturbed conservative system.

The idea of adding a dissipative term, which allows periodic orbits only when the dissipation is zero, is not new; for example, it is used in one of the classical proofs of the Lyapunov Center Theorem where this theorem is shown to correspond to a vertical Hopf bifurcation; see, e.g., [34]. The idea has been used for numerical calculations, for example, in [1, 37].

To be more precise, consider a smooth n-dimensional vector field $g : \mathbb{R}^n \to \mathbb{R}^n$, and assume that the corresponding system

$$\dot{u} = g(u) \qquad (9.1)$$

has a non-trivial first integral $F : \mathbb{R}^n \to \mathbb{R}$. This means that each orbit of (9.1) is contained in a level set of F, and consequently $\nabla F(u) \cdot g(u) \equiv 0$. Let $u_0(t)$ be a periodic solution of (9.1) with initial point $p_0 := u_0(0)$, minimal period $T_0 > 0$, and *monodromy matrix M*. This monodromy matrix is given by $M = V(T_0)$, where $V : \mathbb{R} \to \mathcal{L}(\mathbb{R}^n)$ is the transition matrix for the variational equation $\dot{v} = Dg(u_0(t)) \cdot v$; the eigenvalues of M are the *multipliers* of the periodic solution $u_0(t)$. Assuming that $\nabla F(p_0) \neq 0$ one can show that 1 is an eigenvalue of M with geometric multiplicity $m_g \geq 1$ and algebraic multiplicity

$m_a \geq 2$; full proofs are in [23]. In order to continue the periodic solution $u_0(t)$ we replace equation (9.1) by the extended equation

$$\dot{u} = T \left[g(u) + \alpha \nabla F(u) \right], \tag{9.2}$$

which depends on two (real) parameters T and α. We look for 1-periodic solutions $u(t)$ of (9.2), with $(u(0), T, \alpha)$ near $(p_0, T_0, 0)$. The basic observation is that such solutions can only exist for $\alpha = 0$, since

$$0 = F(u(1)) - F(u(0)) = \int_0^1 \nabla F(u(t)) \cdot \dot{u}(t) \, dt = \alpha T \int_0^1 \| \nabla F(u(t)) \|^2 \, dt.$$

The integral on the right-hand side is different from zero (namely, $\nabla F(u(0))$ is close to $\nabla F(p_0) \neq 0$). We conclude that a 1-periodic solution of (9.2) corresponds (after an appropriate time rescaling) to a T-periodic solution of (9.1).

We denote the flow of (9.2) by $\tilde{u}(t; p, T, \alpha)$, where $p \in \mathbb{R}^n$ is the initial value. To find the 1-periodic solutions we are looking for, we must impose the periodicity condition $\tilde{u}(1; p, T, \alpha) = p$; to avoid phase shifts we must also impose a phase condition. Therefore, we define a mapping

$$G : \mathbb{R}^n \times \mathbb{R} \times \mathbb{R} \to \mathbb{R}^n \times \mathbb{R},$$
$$G(p, T, \alpha) := (\tilde{u}(1; p, T, \alpha) - p, \langle g(p_0), p - p_0 \rangle),$$

where $\langle \cdot, \cdot \rangle$ defines a scalar product on $\mathbb{R}^n$, and we have $G(p_0, T_0, 0) = 0$ by assumption. Now we look for zeros (p, T, α) of G near $(p_0, T_0, 0)$. Using the Implicit Function Theorem in combination with the remark above, one can then prove the following.

Theorem 1. *Let $u_0(t)$ be a periodic solution of the conservative equation (9.1) with initial point $p_0 = u_0(0)$, minimal period $T_0 > 0$ and monodromy matrix M. Assume that $\nabla F(p_0) \neq 0$ and that 1 is an eigenvalue of M with geometric multiplicity $m_g = 1$. Then the solution set of the equation $G(p, T, \alpha) = 0$ consists locally near $(p_0, T_0, 0)$ of a unique smooth curve along which $\alpha \equiv 0$. More precisely, this solution curve can be written in the form $\{(p^*(T), T, 0) \mid T \text{ near } T_0\}$ for some smooth $p^* : \mathbb{R} \to \mathbb{R}^n$ such that $p^*(T_0) = p_0$.*

This theorem is essentially a re-statement of the Cylinder Theorem for conservative systems [17] in a form that is adapted to numerical implementation. It forms the simplest case of a more general continuation result for conservative systems that can be found in [23] and that is described in Sect. 9.2 for the particular case of Hamiltonian systems. If (9.1) has several independent first integrals then the condition $m_g = 1$ of Theorem 1 is not satisfied. In this case the more general theory of [23] is required. This arises in the continuation of the N-body problem, where apart from the Hamiltonian also the components of the total linear momentum and the total angular momentum are first integrals; we refer to Sect. 9.3 for examples.

For the numerical implementation of Theorem 1 (or similar results) the periodicity condition $\tilde{u}(1; p, T, \alpha) = p$ is usually replaced by a boundary value problem for the full solution $\{u(t) \mid 0 \leq t \leq 1\}$, and the Poincaré-type phase condition is replaced by an integral version. Setting $\tilde{u}_0(t) := \tilde{u}(t; p_0, T_0, 0) = u_0(T_0 t)$, this leads to a boundary value problem of the following form:

(**CON-1**) Find $(u(t), T, \alpha)$ near $(\tilde{u}_0(t), T_0, 0)$ such that

$$\begin{cases} \dot{u}(t) = T \left[g(u(t)) + \alpha \nabla F(u(t)) \right], \\ u(1) = u(0), \\ \int_0^1 \langle \dot{\tilde{u}}_0(t)), u(t) - \tilde{u}_0(t) \rangle \, dt = 0. \end{cases} \qquad (9.3)$$

In the following sections we show some results of the numerical implementation of this approach to two particular examples, namely the mathematical pendulum and the Restricted Three-Body Problem (R3BP). While discussing these examples we put some emphasis on numerical issues related to the numerical implementation. In particular, in Sect. 9.1.3 we compare the shooting method with the collocation method from a numerical point of view by computing a 'benchmark' periodic solution of the R3BP taken from exercise 4.12 of [3]. In Sect. 9.1.4 we present partial bifurcation diagrams of the R3BP for two values of the mass ratio that demonstrate the power and versatility of our continuation scheme.

9.1.2 The Mathematical Pendulum

As a simple introductory example we consider the dimensionless mathematical pendulum

$$\begin{aligned} \dot{x}_1 &= x_2, \\ \dot{x}_2 &= -\sin(x_1). \end{aligned} \qquad (9.4)$$

This vector field is a one-degree-of-freedom Hamiltonian system, corresponding to the Hamiltonian $H(x_1, x_2) = \frac{1}{2}x_2^2 + 1 - \cos(x_1)$. The variable $x_1 \in \mathbb{S}^1 = \mathbb{R}/2\pi\mathbb{Z}$ represents the angular displacement from the vertical axis, and $x_2 \in \mathbb{R}$ the angular velocity. The Hamiltonian has been chosen such that the equilibrium at the origin, corresponding to the stable hanging solution, has zero energy.

Equation (9.4) has a family of periodic orbits corresponding to librations of the pendulum; they originate at the origin and terminate at a homoclinic orbit to the saddle point $(\pi, 0)$. Figure 9.1(d) shows some representative periodic orbits of this family, which can be parametrized either by the energy, the period (which increases monotonically from 2π to infinity), or the maximal angular displacement. However, neither of these quantities is explicitly available in (9.4). As we observed before, such behavior (which is nongeneric in dissipative systems) is typical for conservative systems.

In order to calculate the family of periodic solutions we follow the approach outlined in Sect. 9.1.1 and replace (9.4) by (a time rescaled version of) the system

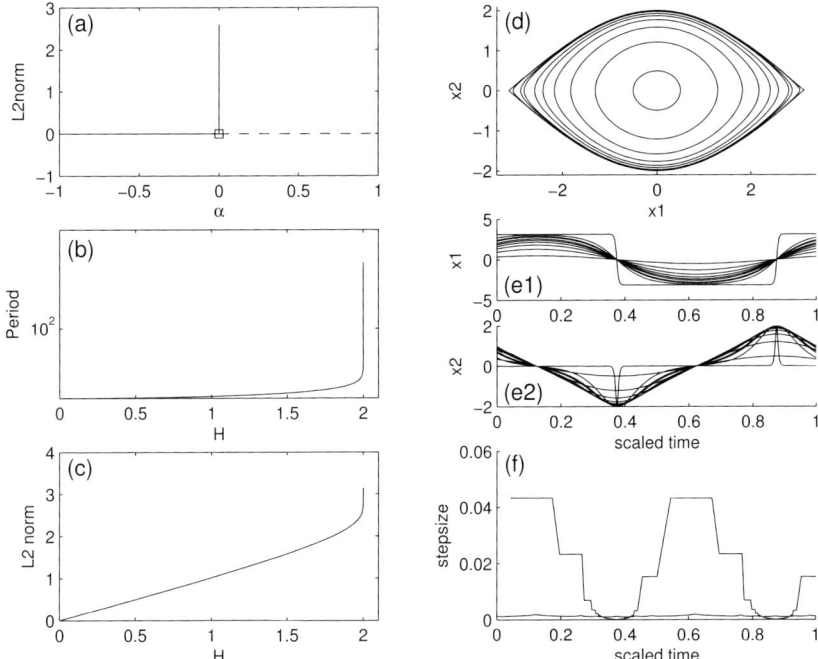

Fig. 9.1. Continuation results for the mathematical pendulum. Panel (a) shows the bifurcation diagram in α with the vertical branch of periodic solutions. Panels (b) and (c) show the period (plotted in a logarithmic scale) and the L_2-norm of the solutions versus the energy H, respectively. Ten representative periodic orbits of the one-parameter family are shown in panel (d). Panels (e1) and (e2) show the time evolutions of x_1 and x_2, respectively, while panel (f) shows the evolution of the time step along the orbit for two solutions, namely one far from homoclinic and the other very close to the homoclinic connection.

$$\dot{x}_1 = x_2 + \alpha \sin(x_1),$$
$$\dot{x}_2 = -\sin(x_1) + \alpha x_2. \qquad (9.5)$$

This system still has two equilibria, the origin which is a stable or an unstable focus depending on the sign of α, and the saddle $(\pi, 0)$. For all non-equilibrium solutions $(x_1(t), x_2(t))$ the function $h(t) := H(x_1(t), x_2(t))$ is strictly decreasing if $\alpha < 0$ or strictly increasing if $\alpha > 0$; this excludes periodic solutions for $\alpha \neq 0$, which agrees with the theoretical results and is also confirmed by a phase plane analysis of (9.5). So periodic solutions are only possible for $\alpha = 0$ in which case (9.5) coincides with (9.4) and we have the family of periodic orbits mentioned before. The bifurcation diagram for periodic orbits of (9.5), therefore, looks as in Fig. 9.1(a); this diagram very much resembles that of a classical Hopf bifurcation, except that in this case the bifurcating branch is completely vertical. It is also clear that α cannot be used to parametrize the family of periodic orbits.

Computationally (9.5) has the desired form, namely, with one external parameter. Starting the computation from, for example, $\alpha = -1$ and with initial point $(x_1, x_2) = (0, 0)$, a software package such as AUTO [7, 9] will locate $(\alpha, x_1, x_2) = (0, 0, 0)$ as a Hopf bifurcation point from the trivial solution and, after switching branches, compute the "vertical branch" of periodic orbits. Along this branch the value of α, computed as part of the solution for each continuation step, appears to be zero (up to numerical precision). The results of such a computation are illustrated in Fig. 9.1; in particular, panels (b) and (c) illustrate that either the energy H or the period can be used to parametrize the family of periodic orbits in panel (d). Some further remarks on the numerical implementation are in order.

1. In the pseudo-arclength continuation technique used by AUTO there is no 'distinguished parameter'. In our particular example α is just one of the quantities that have to be computed at each continuation step. This allows (for example) the computation along folds and, as is illustrated by the example considered here, the continuation of vertical solution branches.

2. Orthogonal collocation with adaptive mesh selection is used in AUTO to solve the boundary value problem at each continuation step. Figure 9.1(f) shows how the time step varies along the orbit: it shrinks at places where the solution varies rapidly, whereas it remains large at slowly varying segments of the orbit. This allows one to compute the family up to orbits with large period, i.e., very close to the homoclinic orbit that terminates the branch; see Fig. 9.1(b).

3. The integral phase condition keeps the segments with a rapid variation of the solution component x_2 at practically the same location when the period becomes large; see Fig. 9.1(e). This allows for bigger continuation steps compared with phase conditions that allow the dip in the profile to move. For further details on this particular aspect of the computations see [8] and Chap. 10.

9.1.3 Collocation Versus Shooting

We now analyze from a numerical point of view a slightly more demanding example taken from exercise 4.12 of [3]. It is very much related to the other examples of this chapter, namely, it concerns a periodic orbit of the R3BP. The aim is to compare a standard shooting method with the orthogonal collocation method of AUTO. Both methods are powerful and versatile enough and have a long tradition in the dynamical system community. In general, both methods behave similarly in terms of efficiency and accuracy.

The R3BP describes the dynamics of a body with negligible mass under the gravitational influence of two massive bodies, called the primaries, which move in circular orbits about their barycenter. Let (x, y, z) denote the position of the negligible-mass body in a rotating barycentric coordinate system, where the x-axis points from the larger to the smaller primary, the z-axis is orthogonal to the orbital plane, and the y-axis completes the orthogonal coordinate system. The units are chosen so that the distance between the primaries, the sum of the masses of the primaries, and the angular velocity of the primaries are all equal to one. The problem then depends on a single external parameter, denoted μ, which is the ratio of the mass of the smaller primary and the total mass. The larger and smaller primaries are then located at $(-\mu, 0, 0)$ and $(1 - \mu, 0, 0)$, respectively, and the equations of motion are given by

$$
\begin{aligned}
\ddot{x} &= 2\dot{y} + x - (1 - \mu)(x + \mu)r_1^{-3} - \mu(x - 1 + \mu)r_2^{-3}, \\
\ddot{y} &= -2\dot{x} + y - (1 - \mu)yr_1^{-3} - \mu y r_2^{-3}, \\
\ddot{z} &= -(1 - \mu)zr_1^{-3} - \mu z r_2^{-3},
\end{aligned}
\tag{9.6}
$$

where

$$
r_1 = \sqrt{(x + \mu)^2 + y^2 + z^2} \quad \text{and} \quad r_2 = \sqrt{(x - 1 + \mu)^2 + y^2 + z^2}.
$$

This dynamical system has one integral of motion, namely, the *Jacobi constant*

$$
E = \frac{1}{2}(\dot{x}^2 + \dot{y}^2 + \dot{z}^2) - U(x, y, z) - \frac{1}{2}\mu(1 - \mu),
$$

where

$$
U = \frac{1}{2}(x^2 + y^2) + \frac{1 - \mu}{r_1} + \frac{\mu}{r_2}.
$$

We have used both a shooting and a collocation method to calculate a particular 'benchmark solution' of (9.6) as described in [3]; it is shown in Fig. 9.2(a). This solution is periodic and planar (observe that the subspace $z = 0$ is invariant under (9.6)), and corresponds to the mass ratio $\mu = 0.01277471$ (which is very close to the Earth-moon case). For the calculations with the shooting method we have used the routine **ode45** of Matlab which is based on an explicit Runge-Kutta (4,5) formula (the Dormand-Prince pair) with decreasing absolute and relative tolerances. For the collocation approach we used AUTO, that is, piecewise-polynomial collocation with Gauss-Legendre collocation points; see Chap. 1. This so-called orthogonal collocation has the desirable property of preserving the symplectic structure of Hamiltonian systems. Furthermore, as implemented in AUTO, it determines at negligible cost the characteristic (Floquet) multipliers as a by-product of the decomposition of the Jacobian of the collocation system. Hence, the stability and bifurcation properties of the calculated periodic solution are available as well.

Figure 9.2 shows the results of the calculations with the shooting algorithm; the collocation approach gives essentially identical results. Panel (a1) and the enlargement (a2) show a near-collision (or flyby) of the negligible, small mass with the small primary at about $t = 6$. This results in a strong acceleration of the negligible mass; see Fig. 9.2(b1). This clearly demonstrates the need for an adaptive mesh so that fast changes in velocity at near-collisions are properly resolved; see panels (c1) and (c2). Indeed fixed step methods should be avoided.

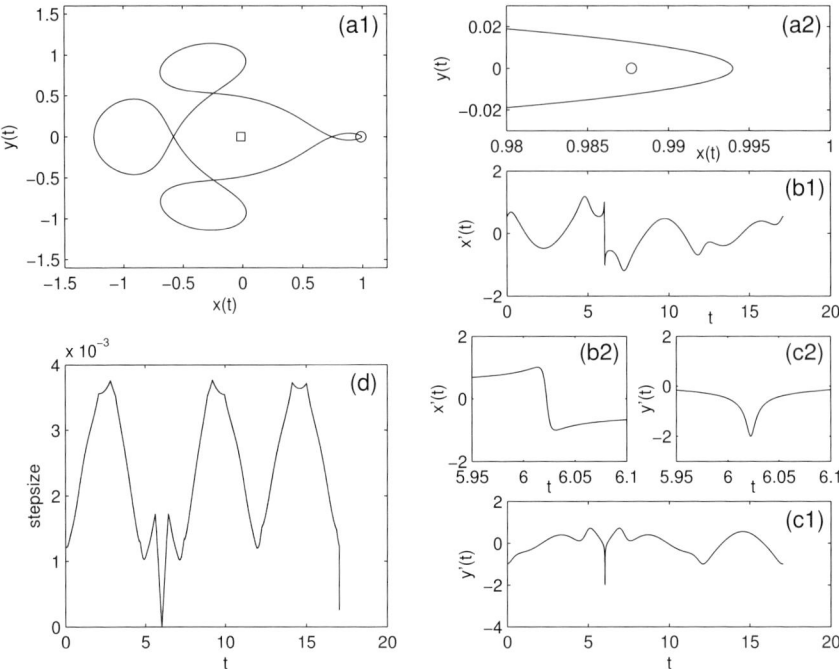

Fig. 9.2. Benchmark periodic orbit of the R3BP with $\mu = 0.01277471$. Panel (a1) shows the orbit in projection onto the (x, y)-plane, and panel (a2) is an enlargement close to the near collision; the primaries are depicted by a circle (for mass μ) and a square (for mass $1 - \mu$). Panel (b1) shows the rapid change of the velocity of the negligible mass, and panel (b2) is an enlargement. Similarly, panels (c1) and (c2) show the time evolution of the velocity component. Panel (d) is a plot of how the adaptive stepsize used by the RK method of Matlab changes along the orbit; the sudden drop of the stepsize around $t = 6$ corresponds to the near collision.

In order to compare the two computational schemes we should study the error, the number of time steps and the CPU-time. The lack of an exact solution and the fact that AUTO is a compiled program, while the shooting calculation was performed with Matlab, only allows us to plot the error in the Cauchy sense. That is, we plot in Fig. 9.3(a) the measure $\max(\|u_k(t) - u_{k-1}(t)\|)$ of the convergence between successive iterations as a function of the number of time steps. From iteration $(k-1)$ to k we increase the number of time intervals (NTST in AUTO) or decrease the tolerances of the ode45 command. As expected, the solution converges more rapidly as the number of time steps is increased. The comparison between the error for the shooting algorithm (squares) and the collocation approach (circles) in Fig. 9.3(a) demonstrates that both methods show quite similar behavior. This might be expected since shooting with Matlab uses a fourth-order Runge-Kutta method and collocation with AUTO was performed with degree-four polynomials (i.e., with $m = 4$

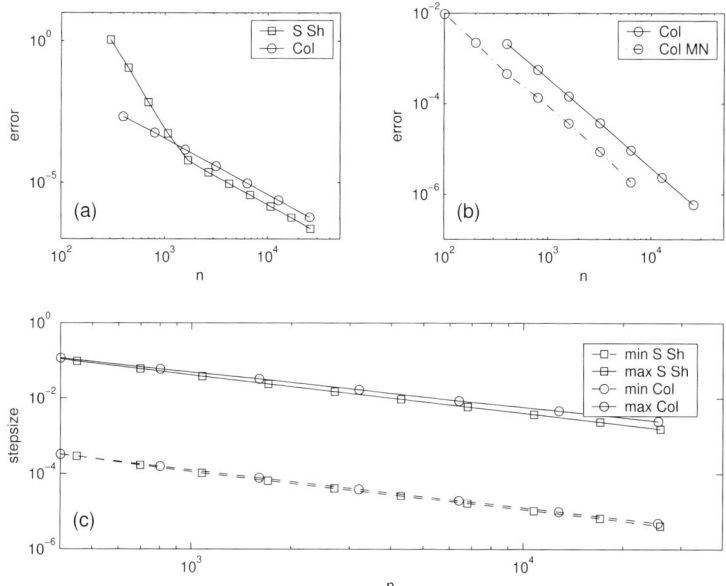

Fig. 9.3. Evolution of error and stepsize as a function of the number of time steps for the benchmark periodic orbit of the R3BP from Fig. 9.2. Panel (a) shows the evolution of the error in the Cauchy sense $(\max(||\mathbf{u}_k(t) - \mathbf{u}_{k-1}(t)||))$ for the shooting algorithm (squares) and for collocation with AUTO (circles). Panel (b) displays the error of the collocation calculation at all mesh points (circles solid line) and only at the main mesh points (circles dashed line); this illustrates the phenomenon of superconvergence at the main mesh points. Both methods use either an adaptive step or an adaptive mesh; panel (c) shows the maximal (solid lines) and minimal (dashed lines) stepsize for both methods. Note that both methods perform comparably well in this example.

collocation points per mesh interval). Indeed, if $u(t)$ is sufficiently smooth then the global accuracy of collocation is known to be of order m. Note from Fig. 9.3(b) that convergence for the collocation results is much better at the main mesh points; this is due to the phenomenon of superconvergence which guarantees an error of order $2m$; see Chap. 1. Finally, panel (c) shows that the mesh adaptation for both methods is very similar. We remark that an exhaustive comparison between shooting and collocation should include the continuation process as well. Furthermore, one should possibly consider multiple shooting for difficult orbits or long time integration.

9.1.4 Families of Lyapunov Solutions in the R3BP

As an illustration of the power and versatility of our computational approach, as outlined in Sect. 9.1.1, we now briefly discuss some numerical results about

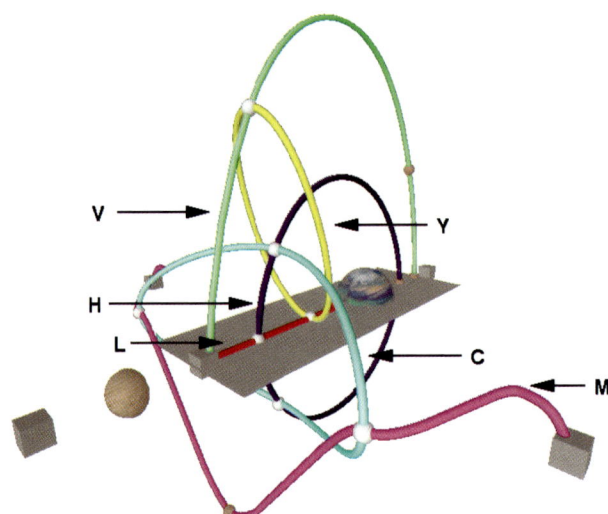

Fig. 9.4. Schematic bifurcation diagram for the R3BP (9.6) of the families of periodic orbits that emanate from the Lagrange point L1 for the mass-ratio $\mu = 0.01215$ of the Earth-moon system. From E.J. Doedel, R. Paffenroth, H. Keller, D. Dichmann, J. Galán-Vioque and A. Vanderbauwhede, Computation of periodic solutions of conservative systems with application to the three-body problem, *Internat. J. Bifur. Chaos Appl. Sci. Engrg.* 13 (2003) 1353–1381 © 2003 by World Scientific Publishing; reprinted with permission.

families of periodic solutions of the R3BP; these results are from [10, 11] where also an extensive bibliography can be found.

It is well known that for each value of μ system (9.6) has five equilibria, called the *Lagrange points* or *libration points*. Three of the Lagrange points, denoted L1, L2 and L3, are collinear with the primary bodies; one of them, L1, lies between the two primaries. Studying the linearization of (9.6) at the collinear Lagrange points and using the Lyapunov Center Theorem one can show that from each of these collinear Lagrange points there emanate two families of periodic orbits: the *Lyapunov family* containing planar orbits in the (x, y)-plane, and the family of so-called *Vertical orbits* that starts off in the vertical z-direction.

Figure 9.4 shows a schematic bifurcation diagram for the two families emanating from L1 and for the secondary families of periodic orbits bifurcating from them; this diagram was calculated for the mass ratio $\mu = 0.01215$ that corresponds to the Earth-moon system. The five libration points are shown as grey cubes. The red line (L) represents the Lyapunov orbits and the green curve (V) represents the Vertical orbits. Any solution branch that intersects the grey plane has a planar periodic solution at the intersection point. Along the (red) Lyapunov family there are two branch points; at the first of these the

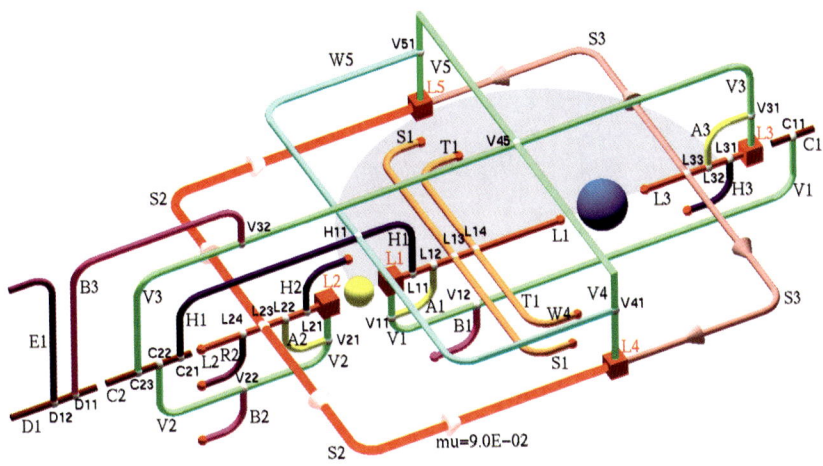

Fig. 9.5. Schematic bifurcation diagram of the families emanating from L1 in the RTBP for mass-ratio $\mu = 0.09$. From E.J. Doedel, V.A. Romanov, R. Paffenroth, H. Keller, D. Dichmann, J. Galán-Vioque and A. Vanderbauwhede, Elemental periodic orbits associated with the Libration Points in the Circular restricted three-body problem, *Internat. J. Bifur. Chaos Appl. Sci. Engrg.* 17 (2007) at press © 2007 by World Scientific Publishing; reprinted with permission.

blue family (H) of so-called *Halo orbits* bifurcates, at the second bifurcation point the yellow family (Y) branches off, connecting the Lyapunov family to the Vertical family. On the blue family of Halo orbits there are two symmetry-related bifurcation points which give rise to the cyan family (C) of orbits. In turn, along this cyan family there are two symmetry-related branching points which give the magenta family (M) that connects the non-collinear Lagrange points L4 and L5; in fact, the magenta family forms the "Vertical family" emanating from L4 and L5. More details on this bifurcation diagram and on the orbits represented can be found in [10].

The picture becomes more complicated and more interesting if the mass-ratio μ is allowed to vary. Detailed computational results for families that emanate from the five libration points in the R3BP for all values of μ are presented in [11]. As a stimulating example of the richness and complexity of the problem we show in Fig. 9.5 the schematic bifurcation diagram for $\mu = 0.09$. The number of families and subfamilies of periodic orbits and their interconnections in this example are a stark reminder of the statement of Poincaré cited above: indeed, periodic orbits play the guiding role in the study of complicated Hamiltonian systems such as the R3BP.

9.2 Theory of Continuation in Hamiltonian Systems

In Hamiltonian systems conserved quantities and symmetries are related by
Noether's theorem, and they result in families of non-isolated periodic orbits.
In this section we describe how the Implicit Function Theorem can be used
in this situation to prove continuation results via a set-up that can also be
exploited numerically; for full details and proofs we refer to [23].

We use the classical set-up for Hamiltonian systems with n degrees of
freedom. The state space is $\mathbb{R}^{2n} = \mathbb{R}^n \times \mathbb{R}^n$, with elements $u = (x, y)$, and with
the standard scalar product $\langle u, \tilde{u} \rangle = \sum_{j=1}^{2n} u_j \tilde{u}_j$. For each smooth function
$H : \mathbb{R}^{2n} \to \mathbb{R}$ the Hamiltonian vector field $X_H : \mathbb{R}^{2n} \to \mathbb{R}^{2n}$ is defined by
$X_H(u) := J\nabla H(u)$, where $J \in \mathcal{L}(\mathbb{R}^{2n})$ is the standard symplectic matrix
given by $J(x, y) := (y, -x)$. We call

$$\dot{u} = X_H(u) \tag{9.7}$$

the *Hamilton equation with Hamiltonian H*, and we denote the flow of (9.7) by
$\varphi_H(t, u) = \varphi_H^t(u)$ (with $t \in \mathbb{R}$ and $u \in \mathbb{R}^{2n}$). A smooth function $F : \mathbb{R}^{2n} \to \mathbb{R}$
is a *first integral* (constant of motion) for (9.7) if and only if $\{F, H\} \equiv 0$,
where the *Poisson bracket* $\{F, H\} : \mathbb{R}^{2n} \to \mathbb{R}$ of F and H is defined by
$\{F, H\}(u) := DF(u) \cdot X_H(u)$. The Poisson bracket is anti-symmetric, that
is, $\{F, H\} = -\{H, F\}$, and, as a consequence, H is a first integral of (9.7).
In case either X_H or X_F has at least one bounded orbit, the condition that
$\{F, H\} \equiv 0$ for F is a first integral of (9.7) is equivalent to the fact that
for all $t, s \in \mathbb{R}$ the symplectic diffeomorphisms φ_F^s and φ_H^t commute. This
is the statement of Noether's theorem: first integrals generate symmetries,
and conversely, (symplectic and continuous) symmetries are generated by first
integrals. We denote the space of all first integrals of (9.7) by

$$\mathcal{F} := \{F \in C^\infty(\mathbb{R}^{2n}; \mathbb{R}) \mid \{F, H\} \equiv 0\}. \tag{9.8}$$

Now assume that $p_0 \in \mathbb{R}^{2n}$ generates a nontrivial periodic orbit $\Gamma_0 :=
\{\varphi_H^t(p_0) \mid t \in \mathbb{R}\}$ of (9.7), with minimal period $T_0 > 0$. In order to continue
this periodic solution we look for solutions (T, p) near (T_0, p_0) of the equation

$$G_0(T, p) := \varphi_H(T, p) - p = 0. \tag{9.9}$$

This equation has the one-dimensional solution curve $\{T_0\} \times \Gamma_0$. If G_0 were
submersive at (T_0, p_0) then this would (locally) give us all solutions. However,
as we will see, this is not the case. To find out about the submersivity prop-
erties of G_0 we calculate the image of $DG_0(T_0, p_0) \in \mathcal{L}(\mathbb{R}^{2n+1}; \mathbb{R}^{2n})$, which is
given by

$$\text{Im } DG_0(T_0, p_0) = \text{Im}(M - I) + \mathbb{R}\, X_H(p_0), \qquad M := D\varphi_H^{T_0} \in \mathcal{L}(\mathbb{R}^{2n}). \tag{9.10}$$

Here M is the *monodromy matrix* of the periodic orbit Γ_0, and its eigenvalues
are the *multipliers* of the periodic orbit Γ_0. Since $X_H(p_0) \in \text{Ker}(M - I)$ there

is always the multiplier 1. We denote by m_g and m_a the geometric and the algebraic multiplicity of the eigenvalue 1 of M. It is not hard to show that M is symplectic, i.e., $M^T J M = J$, and hence m_a is even. One can also prove the following.

Proposition 1. *Let* $W := \{\nabla F(p_0) \mid F \in \mathcal{F}\}$, $\mathcal{Z}_0 := \{G \in \mathcal{F} \mid \{G, F\}(p_0) = 0, \forall F \in \mathcal{F}\}$, *and* $Z_0 := \{X_G(p_0) \mid G \in \mathcal{Z}_0\}$. *Let* $k := \dim W$. *Then*

$$Z_0 \subset JW \subset \operatorname{Ker}(M - I) \quad and \quad \operatorname{Im}(M - I) + Z_0 \subset W^\perp. \tag{9.11}$$

In particular, $m_g \geq k$ *and* $m_a \geq k + \dim Z_0$.

Clearly $H \in \mathcal{Z}_0$ and $X_H(p_0) \in Z_0$; therefore, referring back to (9.10), we see that

$$\operatorname{Im} DG_0(T_0, p_0) = \operatorname{Im}(M - I) + \mathbb{R}X_H(p_0) \subset W^\perp. \tag{9.12}$$

Definition 1. *We say that the periodic orbit* Γ_0 *generated by* p_0 *is normal if in (9.12) we have equality, i.e., if*

$$\operatorname{Im}(M - I) + \mathbb{R}X_H(p_0) = W^\perp. \tag{9.13}$$

This notion of normality is a generalization of the one introduced in [28]; for a discussion of the relationship between the normality property and the submersivity properties of the mapping G_0 see [24]. The following proposition gives conditions for Γ_0 to be normal.

Proposition 2. *The periodic orbit* Γ_0 *of (9.7) is normal if and only if either* $m_g = k$ *or* $m_g = k + 1$ *and* $X_H(p_0) \notin \operatorname{Im}(M - I)$. *In particular,* Γ_0 *is normal if* $m_a = k + 1$.

We remark, in view of Proposition 1, that the condition $m_a = k + 1$ can only be satisfied if k is odd and $\dim Z_0 = 1$.

The usefulness of the concept of a normal periodic orbit becomes clear when one combines the following two observations.

1. If $p_0 \in \mathbb{R}^{2n}$ generates a normal periodic orbit Γ_0 with minimal period T_0 then, by definition of normality, the subspace W is lacking in the image of $DG_0(T_0, p_0)$, and therefore G_0 is not submersive at the point (T_0, p_0). We can 'repair' this lack of submersivity by adding some artificial terms to (9.7). More precisely, we replace (9.7) by

$$\dot{u} = X_H(u) + \sum_{j=1}^{k} \alpha_j \nabla F_j(u), \qquad \alpha = (\alpha_1, \alpha_2, \dots, \alpha_k) \in \mathbb{R}^k, \tag{9.14}$$

where $F_j \in \mathcal{F}$ $(1 \leq j \leq k)$ are chosen such that $\{\nabla F_j(p_0) \mid 1 \leq j \leq k\}$ forms a basis of W (typically one will take $F_1 = H$). Denoting the flow of (9.14) by $\Phi(t, p, \alpha)$ and defining $G : \mathbb{R} \times \mathbb{R}^{2n} \times \mathbb{R}^k \to \mathbb{R}^{2n}$ by $G(T, p, \alpha) := \Phi(T, p, \alpha) - p$ we see that $G(T, p, 0) = G_0(T, p)$, $G(T_0, p_0, 0) = 0$, and one can show that G is submersive at $(T_0, p_0, 0)$. Hence, the solution set of $G(T, p, \alpha) = 0$ forms, locally near $(T_0, p_0, 0)$, a $(k+1)$-dimensional submanifold of $\mathbb{R} \times \mathbb{R}^{2n} \times \mathbb{R}^k$. Since $G(T, p, \alpha) = 0$ means that $\Phi(t, p, \alpha)$ is a T-periodic solution of (9.14), this $(k+1)$-dimensional solution manifold is foliated by a k-parameter family of periodic orbits of (9.14).

2. Suppose that $u(t)$ is a T-periodic solution of (9.14), and let $F(u) := \sum_{j=1}^{k} \alpha_j F_j(u)$. Using $F \in \mathcal{F}$ one then calculates that $d/dt\, F(u(t)) = \|\nabla F(u(t))\|^2$, from which it follows that

$$\int_0^T \|\nabla F(u(t))\|^2 \, dt = F(u(T)) - F(u(0)) = 0 \quad \Rightarrow \quad \nabla F(u(t)) = 0, \ \forall t \in \mathbb{R}.$$

In particular $\nabla F(u(0)) = \sum_{j=1}^{k} \alpha_j \nabla F_j(u(0)) = 0$. Since the vectors $\nabla F_j(p_0)$, $1 \le j \le k$, are linearly independent, the same is true for the vectors $\nabla F_j(u(0))$, $1 \le j \le k$, if $u(0)$ is sufficiently close to p_0. We conclude that (9.14) can only have a periodic orbit near Γ_0 if $\alpha = 0$. In other words, all periodic orbits of (9.14) near Γ_0 are, in fact, periodic orbits of (9.7).

By combining the two observations above, it follows that a normal periodic orbit Γ_0 of (9.7) belongs (locally) to a k-parameter family of normal periodic orbits of the same equation (normality is preserved locally). The question how we can parametrize this family has a straightforward answer only in the simplest possible case, namely if $m_a = k+1$. Then the family can be parametrized by the values of the first integrals F_j, $1 \le j \le k$. Such a parametrization may fail when $m_a > k+1$.

Complementing the periodicity condition $G(T, p, \alpha) = 0$ with appropriate phase conditions also gives an efficient way for actually calculating the periodic orbits. Starting from the point p_0 on Γ_0 we can generate a k-dimensional submanifold of initial points for periodic orbits of (9.7) by applying the flows of the Hamiltonian vector fields X_{F_j}, $1 \le j \le k$. So what we really need is a way to calculate a one-dimensional solution curve in the 'missing direction'; such a curve is given by the following theorem; see [23] for the proof.

Theorem 2. Let $\Gamma_0 = \{\varphi_H(t, p_0) \mid t \in \mathbb{R}\}$ be a normal T_0-periodic solution of $\dot{u} = X_H(u)$. With the notations introduced before, consider the following set of equations for $(T, p, \alpha) \in \mathbb{R} \times \mathbb{R}^{2n} \times \mathbb{R}^k$:

$$G(T, p, \alpha) = 0, \qquad \langle X_{F_j}(p_0), p - p_0 \rangle = 0, \quad 1 \le j \le k. \tag{9.15}$$

Then near $(T_0, p_0, 0)$ the solution set of (9.15) consists of a smooth one-dimensional curve along which $\alpha \equiv 0$. In the case $m_g = k$ this curve can be parametrized by the period T. Projecting the solution curve onto the phase space $\mathbb{R}^{2n}$ and acting on the projection with the flows of the Hamiltonian vector fields X_{F_j}, $1 \le j \le k$, generates a $(k+1)$-dimensional manifold that is invariant under the flow of X_H. Furthermore, this manifold is foliated by a k-parameter family of normal periodic orbits of $\dot{u} = X_H(u)$.

The condition $m_g = k$ appearing in Theorem 2 is typically satisfied, so that the period T can be used as a parameter along the solution curve. For numerical calculations the k phase conditions in (9.15) will usually be replaced by integral versions; see [8] and Chap. 10 for an extensive discussion of such integral phase conditions. Also, instead of looking for T-periodic solutions of (9.14), it is easier to rescale time and to look for 1-periodic solutions of the rescaled equation. The way to implement the theoretical continuation result above then takes the form of the following boundary value problem:

(**CON-2**) Find $(u(t), T, \alpha)$ near $(u_0(t), T_0, 0)$, with $u_0(t) := \varphi_H(T_0 t, p_0)$, such that

$$
\begin{cases}
\dot{u}(t) = T\left(X_H(u(t)) + \sum_{j=1}^{k} \alpha_j \nabla F_j(u(t))\right), \\
u(1) = u(0), \\
\int_0^1 \langle X_{F_j}(u_0(t)), u(t) - u_0(t)\rangle \, dt = 0, \qquad 1 \leq j \leq k.
\end{cases} \tag{9.16}
$$

Such boundary value problems are very well suited for pseudo-arclength continuation as implemented, for example, in AUTO.

In a number of examples (in particular in N-body problems) the Hamiltonian vector field X_H has a scaling property that allows one to obtain new solutions from given ones by appropriate rescalings of time and phase space variables. In this case, Theorem 2 only gives a family of rescaled copies of the starting periodic orbit Γ_0. To obtain a more meaningful result one can fix the period and perform a continuation in an external parameter.

We now give an example of the type of results one can prove in this direction. Consider a Hamiltonian $H_\lambda(u)$ depending smoothly on a real parameter $\lambda \in \mathbb{R}$ and suppose that apart from H_λ the Hamiltonian system

$$
\dot{u} = X_{H_\lambda}(u) \tag{9.17}
$$

has some further independent first integrals F_j, $2 \leq j \leq k$, i.e., $\{H_\lambda, F_j\} \equiv 0$, $2 \leq j \leq k$, for all $\lambda \in \mathbb{R}$. Suppose also that for $\lambda = 0$ equation (9.17) has a periodic orbit $\Gamma_0 = \{\varphi_{H_0}^t(p_0) \mid t \in \mathbb{R}\}$ with minimal period $T_0 > 0$. We set $\mathcal{F}_0 := \{F \in C^\infty(\mathbb{R}^{2n}; \mathbb{R}) \mid \{H_0, F\} \equiv 0\}$ and $W_0 := \{\nabla F(p_0) \mid F \in \mathcal{F}_0\}$. Finally we assume that $\{\nabla H_0(p_0), \nabla F_2(p_0), \ldots, \nabla F_k(p_0)\}$ forms a basis of W_0 and that $m_g = k$. Denote by $\Phi(t, p, \lambda, \alpha)$ the flow of the modified system

$$
\dot{u} = X_{H_\lambda}(u) + \alpha_1 \nabla H_\lambda(u) + \sum_{j=2}^{k} \alpha_j \nabla F_j(u).
$$

Theorem 3. *Under the above assumptions and for each fixed T near T_0 the set of equations*

$$
\Phi(T, p, \lambda, \alpha) = p, \quad \langle X_{F_j}(p_0), p - p_0 \rangle = 0 \ (1 \leq j \leq k, \ F_1 := H_0), \tag{9.18}
$$

has a unique one-dimensional solution branch near $(p, \lambda, \alpha) = (p_0, 0, 0)$. Furthermore, $\alpha \equiv 0$ along this branch, and it can be parametrized by λ.

Theorem 3 lends itself to numerical implementation much like Theorem 2. For an application of Theorem 3 to the three-body problem see Sect. 9.3.

9.2.1 Continuation of Relative Equilibria

As we have seen, Noether's Theorem states a strong relationship between symmetries and first integrals of Hamiltonian systems. We now go a bit deeper into the symmetry aspect; see also Chaps. 10 and 11.

Consider a finite-dimensional subspace $\mathcal{G}$ of $C^\infty(\mathbb{R}^{2n};\mathbb{R})$ with the property that $\{F,G\} \in \mathcal{G}$ for all $F,G \in \mathcal{G}$ (i.e., $\mathcal{G}$ together with the Poisson bracket $\{\cdot,\cdot\}$ forms a Lie algebra). Denote by $\mathfrak{G}$ the (possibly noncompact) Lie group of symplectic diffeomorphisms generated by the one-parameter groups $\{\varphi_F^s \mid s \in \mathbb{R}\}$, with $F \in \mathcal{G}$. If for some Hamiltonian $H : \mathbb{R}^{2n} \to \mathbb{R}$ we have $\{H,F\} \equiv 0$ for all $F \in \mathcal{G}$ (in our earlier notation this means that $\mathcal{G} \subset \mathcal{F}$) then the Hamiltonian system (9.7) is equivariant with respect to the group $\mathfrak{G}$.

Definition 2. *An orbit $\{\varphi_H(t,p_0) \mid t \in \mathbb{R}\}$ of (9.7) is a* relative equilibrium *(with respect to the group $\mathfrak{G}$) if it is contained in the group orbit $\mathfrak{G}(p_0) = \{\gamma(p_0) \mid \gamma \in \mathfrak{G}\}$; this means that there is a mapping $\gamma : \mathbb{R} \to \mathfrak{G}$ such that $\varphi_H(t,p_0) = \gamma(t)(p_0)$.*

It is not hard to show that $p_0 \in \mathbb{R}^{2n}$ generates a relative equilibrium if and only if there is some $F \in \mathcal{G}$ such that $X_H(p_0) = X_F(p_0)$. This means that p_0 is an equilibrium of the Hamiltonian vector field X_{H-F} and that $\varphi_H^t(p_0) = \varphi_F^t(p_0)$. If $F_j \in \mathcal{G}$, $1 \leq j \leq m$, are such that $\{X_{F_j}(p_0) \mid 1 \leq j \leq m\}$ forms a basis of $Y := \{X_F(p_0) \mid F \in \mathcal{G}\}$ then the condition for p_0 to generate a relative equilibrium is that there exist numbers $\Omega_j^0 \in \mathbb{R}$, $1 \leq j \leq m$, such that

$$X_H(p_0) = \sum_{j=1}^m \Omega_j^0 X_{F_j}(p_0). \qquad (9.19)$$

The corresponding relative equilibrium is given by $\varphi_H^t(p_0) = \varphi_F^t(p_0)$, where $F = \sum_{j=1}^m \Omega_j^0 F_j$. Together with $p_0 \in \mathbb{R}^{2n}$ also all the other points on the group orbit $\mathfrak{G}(p_0)$ generate a relative equilibrium; this (m-dimensional) group orbit is then foliated by the X_H-orbits of its elements.

A simple example of relative equilibria appears in the N-body problem where for $\mathfrak{G}$ one can take either the rotation group generated by the components of the total angular momentum or the Euclidean group generated by the components of the total linear momentum or the total angular momentum. For the rotation group the relative equilibria correspond to equilibria in a uniformly rotating frame. Well-known examples are the Lagrange points in the circular restricted three-body problem and the Euler and Lagrange solutions of the three-body problem with three equal masses; see already Sect. 9.3. In each of these examples the relative equilibria are also periodic solutions of the corresponding Hamiltonian system.

For the remaining part of this subsection we make the restrictive assumption that $\mathcal{G} \subset \mathcal{Z} \subset \mathcal{F}$, i.e., $\{F, G\} \equiv 0$ for all $F \in \mathcal{F}$ and all $G \in \mathcal{G}$. This implies, in particular, that $\mathfrak{G}$ is Abelian. As is shown in forthcoming work of Wulff and Schebesch [36], one can cancel this assumption by using some more elaborate symplectic geometry. Assuming (9.19) we complement the $F_j \in \mathcal{G}$, $1 \le j \le m$, with some further first integrals $F_j \in \mathcal{F}$, $m+1 \le j \le k$, such that $\{\nabla F_j(p_0) \mid 1 \le j \le k\}$ forms a basis of $W := \{\nabla F(p_0) \mid F \in \mathcal{F}\}$. We also set

$$L := DX_H(p_0) - \sum_{j=1}^{m} \Omega_j^0 DX_{F_j}(p_0) \in \mathcal{L}(\mathbb{R}^{2n}). \qquad (9.20)$$

Proposition 3. *Under the conditions above we have*

$$Y \subset JW \subset \mathrm{Ker}(L) \quad and \quad \mathrm{Im}(L) + Y \subset W^{\perp}. \qquad (9.21)$$

As a consequence, 0 is an eigenvalue of L with geometric multiplicity $\tilde{m}_g \ge k$ and algebraic multiplicity $\tilde{m}_a \ge k + m$.

Definition 3. *We say that the relative equilibrium generated by p_0 is normal if we have equality in the second inclusion of (9.21), i.e., if*

$$\mathrm{Im}(L) + Y = W^{\perp}. \qquad (9.22)$$

This normality condition is satisfied when either $\tilde{m}_g = k$ or $\tilde{m}_a = k+m$. One can prove the following continuation result for normal relative equilibria [23].

Theorem 4. *Let $\mathcal{G} \subset \mathcal{Z}$, and suppose that $p_0 \in \mathbb{R}^{2n}$ generates a normal relative equilibrium of X_H with respect to $\mathfrak{G}$. Then, in the notation introduced above, the set of equations*

$$X_H(p) = \sum_{j=1}^{m} \Omega_j X_{F_j}(p) + \sum_{j=1}^{k} \alpha_j \nabla F_j(p), \quad \langle X_{F_j}(p_0), p - p_0 \rangle = 0, \ 1 \le j \le k,$$
$$(9.23)$$

has a solution set near $(p, \Omega, \alpha) = (p_0, \Omega^0, 0) \in \mathbb{R}^{2n} \times \mathbb{R}^m \times \mathbb{R}^k$. The solution set is a smooth m-dimensional submanifold, along which $\alpha \equiv 0$. For each $(p, \Omega, 0)$ on this solution manifold we have that $X_H(p) = \sum_{j=1}^{m} \Omega_j X_{F_j}(p)$ and that p generates a normal relative equilibrium of X_H with respect to $\mathfrak{G}$.

In case $\tilde{m}_g = k$ (which is generically satisfied) the solution manifold of (9.23) can be parametrized by $\Omega = (\Omega_1, \Omega_2, \ldots, \Omega_m)$. One can then also fix Ω at any value close to Ω^0 and use a similar set-up to continue the relative equilibrium in external parameters; see the end of the next section for an example.

9.3 Continuation of the Figure-8 Solution

Celestial Mechanics has been at the origin of the theory of dynamical systems and many of the techniques from that theory were developed to analyze the fascinating behavior of a group of massive bodies under the influence of their mutual gravitational interaction.

The spectacular discovery by Chenciner and Montgomery [5] of the existence of a new solution of the Three-Body Problem (3BP) with equal masses in which all three bodies follow the same planar curve with the shape of a 'figure eight' has brought great excitement to the dynamical systems community. The origin of the name of the orbit becomes apparent from a representation in physical space such as in Fig. 9.7 (a). This solution was first discovered (numerically) by Moore [20] in the context of a study of possible braid types associated with the planar N-body problem. The method of proof by Chenciner and Montgomery is based on variational arguments. After some reductions, the action integral is minimized on a restricted set of symmetric arcs to prove the existence of a solution where the three bodies of equal masses chase each other along a single closed trajectory. However, the variational proof is unable to determine the stability of the solution.

Simó [31] computed this remarkable solution numerically with great accuracy and announced elliptic stability; i.e., the non-trivial characteristic multipliers of the periodic orbit are on the unit circle. The precise values of the non-trivial characteristic multipliers (those which are different from one) are given in [31] as $\mu_j = \exp(2\pi i \nu_j)$, $j = 1, 2$, with $\nu_1 = 0.00842272$ and $\nu_2 = 0.29809253$. Note that the smallness of ν_1 indicates that the figure-8 solution is close to a bifurcation. Simó [32] also discovered many (hundreds of) other similar solutions for three equal bodies, as well as for N equal bodies with $3 < N < 799$ [31]. The defining property of these solutions, which are referred to as *choreographies*, is that all bodies follow a single closed curve in phase space, with a fixed time between each of the bodies. From the historical point of view, the solution found by Lagrange in 1772, in which the three bodies form the vertices of an equilateral triangle that rotates with constant angular velocity around its midpoint, can be considered as the first choreography. It has taken more than two hundred years to find the second one.

In this section we apply our continuation scheme to following the figure-8 solution of the 3BP; see also [10, 12] for more details. The original motivation to study this problem was a conjecture by Joe and Herb Keller that it would be possible to connect the two simple Lagrange and figure-8 choreographies in a continuous way by only following periodic orbits. In fact, Marchal [14] has found a family of periodic orbits in a rotating frame (relative periodic orbits) that connects these two highly symmetrical solutions. A further reason for trying to connect the Lagrange and figure-8 choreographies has to do with some controversy about the stability properties of the solution corresponding to the absolute minimizer of the action over certain homotopy classes of loops in configuration space. In Hamiltonian systems with two degrees of freedom such minimizing orbits are always unstable [4]; however, for higher-dimensional systems there are counterexamples. The stable figure-8 solution was obtained by minimizing the action over all loops with a particular symmetry; see [5]. Assuming that the property that 'minimizing orbits are unstable' also holds for the equal mass 3BP there must exist some other unstable periodic orbit in the homotopy class of the figure-8 solution with a lower action. Could this orbit be the equilateral Lagrange choreography? It is unstable, and its action value $(3\pi 3^{2/3} \approx 19.60436$ when the period equals $2\pi)$ is lower than that of the figure-8 solution (≈ 24.37197), but we do not know whether it is in the homotopy class of the figure-8 solution. If it is not then the minimizer over this homotopy class has to be some other unstable periodic orbit which will then probably not be a choreography. Finding some connection between the figure-8 and the Lagrange solutions would at least give some partial answers to the above questions.

We have applied the general continuation scheme of Sect. 9.2 where we used the numerically computed figure-8 solution as the starting solution. The equations of motion of three bodies with masses m_1, m_2 and m_3 under mutual gravitational attraction take the form

$$\ddot{\mathbf{x}}_1 = -m_2 \frac{\mathbf{x}_1 - \mathbf{x}_2}{|\mathbf{x}_1 - \mathbf{x}_2|^3} - m_3 \frac{\mathbf{x}_1 - \mathbf{x}_3}{|\mathbf{x}_1 - \mathbf{x}_3|^3},$$

$$\ddot{\mathbf{x}}_2 = -m_1 \frac{\mathbf{x}_2 - \mathbf{x}_1}{|\mathbf{x}_1 - \mathbf{x}_2|^3} - m_3 \frac{\mathbf{x}_2 - \mathbf{x}_3}{|\mathbf{x}_2 - \mathbf{x}_3|^3}, \qquad (9.24)$$

$$\ddot{\mathbf{x}}_3 = -m_1 \frac{\mathbf{x}_3 - \mathbf{x}_1}{|\mathbf{x}_1 - \mathbf{x}_3|^3} - m_2 \frac{\mathbf{x}_3 - \mathbf{x}_2}{|\mathbf{x}_3 - \mathbf{x}_2|^3}.$$

Here $\mathbf{x}_i = (x_i, y_i, z_i) \in \mathbb{R}^3$ denotes the position of the i-th body, $i = 1, 2, 3$, and the universal gravitational constant has been set to 1. This system can be rewritten as a first-order system of dimension eighteen. There are seven independent conserved quantities: the total energy, the three components of the total linear momentum $\mathbf{P} = \sum_{i=1}^{3} m_i \dot{\mathbf{x}}_i$, and the three components of the total angular momentum $\mathbf{L} = \sum_{i=1}^{3} m_i \mathbf{x}_i \wedge \dot{\mathbf{x}}_i$; they are an immediate consequence of the invariance of the equations under time shifts, translations and rotations.

Additionally, the equations are invariant under the transformation $(t, \mathbf{x}) \mapsto (c^{\frac{3}{2}} t, c \mathbf{x})$, for an arbitrary constant $c > 0$; see [35]. Due to this scaling property there is a trivial continuation of periodic orbits in the period: arbitrarily close to any periodic orbit there is another one with slightly different period, obtained from the first by rescaling. Obviously it has the same stability properties. In order to avoid this trivial continuation we fix the period (say to 2π) and use instead the mass m_1 of the first body as an external continuation parameter. Both other masses are kept equal to 1. It is not difficult to check numerically that the figure-8 solution (corresponding to $m_1 = 1$) is normal (we have $m_g = k = 7$) and, hence, we can apply Theorem 3 with $\lambda = m_1$. As already discussed before, when setting up the continuation algorithm the phase conditions in (9.18) are replaced by integral versions; for the detailed computational formulation we refer to [10].

The calculations were performed with AUTO [7, 9]. While following a one-parameter family of periodic orbits we also monitor the stability of these orbits

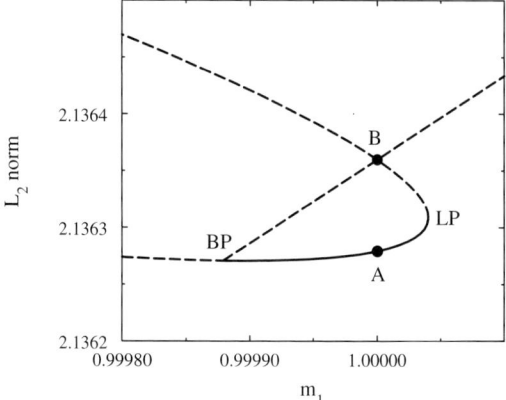

Fig. 9.6. Local bifurcation diagram near the figure-8 solution under variation of the mass m_1. Solutions for the solid section of the curve are stable and those for dashed curves are unstable. Stable solutions appear in a narrow window between a pitchfork bifurcation (BP) and a limit point (LP). Two orbits for the 3BP with equal masses ($m_1 = 1$) are indicated: at A one finds the stable Chenciner-Montgomery figure-8 solution shown in Fig. 9.7(a), and at B one finds the unstable satellite figure-8 solution shown in Fig. 9.7(b). From J. Galán, F.J. Muñoz-Almaraz, E. Freire, E.J. Doedel and A. Vanderbauwhede, Stability and bifurcations of the figure-8 solution of the three-body problem, *Phys. Rev. Lett.* **88** (2002) 241101 © 2002 by the American Physical Society; reprinted with permission.

and the appearance of new branches at bifurcation points; at such bifurcation points AUTO allows one to switch branches and start a new continuation process.

Starting at the figure-8 solution for $m_1 = 1$, the first output of our continuation algorithm are the nontrivial characteristic multipliers of this figure-8 solution; we obtain $\mu_j = \exp(2\pi i \nu_j)$ ($j = 1, 2$), with $\nu_1 = 0.0084227$ and $\nu_2 = 0.2980925$. The good agreement with the results of Simó [31] constitutes a successful practical test for our method.

Figure 9.6 shows the results of the continuation of the figure-8 solution in a small mass-interval around $m_1 = 1$, where we plot the L_2-norm of the solution as a function of m_1. The solution labeled A is the Moore-Chenciner-Montgomery figure-8 solution that is the starting point of our calculation. This special planar orbit is plotted in Fig. 9.7(a). When decreasing m_1 from $m_1 = 1$ we obtain a single solution branch with a pitchfork bifurcation at the point BP. Increasing m_1 from A results in a solution branch that reaches a limit point LP and then continues in the direction of decreasing values of m_1. All solutions along the branch are unstable (hyperbolic), except for those on the section of the branch between BP and LP where the solutions are stable (elliptic). This stable part of the branch corresponds to a very

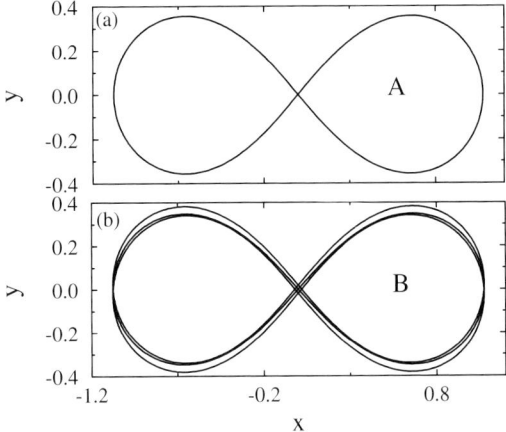

Fig. 9.7. Representation in physical space of the figure-8 solution (a) and of the satellite figure-8 solution (b) of the 3BP with equal masses that correspond to labels A and B in Fig. 9.6. From J. Galán, F.J. Muñoz-Almaraz, E. Freire, E.J. Doedel and A. Vanderbauwhede, Stability and bifurcations of the figure-8 solution of the three-body problem, *Phys. Rev. Lett.* 88 (2002) 241101 © 2002 by the American Physical Society; reprinted with permission.

narrow m_1-interval of the order of 10^{-5}. Continuing the branch beyond the limit point we return to a situation where all three masses are equal, i.e., $m_1 = 1$. The corresponding solution, labeled B in Fig. 9.6, is hyperbolic and by construction in the same homotopy class as the figure-8 solution. However, it has less symmetry and is no longer a choreography: as is shown in Fig. 9.7 (b), the three bodies follow three slightly different figure-8 paths. We refer to this solution as the 'satellite figure-8' solution; it was also found numerically by Simó [32].

At the bifurcation point (BP) there is a pitchfork bifurcation at which the interchange symmetry of the second and third body (which both have the same mass $m_2 = m_3 = 1$) is broken. At this bifurcation two symmetry-related branches are born that are represented by a single curve in Fig. 9.6. Along these branches the solutions are hyperbolic and the mass m_1 increases. Also along these branches one finds a special point where $m_1 = 1$, so that, therefore, all three masses are equal. The corresponding solution is the same as before, namely the satellite figure-8 solution of Fig. 9.7(b) that occurs at the point B; however, the labeling of the three bodies is now different. The fact that all three branches intersect at B is due to the chosen representation.

It is clear from (9.24) that by choosing appropriate units one can always assume that $m_3 = 1$, leaving m_1 and m_2 as (dimensionless) parameters. Since the figure-8 solution (corresponding to $m_1 = m_2 = 1$) is normal it can be continued in both these parameters with a multi-parameter version of The-

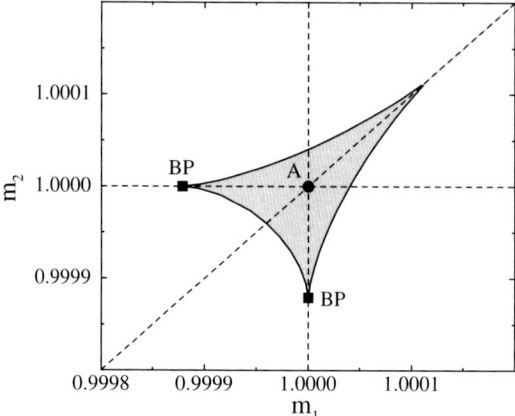

Fig. 9.8. Stability region in the (m_1, m_2)-plane for the continuation of the figure-8 solution. From J. Galán, F.J. Muñoz-Almaraz, E. Freire, E.J. Doedel and A. Vanderbauwhede, Stability and bifurcations of the figure-8 solution of the three-body problem, *Phys. Rev. Lett.* 88 (2002) 241101 © 2002 by the American Physical Society; reprinted with permission.

orem 3. Figure 9.8 shows the result of a continuation of solutions with a numerical determination of their stability. For mass values in the shaded triangular stability region in the (m_1, m_2)-plane the continued figure-8 solution is elliptic; the point labeled A in the center corresponds to the figure-8 solution. Along the border of this stability region (solid curve) the continuation manifold exhibits a fold (LP), and the two points labeled BP correspond to branch points. The diagram is obviously symmetric with respect to the diagonal $m_1 = m_2$, which is a consequence of the invariance of the equations under the symmetry $(\mathbf{x}_1, \mathbf{x}_2, m_1, m_2) \mapsto (\mathbf{x}_2, \mathbf{x}_1, m_2, m_1)$.

In principle one could try to continue the figure-8 solution of the 3BP to a solution of the R3BP by moving along the diagonal and increasing the value of $m_1 = m_2$. However, in practice one finds that, as the two equal masses become larger, the bodies collide.

There is one further remarkable point that came out of our calculations. Chenciner and Montgomery [5] obtained the figure-8 solution A by minimizing the action over a class of loops with some particular symmetry properties. The satellite figure-8 solution B minimizes the action over a much larger class of paths. Therefore, the action corresponding to solution B should be less than or equal to the action corresponding to solution A. This is indeed the case, but in a rather unexpected way: within the precision of our calculations *both solutions have the same action*. Using the standard definition of the action and making the necessary normalizations, the value of the action integral for

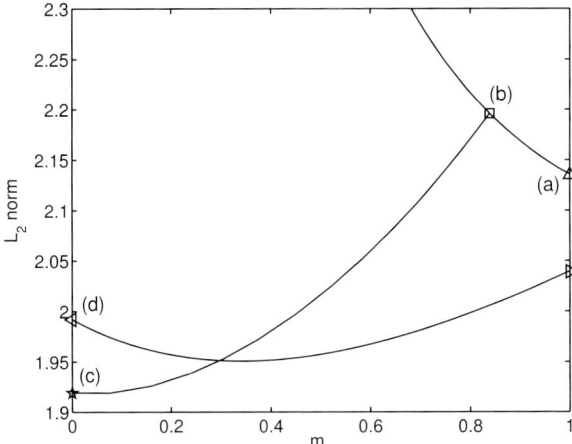

Fig. 9.9. Bifurcation diagram showing branches of periodic orbits that connect the 3BP with equal masses to the R3BP with $\mu = 1/2$. The solutions corresponding to the marked points (a)–(d) along the branches are shown in Fig. 9.10.

both solutions is found to be $S = 24.37197$. This surprising result reveals a degeneracy that deserves further analysis.

Up to now we have restricted the description of our continuation and bifurcation results to a neighborhood of the figure-8 solution. However, nothing prevents us from carrying out the continuation procedure over larger intervals for varying mass m_1. In Fig. 9.9 we show a (partial) global bifurcation diagram spanning the interval $[0, 1]$ for m_1. As before, the plot shows the L_2-norm of some calculated periodic orbits as a function of the mass m_1 of the first body, while the two other masses m_2 and m_3 are equal to 1. The left-hand border at $m_1 = 0$ corresponds to the R3BP with two equal primaries ($\mu = 1/2$, known as the Sitnikov problem), and the right-hand border at $m_1 = 1$ to the 3BP with three equal masses. Figure 9.10 shows representations in physical space of the orbits corresponding to the marked points in the bifurcation diagram of Fig. 9.9.

At the right-hand border of Fig. 9.9 there are two particular solutions of the 3BP with equal masses, namely the figure-8 solution that is also shown in Fig. 9.10 (a) and the equilateral Lagrange solution. (The representation of the Lagrange solution in physical space is simply a circle with radius $3^{-1/6}$ and is not shown in Fig. 9.10). The branch that emerges from the figure-8 solution and crosses the upper-right part of the diagram is the continuation of the lower branch in Fig. 9.6 that contains A and BP. All orbits along this branch are planar; this is an outcome of the calculations, not an imposed condition. The branch ends (just outside of the diagram) in a collision orbit where the smaller body (with mass m_1) collides with one of the larger bodies. Along the branch we have located a bifurcation point, marked (b) in Fig. 9.9 and corresponding

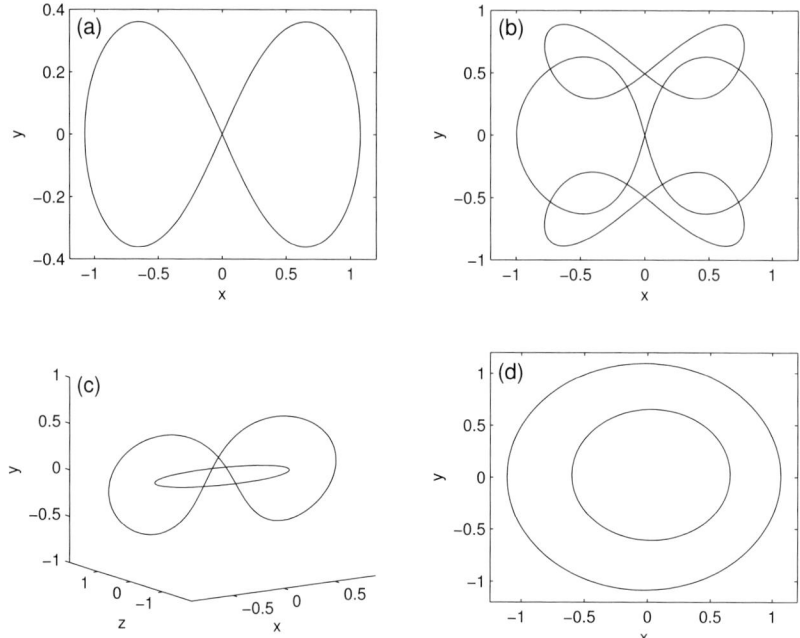

Fig. 9.10. Representation in physical space of solutions (a)–(d) that correspond to the marked points in the global bifurcation diagram of Fig. 9.9.

to the solution depicted in Fig. 9.10(b). For this solution the small mass follows the 8-shaped curve in the middle, while the two other masses each follow one of the curves at the top or the bottom. Along the new branch emanating from this bifurcation point the planar symmetry is broken, i.e., the solutions along this branch have a non-vanishing z-component. This branch of three-dimensional solutions can be continued all the way to $m_1 = 0$. Here we find a special solution (marked (c) in Fig. 9.9) of the so-called Sitnikov problem, that is, of the R3BP with equal masses for the primaries ($\mu = 1/2$). Figure 9.10(c) shows a three-dimensional plot of the solution: the two primaries follow the same (planar) circular orbit, always staying opposite to each other, while the negligible mass traces out the other curve that is intertwined with the circle. We conclude that it is possible, by changing the mass m_1, to connect the figure-8 solution of the 3BP with equal masses to this particular solution of the R3BP with $\mu = 1/2$. This connection, calculated with the implementation of Theorem 3 as described before, was first reported in [10]; further details on the R3BP for $\mu = 1/2$ can be found in [11] and in references therein.

We now turn to the equilateral Lagrange solution and its continuation when the mass m_1 is changed. We choose as our starting orbit the particular Lagrange orbit with period 2π. This solution of the 3BP with equal masses is both a (planar) periodic solution as well as a relative equilibrium under the group $\mathcal{G}$ of rotations generated by the components of the total angular momentum. However, when we try to continue this Lagrange solution we face several difficulties. First, the solution is not normal when considered as a periodic solution. This is mainly due to the fact that it belongs to the so-called homographic family of planar periodic solutions where the three bodies are located at the vertices of an equilateral triangle, each traveling along an appropriate elliptic or circular Kepler orbit; see e.g. [18]. Hence, we cannot apply the continuation results of Sect. 9.2. On the other hand we also cannot apply Theorem 4 (or a variant of it) since $\mathcal{Z}$ contains only the multiples of the Hamiltonian and, hence, $\mathcal{G} \not\subset \mathcal{Z}$. Both difficulties remain when we consider only planar solutions $\mathbf{x}_i \in \mathbb{R}^2$.

The way out of these difficulties lies in a restriction to planar solutions that have a fixed center of mass at the origin. When we set $z_i = 0$, $i = 1, 2, 3$, in (9.24) then the equivalent first order system is Hamiltonian of dimension twelve and has four (independent) first integrals, namely: the Hamiltonian, the two components of the total linear momentum, and the total angular momentum Q. Along the Lagrange solution (which is a relative equilibrium) we have $\nabla H = \nabla Q$ and $X_H = X_Q$. Recall that we have fixed the period at 2π, so that $\Omega^0 = 1$. Since the total linear momentum is a constant of motion, we can choose an inertial frame of reference such that the center of mass remains fixed at the origin. For a general choice of the masses, this gives the condition

$$\sum_{i=1}^{3} m_i \mathbf{x}_i = 0 \quad \Longrightarrow \quad \sum_{i=1}^{3} m_i \dot{\mathbf{x}}_i = 0. \tag{9.25}$$

These equations can be used to eliminate the third body from the equations; see, e.g., [29, §16] where it is shown how this can be done in a canonical way. The resulting reduced system is Hamiltonian of dimension eight and with two remaining first integrals, namely: the Hamiltonian H and the total angular momentum Q, both of course transformed by the elimination process. Since $\{H, Q\} \equiv 0$ we have $\mathcal{Z} = \mathcal{F}$ and, setting $\mathcal{G}$ equal to the multiples of Q, it follows that $\mathcal{G} \subset \mathcal{Z}$. Along the Lagrange relative equilibrium we still have $X_H = X_Q$, so that $k = m = 1$. In [29] the characteristic polynomial of the operator L was explicitly calculated and shown to have the form $p(\lambda) = \lambda^2 (\lambda^2 + 1)(\lambda^4 + \lambda^2 + \gamma)$ with $\gamma = \frac{27}{4}(m_1 m_2 + m_2 m_3 + m_3 m_1)(m_1 + m_2 + m_3)^{-2} = \frac{9}{4}$ for $m_1 = m_2 = m_3 = 1$. This shows that the algebraic multiplicity $\tilde{m}_a$ of the zero eigenvalue equals $2 = k + m$. This was confirmed by a symbolic computation in Mathematica, which also showed that the geometric multiplicity is equal to one, that is, $\tilde{m}_g = 1 = k$. We conclude that for $m_1 = 1$ the Lagrange relative equilibrium is normal (in the restricted setting) and, hence, we can apply the continuation results of Sect. 9.2.1.

In order to avoid rescalings we keep, as before, the period fixed at 2π and use a parameter-dependent version of Theorem 4 to do continuation in the mass m_1 of the first body. Figure 9.9 shows the result of our calculation in the form of a unique branch of relative equilibria that connects the Lagrange solution (lower point at $m_1 = 1$) to a solution of the R3BP with equal primaries (point (d) for $m_1 = 0$). The relative equilibria along this continuation branch are well known: they are homographic solutions where the three bodies are at the vertices of an equilateral triangle, with the two larger bodies (with masses $m_2 = m_3 = 1$) traveling along a circle around the origin, and the smaller body with mass m_1 rotating along a larger circle. The solution of the R3BP at the end point (d) of the branch is nothing else but one of the Lagrange libration points L4 or L5 — which one depends on the direction of rotation of the starting equilateral Lagrange orbit. In a rotating frame, such as used in Sects. 9.1.3 and 9.1.4, this solution is an equilibrium. In a fixed frame the solution looks as shown in Fig. 9.10(d). At each moment the three masses form an equilateral triangle: the two primaries rotate and are opposite to each other on the smaller circle, while the negligible mass rotates along the larger circle.

Along the branch of relative equilibria we find no bifurcations (of relative equilibria). Indeed, the analytical results of Siegel and Moser [29] mentioned before show that the algebraic multiplicity $\tilde{m}_a$ of the zero eigenvalue of the linearization remains equal to two. Therefore, all of these relative equilibria are normal and have unique continuations. We also checked numerically that the geometric multiplicity $\tilde{m}_1$ remains equal to 1 along the full branch. The results of Siegel and Moser also allow us to calculate the multipliers of these relative equilibria when interpreted as periodic orbits of the planar version of (9.24). On top of the multiplier 1 with algebraic multiplicity $m_a = 8$ there is a quadruple of multipliers $(\mu, \bar{\mu}, \mu^{-1}, \bar{\mu}^{-1})$, where $\mu = \exp(2\pi\lambda)$ and λ is any solution of $\lambda^4 + \lambda^2 + \gamma(m_1) = 0$ and $\gamma(m_1) = \frac{27}{4}(2m_1 + 1)(m_1 + 2)^{-2}$.

To conclude this section we observe that, although we have been able to show continuous connections between both the figure-8 and the Lagrange solutions, on the one hand, and specific solutions of the R3BP with $\mu = 1/2$, on the other hand, we have not found a connection from the figure-8 to the Lagrange solution. Therefore, some of the issues discussed earlier remain open.

9.4 Conclusions

We have shown how two-point boundary value problem continuation software can be used to compute families of periodic solutions of symmetric Hamiltonian systems. The theory and the numerical implementations are well developed but not complete. Further work is necessary on the continuation of, first, relative equilibria in the non-Abelian case and, second, of relative periodic orbits; see [36] for some progress in this direction. How to make use of reversibility properties in combination with the Hamiltonian structure also needs further attention; see [22, 24]. An approach for the study of nonholonomic systems based on the ideas in this chapter has been proposed in [19].

In the 3BP and R3BP there are plenty of interesting problems to be investigated and our results can be of some help in this adventure. However, as Poincaré announced, and the last 110 years of Celestial Mechanics have shown, one can spend several lifetimes following periodic orbits in this incredibly complex jungle of trajectories!

Acknowledgements

We are very much indebted to Sebius Doedel, Emilio Freire and Francisco Javier Muñoz-Almaraz. Most of the material of this chapter is the result of our ongoing collaboration over the last 10 years. In particular, we would like to mention that Sebius Doedel was the driving force behind the massive project to obtain a complete picture of families of periodic orbits in the R3BP that was briefly discussed in Sect. 9.1.4. We are very thankful to the publishers of [10], [11] and [12] for the permission to make use of Figs. 9.4, 9.5, 9.6, 9.7 and 9.8. This work has been partially supported by the Spanish Ministry of Education through the grant BFM2003-00336 and MTM2006-00847. A.V. also acknowledges support from the same institution for his sabbatical stay at the University of Seville through grant SAB2005-0188.

References

1. D. G. Aronson, E. J. Doedel, and H. G. Othmer. The dynamics of coupled current-biased Josephson junctions II. *Internat. J. Bifur. Chaos Appl. Sci. Engrg.*, 1: 1–66, 1991.
2. V. I. Arnol'd, V. V. Kozlov, and A. I. Neishtadt. *Dynamical Systems III. Mathematical Aspects of Classical and Celestial Mechanics.* (Springer-Verlag, Berlin, 1993).
3. U. Ascher and L. P. Petzold. *Computer Methods for Ordinary Differential Equations and Differential-Algebraic Equations,* (SIAM, Philadelphia, 1998).
4. G. D. Birkhoff. *Dynamical Systems.* (Amer. Math. Soc., Ann Arbor, 1927).
5. A. Chenciner and R. Montgomery. A remarkable periodic solution of the three-body problem in the case of equal masses. *Annals Math.*, 152: 881–901, 2000.
6. S. N. Chow and J. K. Hale. *Methods of Bifurcation Theory,* Grundlehren der Mathematischen Wissenschaften, 251. (Springer-Verlag, Berlin, 1982).
7. E. J. Doedel. Auto, a program for the automatic bifurcation analysis of autonomous systems. *Congr. Numer.*, 30:265–384, 1981.
8. E. J. Doedel, H. Keller and J.P. Kernévez. Numerical analysis and control of bifurcation problems: II. *Internat. J. Bifur. Chaos Appl. Sci. Engrg.*, 1:745–772, 1991.
9. E. J. Doedel, R. Paffenroth, A. Champneys, F. Fairgieve, Yu. A. Kuznetsov, B. Oldeman, B. Sandstede, and X. Wang. *AUTO2000: Continuation and bifurcation software for ordinary differential equations.* Department of Computer Science, Concordia University, Montreal, Canada, 2000. Available from `http://sourceforge.net/projects/auto2000/`

10. E. J. Doedel, R. Paffenroth, H. Keller, D. Dichmann, J. Galán-Vioque, and A. Vanderbauwhede, Computation of periodic solutions of conservative systems with application to the 3-Body Problem. *Internat. J. Bifur. Chaos Appl. Sci. Engrg.*, 13:1353–1381, 2003.

11. E. J. Doedel, V. A. Romanov, R. Paffenroth, H. Keller, D. Dichmann, J. Galán-Vioque, and A. Vanderbauwhede. Elemental periodic orbits associated with the Libration Points in the Circular restricted three body problem. *Internat. J. Bifur. Chaos Appl. Sci. Engrg.*, in press, 2007.

12. J. Galán, F. J. Muñoz-Almaraz, E. Freire, E. J. Doedel, and A. Vanderbauwhede. Stability and bifurcations of the figure-8 solution of the three-body problem. *Phys. Rev. Lett.*, 88:241101, 2002.

13. M. C. Gutzwiller. *Chaos in Classical and Quantum Mechanics.* (Springer Verlag, New York, 1990).

14. C. Marchal. The family P_{12} of the three-body problem — the simplest family of periodic orbits, with twelve symmetries per period. *Cel. Mech. Dynam. Astron.*, 78:279–298, 2000.

15. K. R. Meyer. Periodic solutions of the N-body problem. *J. Diff. Eqns.*, 39:2–38, 1981.

16. K. R. Meyer. Continuation of periodic solutions in three dimensions. *Physica D*, 112:310–318, 1998.

17. K. R. Meyer. *Periodic Solutions of the N-Body Problem*, Lecture Notes in Mathematics, 1719. (Springer-Verlag, Berlin, 1999).

18. K. R. Meyer and G. R. Hall. *Introduction to Hamiltonian Dynamical Systems and the N-Body Problem.* Applied Mathematical Sciences, 90. (Springer-Verlag, New York, 1992).

19. M. Molina, J. Galán-Vioque, and E. Freire. Generalized Hamiltonian equations of motion for nonholonomic systems. (Preprint, University of Sevilla, 2007).

20. C. Moore. Braids in classical gravity. *Phys. Rev. Lett.*, 70(24):3675–3679, 1993.

21. F. R. Moulton. *Periodic Orbits.* (Carnegie Institutiton of Washington, 1920).

22. F. J. Muñoz-Almaraz. *Continuación y Bifurcaciones de Órbitas Periódicas en Sistemas Hamiltonianos con Simetría*, Ph.D. Thesis, (Universidad de Sevilla, 2003).

23. F. J. Muñoz-Almaraz, E. Freire, J. Galán, E. J. Doedel, and A. Vanderbauwhede. Continuation of periodic orbits in conservative and Hamiltonian systems. *Physica D*, 181:1–38, 2003.

24. F. J. Muñoz-Almaraz, E. Freire, J. Galán, and A. Vanderbauwhede. Continuation of normal doubly symmetric orbits in conservative reversible systems. *Cel. Mech. and Dyn. Astr.*, 97:17–47, 2007.

25. D. Offin. Instability for symmetric periodic solutions of the planar three body problem. Technical report, 2001.

26. H. Poincaré. *Les Méthodes Nouvelles de la Mécanique Céleste.* (Gauthier-Villars, 1892).

27. C. Pugh and C. Robinson. The C^1 closing lemma, including Hamiltonians. *Ergod. Th. Dynam. Sys.*, **3**:261–313, 1983.

28. J. Sepulchre and R. MacKay. Localized oscilations in conservative networks of weakly coupled autonomous oscillators. *Nonlinearity*, 10:679–713, 1997.

29. C. Siegel and J. K. Moser. *Lectures on Celestial Mechanics.* (Springer-Verlag, Berlin, 1971).

30. C. Simó. In Cent ans après les méthodes nouvelles de H. Poincaré, pages 1–23. (Société Mathématique de France, 1996).

31. C. Simó. New families of solutions in n-body problems. In *Proceedings ECM 2000, Barcelona*, 2000.
32. C. Simó. Dynamical properties of the figure-eight solution of the three-body problem. *Contemp. Math.*, 292:209–228, 2002.
33. E. Strömgren. Connaissance actuelle des orbites dans le problème des trois corps. *Bull. Astron. Obs. Copenhagen*, 9(100):87–130, 1935.
34. A. Vanderbauwhede. Families of periodic orbits for autonomous systems. In *Dynamical Systems II*, A. Bednarek and L. Cesari (Eds.), pages 427–44 (Academic Press, 1982).
35. E. T. Whittaker. *A Treatise on the Analytical Dynamics of Particles & Rigid Bodies*. (Cambridge University Press, 1947).
36. C. Wulff and A. Schebesch. Numerical continuation of Hamiltonian relative periodic orbits. Submitted for publication. `http://www.maths.surrey.ac.uk/personal/st/C.Wulff/publicationsframes.html`.
37. J. A. Zufiría. *Symmetry breaking of water waves*, Ph.D. Thesis, (Applied Mathematics, Cal. Inst. Technology, Pasadena, CA, 1987).

Phase Conditions, Symmetries and PDE Continuation

Wolf-Jürgen Beyn and Vera Thümmler

Department of Mathematics, Bielefeld University, Germany

There is a long tradition of making use of continuous symmetries for the analysis of differential equations; see, for example, the monographs [10, 21, 32]. In general, such symmetries are expressed as the equivariance of the differential operator with respect to the action of a Lie group. Solutions of the differential equation then come in group orbits, and this has interesting consequences, for example, it may lead to inherent symmetries of solutions or symmetry-breaking bifurcations. In the theory of equivariant systems one usually tries to reduce the differential equation to the so-called orbit space, the elements of which are equivalence classes created by applying the group action to a single point in phase space. After factoring out the group action in this way, one applies specific results on existence and uniqueness of solutions, on bifurcations, or on asymptotic stability.

Contrary to the situation in the theory, the use of continuous equivariances for efficient numerical computations seems to be rather rare. An early exception is Eusebius Doedel's integral phase condition [13] for the computation of periodic orbits in autonomous ODEs; see also Chap. 1. It is a typical example that shows how a judicious use of symmetry (in this case, equivariance with respect to time shifts) can enhance rather than hamper the efficiency of a numerical method. Namely, less effort is needed for mesh adaptation and larger continuation steps are possible. The ODE example also shows another paradigm of numerical bifurcation analysis. While theory prefers to *reduce* problems, e.g., by Lyapunov-Schmidt or center manifold reduction, it seems advantageous rather to *extend* the problem for numerical purposes (e.g., by choosing unfolding parameters) and then add extra constraints (e.g., normalizing conditions for eigenvectors). In this way one can keep as much structure as possible from the original problem and simultaneously use the normalizing conditions to optimize the conditioning of the extended problem.

In this chapter we discuss the usefulness of phase conditions for the numerical analysis of finite- and infinite-dimensional dynamical systems that have continuous symmetries. Our main topic is the general approach known as the *freezing method*, which was developed in [33] and [7]. It will be presented in an abstract framework for evolution equations that are equivariant with respect to the action of a (not necessarily compact) Lie group. Specifically, we introduce an extra parameter (an element in the associated Lie algebra) that determines the position on the group orbit and impose further constraints or phase conditions such that the point in phase space (e.g., the spatial profile in case of a PDE) varies as little as possible. We show particular applications of phase conditions to periodic, heteroclinic and homoclinic orbits in ODEs, to relative equilibria and relative periodic orbits in PDEs, as well as to time integration of equivariant PDEs.

After reviewing phase conditions that eliminate the time shift in ODEs in Sect. 10.1, we set up in Sect. 10.2 the general freezing method within an abstract framework. We then apply our method to the computation of various spatio-temporal patterns, such as traveling and modulated waves in one, spiral waves in two, and scroll waves in three space dimensions. For problems in one space dimension we also investigate asymptotic stability and discuss the errors introduced by finite boundary conditions.

10.1 Phase Conditions for Orbits in ODEs

Consider a dynamical system generated by an autonomous n-dimensional ordinary differential equation

$$u_t = f(u), \quad u(t) \in \mathbb{R}^n, \quad f : \mathbb{R}^n \to \mathbb{R}^n \text{ smooth.} \tag{10.1}$$

Due to its autonomous character the nonlinear differential operator

$$\mathcal{L}u = u_t - f(u)$$

has a simple equivariance with respect to time shifts. That is, for all $\gamma \in \mathbb{R}$ and for all u in some function space we have

$$[\mathcal{L}u](\cdot - \gamma) = \mathcal{L}[u(\cdot - \gamma)]. \tag{10.2}$$

Depending on the application, appropriate function spaces may be chosen, such as the Sobolev space $\mathcal{H}^1(\mathbb{R}, \mathbb{R}^n)$, the space of bounded uniformly continuous C^1-functions $C^1_{\mathrm{unif}}(\mathbb{R}, \mathbb{R}^n)$ or the space of one-periodic functions $C^1_{\mathrm{per}}(\mathbb{R}, \mathbb{R}^n)$.

10.1.1 Periodic Orbits

In order to determine a periodic orbit of (10.1) we should find a period $T > 0$ and a solution $u(t)$ of the boundary value problem

$$u_t = f(u), \ t \in [0, T], \ u(0) = u(T).$$

Introducing the scaled function $v(t) = u(tT)$, $t \in [0,1]$ the boundary value problem for $(v,T) \in C^1([0,1], \mathbb{R}^n) \times \mathbb{R}$ now reads

$$v_t = Tf(v), \ t \in [0,1], \quad v(0) = v(1). \tag{10.3}$$

Due to equivariance (10.2) the solutions of (10.3) are only determined up to a phase shift and a further condition is needed to make the solution unique. In the first publication on the continuation package AUTO [13] Doedel suggested to use an 'anchor equation' (as it was called in [13]) that tries to minimize the $\mathcal{L}_2$ distance to some template function $\hat{v} \in C^1_{\mathrm{per}}(\mathbb{R}, \mathbb{R}^n)$, i.e., tries to minimize

$$\rho(v,\gamma) = \int_0^1 ||v(t-\gamma) - \hat{v}(t)||_2^2 \, dt = ||v(\cdot - \gamma) - \hat{v}||^2_{\mathcal{L}_2}. \tag{10.4}$$

By differentiating with respect to γ, a necessary condition for a local minimum is

$$\int_0^1 (v(t-\gamma) - \hat{v}(t))^T \hat{v}_t(t) \, dt = 0. \tag{10.5}$$

A more formal statement is contained in the following lemma; see [4] for a proof.

Lemma 1. *Suppose that $\hat{v} \in C^1_{\mathrm{per}}(\mathbb{R}, \mathbb{R}^n)$ is a nonconstant 1-periodic function. Then there exist neighborhoods U of $\hat{v}$ in the C^1-topology and $\Gamma \subset \mathbb{R}$ of 0 such that for any $v \in U$ the $\mathcal{L}_2$-distance from (10.4) has a unique minimum at $\gamma = \gamma(v) \in \Gamma$ where $\gamma : U \to V$ is a C^1-mapping satisfying $\gamma(\hat{v}) = 0$ and condition (10.5).*

During computations one selects v such that the phase condition (10.5) holds at $\gamma = 0$, i.e.,

$$\int_0^1 (v(t) - \hat{v}(t))^T \hat{v}_t(t) \, dt = 0. \tag{10.6}$$

This condition has several advantages over a Poincaré-type condition such as

$$(v(0) - \hat{v}(0))^T f(\hat{v}(0)) = 0. \tag{10.7}$$

If $\hat{v}$ is a good approximation of v obtained from continuation along a branch, then condition (10.6) tries to keep a steep front or a peak of the solution in the same place. Usually this facilitates mesh adaptation and simultaneously allows for larger step sizes along branches. This phase condition is now built into standard continuation packages, such as AUTO (with HOMCONT) [17], CONTENT [25] and MATCONT [12]; see also Chap. 2. It has proved to be most reliable in many applications.

For an illustration we take the model example from [13], namely

$$\begin{pmatrix} u_1 \\ u_2 \end{pmatrix}_t = \begin{pmatrix} (1-\lambda)u_1 - u_2 \\ u_1 + u_1^2 \end{pmatrix}.$$

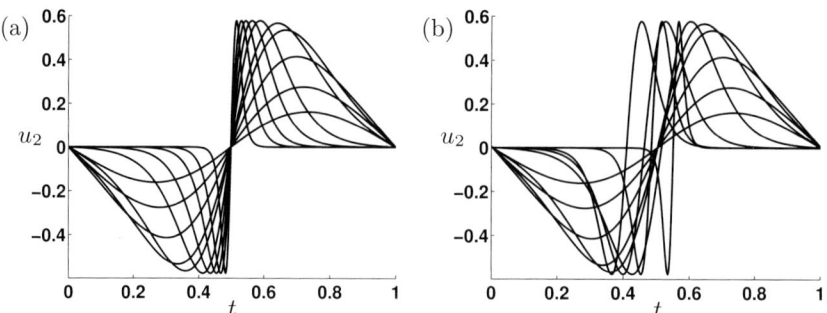

Fig. 10.1. Effect of an integral phase condition (a) and of a point phase condition (b).

This system is, in fact, Hamiltonian at $\lambda = 1$ and a continuous family of periodic orbits bifurcating from the origin and ending in a homoclinic orbit exists. For a detailed treatment of the Hamiltonian case we refer to Chap. 9. Figure 10.1 shows the result of continuing the periodic orbits with both phase conditions (10.6) and (10.7).

We finally mention that conditions (10.6) and (10.7) are special cases of the general form $\psi(v) = 0$, where $\psi : C^1([0,1], \mathbb{R}^n) \to \mathbb{R}$ is a C^1-mapping. Following [4, 16] one can characterize the admissible phase conditions that lead to a regular solution $(v, T) \in C^1([0,1], \mathbb{R}^n) \times \mathbb{R}$ of the operator equation

$$F(v, T) = (v_t - Tf(v), v(0) - v(1), \psi(v)) = 0 \qquad (10.8)$$

as follows. Let $u(t)$ be a T-periodic solution of (10.1) such that $v(t) = u(tT)$ satisfies $\psi(v) = 0$. Then the pair (v, T) is a regular solution of (10.8) if, and only if, 1 is a simple Floquet multiplier and $D\psi(v)v_t \neq 0$ where $D\psi$ denotes the Fréchet derivative of ψ. An easy calculation shows, that the latter condition requires $\langle v_t, \hat{v}_t \rangle_{\mathcal{L}_2} \neq 0$ for (10.6) and $v_t(0)^T f(\hat{v}(0)) \neq 0$ for (10.7). One may call (10.8) a defining equation for an isolated periodic orbit. The recent paper [15] provides a considerable extension of this general approach to defining equations for all codimension-one bifurcations of periodic orbits, namely: fold (saddle-node), flip (period-doubling) and Neimark-Sacker bifurcations; see also [26] and Chap. 2.

10.1.2 Homoclinic and Heteroclinic Orbits

It is natural to extend the numerical methods for periodic orbits to orbits that connect stationary points in infinite time. Such orbits typically occur in parametrized systems

$$u_t = f(u, \lambda), \quad u(t) \in \mathbb{R}^n, \quad f : \mathbb{R}^n \times \mathbb{R}^p \to \mathbb{R}^n \text{ smooth.} \qquad (10.9)$$

Connecting orbits have numerous applications, in particular, they appear as traveling waves of PDEs; see Sect. 10.2.1.

Definition 1. *A pair $(\bar{u}, \bar{\lambda}) \in C^1(\mathbb{R}, \mathbb{R}^n) \times \mathbb{R}^p$ is called a connecting orbit pair of system (10.9) if $\bar{u}$ is a solution at $\lambda = \bar{\lambda}$ and if the limits*

$$\lim_{t \to \infty} \bar{u}(t) = \bar{u}_+, \qquad \lim_{t \to -\infty} \bar{u}(t) = \bar{u}_- \tag{10.10}$$

exist. The connecting orbit is called homoclinic if $\bar{u}_+ = \bar{u}_-$ and heteroclinic otherwise.

From (10.10) we infer that $\bar{u}_\pm$ are stationary points, i.e., $f(\bar{u}_\pm, \bar{\lambda}) = 0$, and that $\bar{u} \in C_b^1(\mathbb{R}, \mathbb{R}^n)$, i.e., C^1 and bounded on $\mathbb{R}$. Nondegeneracy of a connecting orbit may be defined as follows; cf. [3].

Definition 2. *A connecting orbit pair $(\bar{u}, \bar{\lambda})$ is called nondegenerate if the following conditions hold:*

(i) The matrices $f_u(\bar{u}_\pm, \bar{\lambda}) \in \mathbb{R}^{n,n}$ are hyperbolic with $n_{\pm s}$ eigenvalues of negative real part and $n_{\pm u} = n - n_{\pm s}$ eigenvalues of positive real part;
(ii) $p = n_{-s} - n_{+s} + 1$;
(iii) If $u \in C_b^1(\mathbb{R}, \mathbb{R}^n), \lambda \in \mathbb{R}^p$ satisfies the variational equation
$u_t = f_u(\bar{u}, \bar{\lambda})u + f_\lambda(\bar{u}, \bar{\lambda})\lambda$, *then $\lambda = 0$ and $u = c\bar{u}_t$ for some $c \in \mathbb{R}$.*

Conditions (i) and (ii) ensure that the dimension $n_{-u}+p$ of the center-unstable manifold of $(\bar{u}_-, \bar{\lambda})$ in the extended phase space $\mathbb{R}^n \times \mathbb{R}^p$ and the dimension $n_{+s}+p$ of the center-stable manifold of $(\bar{u}_+, \bar{\lambda})$ add up to $n+p+1$, which is one plus the dimension of the extended phase space $\mathbb{R}^n \times \mathbb{R}^p$. Condition (iii) then guarantees that these two manifolds intersect transversely in the connecting orbit $\{(\bar{u}(t), \bar{\lambda}) : t \in \mathbb{R}\}$. Similar to the periodic case, one can characterize connecting orbit pairs as regular solutions of an operator equation

$$F(u, \lambda) = (u_t - f(u, \lambda), \psi(u, \lambda)) = 0, \tag{10.11}$$

where a smooth map $\psi : C_b^1(\mathbb{R}, \mathbb{R}^n) \times \mathbb{R}^p \to \mathbb{R}$ defines the phase condition; see [3] for a proof.

Proposition 1. *Let $(\bar{u}, \bar{\lambda})$ be a connecting orbit pair satisfying $\psi(\bar{u}, \bar{\lambda}) = 0$ and conditions (i) and (ii) of Definition 2. Then $(\bar{u}, \bar{\lambda})$ is a regular solution of (10.11) if, and only if, the orbit pair is nondegenerate and $\psi_u(\bar{u}, \bar{\lambda})\bar{u}_t \neq 0$.*

The analogue of the functional (10.4) to be minimized is

$$\rho(u, \gamma) = \int_{-\infty}^{\infty} ||u(t - \gamma) - \hat{u}(t)||_2^2 \, dt = ||u(\cdot - \gamma) - \hat{u}||_{\mathcal{L}_2}^2 = ||u - \hat{u}(\cdot + \gamma)||_{\mathcal{L}_2}^2,$$

where we take $u \in \hat{u} + \mathcal{H}^1$ and assume that $\hat{u} \in C_b^2(\mathbb{R}, \mathbb{R}^n)$ is a template function that satisfies $\hat{u}_t \in \mathcal{H}^1$. Then the phase condition is again obtained from the necessary condition of a minimum $\langle u - \hat{u}, \hat{u}_t \rangle_{\mathcal{L}_2} = 0$. In applications it may be unrealistic to assume that such a template function is known, because this essentially requires one to know $\bar{u}_\pm$ beforehand and to choose $\hat{u}$ such that $\|\hat{u}(t)-\bar{u}_\pm\| = \mathcal{O}(e^{-\alpha|t|})$. In general, $\bar{u}_\pm$ will depend on λ and be determined by $f(\bar{u}_\pm, \lambda) = 0$. Therefore, Doedel and Friedman [14, 20] suggested to minimize $\|u_t - \hat{u}_t(\cdot - \gamma)\|_{\mathcal{L}_2}$, which leads to the phase condition

$$\langle u_t - \hat{u}_t, \hat{u}_{tt}\rangle_{\mathcal{L}_2} = 0. \tag{10.12}$$

There are several ways to solve the boundary value problem (10.11) on the infinite line. One may discretize it by using globally-defined Galerkin functions or transform the domain $\mathbb{R}$ to a bounded interval and then devise methods that handle the resulting singularities; see [27, 30, 31]. Perhaps the simplest method that allows one to employ existing boundary value solvers is to approximate (10.11) by a finite boundary value problem on some large interval $J = [T_-, T_+]$. This approach was proposed and analyzed in [3, 14, 20] and implemented in the HOMCONT part of AUTO[17].

For $u \in C^1(J, \mathbb{R}^n), \lambda \in \mathbb{R}^p$ we consider the finite boundary value problem

$$F_J(u, \lambda) = (u_t - f(u, \lambda), B(u(T_-), u(T_+), \lambda), \psi_J(u, \lambda)) = 0, \tag{10.13}$$

where the smooth maps $B : \mathbb{R}^{2n+p} \to \mathbb{R}^{n+p-1}, (u_-, u_+, \lambda) \mapsto B(u_-, u_+, \lambda)$ and $\psi_J : C^1(J, \mathbb{R}^n) \times \mathbb{R}^p \to \mathbb{R}$ determine the boundary condition and the approximate phase condition, respectively. The error introduced by this approximation can be estimated as follows; see [3, 20, 40].

Theorem 1. *Let $(\bar{u}, \bar{\lambda})$ be a nondegenerate connecting orbit pair of (10.9) such that*

(i) $B(\bar{u}_-, \bar{u}_+, \bar{\lambda}) = 0$ and the matrix

$$\left(\frac{\partial B}{\partial u_-}(\bar{u}_-, \bar{u}_+, \bar{\lambda})X_{-s} \quad \frac{\partial B}{\partial u_+}(\bar{u}_-, \bar{u}_+, \bar{\lambda})X_{+u} \right) \in \mathbb{R}^{(n+p-1)\times(n+p-1)}$$

is nonsingular, where the columns of $X_{-s} \in \mathbb{R}^{n\times n-s}$ and $X_{+u} \in \mathbb{R}^{n\times n+u}$ form a basis of the stable subspace of $f_u(\bar{u}_-, \bar{\lambda})$ and of the unstable subspace of $f_u(\bar{u}_+, \bar{\lambda})$, respectively;
(ii) $\psi(\bar{u}, \bar{\lambda}) = 0, \psi_J(\bar{u}_{|J}, \bar{\lambda}) \to 0$ as $J \to \mathbb{R}$, the derivatives $D\psi_J$ are equicontinuous in a uniform neighborhood of $(\bar{u}_{|J}, \bar{\lambda})$ and $|D\psi_J(\bar{u}_{|J}, \bar{\lambda})\bar{u}'_{|J}| \geq \delta > 0$ for some $\delta > 0$.

Then there exist constants $\rho, K > 0$ and an interval $J_0 \subset \mathbb{R}$ with the following properties. For all $J_0 \subset J$ the boundary value problem (10.13) has a unique solution (u_J, λ_J) in a C^1-ball of radius ρ and center $(\bar{u}_{|J}, \bar{\lambda})$. Furthermore, there is a unique phase shift γ_J near zero such that $\tilde{u} = \bar{u}(\cdot - \gamma_J)$ satisfies $\psi_J(\tilde{u}_{|J}, \lambda_J) = 0$ and the following estimate holds

$$\|\tilde{u}_{|J} - u_J\|_{C^1} + \|\bar{\lambda} - \lambda_J\| \leq C\|B(\tilde{u}(T_-), \tilde{u}(T_+), \bar{\lambda})\|. \tag{10.14}$$

In view of (10.12) and (10.13) it is natural to take the phase conditions

$$\psi_J(u, \lambda) = \langle u - \hat{u}, \hat{u}_t \rangle_{\mathcal{L}_2(J)} \quad \text{or} \quad \psi_J(u, \lambda) = \langle u_t - \hat{u}_t, \hat{u}_{tt} \rangle_{\mathcal{L}_2(J)}.$$

The most natural choice for boundary conditions are so-called projection boundary conditions that force the end points $u(T_-), u(T_+)$ to lie in the tangent spaces of the unstable manifold at $\bar{u}_-$ and of the stable manifold at $\bar{u}_+$. These conditions may be written as

$$B(u_-, u_+, \lambda) = \begin{pmatrix} Y_{-s}^T(\lambda)(u_- - u_-(\lambda)) \\ Y_{+u}^T(\lambda)(u_+ - u_+(\lambda)) \end{pmatrix}, \tag{10.15}$$

where $f(u_\pm(\lambda), \lambda) = 0$ and the columns of $Y_{-s}(\lambda) \in \mathbb{R}^{n \times n_{-s}}$ and $Y_{+u}(\lambda) \in \mathbb{R}^{n \times n_{+u}}$ form a basis of the stable subspace of $f_u^T(u_-(\lambda), \lambda)$ and of the unstable subspace of $f_u^T(u_+(\lambda), \lambda)$, respectively. Note that, by Definition 2, (10.15) imposes $n_{-s} + n_{+u} = n + p - 1$ boundary conditions. Methods to compute these matrices such that they depend smoothly on the parameter λ were proposed in [3] and, more recently, via a smooth block Schur decomposition in [11]. For numerous computations that apply this approach to specific examples we refer to [3, 14, 17, 20].

We finally notice that projection boundary conditions imply exponential decay of the term on the right-hand side of (10.14). More precisely, we have

$$||\tilde{u}_{|J} - u_J||_{C^1} + ||\bar{\lambda} - \lambda_J|| = \mathcal{O}(e^{2\alpha_- T_-} + e^{-2\alpha_+ T_+}),$$

where $0 < \alpha_- < \text{Re}(\mu)$ for all eigenvalues μ of $f_u(\bar{u}_-, \lambda)$ with positive real part and $\text{Re}(\mu) < -\alpha_+ < 0$ for all eigenvalues of $f_u(\bar{u}_+, \bar{\lambda})$ with negative real part. For the parameter a superconvergence behavior was observed in [3] and a corresponding estimate proved in [34], namely:

$$||\bar{\lambda} - \lambda_J|| = \mathcal{O}(e^{(2\alpha_- + \alpha_+)T_-} + e^{-(2\alpha_+ + \alpha_-)T_+}).$$

10.2 Phase Conditions and Equivariant PDEs

In this section we consider time-dependent PDEs that have continuous symmetries in the spatial operator. Therefore, we will be concerned with phase conditions that act on the spatial variables of the solutions. First, we introduce the method of freezing that employs phase conditions in order to decompose a time-dependent solution into a time-dependent group orbit and a spatial profile that varies as little as possible. Second, this method will be used to compute relative equilibria, i.e., spatial profiles of which the group orbits are invariant under the PDE flow. The underlying general approach was developed independently in [33] and in [7].

10.2.1 Traveling Waves

A special class of relative equilibria in one space dimension are traveling wave solutions $u(x,t) = \bar{v}(x - \bar{\lambda}t)$ of parabolic PDEs

$$u_t = Au_{xx} + f(u), \quad u(\cdot,0) = u^0, \qquad x \in \mathbb{R}, \; u(x,t) \in \mathbb{R}^m, \qquad (10.16)$$

where $A \in \mathbb{R}^{m \times m}$ is positive definite, $\bar{v}$ denotes the profile of the wave and $\bar{\lambda} \in \mathbb{R}$ its velocity.

These solutions are stationary in the moving coordinate system which is obtained via the transformation $v(\xi,t) = u(x,t)$, $\xi = x - \bar{\lambda}t$, i.e., $\bar{v}$ and $\bar{\lambda}$ solve

$$0 = A\bar{v}_{xx} + f(\bar{v}) + \bar{\lambda}\bar{v}_x. \qquad (10.17)$$

Given a stationary solution $\bar{v}$, each shifted version $\bar{v}_\gamma = \bar{v}(\cdot - \gamma)$ is also a solution of (10.17). As in Sect. 10.1 we add a phase condition defined by some functional ψ in order to obtain a well-posed boundary value problem for (v, λ), namely:

$$0 = Av_{xx} + f(v) + \lambda v_x,$$
$$0 = \psi(v, v_x, \lambda).$$

The natural choice for ψ stems from the phase condition discussed in Sect. 10.1. One minimizes the $\mathcal{H}^1$-distance or the $\mathcal{L}_2$-distance to a template function $\hat{v}$. This leads to the functional $\psi(v) = \langle \hat{v}_x, v - \hat{v} \rangle_{\mathcal{H}^1}$ or

$$\psi(v) = \langle \hat{v}_x, v - \hat{v} \rangle_{\mathcal{L}_2}. \qquad (10.18)$$

Transforming to a first-order system, we can apply the results from Sect. 10.1 for studying well-posedness (Proposition 1) and approximation (Theorem 1).

In our next step we are going to use phase conditions also for the non-stationary case. Now we let the transformation into the moving frame depend on time in the following way

$$u(x,t) = v(x - \gamma(t), t), \qquad (10.19)$$

where $\gamma(0) = 0$ and we define $\lambda(t) = \dot{\gamma}(t)$. In this setting, (10.16) together with the phase condition transforms into a partial differential algebraic equation (PDAE) for (v, λ), namely:

$$v_t = Av_{xx} + f(v) + \lambda v_x, \quad v(\cdot,0) = u^0,$$
$$0 = \psi(v, v_x, \lambda). \qquad (10.20)$$

Note that the initial value $\lambda(0)$ is not prescribed but, as usual with DAEs, is determined by differentiating the constraint $\psi = 0$ with respect to time and using the differential equation. In Sect. 10.2.3 we will discuss possible choices for the phase condition that lead to PDAEs of different index. System (10.20) can be completed by the simple ODE $\dot{\gamma} = \lambda(t)$, $\gamma(0) = 0$ (called the reconstruction equation in [33]). The traveling wave $(\bar{v}, \bar{\lambda})$ now appears as a stationary solution of system (10.20) and, in case of stability, we expect the solution of (10.20) to converge to $(\bar{v}, \bar{\lambda})$ during time evolution; see Sect. 10.3.

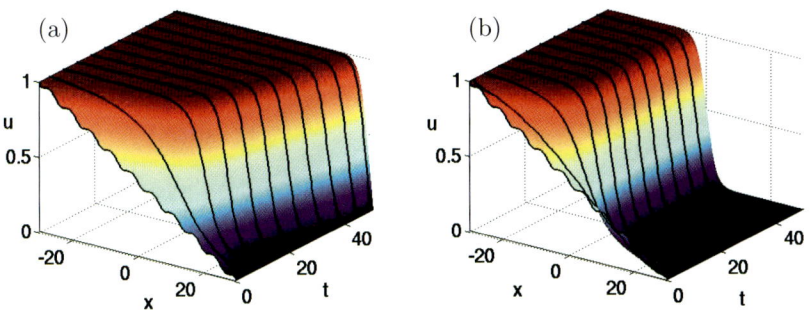

Fig. 10.2. Calculation of a wave in the Nagumo equation (10.21): traveling wave (a) and frozen wave (b).

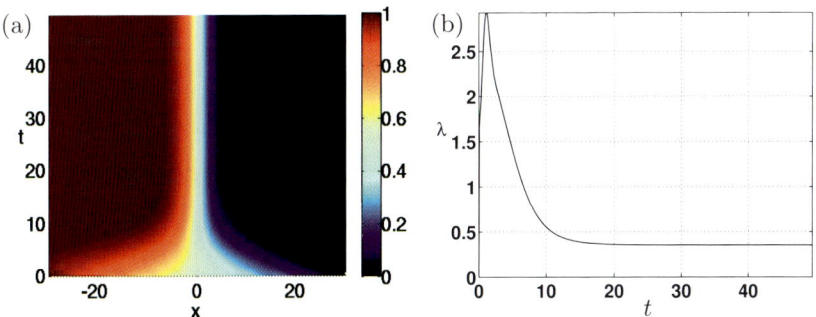

Fig. 10.3. Calculation of a wave in the Nagumo equation (10.21): (t,x)-plot of the frozen wave (a), and time evolution of λ (b).

Example 1. The standard toy example of a traveling wave is a heteroclinic orbit between two metastable states in the Nagumo equation [23]

$$u_t = u_{xx} + u(1-u)(u-a), \quad u(x,t) \in \mathbb{R}, \; x \in \mathbb{R}, \; t > 0, \qquad (10.21)$$

where $a \in (0, \frac{1}{2})$. An explicit traveling wave connecting the stationary points $u_- = 0$, $u_+ = 1$ is

$$\bar{v}(x) = \left(1 + e^{\frac{-x}{\sqrt{2}}}\right)^{-1}, \quad \bar{\lambda} = -\sqrt{2}\left(\tfrac{1}{2} - a\right). \qquad (10.22)$$

In Fig. 10.2 we show the results of a numerical computation for $a = 0.25$ with finite differences in space ($\Delta x = 0.1$) and the implicit Euler method in time ($\Delta t = 0.1$). Panel (a) is for the non-frozen system (10.21) and panel (b) for the frozen system (10.20). In both cases the spatial interval is $J = [-30, 30]$ and we use Dirichlet boundary conditions. Similar to Sect. 10.1, the frozen system has the advantage that steep gradients stay in approximately the same place and the front does not leave the computational domain in finite time. In

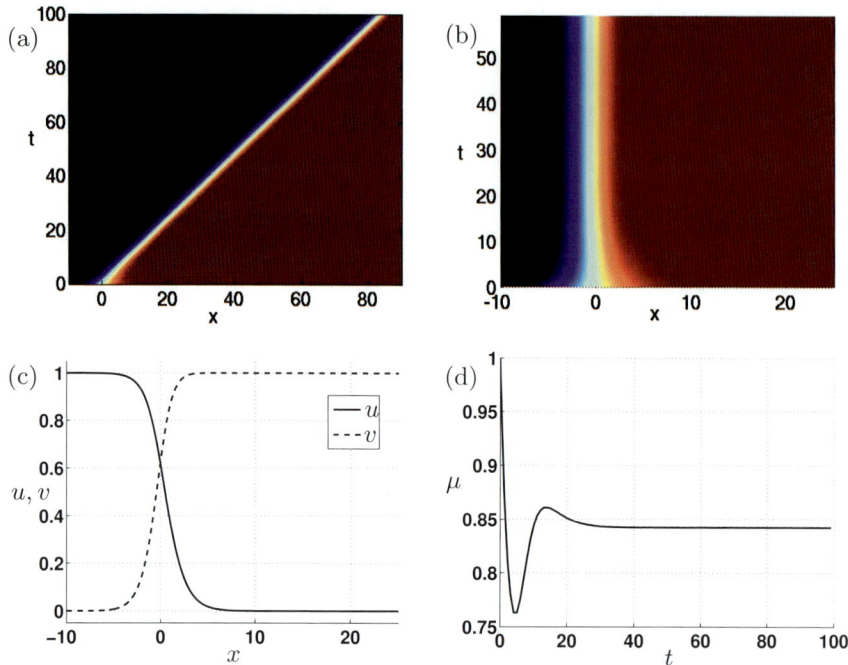

Fig. 10.4. Calculation of a wave in the autocatalytic system (10.23): traveling wave (a), frozen wave (b), u- and v-components at $t = 100$ (c), and time evolution of μ (d).

Fig. 10.3 we show another representation of the frozen wave as a color-coded (t, x)-plot in panel (a), while panel (b) is the time evolution of the velocity λ. This type of figure will be used throughout this chapter.

Example 2. We consider an autocatalytic system [1, 28] as given by

$$
\begin{aligned}
u_t &= au_{xx} - uf(v), \quad a > 0, \ u, v : \mathbb{R} \to \mathbb{R}, \\
v_t &= v_{xx} + uf(v),
\end{aligned}
\tag{10.23}
$$

where $f(v) = v^m$ for $v \geq 0$ and zero otherwise. This system has traveling wave solutions if the parameter $m \geq 2$ is not too large. As in [1, 28] we choose limit values $(u_-, v_-) = (0, 1)$, $(u_+, v_+) = (1, 0)$ in order to eliminate a scaling invariance.

Figure 10.4(a) and (b) show the solution of the original and of the frozen system, respectively, in an interval of length 100 for the original system and of length 30 for the frozen system. Here $a = 0.1$, $m = 2$, and we use the Crank-Nicholson method ($\Delta x = 0.1$, $\Delta t = 0.1$) and Dirichlet boundary conditions. Figure 10.4(c) and (d) show the u- and v-components of the frozen system and the time evolution of μ. Again, the example shows how the method of freezing allows one to observe phenomena that become visible only after a transient phase, while in a direct numerical simulation the solution may leave the finite domain before the steady profile appears.

10.2.2 Freezing Solutions of Equivariant PDEs

Let M be a manifold modeled over some Banach space X and let N be a submanifold modeled over some dense subspace $Y \subset X$ [9]. Consider an evolution equation

$$u_t = F(u), \quad u(0) = u^0, \tag{10.24}$$

for a vector field $F : N \to TM$ where TM denotes the tangent bundle of M. We assume that (10.24) is equivariant with respect to a finite-dimensional (possibly noncompact) Lie group G acting on M via

$$a : G \times M \to M, \ (\gamma, v) \mapsto a(\gamma, v),$$

with the property

$$a(\gamma_1 \circ \gamma_2, v) = a(\gamma_1, a(\gamma_2, v)), \quad a(\mathbb{1}, v) = v, \quad \mathbb{1} = \text{unit element in } G.$$

By equivariance we mean that the following relation holds:

$$a(\gamma, N) \subset N \quad \forall \gamma \in G,$$
$$F(a(\gamma, u)) = Ta(\gamma, u)F(u), \quad \forall u \in N, \gamma \in G,$$

where $Ta : G \times TM \to TM$ denotes the tangent action of a. We assume that the linear map

$$Ta(\gamma, v) : T_v M \to T_{a(\gamma, v)} M, \quad w \mapsto Ta(\gamma, v)w$$

is a homeomorphism for each $v \in M$. (Note that $a(g, \cdot)$ corresponds to $\Phi_g : N \to N$ in [29, 33] and $Ta(g, \cdot)$ corresponds to $\Psi_g : TM \to TM$.) Furthermore, we assume that for any $v \in M$ the map

$$a(\cdot, v) : G \to M, \quad \gamma \mapsto a(\gamma, v)$$

is continuous and that it is continuously differentiable for any $v \in N$ with derivative denoted by

$$da(\gamma, v) : T_\gamma G \to T_{a(\gamma)v} M, \quad \lambda \mapsto da(\gamma, v)\lambda.$$

For the construction of some spaces that satisfy this smoothness requirement we refer to [7]. Finally, we denote by $L_\gamma : G \to G$, $g \mapsto \gamma \circ g$ the multiplication by $\gamma \in G$ from the left and by $dL_\gamma(g) : T_g G \to T_{\gamma \circ g} G$ its derivative. Then we define the exponential $\exp(t\mu)$ for μ in the Lie algebra $T_\mathbb{1} G$ as the solution of

$$\dot{\gamma} = dL_\gamma(\mathbb{1})\mu.$$

The evolution of $\gamma(t)$ describes the motion on the group. Other equivalent definitions of exp are in common use [9, 10, 29].

Generalizing ansatz (10.19) to $u(t) = a(\gamma(t), v(t))$, (10.24) can be transformed into a system for the unknowns $v(t) \in M$, $\gamma(t) \in G$, $\mu(t) \in T_\mathbb{1}G$ as follows (cf. [7, 33]):

$$v_t = F(v) - da(\mathbb{1}, v)\mu, \qquad v(0) = u^0, \tag{10.25a}$$
$$\dot{\gamma} = dL_\gamma(\mathbb{1})\mu, \qquad \gamma(0) = \mathbb{1}. \tag{10.25b}$$

Lemma 2. *For some $T > 0$ let $u \in C^1((0, T], M) \cap C([0, T], N)$ be a solution of (10.24) and let $\gamma \in C^1([0, T], G)$ be arbitrary with $\gamma(0) = \mathbb{1}$. Then $v(t)$ defined by $u(t) = a(\gamma(t), v(t))$ and $\mu(t)$ defined by (10.25b) are solutions of (10.25a). Conversely, assume that $v \in C^1((0, T], M) \cap C([0, T], N)$ and $\mu \in C^1([0, T], T_\mathbb{1}G)$ solve (10.25a) and define $\gamma \in C^1([0, T], G)$ as the solution of (10.25b). Then $u(t) = a(\gamma(t), v(t))$ solves (10.24) on $[0, T]$.*

Proof. Insert the ansatz $u(t) = a(\gamma(t), v(t))$ into (10.24) and use equivariance to obtain

$$da(\gamma, v)\dot{\gamma} + Ta(\gamma, v)v_t = u_t = F(u) = F(a(\gamma, v)) = Ta(\gamma, v)F(v). \tag{10.26}$$

Differentiating the relation $a(\gamma, a(g, v)) = a(\gamma \circ g, v)$, $g, \gamma \in G$, $v \in N$ with respect to g at $g = \mathbb{1}$ leads to

$$Ta(\gamma, v)da(\mathbb{1}, v)\mu = da(\gamma, v)dL_\gamma(\mathbb{1})\mu, \quad \forall \mu \in T_\mathbb{1}G. \tag{10.27}$$

Finally, define $\mu(t)$ by $\dot{\gamma}(t) = dL_\gamma(\mathbb{1})\mu$ and combine (10.26), (10.27) to find

$$Ta(\gamma, v)\left[v_t - F(v) + da(\mathbb{1}, v)\mu\right] = 0$$

and, hence, (10.25a) by the invertibility of $Ta(\gamma, v)$. The converse is proved in a similar way. $\square$

Lemma 2 shows that system (10.25) does not have a unique solution (v, μ, γ). Rather we have $p = \dim G$ additional degrees of freedom that will be fixed by a phase condition $\psi : N \times T_\mathbb{1}G \to \mathbb{R}^p$. The phase condition together with (10.25a) yields a PDAE for v and μ, namely:

$$v_t = F(v) - da(\mathbb{1}, v)\mu, \tag{10.28}$$
$$0 = \psi(v, \mu).$$

Equation (10.25b) is called the *reconstruction equation* in [33]. It is decoupled from system (10.28) and can be solved by an a-posteriori process.

The traveling waves in Examples 1 and 2 easily fit into the abstract framework.

Example 3. For the Lie group $G = \mathbb{R}$ consider the shift action $a(\gamma, u)(x) = u(x - \gamma)$. There are different possibilities for the choice of spaces M and N. Either take $M = C_{\text{unif}}$, $N = C_{\text{unif}}^2$ or $M = w + \mathcal{L}_2 \supset N = w + \mathcal{H}^2$ where $w \in C_b^2(\mathbb{R}, \mathbb{R}^2)$ satisfies $w_x, w_{xx} \in \mathcal{L}_2$ and has the correct limit behavior, e.g., $w(x) = \bar{u}_\pm + \mathcal{O}(e^{-\alpha|x|})$ as $x \to \pm\infty$. For the last choice we actually use the manifold structure of M and N. In both cases we have $da(\mathbb{1}, v)\mu = -\mu v_x$ and using a template function $\hat{v} \in N$ the system (10.28) is given by (10.20) with ψ given in (10.18).

Example 4. Consider a system (10.16) of dimension $m = 2$ such that the nonlinearity is equivariant with respect to rotations, i.e.,

$$f(R_\rho v) = R_\rho f(v) \ \forall v \in \mathbb{R}^2, \ \rho \in \mathbb{R}, \ \text{where} \ R_\rho = \begin{pmatrix} \cos \rho & -\sin \rho \\ \sin \rho & \cos \rho \end{pmatrix}.$$

Equations of this type arise as real versions of complex-valued systems, such as the Ginzburg-Landau equation. The Lie group is $G = \mathbb{S}^1 \times \mathbb{R}$ and the action $a : G \times \mathcal{L}_2 \to \mathcal{L}_2$ on $u : \mathbb{R} \to \mathbb{R}^2$ at $\gamma = (\rho, \tau)$ is given by

$$a(\gamma, u)(x) = R_{-\rho} u(x - \tau).$$

With $M = \mathcal{L}_2$, $N = \mathcal{H}^2$ we obtain $da(\mathbb{1}, v)(\mu_\tau, \mu_\rho) = -v_x \mu_\tau - R_{\frac{\pi}{2}} v \mu_\rho$ and (10.28) has the form

$$v_t = Av_{xx} + \mu_\tau v_x + \mu_\rho R_{\frac{\pi}{2}} v + f(v),$$
$$0 = \langle \hat{v}', v - \hat{v} \rangle_{\mathcal{L}_2}, \quad 0 = \langle R_{\frac{\pi}{2}} v, v - \hat{v} \rangle_{\mathcal{L}_2}.$$

The reconstruction equations read $\dot{\tau} = \mu_\tau$, $\tau(0) = 0$ and $\dot{\rho} = \mu_\rho$, $\rho(0) = 0$.

10.2.3 Fixed Versus Minimizing Phase Conditions

In the abstract setting of Sect. 10.2.2 assume that M is a Banach space in which we have a continuous inner product $\langle \cdot, \cdot \rangle_2$ with associated norm $\|v\|_2$. One way to set up a phase condition (in the spirit of Sect. 10.1) is to minimize the distance of the frozen solution v from the group orbit $\mathcal{O}(\hat{v}) = \{a(\gamma, \hat{v}) : \gamma \in G\}$ of a template function $\hat{v}$; see Fig. 10.5(a). The necessary condition for a minimum of $\|a(\gamma, \hat{v}) - v\|_2$ to occur at $\gamma = \mathbb{1}$ is

$$\psi_{\text{fix}}(v)\mu = \langle da(\mathbb{1}, \hat{v})\mu, v - \hat{v} \rangle_2 = 0 \quad \forall \mu \in T_{\mathbb{1}}G. \tag{10.29}$$

In the beginning one may choose as template the initial value $\hat{v} = u^0$. Note that ψ in (10.29) maps into the dual $T_{\mathbb{1}}^* G$ of the Lie algebra, which is isomorphic to $\mathbb{R}^p$.

Another possibility is to minimize the temporal change $\|v_t\|_2$ at each time instance which leads to the condition

$$\psi_{\text{min}}(v)\mu = \langle da(\mathbb{1}, v)\mu, v_t \rangle_2 = 0 \quad \forall \mu \in T_{\mathbb{1}}G. \tag{10.30}$$

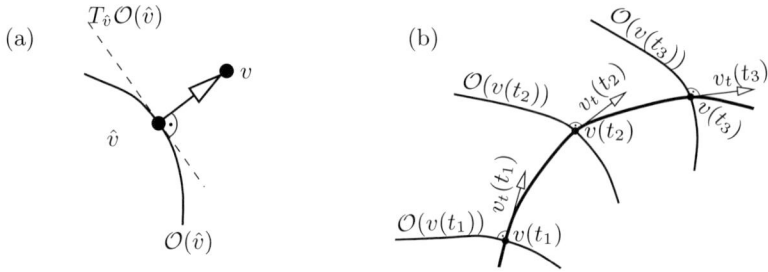

Fig. 10.5. Conditions ψ_{fix} (a) and ψ_{min} (b).

As is illustrated in Fig. 10.5(b) this condition requires the frozen trajectory $v(t)$ to be orthogonal to the group orbit of $v(t)$ at all times.

For the case of traveling waves we show how to transform solutions of the PDAE for both phase conditions into each other, i.e., we transform solutions (v, λ) of

$$v_t = v_{xx} + f(v) + \lambda v_x$$
$$0 = \langle \bar{v}_x, v - \bar{v} \rangle_{\mathcal{L}_2} \tag{10.31}$$

into solutions (w, μ) of

$$w_t = w_{xx} + f(w) + \lambda w_x$$
$$0 = \langle w_x, w_t \rangle_{\mathcal{L}_2}. \tag{10.32}$$

The following Lemma will be used for the stability analysis in Sect. 10.3.3.

Lemma 3. *Let (v, λ) be a solution of (10.31). Then (w, μ) defined by*

$$w(x, t) = v(x - \eta(t), t), \quad \mu = \lambda - \dot{\eta},$$

and

$$\eta(t) = \int_0^t \frac{\langle v_x(\cdot, \tau), v_t(\cdot, \tau) \rangle_{\mathcal{L}_2}}{\|v_x(\cdot, \tau)\|_{\mathcal{L}_2}^2} \, d\tau$$

is a solution of system (10.32).

Proof. We have

$$w_t = v_t - v_x \dot{\eta} = v_{xx} + f(v) + (\lambda - \dot{\eta})v_x = w_{xx} + f(w) + \mu w_x.$$

With the shift invariance of $\langle \cdot, \cdot \rangle_{\mathcal{L}_2}$ and the definition of η we get

$$\langle w_t, w_x \rangle_{\mathcal{L}_2} = \langle v_t - \dot{\eta} v_x, v_x \rangle_{\mathcal{L}_2} = 0. \qquad \square$$

10.3 Relative Equilibria and Stability

In this section we study relative equilibria of equivariant evolution equations. In particular, we consider relative equilibria of parabolic systems in one space dimension. We use phase conditions to approximate relative equilibria on finite intervals and study their asymptotic stability in the Lyapunov sense via the freezing method.

10.3.1 Relative Equilibria

We seek solutions of (10.24) that have the special form $u(t) = a(\gamma(t), \bar{v})$ for some time-independent function $\bar{v}$.

Definition 3. *A solution $\bar{u}$ of (10.24) is called a relative equilibrium if it has the form $\bar{u}(t) = a(\bar{\gamma}(t), \bar{v})$ for some $\bar{v} \in N$ and for some function $\bar{\gamma} \in C^1([0, \infty), G)$.*

Without loss of generality we can assume that $\bar{\gamma}(0) = \mathbb{1}$. Usually the whole group orbit $\mathcal{O}(\bar{v}) = \{a(\gamma, \bar{v}), \ \gamma \in G\}$ is called a relative equilibrium if it is invariant under the semi-flow; see [10, 29]. We found the equivalent constructive definition above more convenient from a numerical point of view [7], because it explicitly includes the orbit $\bar{\gamma}(t)$ on the group. The following lemma shows the connection between $\bar{u}$, $\bar{\gamma}$ and $\bar{v}$.

Lemma 4. *Let $\bar{u}(t) = a(\bar{\gamma}(t), \bar{v})$ be a relative equilibrium with trivial stabilizer $\mathcal{S}_{\bar{v}} = \{\gamma \in G : \ a(\gamma, \bar{v}) = \bar{v}\}$. Then there exists $\bar{\mu} \in T_{\mathbb{1}}G$ such that $(\bar{v}, \bar{\mu})$ solve*

$$0 = F(\bar{v}) - da(\mathbb{1}, \bar{v})\bar{\mu} \tag{10.33}$$

and $\dot{\bar{\gamma}} = dL_{\bar{\gamma}}(\mathbb{1})\bar{\mu}$, $\gamma(0) = \mathbb{1}$.
Conversely, if (10.33) holds for $(\bar{v}, \bar{\mu})$ then $\bar{u}(t) = a(\bar{\gamma}(t), \bar{v})$ with $\bar{\gamma} = \exp(t\bar{\mu})$ is a relative equilibrium of (10.24).

Proof. The orbit $\mathcal{O}(\bar{v})$ has tangent space $T_{\bar{v}}\mathcal{O}(\bar{v}) = \mathrm{range}(da(\mathbb{1}, \bar{v}))$, and it is well known [10, Lemma 4.10.4] that $\dim T_{\bar{v}}\mathcal{O}(\bar{v}) = \dim G - \dim \mathcal{S}_{\bar{v}}$. Hence, the stabilizer is trivial if, and only if, $da(\mathbb{1}, \bar{v})$ is one-to-one. By Lemma 2 we find that $\bar{\mu}(t) = dL_{\gamma}(\mathbb{1})^{-1}\dot{\bar{\gamma}} \in T_{\mathbb{1}}G$ is continuous and satisfies (10.33). Since $\bar{v}$ is independent of t and $da(\mathbb{1}, \bar{v})$ is one-to-one, we obtain that $\bar{\mu}$ is independent of t as well. $\square$

Remark 1. If $\bar{v}$ has nontrivial stabilizer then one can still write $\bar{\gamma}$ as an exponential in terms of the Lie algebra of the stabilizer and its normalizer; see [10, Th. 7.2.4].

Choosing a basis $\{e^1, \ldots, e^p\}$ in $T_{\mathbb{1}}G$ we can identify the Lie algebra with $\mathbb{R}^p$ via $\mu = \sum_{i=1}^{p} \mu_i e^i$. Further, setting $S^i(v) = -da(\mathbb{1}, v)e^i$, we find from (10.28) and Lemma 4 the equation to be solved for $(\bar{v}, \bar{\mu})$, namely:

$$0 = F(v) + S(v)\mu, \quad \text{where } S(v)\mu = \sum_{i=1}^{p} \mu_i S^i(v),$$

$$0 = \psi(v, \mu).$$

10.3.2 Approximation of Relative Equilibria on Finite Intervals

We now treat the special case when the evolution equation (10.24) is a parabolic system of the form (10.16). We assume that the operators S^i are differential operators $S^i(v)(x) = S_0^i v(x) + S_1^i v_x(x)$ for suitable matrices $S_0^i, S_1^i \in \mathbb{R}^{m \times m}$.

For the numerical computation of relative equilibria of (10.16) we solve a boundary value problem on a finite interval $J = [x_-, x_+]$, namely:

$$0 = A v_{xx} + S(v)\mu + f(v), \quad x \in [x_-, x_+], \tag{10.34a}$$

$$\eta = \mathcal{B}v, \tag{10.34b}$$

$$0 = \langle S^i(\hat{v})|_J, v - \hat{v}|_J \rangle_J, \quad i = 1, \dots, p. \tag{10.34c}$$

Here $\hat{v}$ is a template function and $\mathcal{B}$ is the two-point boundary operator

$$\mathcal{B}v = P_- v(x_-) + Q_- v_x(x_-) + P_+ v(x_+) + Q_+ v_x(x_+), \quad P_\pm, Q_\pm \in \mathbb{R}^{2m \times m}.$$

The linearization of (10.34a) with respect to v at $(\bar{v}, \bar{\mu})$ is given by

$$\Lambda u = A u_{xx} + B u_x + C u, \quad B = \sum_{i=1}^{p} \bar{\mu}_i S_1^i, \ C(x) = f'(\bar{v}(x)) + \sum_{i=1}^{p} \bar{\mu}_i S_0^i. \tag{10.35}$$

If $\lim_{x \to \pm\infty} \bar{v}(x) = v_\pm$ and $\lim_{x \to \pm\infty} \bar{v}_x(x) = 0$ then Λ turns for $x \to \pm\infty$ into the constant-coefficient operator

$$\Lambda_\pm v = A v_{xx} + B v_x + C_\pm v, \quad C_\pm = \lim_{x \pm \infty} C(x).$$

The main spectral assumptions on Λ are the following:

Hypothesis 1 (spectral condition) *The eigenvalue 0 lies in the connected component of $\mathbb{C} \setminus \{\Sigma_+ \cup \Sigma_-\}$ that contains a right half-plane, where*

$$\Sigma_\pm = \{s \in \mathbb{C} : \det(-\kappa^2 A + i\kappa B + C_\pm - sI) = 0, \ \text{for some } \kappa \in \mathbb{R}\}.$$

Hypothesis 2 (eigenvalue condition) *The functions $S^i(\bar{v}) = -da(1, \bar{v})e^i$, $i = 1, \dots, p$ lie in $\mathcal{H}^2$, are linearly independent and span the nullspace of $\Lambda : \mathcal{H}^2 \to \mathcal{L}_2$, i.e., $\ker(\Lambda) = \mathrm{span}\{S^1(\bar{v}), \dots, S^p(\bar{v})\}$. Moreover, the algebraic and the geometric multiplicity of zero are both equal to p.*

Hypothesis 1 guarantees that the quadratic eigenvalue problem associated with Λ_+ has m stable and m unstable eigenvalues; cf. [6]. In view of condition (i) of Theorem 1 we consider the determinant (cf. [6])

$$\mathcal{D} = \det \left((P_- \; Q_-) \begin{pmatrix} Y_-^s \\ Y_-^s \Sigma_-^s \end{pmatrix} (P_+ \; Q_+) \begin{pmatrix} Y_+^u \\ Y_+^u \Sigma_+^u \end{pmatrix} \right), \qquad (10.36)$$

where (Σ_-^s, Y_-^s), $(\Sigma_+^u, Y_+^u) \in \mathbb{R}^{m \times m} \times \mathbb{R}^{m \times m}$ solve the quadratic eigenvalue problems

$$AY\Sigma^2 + BY\Sigma + C_{\pm}Y = 0$$

with $\operatorname{Re}\sigma(\Sigma_-^s) < 0$ and $\operatorname{Re}\sigma(\Sigma_+^u) > 0$. Then we can formulate the determinant condition and a consistency assumption for the boundary conditions.

Hypothesis 3 (boundary conditions) *The boundary condition (10.34b) is satisfied at the stationary points $\bar{v}_{\pm}$, i.e., $\eta = P_- \bar{v}_- + P_+ \bar{v}_+$ and the determinant $\mathcal{D}$ defined in (10.36) is nonzero.*

As in Sect. 10.1, the boundary conditions have to control the terms that grow in forward time on the positive axis and in backward time on the negative axis. These are given by the stable or unstable manifolds of the stationary points. Note that the determinant condition is satisfied for Dirichlet, Neumann and periodic boundary conditions; cf. [6].

For simplicity we first formulate the theorem for pulses, i.e., we use $M = \mathcal{L}_2$, $N = \mathcal{H}^2$. In order to generalize this to fronts one needs the additional condition $a(\gamma, \hat{v}) - \bar{v} \in \mathcal{H}^2$ for all $\gamma \in G$.

Hypothesis 4 (phase condition) *The phase condition is satisfied by $\bar{v}$, i.e., $\langle S(\hat{v}), \bar{v} - \hat{v} \rangle_{\mathcal{L}_2} = 0$, $\bar{v} - \hat{v} \in \mathcal{H}^1$, $S(\hat{v}) \in \mathcal{L}_2$ and the matrix*

$$\langle S(\hat{v}), S(\bar{v}) \rangle_{\mathcal{L}_2} = \left(\int_{\mathbb{R}} [S^i(\hat{v})](x)^T [S^j(\bar{v})](x) dx \right)_{i,j=1}^p \in \mathbb{R}^{p \times p}$$

is nonsingular.

The following approximation result is an adaptation of Theorem 1 to the current situation; see [39] for a proof.

Theorem 2 (Approximation of relative equilibria on finite intervals). *Assume Hypotheses 1– 4 hold. Then there exist $\varrho > 0$, $T > 0$, such that for $\min\{-x_-, x_+\} > T$ the boundary value problem (10.34) has a unique solution (v_J, μ_J) in a ball $B_\varrho(\bar{v}_{|J}, \bar{\mu}) = \{(v, \mu) \in \mathcal{H}^2(J, \mathbb{R}^m) \times \mathbb{R}^p : \|\bar{v}_{|J} - v\|_{\mathcal{H}^2} + \|\bar{\mu} - \mu\| < \varrho\}$. Further, there exist group elements $\gamma_J \in G$ such that $\tilde{v} = a(\gamma_J, \bar{v})$ satisfies the following estimate for some $\alpha > 0$*

$$\|v_J - \tilde{v}_{|J}\|_{\mathcal{H}^2} + \|\mu_J - \bar{\mu}\| \leq \text{const } e^{-\alpha \min\{-x_-, x_+\}}.$$

A similar version for a full discretization with finite differences can be found in [43]. In that case one obtains an error estimate on the grid $J_h = \{hn, n_- \leq n \leq n_+\}$ for the approximate solution (v_h, μ_h), namely:

$$\|v_h - \bar{v}_{|J_h}\|_{\mathcal{H}_h^2} + \|\mu_h - \bar{\mu}\| \leq \text{const } (h^2 + e^{-\alpha h \min\{-n_-, n_+\}}), \qquad (10.37)$$

where $\|\cdot\|_{\mathcal{H}_h^2}$ is the discrete analog of the $\mathcal{H}^2$ norm. A similar result holds for the norm $\|\cdot\|_\infty$.

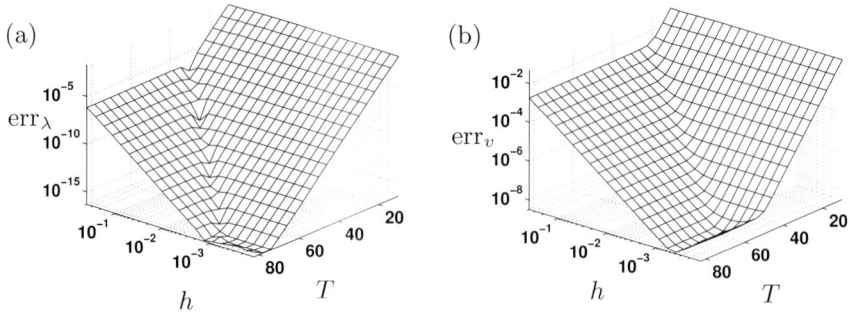

Fig. 10.6. Approximation errors $\mathrm{err}_\lambda = |\lambda_h - \bar\lambda|$ (a) and $\mathrm{err}_v = \|v_h - \bar v_{|J_h}\|_{\mathcal{H}^2_h}$ (b).

Example 5. In the case of the Nagumo equation (10.21) from Example 1 we can compare the approximation with the exact solution. Figure 10.6 shows the approximation errors of the traveling wave for Dirichlet boundary conditions. The grid size h was varied logarithmically from 10^{-4} to 10^{-1} and the size of the symmetric interval $[-T, T]$ linearly from 20 to 80. We observe that the convergence of $\tilde v$ and $\tilde\lambda$ to the exact solution $\bar v$ and $\bar\lambda$ given in (10.22) is exponential in T and quadratic in h. This is in good agreement with the approximation result (10.37).

10.3.3 Stability of Relative Equilibria in One Space Dimension

Stability results for traveling waves on the real line or, more generally, for relative equilibria are well known for parabolic systems [22, 44]. Here the notion of asymptotic stability with asymptotic phase is used. By the freezing method this notion is converted into the usual asymptotic (Lyapunov-) stability. We now present a stability result for relative equilibria in the frozen setting. To this end, the spectral assumptions 1 and 2 have to be tightened as follows.

Hypothesis 5 *The curves $\Sigma_+ \cup \Sigma_-$ lie in the open left half-plane and zero is the only eigenvalue with real part greater equal zero.*

Theorem 3. *Assume Hypotheses 1, 2, 4 and 5 hold. Then there exist $\varepsilon, \nu > 0$ such that for all $u_0 \in \hat v + \mathcal{H}^1(\mathbb{R})$ with $\|u_0 - \bar v\|_{\mathcal{H}^1} \le \varepsilon$ the system*

$$v_t = Av_{xx} + f(v) + S(v)\mu, \quad v(\cdot, 0) = u_0,$$
$$0 = \langle S(\hat v), v - \hat v\rangle_{\mathcal{L}_2}$$

has a unique solution $v \in C^1((0, \infty), \hat v + \mathcal{H}^1(\mathbb{R})) \cap C([0, \infty), \hat v + \mathcal{H}^1(\mathbb{R}))$ and $\mu \in C([0, \infty), \mathbb{R}^p)$. Moreover, this solution satisfies

$$\|v(\cdot, t) - \bar v\|_{\mathcal{H}^1} + \|\mu(t) - \bar\mu\| \le \mathrm{const}\ e^{-\nu t}\|v^0 - \bar v\|_{\mathcal{H}^1} \quad \forall t \ge 0.$$

Remark 2. For the case of traveling waves a proof of this theorem can be found in [41]. The generalization to arbitrary groups is straightforward by the techniques used for Theorem 2. It is also shown in [41] that one can allow more general nonlinearities $f(v, v_x)$ of the form

$$f(u, v) = f_1(u)v + f_2(u), \quad f_1 \in C^1(\mathbb{R}^m, \mathbb{R}^{m \times m}), f_2 \in C^1(\mathbb{R}^m, \mathbb{R}^m),$$

where f_1, f_2, f_1', f_2' are globally Lipschitz. This includes the case of the nonlinearity uu_x in Burgers equation.

An analogous result for a spatial discretization with finite differences is given in [41] for traveling waves and in [42] for general relative equilibria in one space dimension.

Remark 3. We note that a general stability theorem for finite-dimensional equivariant systems is given in [10, Th. 7.4.2].

For the special case of stationary solutions of (10.31) the local stability estimate reads

$$\|v(\cdot, t) - \bar{v}\|_{\mathcal{H}^1} + |\lambda(t) - \bar{\lambda}| \leq \text{const } e^{-\alpha t} \|v^0 - \bar{v}\|_{\mathcal{H}^1} \quad \forall t \geq 0. \qquad (10.38)$$

Using the transformation between the different phase conditions ψ_{fix} and ψ_{min} in Lemma 3, we will show how stability transfers to the ψ_{min}-case.

We define the bilinear form $b : \mathcal{H}^1 \times \mathcal{H}^1 \to \mathbb{R}$ via

$$b(u, v) = \int_{\mathbb{R}} -u_x(x)^T A v_x(x) + u(x)^T (B v_x(x) + C(x)v(x)) \, dx$$

where $A, B, C(\cdot)$ are the bounded matrix functions defined in (10.35). Via integration by parts we then get

$$b(\hat{v}_x, v) = \langle \hat{v}_x, \Lambda v \rangle_{\mathcal{L}_2} \quad \text{for} \quad v \in \mathcal{H}^2 \quad \text{and} \quad |b(\hat{v}_x, v)| \leq \text{const } \|v\|_{\mathcal{H}^1}.$$

We define the projector P onto $\hat{v}_x^{\perp}$ along $\bar{v}_x$ and the projected differential operator Λ_P through

$$Pv = v - \hat{v}_x \langle \hat{v}_x, \bar{v}_x \rangle_{\mathcal{L}_2}^{-1} \langle \hat{v}_x, v \rangle_{\mathcal{L}_2}, \quad \Lambda_P = P\Lambda_{|range(P)}.$$

The following lemma gives the main estimate for solutions of the nonautonomous PDAE

$$\begin{aligned} v_t &= \Lambda v + \mu \bar{v}_x + g(t, v, \mu), \quad v(0) = v^0, \\ 0 &= \langle \hat{v}_x, v \rangle. \end{aligned} \qquad (10.39)$$

Lemma 5. *Assume that g satisfies*

$$\|g(t, v, \mu)\| \leq \text{const } e^{-\beta t}(\|v\|_{\mathcal{H}^1} + \|\mu\|), \quad \beta > 0. \qquad (10.40)$$

Then there exist $\rho > 0$ and $\nu \in (0, \beta)$ such that any solution (v, μ) of (10.39) with $\|v^0\|_{\mathcal{H}^1} < \rho$ obeys the exponential estimate

$$\|v(t)\|_{\mathcal{H}^1} + \|\mu(t)\| \leq \text{const } e^{-\nu t} \|v^0\|_{\mathcal{H}^1}. \qquad (10.41)$$

Proof. The proof relies on the estimates for $r \in \text{range}(P)$ and some $\alpha > 0$

$$\|e^{\Lambda_P t} r\|_{\mathcal{L}_2} \leq K e^{-\alpha t} \|r\|_{\mathcal{L}_2}, \quad \|e^{\Lambda_P t} r\|_{\mathcal{H}^1} \leq K e^{-\alpha t} t^{-\frac{1}{2}} \|r\|_{\mathcal{L}_2}, \qquad (10.42)$$

which follow from the fact that the eigenvalue 0 has been eliminated from the spectrum of Λ_P; cf. [41, Lemma 1.24]. By the variation of constants formula the PDAE (10.39) can be written equivalently as

$$v(t) = e^{\Lambda_P t} v^0 + \int_0^t e^{\Lambda_P (t-s)} P\, g(s, v(s), \mu(s))\, ds,$$
$$\mu(t) = -\langle \hat{v}_x, \bar{v}_x \rangle_{\mathcal{L}_2}^{-1} [b(\hat{v}_x, v(t)) + \langle \hat{v}_x, g(t, v(t), \mu(t)) \rangle_{\mathcal{L}_2}]. \qquad (10.43)$$

Using this form, one first shows via Gronwall estimates as in [41] a global bound

$$\|v(t)\|_{\mathcal{H}^1} + \|\mu(t)\| \leq C \|v^0\|_{\mathcal{H}^1} \quad \forall t \geq 0. \qquad (10.44)$$

From the second equation in (10.43) we find with (10.40) that

$$\|\mu(t)\| \leq C[\|v(t)\|_{\mathcal{H}^1} + e^{-\beta t} (\|v(t)\|_{\mathcal{H}^1} + \|\mu(t)\|)].$$

Choose $T > 0$ such that $C e^{-\beta T} \leq \frac{1}{2}$ and obtain

$$\|\mu(t)\| \leq C \|v(t)\|_{\mathcal{H}^1} \quad \forall t \geq T. \qquad (10.45)$$

Now choose $0 < \nu < \min(\alpha, \beta)$ and use (10.42), (10.44) and (10.45) in the first equation of (10.43) to obtain

$$n(t) = \|v(t)\|_{\mathcal{H}^1} e^{\nu t} \leq C\Big(e^{(\nu-\alpha)t} \|v^0\|_{\mathcal{H}^1}$$
$$+ \int_0^t \frac{e^{(\nu-\alpha)(t-s)}}{\sqrt{t-s}} e^{\nu s} e^{-\beta s} (\|v(s)\|_{\mathcal{H}^1} + \|\mu(s)\|)\, ds \Big)$$
$$\leq C\Big(e^{(\nu-\alpha)t} \|v^0\|_{\mathcal{H}^1} + e^{(\nu-\alpha)(t-T)} \int_0^T \frac{e^{(\nu-\alpha)(T-s)}}{\sqrt{t-s}} \|\mu(s)\|\, ds$$
$$+ \int_0^t \frac{e^{(\nu-\alpha)(t-s)}}{\sqrt{t-s}} e^{\nu s} \|v(s)\|_{\mathcal{H}^1}\, ds \Big)$$
$$\leq C\Big(e^{(\nu-\alpha)t} \|v^0\|_{\mathcal{H}^1} + \int_0^t \frac{e^{(\nu-\alpha)(t-s)}}{\sqrt{t-s}} n(s)\, ds \Big).$$

The Gronwall inequality with weak singularities (cf. [22, Lemma 7.1.1]) yields the assertion. □

Lemma 6. *Let the assumptions of Theorem 3 be satisfied for a nonconstant solution* $(\bar{v}, \bar{\lambda}) \in C_b^2 \times \mathbb{R}$ *of* (10.31). *Then there exists a shift* $\gamma \in \mathbb{R}$ *such that* $(\bar{v}(\cdot + \gamma), \bar{\lambda})$ *is an asymptotically stable solution of* (10.32).

Proof. Note that $w = v_t$ solves

$$w_t = w_{xx} + f'(v)w + w_x\lambda + v_x\dot{\lambda},$$
$$0 = \langle \hat{v}_x, w \rangle,$$

and that the first equation with $\mu := \dot{\lambda}$ is equivalent to

$$w_t = \Lambda w + \bar{v}_x\mu + (f'(v) - f'(\bar{v}))w + (\lambda - \bar{\lambda})w_x + (v_x - \bar{v}_x)\mu.$$

Now we apply Lemma 5 for small $\|v^0\|_{\mathcal{H}^1}$ with $\beta = \alpha$ and

$$g(t, w, \mu) = (f'(v(t)) - f'(\bar{v}))w + (\lambda(t) - \bar{\lambda})w_x + (v_x(t) - \bar{v}_x)\mu.$$

Note that the exponential decay (10.40) follows from the stability estimate (10.38). Since g is linear in w and μ, we obtain for all v^0 from (10.41) the estimate

$$\|v_t\| + \|\dot{\lambda}\| \le \text{const } e^{-\nu t}\|v^0\|_{\mathcal{H}^1} \quad \forall t \ge 0.$$

By this estimate the integral

$$\eta_\infty = \int_0^\infty \frac{\langle v_x(\cdot, \tau), v_t(\cdot, \tau)\rangle}{\|v_x(\cdot, \tau)\|^2} \, d\tau$$

exists and we have

$$|\eta(t) - \eta_\infty| \le \int_t^\infty \frac{|\langle v_x(\cdot, \tau), v_t(\cdot, \tau)\rangle|}{\|v_x(\cdot, \tau)\|^2} \, d\tau \le \int_t^\infty \frac{\|v(\cdot, \tau)\|_{\mathcal{H}^1}\|v_t(\cdot, \tau)\|}{\|v_x(\cdot, \tau)\|_{\mathcal{L}_2}^2} \, d\tau$$

$$\le \int_t^\infty \frac{(\|\bar{v}\|_{\mathcal{H}^1} + C\delta)Ce^{-\nu\tau}}{(\|\bar{v}_x\|_{\mathcal{L}_2} - C\delta)^2} \, d\tau \le \text{const } e^{-\nu t}.$$

Together with the local stability estimate (10.38) this leads to

$$\|w(\cdot, t) - \bar{v}(\cdot - \eta_\infty)\|_{\mathcal{H}^1} + |\mu(t) - \bar{\lambda}|$$
$$\le \|v(\cdot - \eta(t), t) - \bar{v}(\cdot - \eta_\infty)\|_{\mathcal{H}^1} + |\lambda(t) - \dot{\eta}(t) - \bar{\lambda}|$$
$$\le \|v(\cdot, t) - \bar{v}\|_{\mathcal{H}^1} + \|\bar{v} - \bar{v}(\cdot + \eta(t) - \eta_\infty)\|_{\mathcal{H}^1} + |\lambda(t) - \bar{\lambda}| + |\dot{\eta}(t)|$$
$$\le \text{const } e^{-\nu t}. \qquad \square$$

10.4 Spiral Waves and Beyond

Embedding spiral waves of parabolic systems in $\mathbb{R}^2$ into the abstract framework of Sect. 10.2.2 is a considerable task [36, 46]. Therefore we do not pursue this in detail here.

10.4.1 Spiral Waves in Two Space Dimensions

Consider a PDE in two space dimensions

$$u_t = \Delta u + f(u), \quad t \geq 0,$$
$$u(x,0) = u_0(x), \quad x \in \mathbb{R}^2. \tag{10.46}$$

This equation is equivariant with respect to the Euclidean group $SE(2) = \mathbb{S}^1 \ltimes \mathbb{R}^2 \ni (\phi, \tau)$ with action $a(\gamma, v)(x) = v(R_{-\phi}(x - \tau))$ and group multiplication

$$(\phi_1, \tau_1) \circ (\phi_2, \tau_2) = (\phi_1 + \phi_2, \tau_1 + R_{\phi_1}\tau_2),$$

where R_ϕ again denotes rotations; cf. Example 4.

Take, for example, $v \in C_{\text{unif}} = M$ and $N = C_{\text{unif}}^2$ or the subspace $C_{\text{eucl}} \subset C_{\text{unif}}$ on which $SE(2)$ acts continuously [46]. Then the infinitesimal generators $da(\mathbb{1}, v)e^i$, $i = 1, 2, 3$ read $da(\mathbb{1}, v)e^1 = x_2 v_{x_1} - x_1 v_{x_2}$, $da(\mathbb{1}, v)e^2 = -v_{x_1}$, $da(\mathbb{1}, v)e^3 = -v_{x_2}$. For a relative equilibrium $u(x,t) = \bar{v}(R_{-\phi}(x - \tau))$ the wave form $\bar{v}$ is a solution of

$$0 = \Delta v + f(v) + \bar{\mu}_1(x_1 v_{x_2} - x_2 v_{x_1}) + \bar{\mu}_2 v_{x_1} + v_{x_2}\bar{\mu}_3$$

for some $\bar{\mu} \in se(2)$. The motion on the group orbit is given by

$$\dot{\gamma} = dL_\gamma(\mathbb{1})\bar{\mu} = \begin{pmatrix} 1 & 0 \\ 0 & R_\phi \end{pmatrix}\bar{\mu},$$

with solution

$$\bar{\gamma}(t) = \begin{pmatrix} \bar{\mu}_1 t \\ (I - R_{\bar{\mu}_1 t})\xi \end{pmatrix}, \quad \text{where} \quad \xi = \frac{1}{\bar{\mu}_1}\begin{pmatrix} -\bar{\mu}_3 \\ \bar{\mu}_2 \end{pmatrix}.$$

Then we can represent the relative equilibrium as follows:

$$\bar{u}(x,t) = \bar{v}(R_{-\bar{\mu}_1 t}(x + (R_{\bar{\mu}_1 t} - I)\xi)) = \bar{v}(R_{-\bar{\mu}_1 t}(x - \xi) + \xi).$$

A fixed reference point $\bar{x} \in \mathbb{R}^2$ with value $\bar{v}(\bar{x})$ traces the curve

$$x(t) = R_{\bar{\mu}_1 t}(\bar{x} - \xi) + \xi, \tag{10.47}$$

which is a circle of radius $\|\bar{x} - \xi\|$ and center ξ. If one uses a geometric definition for the tip x_{tip} of a spiral wave (see e.g. [8, 46]) then the tip moves on a circle of radius $r_{\text{tip}} = \|x_{\text{tip}} - \xi\|$. A special case are rigidly rotating Archimedean spirals which can be written in polar coordinates as $\bar{v}(x) = w(r, \phi)$ with $w(r, \phi) \to w_\infty(kr + \phi)$ as $r \to \infty$ for some periodic function w_∞.

Example 6. We use a diffusive version of Barkley's system [2] as an example, namely:

$$u_t = \Delta u + \frac{1}{\epsilon}u(1 - u)(u - \frac{1}{a}(v + b)),$$
$$v_t = D_v\Delta v + u - v.$$

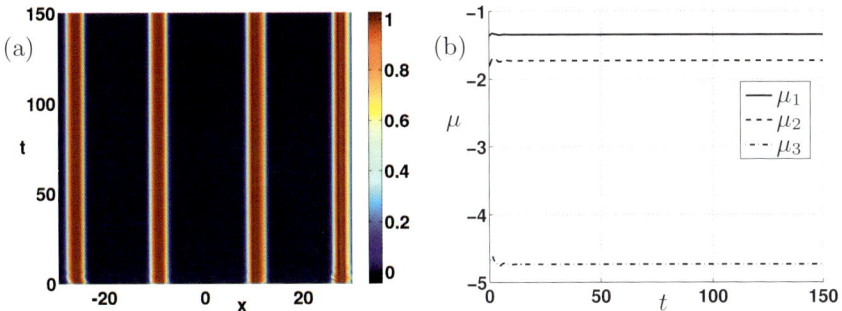

Fig. 10.7. Frozen spiral wave of PDE (10.46): time evolution of the u-component (a) and time evolution of μ (b).

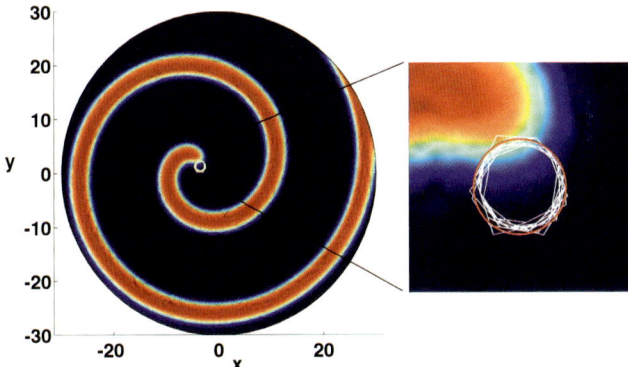

Fig. 10.8. Prediction of the tip-motion of the spiral wave from Fig. 10.7 via (10.47) (red circle) and tip-motion of the non-frozen spiral starting from the same initial condition (white trace).

We solve the corresponding PDAE

$$u_t = \Delta u + \tfrac{1}{\epsilon} u(1-u)(u - \tfrac{1}{a}(v+b)) + \lambda_1(y u_x - x u_y) + \lambda_2 u_x + \lambda_3 u_y$$
$$v_t = D_v \Delta v + u - v + \lambda_1(y v_x - x v_y) + \lambda_2 v_x + \lambda_3 v_y$$
$$0 = \langle y\hat{u}_x - x\hat{u}_y, u - \hat{u} \rangle_{\mathcal{L}_2}, \quad 0 = \langle \hat{u}_x, u - \hat{u} \rangle_{\mathcal{L}_2}, \quad 0 = \langle \hat{u}_y, u - \hat{u} \rangle_{\mathcal{L}_2}$$
$$0 = \langle y\hat{v}_x - x\hat{v}_y, v - \hat{v} \rangle_{\mathcal{L}_2}, \quad 0 = \langle \hat{v}_x, v - \hat{v} \rangle_{\mathcal{L}_2}, \quad 0 = \langle \hat{v}_y, v - \hat{v} \rangle_{\mathcal{L}_2}$$

with $(\hat{u}, \hat{v}) = (u^0, v^0)$ numerically for the parameters $D_v = 0.5$, $a = 0.5$, $b = 0.05$, $\epsilon = \tfrac{1}{50}$ by using the Finite Element package Comsol Multiphysics^{TM}.

In Fig. 10.7 the time evolutions of the u-component and the parameter μ are displayed. In Fig. 10.8 the prediction of the motion of the tip via (10.47) (red circle) is compared to the tip motion of the non-frozen spiral starting

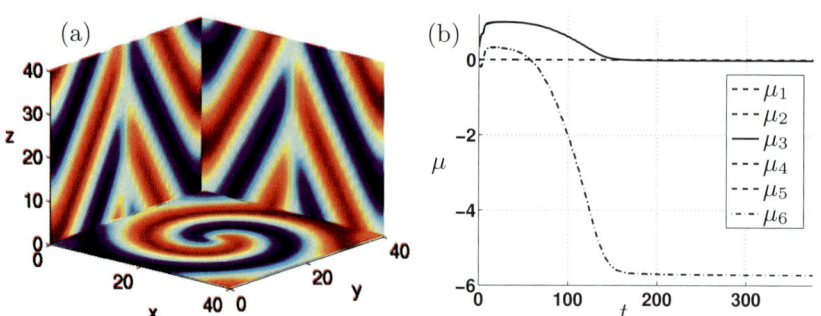

Fig. 10.9. Three-dimensional scroll wave of (10.48), namely: slices of the development of u (a) and time evolution of μ (b).

from the same initial conditions (white trace). For the definition of x_{tip} we used the condition $u = \frac{1}{2}$, $v = \frac{a}{2} - b$ from [2].

10.4.2 A Scroll Wave in Three Space Dimensions

In three space dimensions (10.46) is equivariant with respect to $G = SO(3) \ltimes \mathbb{R}^3 = SE(3)$ with action $a(\gamma, v)(x) = v(\mathcal{R}^{-1}(x - \tau))$, $\gamma = (\mathcal{R}, \tau)$, $\tau = (\tau_1, \tau_2, \tau_3)$ and group operation $\gamma \circ \tilde{\gamma} = (\mathcal{R}\tilde{\mathcal{R}}, \tau + \mathcal{R}\tilde{\tau})$. We denote the rotations about the x_1, x_2 and x_3 axes by R_{x_i}, and find by differentiating with respect to x_1, x_2 and x_3 the formula

$$-da(\mathbb{1}, \nu)\mu = \mu_1(v_{x_2}x_3 - v_{x_3}x_2) + \mu_2(v_{x_3}x_1 - v_{x_1}x_3) + \mu_3(v_{x_1}x_2 - v_{x_2}x_1)$$
$$+ \mu_4 v_{x_1} + \mu_5 v_{x_2} + \mu_6 v_{x_3}.$$

Example 7. We consider the following λ-ω system in complex form

$$u_t = \Delta u + 1 - |u|^2 - i|u|^2 u, \quad x \in \mathbb{R}^3, \ u(x, t) \in \mathbb{C}, \tag{10.48}$$

for which rigidly rotating waves exist [24].

We use an adapted version of the code EZSCROLL [18] and start in a box of length 40 with $\Delta x = 0.1$ from an initial function given in cylindrical coordinates as

$$u_0(r, \varphi, z) = e^{\frac{iz}{2\pi}}\frac{r}{40}(\cos(\varphi) + i\sin(\varphi)),$$

which ensures that in each z-slice a rotating spiral develops. We use periodic boundary conditions on the z-faces and Neumann boundary conditions on the x- and y-faces. Therefore, the initial function initiates a scroll wave twisted once in the z-direction; see [18, 19] for more information on scroll waves and scroll rings.

Figure 10.9 shows the real part of the solution of the frozen system at the final time instance, as well as the time evolution of μ. The solution in panel (a) is shown in the form of slices in x-,y-,z-directions through the origin $(0,0,0)$ which have been projected to the boundaries to increase visibility. From panel (b) one can see that first the rotation around and the translations along the z-axis are active. However, after some transient time only the z-translation is used to freeze the solution.

We expect that this solution has a nontrivial stabilizer since vertical motions in the z-direction and rotations about the z-axis can be exchanged. Nevertheless, our method seems to work. In fact, system (10.48) is actually equivariant with respect to the seven-dimensional group $\mathbb{S}^1 \times SE(3)$, where $\theta = e^{i\rho} \in \mathbb{S}^1$ acts as in Example 4. Including this symmetry in the computations leads to ill-conditioned systems when resolving the phase conditions for the seven parameters.

10.4.3 Relative Periodic Orbits

For relative periodic orbits we have a similar definition as for relative equilibria. We seek solutions of (10.24) that have the special form $\bar{u}(t) = a(\bar{\gamma}(t), \bar{v}(t))$ for some time periodic function $\bar{v}$.

Definition 4. *A solution $\bar{u}$ of (10.24) is called a relative periodic orbit if it has the form $\bar{u}(t) = a(\bar{\gamma}(t), \bar{v}(t))$ where $\bar{\gamma} : \mathbb{R} \to G$ is a smooth curve satisfying $\bar{\gamma}(0) = \mathbb{1}$ and $\bar{v} : \mathbb{R} \to N$ satisfies $\bar{v}, \bar{v}_t, F(\bar{v}) \in C(\mathbb{R}, M)$, is nonconstant and time periodic, i.e., $\bar{v}(\cdot + T) = \bar{v}$ for some period $T > 0$.*

As for relative equilibria (see Lemma 3) we can relate $\bar{u}$, $\bar{\gamma}$ and $\bar{v}$.

Lemma 7. *Let $\bar{u}(t) = a(\bar{\gamma}(t), \bar{v}(t))$ be a relative periodic orbit with trivial stabilizer $\mathcal{S}_{\bar{v}(t)}$, $t \in [0, T]$. Then there exists a T-periodic function $\bar{\mu} \in C(\mathbb{R}, T_{\mathbb{1}}G)$ such that for all $t \in \mathbb{R}$*

$$\bar{v}_t = F(\bar{v}) - da(\mathbb{1}, \bar{v})\bar{\mu} \tag{10.49}$$
$$\bar{\gamma}_t = dL_\gamma(\mathbb{1})\bar{\mu}, \quad \gamma(0) = \mathbb{1}. \tag{10.50}$$

Conversely, if $\bar{\mu} \in C(\mathbb{R}, T_{\mathbb{1}}G)$ and $\bar{v} \in C^1(\mathbb{R}, M)$ with $\bar{v}(t) \in N \ \forall t \in \mathbb{R}$ are T-periodic and solve (10.49) then $a(\bar{\gamma}(t), \bar{v})$ with $\bar{\gamma}$ defined by (10.50) is a relative periodic orbit of (10.24).

Proof. The proof is similar to that of Lemma 4. For the first assertion define $\bar{\mu}(t)$ by (10.50) and obtain (10.49) from Lemma 2. Equation (10.49) then shows that $\bar{\mu}$ is T-periodic. The converse follows in a similar manner. $\square$

Scaling with the period T as in Sect. 10.1, we find that $v(t) = \bar{v}(tT)$, $\mu(t)$ solve

$$v_t = T[F(v) - da(\mathbb{1}, v)\mu], \quad t \in [0, 1], \quad v(0) = v(1).$$

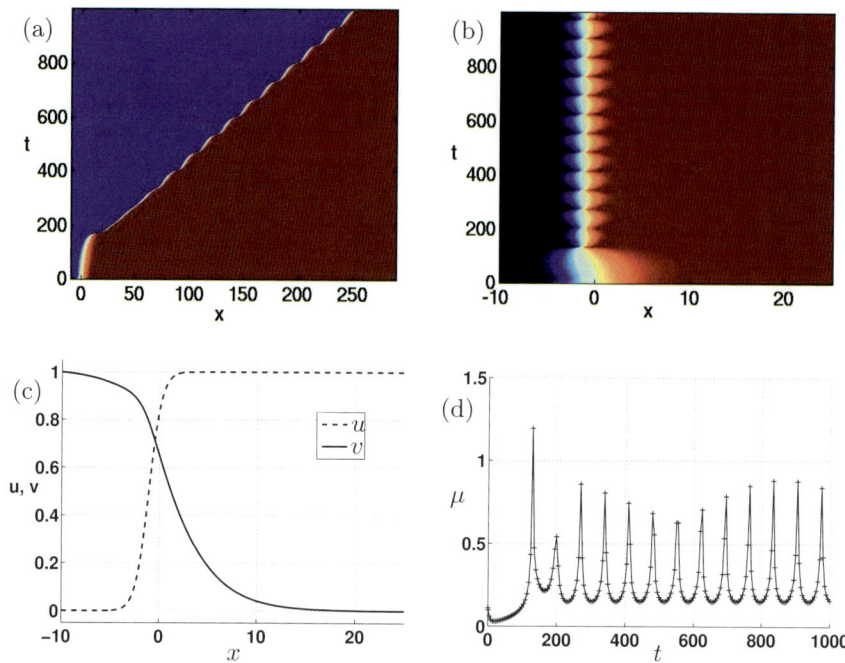

Fig. 10.10. Calculation of a modulated traveling wave in the autocatalytic system (10.23) for $\theta = \frac{1}{2}$: traveling wave (a), frozen wave (b), u- and v-components at $t = 1000$ (c), and time evolution of μ (d).

Example 8. Relative periodic orbits in one space dimension are modulated traveling waves [37]. These have the form $\bar{u}(x,t) = \bar{v}(x - \bar{\lambda}t, t)$ with $\bar{v}(x,t) = \bar{v}(x, t + T)$. Such solutions occur, for example, in the autocatalytic system (10.23) that was already considered in Example 2. We solve the system for $a = 0.1$, $m = 9$ on an interval of length 35 in the frozen case, and of length 300 for the direct simulation with $\Delta x = 0.1$, with Dirichlet boundary conditions and with a θ-method with $\Delta t = 0.1$. In Fig. 10.10(a) and (b) the numerical solutions for $\theta = \frac{1}{2}$ of the PDE (10.23) and of the corresponding PDAE (defining the frozen system) are shown. Panel (c) shows the solution at the last time instance, while panel (d) displays the time evolution of μ. The periodicity of the wave and the velocity can clearly be seen.

Figure 10.11 shows the modulated traveling wave and the frozen wave when the implicit Euler method is used, that is, $\theta = 1$. Note that the oscillations are strongly damped for the standard simulation in panel (a), whereas they are clearly visible in the frozen case in panel (b).

Combining the principles from Sects. 10.1 and 10.2 for the computation of relative periodic orbits, we arrive at a boundary value problem (in space and time) for $v \in C([0,1], M)$, $\mu \in C([0,1], T_{\mathbb{1}}G)$ and $T \in \mathbb{R}$ as follows:

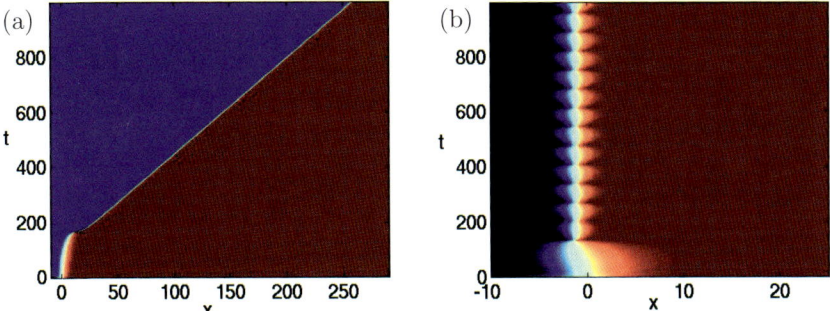

Fig. 10.11. Calculation of a modulated traveling wave in the autocatalytic system (10.23) for $\theta = 1$ (implicit Euler method): traveling wave (a), frozen wave (b).

$$v_t = T[F(v) - da(\mathbb{1}, v)\mu], \quad v(0) = v(1),$$

$$0 = \psi(v)\lambda = \int_0^1 \langle da(\mathbb{1}, \hat{v})\lambda, v - \hat{v} \rangle \, dt \quad \forall \lambda \in T_{\mathbb{1}}G,$$

$$0 = \phi(v) = \int_0^1 \langle \hat{v}_t, v - \hat{v} \rangle \, dt.$$

Here $\langle \cdot, \cdot \rangle$ denotes an inner product on M and $\hat{v} \in C([0,1], M)$ is a suitable template function.

10.5 Conclusions and Perspectives

Phase conditions are an effective tool in selecting specific orbits in equivariant evolution equations. When based on minimization principles they facilitate mesh adaptation and speed up continuation along branches. In many applications the underlying symmetry is induced by the Euclidean group $SE(d)$ acting on functions defined on the whole space $\mathbb{R}^d$. For numerical computations one has to truncate to bounded domains and use asymptotic boundary conditions. Truncation in combination with the method of freezing spatio-temporal patterns in a co-moving frame raises several numerical as well as theoretical problems. Only a few of them have been tackled in this chapter, mainly for parabolic systems in one space dimension.

Considerable challenges remain, and we expect the further development of the field to address theoretical and numerical issues, including the following:

1. Computation and continuation of relative equilibria and relative periodic orbits in equivariant systems, the detection of bifurcation points and branch switching at symmetry breaking bifurcations; see, for example, the recent progress by Wulff and Schebesch [45].

2. Adaptation of the freezing method to relative equilibria with nontrivial stabilizers.
3. Consideration of linear versus nonlinear stability for spatio-temporal patterns in space dimensions ≥ 2. There are extensive studies of the spectra associated with systems linearized about spiral waves and their truncation to bounded domains (see [35, 36, 38]), but a result on nonlinear stability still seems to be lacking.
4. Development of (implementable) asymptotic boundary conditions for spiral waves, scroll waves and the like.
5. Application of the freezing method to viscous conservation laws. As for modulated or spiral waves, this case is difficult because of the fact that the essential spectrum has a quadratic tangency with the imaginary axis; see [5].

References

1. N. J. Balmforth, R. V. Craster, and S. J. A. Malham. Unsteady fronts in an autocatalytic system. *Proc. Royal Soc. A*, 455(1984):1401–1433, 1999.
2. D. Barkley. A model for fast computer simulation of waves in excitable media. *Physica D*, 49:61–70, 1991.
3. W.-J. Beyn. The numerical computation of connecting orbits in dynamical systems. *IMA J. Numer. Anal.*, 10(3):379–405, 1990.
4. W.-J. Beyn. Numerical methods for dynamical systems. In *Advances in Numerical Analysis, Vol. I (Lancaster, 1990)*, Oxford Sci. Publ., pages 175–236. (Oxford University Press, 1991).
5. W.-J. Beyn and J. Lorenz. Stability of viscous profiles: proofs via dichotomies. *J. Dynam. Diff. Eqns.*, 18:141–195, 2006.
6. W.-J. Beyn and J. Lorenz. Stability of traveling waves: dichotomies and eigenvalue conditions on finite intervals. *Numer. Funct. Anal. Optim.*, 20(3-4):201–244, 1999.
7. W.-J. Beyn and V. Thümmler. Freezing solutions of equivariant evolution equations. *SIAM J. Appl. Dyn. Sys.*, 3(2):85–116, 2004.
8. M. Braune and H. Engel. Compound rotation of spiral waves in a lightsensitive Belousov-Zhabotinsky medium. *Chem. Phys. Lett.*, 204(3-4):257–264, 1993.
9. P. R. Chernoff and J. E. Marsden. *Properties of Infinite Dimensional Hamiltonian Systems*. Lecture Notes in Mathematics, 425. (Springer-Verlag, Berlin, 1974).
10. P. Chossat and R. Lauterbach. *Methods in Equivariant Bifurcations and Dynamical Systems*, Advanced Series in Nonlinear Dynamics 15. (World Scientific Publishing, Singapore, 2000).
11. J. W. Demmel, L. Dieci, and M. J. Friedman. Computing connecting orbits via an improved algorithm for continuing invariant subspaces. *SIAM J. Sci. Comput.*, 22(1):81–94 (electronic), 2000.
12. A. Dhooge, W. Govaerts, and Yu. A. Kuznetsov. MATCONT: a MATLAB package for numerical bifurcation analysis of ODEs. *ACM Trans. Math. Software*, 29(2):141–164, 2003.

13. E. J. Doedel. AUTO: a program for the automatic bifurcation analysis of autonomous systems. In *Proceedings of the Tenth Manitoba Conference on Numerical Mathematics and Computing, Vol. I (Winnipeg, Man., 1980)*, volume 30, pages 265–284, 1981.

14. E. J. Doedel and M. J. Friedman. Numerical computation of heteroclinic orbits. *J. Comput. Appl. Math.*, 26(1-2):155–170, 1989.

15. E. J. Doedel, W. Govaerts, and Yu. A. Kuznetsov. Computation of periodic solution bifurcations in ODEs using bordered systems. *SIAM J. Numer. Anal.*, 41(2):401–435, 2003.

16. E. J. Doedel, A. D. Jepson, and H. B. Keller. Numerical methods for Hopf bifurcation and continuation of periodic solution paths. In *Computing Methods in Applied Sciences and Engineering, VI (Versailles, 1983)*, pages 127–138. (North-Holland, 1984).

17. E. J. Doedel, R. C. Paffenroth, A. R. Champneys, T. F. Fairgrieve, Yu. A. Kuznetsov, B. E. Oldeman, B. Sandstede, and X. J. Wang. AUTO2000: Continuation and bifurcation software for ordinary differential equations (with HOMCONT). Available via `http://cmvl.cs.concordia.ca/`.

18. M. Dowle, R. M. Mantel, and D. Barkley. Fast simulations of waves in three-dimensional excitable media. *Internat. J. Bifur. Chaos Appl. Sci. Engrg.*, 7(11):2529–2545, 1997.

19. B. Fiedler and R. M. Mantel. Crossover collision of scroll wave filaments. *Doc. Math.*, 5:695–731 (electronic), 2000.

20. M. J. Friedman and E. J. Doedel. Numerical computation and continuation of invariant manifolds connecting fixed points. *SIAM J. Numer. Anal.*, 28(3):789–808, 1991.

21. M. Golubitsky and I. Stewart. *The Symmetry Perspective*. Progress in Mathematics, 200. (Birkhäuser, 2002).

22. D. Henry. *Geometric Theory of Semilinear Parabolic Equations* Lecture Notes in Mathematics, 840. (Springer-Verlag, Berlin, 1981).

23. S. Arimoto J. Nagumo and S. Yoshizawa. An active pulse transmission line simulating nerve axon. In *Proceedings of the IRE*, volume 50, pages 2061–2070, 1962.

24. Y. Kuramoto and S. Koga. Turbulized rotating chemical waves. *Progr. Theor. Phys.*, 66(3):1081–1085, 1981.

25. Yu. A. Kuznetsov, V. V. Levitin, and A. R. Skovoroda. Continuation of stationary solutions to evolution problems in CONTENT. Technical Report AM-R961, Centrum voor Wiskunde en Informatica, Amsterdam, 1996.

26. Yu. A. Kuznetsov. *Elements of Applied Bifurcation Theory*, third edition. Applied Mathematical Sciences, 112. (Springer-Verlag, Berlin, 2004).

27. L. Liu, G. Moore, and R. D. Russell. Computation and continuation of homoclinic and heteroclinic orbits with arclength parameterization. *SIAM J. Sci. Comput.*, 18(1):69–93, 1997.

28. S. J. A. Malham and M. Oliver. Accelerating fronts in autocatalysis. *Proc. Royal Soc. A*, 456(1999):1609–1624, 2000.

29. J. E. Marsden and T. S. Ratiu. *Introduction to Mechanics and Symmetry*, volume 17 of Texts in Applied Mathematics. (Springer-Verlag, second edition, 1999).

30. G. Moore. Computation and parametrization of periodic and connecting orbits. *IMA J. Numer. Anal.*, 15(2):245–263, 1995.

31. G. Moore. Laguerre approximation of stable manifolds with application to connecting orbits. *Math. Comp.*, 73(245):211–242, 2004.
32. P. J. Olver. *Applications of Lie Groups to Differential Equations*, second edition. Graduate Texts in Mathematics, 107. (Springer-Verlag, Berlin, 1993).
33. C. W. Rowley, I. G. Kevrekidis, J. E. Marsden, and K. Lust. Reduction and reconstruction for self-similar dynamical systems. *Nonlinearity*, 16(4):1257–1275, 2003.
34. B. Sandstede. Convergence estimates for the numerical approximation of homoclinic solutions. *IMA J. Numer. Anal.*, 17(3):437–462, 1997.
35. B. Sandstede and A. Scheel. Absolute and convective instabilities of waves on unbounded and large bounded domains. *Physica D*, 145(3-4):233–277, 2000.
36. B. Sandstede and A. Scheel. Absolute versus convective instability of spiral waves. *Phys. Rev. E*, 62(6):7708–7714, 2000.
37. B. Sandstede and A. Scheel. On the structure of spectra of modulated traveling waves. *Math. Nachr.*, 232:39–93, 2001.
38. B. Sandstede and A. Scheel. Curvature effects on spiral spectra: Generation of point eigenvalues near branch points. *Phys. Rev. E*, 73(1):016217, 8, 2006.
39. S. Selle. *Approximation von relativen Gleichgewichten auf endlichen Intervallen.* Master's thesis, (Dept. of Mathematics, Bielefeld University, 2006).
40. V. Thümmler. *Numerische Stabilitätsindikatoren für wandernde Wellen.* Master's thesis, (Dept. of Mathematics, Bielefeld University, 1998).
41. V. Thümmler. *Numerical Analysis of the Method of Freezing Traveling Waves.* PhD thesis, (Dept. of Mathematics, Bielefeld University, 2005).
42. V. Thümmler. Asymptotic stability of frozen relative equilibria. Preprint no. 06-30 of the CRC 701, Bielefeld University, 2006.
43. V. Thümmler. Numerical approximation of relative equilibria for equivariant PDEs. Preprint no. 05-017 of the CRC 701, Bielefeld University, 2006.
44. A. I. Volpert, V. A. Volpert, and V. A. Volpert. *Traveling Wave Solutions of Parabolic Systems.* Translations of mathematical monographs, 140. (American Mathematical Society, 1994).
45. C. Wulff and A. Schebesch. SYMPERCON (continuation for symmetric periodic orbits). Technical report, University of Surrey, Department of Mathematics and Statistics, 2006.
46. C. Wulff. Transitions from relative equilibria to relative periodic orbits. *Doc. Math.*, 5:227–274, 2000.

11

Numerical Computation of Coherent Structures

Alan R Champneys[1] and Björn Sandstede[2]

[1] Department of Engineering Mathematics, University of Bristol, United Kingdom
[2] Department of Mathematics, University of Surrey, United Kingdom

In many applications one is interested in finding solutions to nonlinear evolution equations with a particular spatial and temporal structure. For instance, solitons in optical fibers and wave guides or buckling modes of long structures can be interpreted as localized traveling or standing waves of an appropriate underlying partial differential equation (PDE) posed on an unbounded domain. Spiral waves or other defects in oscillatory media are time-periodic waves with an asymptotic spatially periodic structure. All of these examples are referred to as *coherent structures*. They represent relative equilibria, that is, their temporal evolution is determined by a symmetry of the underlying PDE: namely, translational symmetry for traveling waves, rotational symmetry for spiral waves, and phase symmetry for oscillatory structures.

Given the complexity of typical PDE models, these nonlinear waves are in general accessible only through numerical computations. One possible approach is via direct simulation which is, however, expensive and fails to capture solutions that are either unstable or may have a small basin of attraction. Simulation therefore often fails to provide valuable information on how branches of solutions are organized in parameter space.

In this chapter, we give an overview of boundary value problem formulations for coherent structures which provide a robust and less expensive alternative to simulation. Moreover, setting up well-posed boundary value problems allows us to continue solutions in parameter space, investigate their spectral stability directly, and continue branches of solutions efficiently as parameters vary.

In the next section we outline how PDEs can be supplemented by phase conditions that allow us to compute nonlinear waves as regular zeros of the resulting nonlinear system. In the remaining sections, we treat different kinds of coherent structures, namely traveling waves, time-periodic structures, and planar localized patterns. In each case we explain how to set up a well-posed boundary value problem and illustrate the theory with the results of an example computation.

11.1 Symmetries and Phase Conditions

It is desirable to formulate the problem of computing nonlinear waves in such a way that solutions correspond to regular zeros of an appropriate smooth function. Indeed, Newton's method is then applicable and gives a robust and efficient way of computing zeros. When computing coherent structures, solutions are often not unique due to the presence of translational or other symmetries in the underlying partial differential equation. In other words, symmetries may give rise to families of solutions. To make Newton's method work, we need to single out one particular solution. This can be done by adding extra functionals to the problem and identifying additional parameters that compensate for the added functional to make the overall problem invertible; see also Chaps. 9 and 10. In this section, we recall how the extra functionals, which we refer to as phase conditions, can be constructed and the additional parameters be identified in systems with symmetries.

Thus, consider the equation

$$u_t = f(u), \qquad u \in \mathcal{X}, \tag{11.1}$$

where $\mathcal{X}$ is a Banach space. For ease of exposition, we assume that f maps $\mathcal{X}$ into itself and is smooth; we remark that the first property fails if (11.1) represents a partial differential equation, but the results that follow actually hold for PDEs nevertheless; see [39].

The key assumption is that (11.1) is equivariant under the action of a finite-dimensional Lie group G on $\mathcal{X}$ so that

$$g f(u) = f(gu), \qquad \forall g \in G, \ \forall u \in \mathcal{X}.$$

We denote by

$$\exp : \mathrm{alg}(G) \longrightarrow G, \quad \xi \longmapsto \exp(\xi)$$

the exponential map of the Lie algebra $\mathrm{alg}(G)$ of G into G.

Example 1. Take $\mathcal{X} = C^1(\mathbb{R}, \mathbb{R}^n)$ and consider the action of $G = \mathbb{R}$ defined by $g : \mathcal{X} \to \mathcal{X}$, $u(\cdot) \mapsto u(\cdot + g)$ for all $g \in G = \mathbb{R}$; then $\exp = \mathrm{id}$.

Particularly interesting solutions $u(t)$ of (11.1) are relative equilibria which are equilibria when the group action is factored out. More precisely, u_* is a relative equilibrium if there exists $\xi_* \in \mathrm{alg}(G)$ so that

$$u(t) = \exp(\xi_* t) u_* \qquad \forall t \in \mathbb{R}.$$

Note that $u(t) \in G u_*$ for all t. We emphasize that u_* satisfies

$$\xi_* u_* = f(u_*), \tag{11.2}$$

where we identify an element $\xi \in \mathrm{alg}(G)$ with the generator

$$\xi = \frac{\mathrm{d}}{\mathrm{d}t}\exp(\xi t)\Big|_{t=0} : \mathcal{X} \longrightarrow \mathcal{X}$$

of the one-parameter group $\exp(\xi t)$.

Example 2 (Example 1 continued). Relative equilibria of the group action $g : \mathcal{X} \to \mathcal{X}$, $u(\cdot) \mapsto u(\cdot + g)$ for $g \in G = \mathbb{R}$ on $\mathcal{X} = C^1(\mathbb{R}, \mathbb{R}^n)$ are of the form $u(x,t) = u_*(x + c_0 t)$ for $c_0 \in \mathbb{R}$. Therefore, they are traveling waves and we have

$$\xi_* u_* := \frac{\mathrm{d}}{\mathrm{d}t}\exp(c_0 t)u_*\Big|_{t=0} = \frac{\mathrm{d}}{\mathrm{d}t}u_*(\cdot + c_0 t)\Big|_{t=0} = c_0 u'_*(\cdot + c_0 t)\Big|_{t=0} = c_0 u'_*,$$

and (11.2) becomes the traveling wave equation $c_0 u'_* = f(u_*)$.

If u_* is a relative equilibrium then the entire group orbit Gu_* consists of relative equilibria since

$$gu(t) = g\exp(\xi_* t)g^{-1}gu_* = \exp(\mathrm{Ad}_g\xi_* t)gu_*$$

is again a relative equilibrium, where

$$\mathrm{Ad}_g\xi := \frac{\mathrm{d}}{\mathrm{d}t}g\exp(\xi t)g^{-1}\Big|_{t=0} \in \mathrm{alg}(G).$$

If G is Abelian, then Ad_g is the identity, and the velocity ξ_* is the same for all relative equilibria in the group orbit Gu_*.

Example 3 (Example 1 continued). With $u(t) = u_*(\cdot + c_0 t)$, we find that $u_*(\cdot + g + c_0 t)$ is a traveling wave for each $g \in G = \mathbb{R}$.

Thus, to compute or continue relative equilibria, we need to solve

$$f(u) - \xi u = 0$$

for $(u, \xi) \in \mathcal{X} \times \mathrm{alg}(G)$ and expect to find a family $\{(gu_*, \mathrm{Ad}_g\xi_*); \ g \in G\}$ of solutions. To find (u, ξ) as a regular zero of an appropriate function, we need to add phase conditions that single out a relative equilibrium (u_*, ξ_*) among the family of solutions.

For simplicity, we assume from now on that the Lie group G is Abelian. The family of relative equilibria is then given by (gu_*, ξ_*) for $g \in G$, and we have $f(gu_*) - \xi_* gu_* = 0$ for all $g \in G$. Choosing $g = \exp(\xi t)$ and taking the derivative with respect to t at $t = 0$, we obtain

$$f_u(u_*)\xi u_* - \xi_* \xi u_* = 0$$

for all $\xi \in \mathrm{alg}(G)$. Thus, if we assume that the isotropy group $\{g \in G; \ gu_* = u_*\}$ is discrete, we can conclude that the linearization $f_u(u_*) - \xi_*$ of (11.2) has a null space of dimension at least $\dim G$ with elements ξu_* for $\xi \in \mathrm{alg}(G)$.

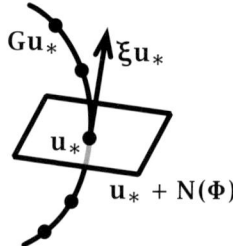

Fig. 11.1. The translated null space $u_* + N(\Phi)$ of an admissible phase condition is transverse to the group orbit Gu_* at the relative equilibrium u_*. Imposing the condition $\Phi(u - u_*) = 0$ restricts u to $u_* + N(\Phi)$ and, therefore, singles out the relative equilibrium u_*.

Hypothesis 1 *Assume that G is Abelian with $\dim G = m$, and (u_*, ξ_*) is a relative equilibrium such that $f_u(u_*) - \xi_*$ is a Fredholm operator with index zero. Furthermore, assume that $f_u(u_*) - \xi_*$ has an eigenvalue $\lambda = 0$ with geometric and algebraic multiplicity m and that its null space is given by $\{\xi u_*\}_{\xi \in \mathrm{alg}(G)}$.*

We choose linear phase conditions of the form

$$\Phi : \mathcal{X} \longrightarrow \mathbb{R}^m, \quad u \longmapsto \Phi u$$

and say that Φ is admissible if the linear operator

$$\mathrm{alg}(G) \longrightarrow \mathbb{R}^m, \quad \xi \longmapsto \Phi \xi u_*$$

is invertible. The following theorem is illustrated in Fig. 11.1.

Theorem 1. *Assume that Hypothesis 1 holds and that Φ is an admissible phase condition. Then the relative equilibrium (u_*, ξ_*) is a regular zero of the map*

$$\mathcal{F} : \mathcal{X} \times \mathrm{alg}(G) \longrightarrow \mathcal{X} \times \mathbb{R}^m, \quad (u, \xi) \longmapsto (f(u) - \xi u, \Phi(u - u_*)).$$

Proof. Clearly, $\mathcal{F}(u_*, \xi_*) = 0$. The linearization of $\mathcal{F}$ about (u_*, ξ_*) is given by

$$\mathcal{L} : (u, \xi) \longmapsto ([f_u(u_*) - \xi_*]u - \xi u_*, \Phi u).$$

The bordering lemma [3, Lemma 2.3] shows that this operator is Fredholm with index zero and, therefore, it suffices to show that its null space is trivial. Since the geometric and algebraic multiplicities of the eigenvalue $\lambda = 0$ of $f_u(u_*) - \xi_*$ are equal to each other, there is a spectral decomposition

$$\mathcal{X} = \tilde{\mathcal{X}} \oplus \{\xi u_*; \ \xi \in \mathrm{alg}(G)\}$$

so that $f_u(u_*) - \xi_*$ is an isomorphism from $\tilde{\mathcal{X}}$ into itself. Writing $u = \tilde{u} + \zeta u_*$, we find

$$\mathcal{L}(u, \xi) = \mathcal{L}(\tilde{u} + \zeta u_*, \xi) = ([f_u(u_*) - \xi_*]\tilde{u} - \xi u_*, \Phi(\tilde{u} + \zeta u_*)).$$

Using again the decomposition of $\mathcal{X}$, we see that $\tilde{u} = \xi = 0$, and the remaining equation $\Phi \zeta u_* = 0$ finally gives $\zeta = 0$ since Φ is admissible. $\square$

Remark 1. Suppose that $\mathcal{X}$ is a Hilbert space with scalar product $\langle \cdot, \cdot \rangle$. Choose a basis $\{\xi_j\}_{j=1,\dots,m}$ of $\mathrm{alg}(G)$. Then the phase condition

$$\Phi : \mathcal{X} \longrightarrow \mathbb{R}^m, \qquad u \longmapsto (\langle \xi_j u_*, u \rangle)_{j=1,\dots,m}$$

is admissible.

Example 4 (Example 1 continued). We wish to find a functional $\Phi : C^1(\mathbb{R}, \mathbb{R}^n) \to \mathbb{R}$ such that $\Phi u_*' \neq 0$. The space $C^1(\mathbb{R}, \mathbb{R}^n)$ is not a Hilbert space, but if u_*' decays exponentially then

$$\Phi u = \int_{\mathbb{R}} \langle u_*'(x), u(x) \rangle \, dx$$

is an admissible phase condition.

11.2 Traveling Waves: Pulses and Fronts

We consider systems of ordinary differential equations (ODEs) of the form

$$u_x = f(u; \alpha), \quad x \in \mathbb{R}, \quad u \in \mathbb{R}^n, \quad f \in C^1(\mathbb{R}^n \times \mathbb{R}^p, \mathbb{R}^n), \quad \alpha \in \mathbb{R}^p. \quad (11.3)$$

Such problems typically arise as steady state or traveling wave reductions of PDEs of the form

$$U_t = D U_{XX} + F(U; \beta), \qquad U \in \mathbb{R}^N.$$

Here, for a traveling wave $U(X, t) = U(X - ct)$ of speed c, we have $x = X - ct$, $u = (U, V) = (U, U_X)$, $\alpha = (c, \beta)$, and $f(u; \alpha) = (V, -D^{-1}[F(U; \beta) + cU])$.

Front solutions $q(x)$ of (11.3) at parameter values α_* correspond to heteroclinic connections between equilibria, and solve the following boundary value problem on the real line

$$\begin{cases} q_x = f(q(x); \alpha_*), \\ q(x) \to u_0 \text{ as } x \to -\infty, \quad f(u_0; \alpha_*) = 0, \\ q(x) \to u_1 \text{ as } x \to \infty, \quad f(u_1; \alpha_*) = 0. \end{cases} \quad (11.4)$$

In the special case that $u_0 = u_1$, the pulse $q(x)$ corresponds to a homoclinic orbit of the ODE.

To set up a well-posed numerical boundary value problem for solutions of (11.4), we need to deal with the infinite interval $x \in (-\infty, \infty)$ in some way, to replace the asymptotic boundary conditions by some regular condition, and to deal with the translation symmetry $q(x) \to q(x+g)$. Several approaches for performing this task have been proposed, including shooting methods [25, 37], computation of large-period periodic orbits [40], and mapping the infinite domain to a finite interval; see, e.g., [26]. We focus here on methods, owing to Beyn [3] and Friedman and Doedel [15, 18], that involve truncation to a finite domain $[-L, L]$ and use projection boundary conditions.

Suppose that the matrices $f_u(u_j; \alpha)$ have n_j^s eigenvalues (counting multiplicities) with negative real part, n_j^c eigenvalues with zero real part, and n_j^u eigenvalues with positive real part, so that $n_j^s + n_j^c + n_j^u = n$ for $j = 1, 2$. To set up a numerical boundary value problem we then look for a solution $u(x)$ with $x \in [-L, L]$ subject to the boundary conditions

$$P^s(u_0; \alpha)(u(-L) - u_0) = 0, \qquad P^u(u_1; \alpha)(u(L) - u_1) = 0. \qquad (11.5)$$

Here $P^s(u_0, \alpha)$ is an $n_0^s \times n$ matrix whose rows form a basis for the stable eigenspace of $f_u^T(u_0; \alpha)$. Accordingly, $P^u(u_1; \alpha)$ is an $n_1^u \times n$ matrix, whose rows form a basis for the unstable eigenspace of $f_u^T(u_1; \alpha)$. The boundary conditions (11.5) thus place the boundary points in the center-unstable eigenspace of $f_u(u_0; \alpha)$ and the center-stable eigenspace of $f_u(u_1; \alpha)$, respectively.

If each of the equilibria u_j is hyperbolic, then $n_j^c = 0$, and by simply counting dimensions of the corresponding stable and unstable manifolds under generic hypotheses on transversality, we find that the codimension of the parameter set in which there exists a heteroclinic connection is equal to $m = n - (n_0^u + n_1^s) + 1$. If this number is negative then additional internal boundary conditions need to be set up to choose a member of the $|m|$-dimensional continuum of connections. From now on, we assume that m is non-negative. Hence, we need to free m parameters, say $(\alpha_1, \ldots, \alpha_m)$, to find a regular solution to the boundary value problem (11.3) and (11.5). However, (11.5) represents only

$$n_0^s + n_1^u = (n - n_0^u) + (n - n_1^s) = n + m$$

conditions for the $n+m+1$ unknowns $u \in \mathbb{R}^n$ and $(\alpha_1, \ldots, \alpha_m)$, so that we need an extra condition. At the same time, we see that the original problem respects a translation symmetry since if $u(x)$ solves (11.4), then so does $u(x + g)$ for any $g \in \mathbb{R}$. So, according to the theory of Sect. 11.1, we need a phase condition that is transverse to the generator $\xi u = u_x$. Such a condition can be posed at a specific point in the domain, for example, at the left-hand boundary

$$\Phi(u) := \langle v, u(-L) \rangle = 0 \quad \text{where } v \text{ is such that} \quad \langle v, u_x(-L) \rangle \neq 0. \qquad (11.6)$$

This is equivalent to choosing the left-hand boundary point to be in the Poincaré section $\Sigma = \{u \in \mathbb{R}^n; \langle v, u \rangle = 0\}$, with the inner product condition implying that the corresponding trajectory of (11.3) crosses Σ transversally. However, for practical implementation in continuation software, one cannot guarantee a priori that all solutions along a branch will remain transverse to Σ for all parameter values α. Therefore, it is better to take as phase condition the integral condition

$$\Phi(u) := \int_{-L}^{L} \langle u'_*(x), u(x) \rangle \, \mathrm{d}x \tag{11.7}$$

where u_* is a reference solution (for example the previously computed solution along a continuation branch); see Example 4 and also Chaps. 1, 9 and 10.

Thus we solve the two-point boundary value problem

$$\begin{pmatrix} u_x = f(u; \alpha) \\ P^s(u_0; \alpha)(u(-L) - u_0) \\ P^u(u_1; \alpha)(u(L) - u_1) \\ \int_{-L}^{L} \langle u'_*(x), u(x) - u_*(x) \rangle \, \mathrm{d}x \end{pmatrix} = 0. \tag{11.8}$$

Under certain nondegeneracy conditions, the existence of a homoclinic solution to the original problem (11.4) on the infinite interval implies the existence of a unique solution to the truncated problem (11.8) provided L is sufficiently large; see [3, 18, 38]. Furthermore, the error involved in the truncation scales exponentially with L for both the parameter and the solution.

There have been a number of implementations of the above algorithm for computing homoclinic and heteroclinic connections. For example, the routines HOMCONT [7] are part of the package AUTO [14]. Some extensions to this basic algorithm allow one to deal with special situations as we now outline:

Reversibility

Many ODEs arise from reductions of even-order PDEs that contain only even spatial derivatives. In this case, the ODE can be written as an even-dimensional reversible system with respect to an involution $\mathcal{R}$ so that

$$f(\mathcal{R}u; \alpha) = -\mathcal{R}f(u; \alpha),$$

with $\mathcal{R}^2 = \mathrm{id}$ and $\dim \mathrm{Fix}(\mathcal{R}) = n/2$. Standing symmetric pulses, which are invariant under $x \mapsto -x$, correspond to reversible homoclinic orbits q to a reversible hyperbolic equilibrium u_0, and are of codimension zero in parameter space [6, 12]. For such an orbit we can replace the right-hand boundary condition of (11.5) by the condition $u(0) \in \mathrm{Fix}(\mathcal{R})$. Note that, provided the solution $q(x)$ intersects $\mathrm{Fix}(\mathcal{R})$ transversally, this boundary condition also breaks the translation invariance of the system. Therefore, there is no need for a phase condition. This is good since we require one condition less, which agrees with having one free parameter less compared to the non-reversible case. This approach has been used extensively to compute so-called *gap solitons* in nonlinear optics; see e.g. [11].

Hamiltonian Systems

Muñoz-Almaraz et al. [33] consider how to compute periodic orbits in Hamiltonian systems; see also Chap. 9. Their approach naturally extends to heteroclinic connections under generic hypotheses on the connection. Hamiltonian systems conserve a first integral, the Hamiltonian itself. Noether's theorem says that any other independent (in the sense of a Poisson bracket) conserved quantity $\mathcal{C}(u)$ corresponds to a symmetry of the underlying system. Thus each conserved quantity (including the Hamiltonian) reduces the codimension of the solution one is trying to compute. Thus we need to remove one boundary condition for each additional conserved integral, or introduce an extra artificial parameter. The approach used in [33], for a system with k independent conserved smooth scalar functionals $\mathcal{C}_j : \mathbb{R}^n \times \mathbb{R}^p \to \mathbb{R}$ for $j = 1, \ldots, k$, is to introduce artificial parameters $\tilde{\alpha} \in \mathbb{R}^k$ and solve

$$u_x = f(u; \alpha) + \sum_{j=1}^{k} \tilde{\alpha}_j \partial_u \mathcal{C}_j(u; \alpha). \tag{11.9}$$

It has been proved rigorously that this leads to well-posed boundary value problems under appropriate generic hypotheses. We record that $\tilde{\alpha} = 0$ for the true solution to the boundary value problem and that $\partial_u \mathcal{C}_j(u; \alpha)$ is precisely orthogonal to the level set of $\mathcal{C}_j$, which is in keeping with the theory of Sect. 11.1 above. Note that we still need to retain exactly the same projection boundary conditions (11.5) and (11.7), but now the free parameters are replaced with the artificial parameters $\tilde{\alpha} \in \mathbb{R}^k$.

Nonhyperbolic Equilibria

If $u_0 = u_1$ is a nonhyperbolic equilibrium, then one has to make special considerations to determine the codimension of the homoclinic connection and, hence, the number of free parameters that is needed. We consider two cases separately.

Our first case of interest is that when $u_0 = u_1$ and $n^c = 2$ with a pair of purely imaginary eigenvalues in the spectrum. In Hamiltonian or reversible systems such a linearization is generic, and such points are called saddle-centers if $n^u = n^s > 0$: in this situation, it is then possible to form homoclinic connections with exponentially decaying tails that lie in the stable and unstable manifolds of the equilibrium. Such solutions in reversible systems correspond to so-called *embedded solitons* of nonlinear wave problems [46]. To compute these objects, we must replace the boundary condition (11.5) with

$$P^{cs}(u_0; \alpha)(u(-L) - u_0) = 0, \qquad u(0) \in \mathrm{Fix}(\mathcal{R}),$$

where $P^{cs}(u_0; \alpha)$ is the spectral projection onto the center-stable eigenspace of u_0 whose rows, therefore, form a basis for the center-stable eigenspace of $f_u^T(u_0; \alpha)$.

The second case we consider is that $u_0 = u_1$ is a saddle node. Then pulse solutions can exist with algebraic decay as $x \to \pm\infty$. Besides the condition that u_0 is a saddle-node equilibrium, which can be taken to be $\det f_u(u_0; \alpha) = 0$, we provide the $n - 1$ boundary conditions (11.5) plus the phase condition (11.7). In particular, these solutions have codimension one in the parameter space; see, e.g., [24, §7.1.2].

Saddle-Homoclinic Orbits and Their Orientation

Homoclinic bifurcations of various kinds can be treated by freeing an additional parameter α_j and adding appropriate conditions whose regular zeros define the extra degeneracy along a branch of homoclinic solutions; see [7] for a number of examples.

We discuss here a particular kind of codimension-two bifurcation that is known to lie at the heart of the creation of multi-pulse solutions from a branch of one-pulses, namely the so-called *inclination-flip* homoclinic bifurcation. It is caused by a degeneracy in the global twistedness of the stable and unstable manifolds $W^{s,u}(u_0)$ of a saddle u_0 around its homoclinic orbit $q(x)$; see e.g. [36] and references therein. At each point $q(x)$ along the homoclinic orbit, the normal bundle

$$Z(x) = [T_{q(x)}W^s(u_0) + T_{q(x)}W^u(u_0)]^{\perp}$$

of the tangent spaces of stable and unstable manifolds can be defined. Generically, these tangent spaces intersect along the one-dimensional subspace spanned by $q'(x)$ and, therefore, $Z(x)$ is one dimensional for each x. The collection of $Z(x)$ generically forms a one-dimensional bundle along the set $\mathbb{S}^1 \cong \{u_0\} \cup \{q(x); \ x \in \mathbb{R}\}$ which can be orientable (homeomorphic to a cylinder) or non-orientable (homeomorphic to a Möbius band). Inclination-flips occur at codimension-two points in parameter space where this bundle changes from orientable to non-orientable (or vice versa). An efficient way to detect inclination-flip points is to compute normal vectors $w(x) \in Z(x)$ as solutions to the adjoint variational problem

$$w_x = -f_u^T(q(x); \alpha)w, \qquad w \to 0 \text{ as } x \to \pm\infty. \tag{11.10}$$

If we truncate, then we can approximate $w(x)$ on the same finite interval $[-L, L]$ as $q(x)$ by defining boundary conditions

$$Q^u(u_0; \alpha)w(-L) = 0, \qquad Q^s(u_1; \alpha)w(L) = 0, \tag{11.11}$$

where $Q^u(u_0; \alpha)$ is an $n_0^u \times n$ matrix whose rows form a basis for the unstable eigenspace of $f_u(u_0; \alpha)$. Similarly, $Q^s(u_1; \alpha)$ is an $n_1^s \times n$ matrix, whose rows form a basis for the stable eigenspace of $f_u(u_1; \alpha)$. Note that a solution to (11.10) is only defined up to scalar multiplication $w \to gw$. Therefore, we need a phase condition that fixes the amplitude which we may take to be

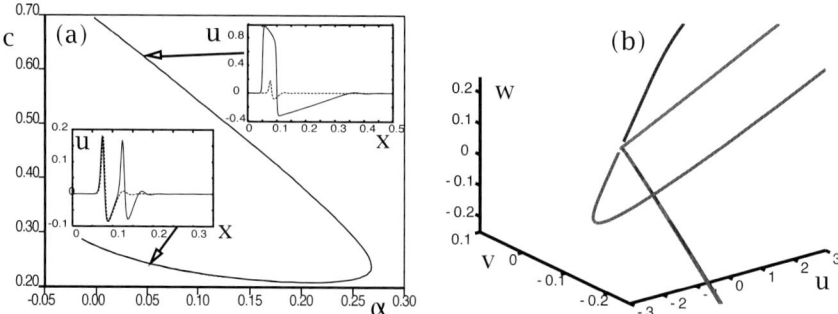

Fig. 11.2. Panel (a) shows a continuation of the branch of single-pulses to the FitzHugh-Nagumo system (11.15) for $\gamma = 0$ and $\varepsilon = 0.025$; the insets show the fast and the slow pulse for $\alpha = 0.02$ and a two-pulse orbit that is close to the slow pulse. Panel (b) is an enlargement near the origin of the solution of the adjoint variational equation around the pulse at the inclination-flip point and at two points that are parameter perturbations of $O(10^{-3})$ along the branch of homoclinic solutions.

$$\Phi(w) = \int_{-L}^{L} \langle w_*(x), w(x) - w_*(x)\rangle \, \mathrm{d}x = 0, \tag{11.12}$$

where $w_*(x)$ is a reference solution (for instance, the solution computed at a previous continuation step). Note though that (11.12) and (11.11) provides $n+1$ boundary conditions, so we need to free an additional parameter. In this case, it is convenient to include a parameter α_0 that unfolds the orthogonality of w and u_x:

$$w_x = -f_u^T(u(x); \alpha)w + \alpha_0 f(u(x); \alpha), \tag{11.13}$$

with the true solution to (11.13), (11.11), and (11.12) satisfying $\alpha_0 = 0$.

We finish with an example, namely, the FitzHugh-Nagumo system; see also [14] for a demo on switching branches to multi-pulse orbits in this system, during continuation of a single pulse, and also for references. It is given by the PDE

$$u_t = u_{yy} - f_\alpha(u) - w, \qquad f_\alpha(u) = u(u - \alpha)(u - 1), \tag{11.14}$$
$$w_t = \varepsilon(u - \gamma w),$$

for the functions $(u, w)(y, t)$. Looking for traveling waves by letting $x = y + ct$, we obtain the ODE system

$$u_x = v,$$
$$v_x = cv + f_\alpha(u) + w, \tag{11.15}$$
$$w_x = \frac{\varepsilon}{c}(u - \gamma w).$$

Figure 11.2 shows the result of numerical continuation of the single-pulse solution in the (α, c)-plane for $\gamma = 0$ and $\varepsilon = 0.025$. Note that for most values of α there are either none or two wave speeds c for which there exists a traveling pulse. The faster wave is the one that is generally stable. Note that both the upper and the lower branch appear to end 'in mid air' as α is reduced. In fact, the branches fold back on themselves and the pulses return as their own two-pulse orbits at very nearby parameter values to those at which the one-pulse exists; see the dashed curve in the inset to Fig. 11.2(a).

For α sufficiently large ($\alpha > 0.1318124$ on the lower branch and $\alpha > 0.107652$ on the upper branch) the tail of the pulse has monotonic decay because the origin of the ODE system (11.15) is a real saddle. In such circumstances, the orientability of the normal bundle to the stable and unstable manifolds along the homoclinic orbit is well defined. Indeed we find that at $(\alpha, c) = (0.240314, 0.211443)$ there is an inclination flip with respect to the stable manifold; see Fig. 11.2(b). Theoretical and numerical work for inclination flips in the FitzHugh-Nagumo system can also be found in [23].

11.3 Traveling Waves with Spatially Periodic Asymptotics

Dieci and Rebaza [13] considered a general boundary value approach for computing heteroclinic connections between hyperbolic periodic orbits and from equilibria to periodic orbits for ODEs. These correspond to traveling waves whose profile becomes spatially periodic as $x \to \infty$ and/or $x \to -\infty$. For the special case where the periodic solutions in the two tails are the same (up to a phase shift), a popular approach for computing such generalized pulses is to study the problem in the setting of a Poincaré map and to compute homoclinic connections to hyperbolic fixed points of the map; see, for example, [1, 5, 21] and also the AUTO implementation HOMMAP due to Yagasaki [45]. Instead, we present here the approach from [13] (see also [22]) that relies on setting up a coupled boundary value problem for the periodic orbits and the homoclinic orbit.

Suppose a connection $q(x)$ is sought between two separate hyperbolic periodic solutions $p_j(x; \alpha)$ with periods L_j for $j = 0, 1$ which have n_j^s Floquet multipliers inside the unit circle and $n_j^u = n - n_j^s + 1$ outside. Note that $n_j^c = 1$ due to the presence of the trivial Floquet multiplier 1. We then seek a solution to the boundary value problem

$$\begin{pmatrix} u' - f(x; \alpha) \\ v_{0,1}' - f(v^{0,1}; \alpha) \\ v_{0,1}(0) - v_{0,1}(L_{0,1}) \\ P^{cs}(v_0; \alpha)(u(-L) - v_0(0)) \\ P^{cu}(v_1; \alpha)(u(L) - v_1(0)) \\ \Phi_{0,1}(v_{0,1} - v_{0,1}^*) \end{pmatrix} = 0, \qquad (11.16)$$

with solution (u, v_0, v_1) near (q, p_0, p_1). Here, the projection boundary condition $P^{cs}(v_0; \alpha)$ is an $(n_0^s + 1) \times n$ matrix whose columns are orthogonal

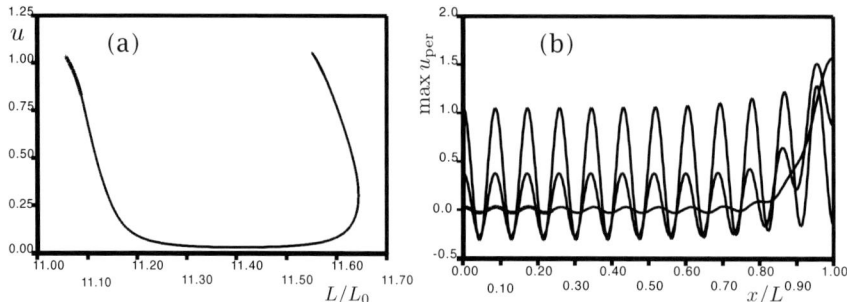

Fig. 11.3. Homoclinic orbits to periodic orbits of (11.17) computed on the half-interval $[-L, 0]$ (a), and their locus in the phase shift versus tail amplitude plane (b). From A.R. Champneys and G.J. Lord, Computation of homoclinic solutions to periodic orbits in a reduced water-wave problem, *Physica D*, 102 (1997) 101–124 © 1997 by Elsevier Science; reprinted with permission.

to the unstable eigenspace of the monodromy matrix associated with v_0 at $x = 0$. In other words, the rows of $P^{cs}(v_0; \alpha)$ are formed by the center-stable Floquet eigenspace of the adjoint variational problem $w_x = f_u^T(v_0(x); \alpha)w$ for $x \in [0, L_0]$. Note the importance of solving the *adjoint* problem in order that the projection represented by P^{cs} is orthogonal to the unstable Floquet eigenspace, rather than being along the center-stable Floquet directions (this detail was inadvertently omitted in [13]). Similarly, P^{cu} sets the component orthogonal to the stable Floquet eigenspace to zero. The phase conditions $\Phi_{0,1}$ are chosen to factor out the translation symmetry of the periodic orbits, and may be taken to be

$$\Phi_j(v) = \int_0^{L_j} \langle \partial_x v_j^*(x), v(x) \rangle \, dx, \qquad j = 1, 2,$$

where $v_j^*(x)$ are reference periodic solutions (we may again take v_j^* to be p_j computed at a previous point along a continuation branch). Note that there is no phase condition on the heteroclinic orbit $u(x)$ since the translation symmetry is broken by fixing the phase of the two periodic orbits. Moreover, the intervals L_0, L_1 and L should all be taken as unknowns. This is because L_0 and L_1 are the a priori unknown periods of the periodic orbits p_0 and p_1, and L must be taken to be free in order to find the unknown *phase shift* between the periodic orbits p_0 and p_1.

Practical implementation details of the boundary value problem (11.16) are discussed in [13], which also contains a convergence proof based on the methods of [4]. In particular, care has to be taken in order to evaluate the matrices $P^{cs}(v_0; \alpha)$ and $P^{cu}(v_1; \alpha)$ which themselves depend on the solutions $v_j(x)$.

We illustrate the approach outlined above by considering an application to a fourth-order water-wave problem. Champneys and Lord [9] consider the system

$$\varepsilon^2 u^{iv}(x) + u''(x) - u(x) + u(x)^2 = 0, \quad x \in \mathbb{R}, \tag{11.17}$$

where $u(x)$ represents the amplitude of water waves in the presence of surface tension for a model equation that is only valid for small-amplitude waves close to the critical point of Bond number $1/3$ and Froude number 1. Equation (11.17) can easily be shown to be equivalent to a fourth-order Hamiltonian system, with total energy given by

$$\mathcal{H} = \varepsilon^2 u' u''' + \frac{[u']^2}{2} - \frac{\varepsilon^2}{2}[u'']^2 - \frac{u^2}{2} + \frac{u^3}{3}.$$

The problem is also reversible with respect to the involution

$$\mathcal{R}: \quad (u, u', u'', u''') \longmapsto (u, -u', u'', -u''').$$

Figure 11.3 shows the results of computations of reversible homoclinic orbits to reversible periodic orbits, where the right-hand boundary conditions in (11.16) are replaced with $u(0) \in \mathrm{Fix}(\mathcal{R})$ and $v_0(L_0/2) \in \mathrm{Fix}(\mathcal{R})$.

Similar results were found for the full water-wave problem in the presence of surface tension using boundary-integral methods [10]; see Fig. 11.4. These careful computations helped to settle a conjecture (in the negative) of whether true solitary waves of elevation exist in the classical water wave problem; see also [41].

11.4 Moving Discrete Breathers in $\mathbb{R} \times \mathbb{Z}$

The discrete nonlinear Schrödinger equation with a saturable nonlinear term, representing the effects of a photorefractive crystal lattice, may be written in the form

$$i\dot{u}_n(t) = -\varepsilon[\Delta_2 u(t)]_n + \frac{u_n(t)}{1 + |u_n(t)|^2}, \quad n \in \mathbb{Z}, \tag{11.18}$$

where Δ_2 is the standard second-order spatial difference operator $[\Delta_2 u]_n = u_{n+1} - 2u_n + u_{n-1}$, and ε represents the dimensionless coupling strength between each lattice site. Localized time-periodic solutions to such equations have been given the name *discrete breathers*.

Standing discrete breathers are relatively easy to compute since, after making a phase transformation $u_n(t) = e^{i\Lambda t} U_n$, one has to solve for homoclinic orbits of the discrete map

$$i\Lambda U_n = -\varepsilon[\Delta_2 U]_n + \frac{U_n}{1 + |U_n|^2}, \quad n \in \mathbb{Z},$$

which are structurally stable (and indeed many of which are stable and may be found as solutions of the initial value problem (11.18)). However, if one

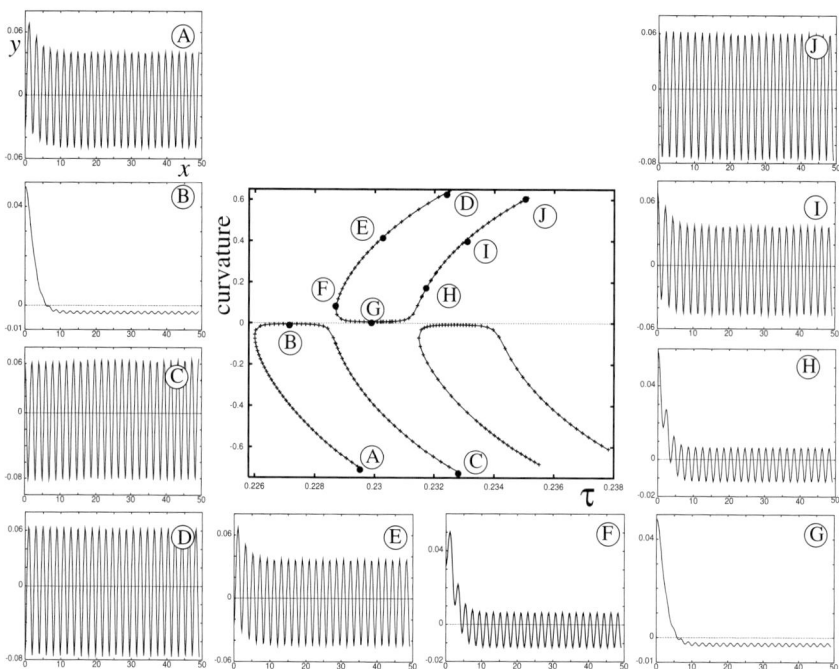

Fig. 11.4. Equivalent diagram to Fig. 11.3 for the Euler-equation formulation of water waves in the presence of surface tension plotted as Bond number τ against a signed measure of the tail amplitude for fixed Froude number $F = 1.002$ and for fixed domain size $L = 98.33$. From A.R. Champneys, J.M. Vanden-Broeck and G.J. Lord, Do true elevation gravity-capillary solitary waves exist? A numerical investigation, *J. Fluid Mech.*, 454 (2002) 403–417 © 2002 by Cambridge University Press; reprinted with permission.

tries to make such structures move with wave speed $c \neq 0$, they typically shed radiation and eventually stop or cease to survive as coherent structures.

Moving discrete breathers can be sought by making the substitution

$$u_n(t) = \psi(z)\mathrm{e}^{-\mathrm{i}\Lambda t} \qquad \text{with } z = n - ct \in \mathbb{R},$$

which gives the advanced-retarded equation

$$-\mathrm{i}c\psi'(z) = (2\varepsilon - \Lambda)\psi(z) - \varepsilon(\psi(z+1) + \psi(z-1)) + \frac{\psi(z)}{1 + |\psi(z)|^2}, \quad (11.19)$$

for which we seek homoclinic solutions $\psi(z) \to 0$ as $z \to \pm\infty$. Simple spectral analysis shows that the spectrum of the problem linearized about $\psi = 0$ is symmetric with respect to the imaginary axis, so that the dimension n^c of the center manifold is even; moreover, $n^c \geq 2$ for all $c, \varepsilon > 0$. Counting the codimension of the stable and unstable manifolds shows that the best we can hope for is that $n^c = 2$, and then homoclinic solutions to (11.19) should

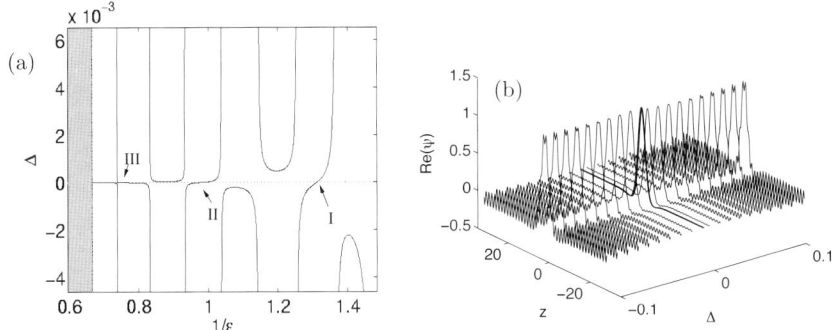

Fig. 11.5. Continuation of breathers on a periodic background for various values of $1/\varepsilon$ showing three zeros in δ for $c = 0.7$, $\Lambda = 0.5$, and $L = 60$ (a); the shaded region represents the spectral band where any embedded solitons would be of codimension two. Panel (b) shows solutions on the branch with a zero of δ near $\varepsilon = 1.02$. From T.R.O. Melvin, A.R. Champneys, P.G. Kevrekidis and J. Cuevas, Radiation-less traveling waves in saturable nonlinear Schrödinger lattices, *Phys. Rev. Lett.* 97 (2006) 124101 © 2006 by the American Physical Society; reprinted with permission.

exist along codimension-one lines in the parameter plane. An efficient way to compute such solutions is to seek solutions on a large interval $z \in [-L, L]$ with appropriate boundary conditions. Note that (11.19) is invariant under the reversing transformation

$$\mathcal{R}: \quad z \longmapsto -z, \quad (\operatorname{Re}\psi, \operatorname{Im}\psi) \longmapsto (-\operatorname{Re}\psi, \operatorname{Im}\psi).$$

We can find $\mathcal{R}$-symmetric periodic solutions by seeking solutions with $\psi(-L) \in \operatorname{Fix}(\mathcal{R})$, $\psi(0) \in \operatorname{Fix}(\mathcal{R})$. The use of these boundary conditions effectively fixes the phase symmetry $\phi \mapsto e^{\mathrm{i}g}\phi$ according to the theory of Sect. 11.1.

An efficient way to solve such boundary value problems is to use a pseudo-spectral method [16] by making the substitution

$$\psi(z) = \sum_{j=1}^{N} a_j \cos(\pi j z/L) + \mathrm{i}\, b_j \sin(\pi j z/L) \tag{11.20}$$

for the unknown coefficients (a_j, b_j). Substitution of (11.20) into (11.19) and evaluating at the collocation points $z_j = \frac{jL}{-2(N+1)}$ leads to a regular system of $2N$ algebraic equations that can be solved numerically.

In particular, we can use this method to compute periodic solutions of fixed large period that approximate *quasi-solitons* made up of an exponentially localized core ψ_{core} and a nonvanishing oscillatory background ψ_{tail}:

$$\psi(z) = \psi_{\mathrm{core}}(z) + \psi_{\mathrm{tail}}(z).$$

At a sufficient distance from the center of the breather ψ_{core} is zero due to its exponential localization and, therefore, we are left with the tail which, if

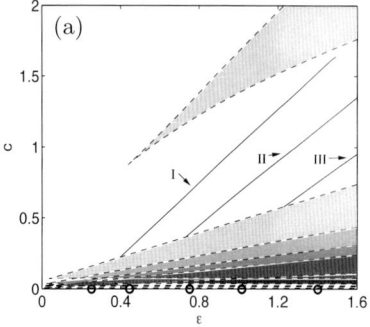
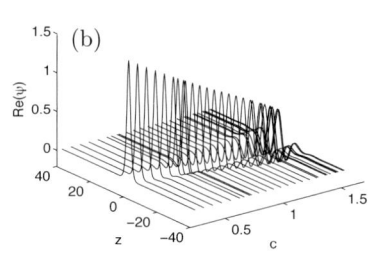

Fig. 11.6. Continuation of the three zeros of δ shown in Fig. 11.5, varying ε and c with $\Lambda = 0.5$ (a); the depth of shading represents the number of pairs $n^c/2$ of imaginary eigenvalues of the linearized problem in each parameter region: white $n^c = 2$, lightest gray $n^c = 4$, and so on; circles on the $\{c = 0\}$ axis indicate the transparent points at which the energy barrier for steady breathers vanishes. Panel (b) shows the real parts of solutions on the first branch, whose amplitude goes to zero as the upper grey wedge is approached. From T.R.O. Melvin, A.R. Champneys, P.G. Kevrekidis and J. Cuevas, Radiationless traveling waves in saturable nonlinear Schrödinger lattices, *Phys. Rev. Lett.* 97 (2006) 124101 © 2006 by the American Physical Society; reprinted with permission.

it is of small amplitude, is in the center manifold and hence approximately sinusoidal. Because of the way $\psi(z)$ has been approximated in (11.20) we know that the imaginary part of $\psi_{\text{tail}}(z)$ is odd around $z = 0$ and the quantity

$$\delta = \text{Im}\,\psi(L/2)$$

can be used as a signed measure of the amplitude of the tail. Regular zeros of δ can be followed in multiple parameters, and these give the traveling discrete breathers that we are seeking. These moving structures with zero tails were found to be remarkably stable as solutions to the initial value problem.

Figure 11.5 shows the results of computations of solutions with periodic tails for fixed c and Λ, but with ε allowed to vary. Note that the U- and N-shaped nature of the branches is qualitatively the same as in Fig. 11.4, except here we see the occasional S-shaped dislocation of the Us and Ns. These lead to values of c at which the measure δ of the tail amplitude undergoes a regular zero. Hence, we have found a truly localized solution that can be continued in two parameters. Three such zeros are found in Fig. 11.5, and they lead to the three branches labeled I, II and III in Fig. 11.6. Note that these branches terminate when they reach the boundary of the white region in Fig. 11.6(a), when $n^c = 2$ is no longer satisfied. More details can be found in [32].

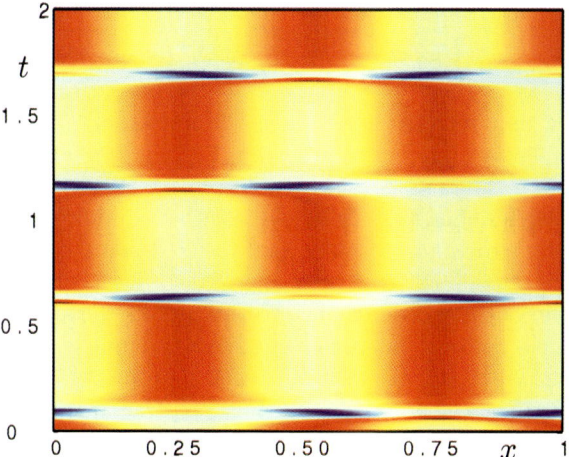

Fig. 11.7. A space-time contour plot of a typical heteroclinic cycle of (11.21) is shown for $\alpha = 70$, $\nu = 1$, and $\mu = -4$, where dark color corresponds to smaller and light color to larger values of the amplitude $|u(x,t)|$. From D.J.B. Lloyd, A.R. Champneys and R.E. Wilson, Robust heteroclinic cycles in the one-dimensional complex Ginzburg-Landau equation, *Physica D* 204 (2005) 240–268 © 2005 by Elsevier Science; reprinted with permission.

11.5 Computing Robust Heteroclinic Cycles

Heteroclinic cycles are formed of trajectories that connect equilibria or periodic orbits. This behavior is typically structurally unstable in generic dynamical systems. However, in systems with symmetry, robust connections between saddles can occur if each saddle is actually a sink within an invariant subspace generated by a discrete symmetry.

As an example, we consider the one-dimensional complex Ginzburg–Landau equation (CGL)

$$u_t = (1 + \mathrm{i}\nu)u_{xx} + \alpha u - (1 + \mathrm{i}\mu)|u|^2 u \tag{11.21}$$

on a periodic domain $x \in \mathbb{S}^1 \cong [0,1]/\sim$, where $u \in \mathbb{C}$ and $\nu, \mu, \alpha \in \mathbb{R}$. The CGL (11.21) is invariant under the action of the $\mathbb{S}^1$ phase symmetry

$$u(x) \longmapsto \mathrm{e}^{\mathrm{i}\phi}u(x)$$

and the O(2)-symmetry

$$u(x) \longmapsto u(-x), \qquad u(x) \longmapsto u(x - y)$$

for $y \in \mathbb{S}^1 \cong [0,1]/\sim$.

A typical heteroclinic cycle of (11.21), obtained by a direct integration, is shown in Fig. 11.7. We observe long dormant behavior followed by bursts of spatio-temporal activity. The dormant states are relative equilibria of (11.21) of the form

$$u(x,t) = U(x)e^{i\omega t} \qquad (11.22)$$

that are reflection symmetric about $x = 0.5$ and satisfy $U(x + 1/2) = U(x)$ for all x. The relative equilibria observed in Fig. 11.7 are all related by a 1/4-spatial translation. The bursts are characterized by initial symmetry breaking with spatial wavenumber $k = 2\pi$ of the relative equilibria (11.22), and the corresponding heteroclinic orbits lie in the fixed-point space generated by reflection across $x = 0$, $x = 1/4$, $x = 1/2$ or $x = 3/4$. The trajectory shown in Fig. 11.7 follows these different heteroclinic orbits seemingly randomly.

Since we have periodic boundary conditions, it is natural to use Fourier modes and, following [31], we, therefore, seek solutions of the form

$$u(x,t) = e^{i\phi(t)} \sum_{n\in\mathbb{Z}} W_n(t)e^{2\pi i n x}, \qquad (11.23)$$

where we choose $\phi(t)$ so that $W_0(t)$ is real-valued, while $W_n(t)$ is complex-valued for $n \neq 0$, to factor out the $\mathbb{S}^1$ phase symmetry. Substituting (11.23) into (11.21) yields the infinite-dimensional system

$$i\dot\phi(t)W_n + \dot W_n = [\alpha - (2\pi n)^2(1+i\nu)]W_n - (1+i\mu)\sum_{k-l+m=n} W_k \overline{W}_l W_m, \quad (11.24)$$

which governs the evolution of W_n, where the phase velocity $\dot\phi$ is expressed explicitly by

$$\dot\phi = \mathrm{Im}\left[-\frac{1+i\mu}{W_0} \sum_{k-l+m=0} W_k \overline{W}_l W_m \right]. \qquad (11.25)$$

If we wish to compute the heteroclinic orbit in the fixed-point space of functions invariant under the reflection across $x = 0$, we would restrict ourselves to functions that are even in x and set $W_n = W_{-n}$ for all $n \in \mathbb{Z}$. This would then allow us to obtain the desired heteroclinic orbit as a codimension-zero saddle-sink connection using again a boundary value formulation.

Instead, we now describe the results obtained in [28] for the $N = 2$ Fourier truncation of (11.24) without assuming any relation between W_n and W_{-n}; in particular, (11.24) is five dimensional for $N = 2$. The results are shown in Fig. 11.8. The spatial average mode W_0, shown as a solid curve, exhibits the standard heteroclinic cycle behavior. The quantities $\mathrm{Re}(W_{-1} - W_1)$ and $\mathrm{Re}(W_{-2} - W_2)$ measure how far we are away from the fixed-point space of even functions. We find that the quantity $\mathrm{Re}(W_{-2} - W_2)$ vanishes for all times even though $W_2, W_{-2} \neq 0$. This implies, and can be checked numerically, that $W_{-2}(t) = W_2(t)$ is a spatial symmetry of the heteroclinic cycles. The other quantity $\mathrm{Re}(W_{-1} - W_1)$ is plotted as the dashed curve in Fig. 11.8 and turns out to vanish for approximately half the heteroclinic connections. Hence, roughly half of the heteroclinic connections lie in the invariant subspace

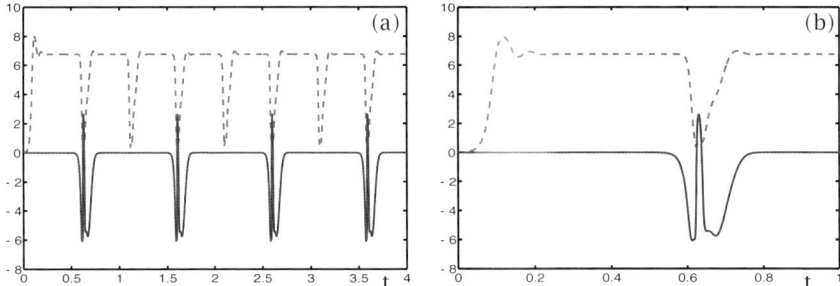

Fig. 11.8. Heteroclinic cycle in Fourier space (a) found by direct simulation of (11.24) with $N = 2$ for $\alpha = 70$, $\mu = -4$, and $\nu = 1$; the solid curve shows the evolution of W_0, while the dashed curve gives $\mathrm{Re}(W_{-1} - W_1)$. Panel (b) shows an enlargement of the cycle. From D.J.B. Lloyd, A.R. Champneys and R.E. Wilson, Robust heteroclinic cycles in the one-dimensional complex Ginzburg-Landau equation, *Physica D* 204 (2005) 240–268 © 2005 by Elsevier Science; reprinted with permission.

$\{W_{-n} = W_n; \; n \neq 0\}$ corresponding to even functions. If we plotted $\mathrm{Re}(W_{-1} + W_1)$, then we would see that this quantity is also zero half of the time but for precisely the other half of heteroclinic connections. This then gives us the invariant subspace of odd functions. Both subspaces intersect at the equilibria where $W_{-1} = W_1 = 0$ allowing for structurally stable heteroclinic cycles to exist to such equilibria within either subspace.

11.6 Traveling Waves on Cylinders $\mathbb{R} \times \mathbb{S}^1$

We now treat the case of PDEs on multi-dimensional domains with a single unbounded direction. For simplicity we shall restrict to two spatial dimensions and assume periodic boundary conditions in the other direction. Rather than presenting the general theory, which can be found in [30], we treat a specific example which is concerned with localized buckling of cylindrical shells [29].

Consider an infinitely long cylinder of radius R and shell thickness t. The equilibrium for the in-plane stress function ϕ and displacement w in the post-buckling regime of the cylinder is governed by the von Kármán–Donnell equations:

$$\kappa^2 \nabla^4 w + \lambda w_{xx} - \rho \phi_{xx} = w_{xx}\phi_{yy} + w_{yy}\phi_{xx} - 2w_{xy}\phi_{xy}, \quad (11.26)$$

$$\nabla^4 \phi + \rho w_{xx} = (w_{xy})^2 - w_{xx}w_{yy}, \quad (11.27)$$

where ∇^4 denotes the two-dimensional bi-harmonic operator, $x \in \mathbb{R}$ is the axial and $y \in [0, 2\pi R)$ the circumferential coordinate. Furthermore, $\rho := 1/R$ is the curvature, $\kappa^2 := t^2/12(1 - \nu^2)$, ν is Poisson's ratio, and the bifurcation parameter is $\lambda := P/Et$ where P is the compressive axial load (force per unit length), and E is Young's modulus. The form of solutions sought are localized in the axial length, but circumferentially periodic, suggesting that equations (11.26)–(11.27) should be supplemented with periodic boundary conditions in y and asymptotic boundary conditions in the axial direction x. Simplifications can be made by considering two types of symmetry conditions in the mid plane of the cylinder, namely

$$\begin{aligned} \mathcal{R}: \quad & w(x,y) = w(-x,y), \quad \phi(x,y) = \phi(-x,y), \\ \mathcal{S}: \quad & w(x,y) = w(-x, y + \pi R/s), \quad \phi(x,y) = \phi(-x, y + \pi R/s), \end{aligned}$$

where s is the axial wavenumber of the post-buckling pattern sought.

The numerical approach taken in [29] is to use a spectral decomposition circumferentially for each value of s of the form

$$w(x,y) = \sum_{j=0}^{N} a_j(x) \cos(js\rho y), \qquad \phi(x,y) = \sum_{j=0}^{N} b_j(x) \cos(js\rho y),$$

allied to Galerkin projection, to reduce the PDEs (11.26)–(11.27) to a system of $2N$ fourth-order ODEs, each akin to (11.17), for the amplitudes $(a_j, b_j)(x)$. The symmetries $\mathcal{R}$ and $\mathcal{S}$ provide two different reversing transformations of the ODEs, and so the projection boundary condition at one end can be replaced by using the $\mathrm{Fix}(\mathcal{R})$ or $\mathrm{Fix}(\mathcal{S})$ boundary conditions depending on which symmetric solution is sought.

Figure 11.9(a) and (b) shows numerical continuation results for both $\mathcal{R}$-symmetric and $\mathcal{S}$-symmetric localized modes for a typical experimentally amenable geometry, but for different circumferential wavenumber s. The horizontal axis in these plots is the overall end shortening implied by the deformation; see [29] for the details. The trend is the same for all solutions, namely that they bifurcate from the trivial solution subcritically (the load λ reduces sharply). They then reach a turning point with respect to λ at which point the branch would become stable in a controlled end-shortening experiment. Figure 11.9(c) shows an example of a localized buckle shape near such a fold.

11.7 Defects in Oscillatory Media

In the preceding section we discussed the computation of localized stationary solutions of PDEs posed on the cylinder $\mathbb{R} \times \mathbb{S}^1$. Here we illustrate that the computation of (not necessarily localized) time-periodic solutions of reaction-diffusion systems

$$u_t = u_{xx} + f(u), \qquad x \in \mathbb{R}, \ u \in \mathbb{R}^n \tag{11.28}$$

proceeds similarly. We denote the temporal frequency of a time-periodic solution by ω and use the rescaled time variable $\tau = \omega t$. Time-periodic solutions of (11.28) are then in one-to-one correspondence with solutions of

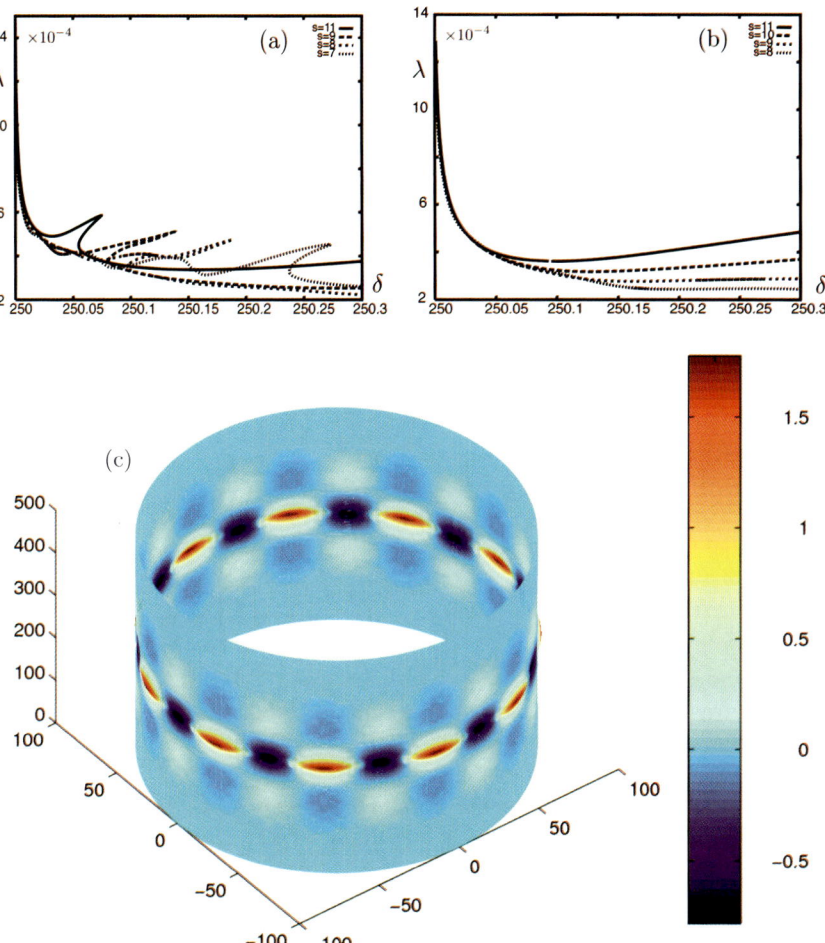

Fig. 11.9. Bifurcation diagrams for (a) $\mathcal{R}$-symmetric and (b) $\mathcal{S}$-symmetric solutions to (11.26)–(11.27) for different seed modes s with $\rho = 0.01$, $t = 0.247$, $\nu = 0.3$, $E = 5.56$, and $N = 6$. Panel (c) shows the displacement $w(x, y)$ of the $\mathcal{R}$-symmetric mode for $s = 11$ with λ at its local minimum value. From G.J. Lord, A.R. Champneys and G.W. Hunt, Computation of homoclinic orbits in partial differential equations: an application to cylindrical shell buckling, *SIAM J. Sci. Comp* 21 (1999) 591–619 © 1999 by the Society for Industrial and Applied Mathematics; reprinted with permission.

$$\omega u_\tau = D u_{xx} + f(u), \qquad (x, \tau) \in \mathbb{R} \times \mathbb{S}^1, \qquad (11.29)$$

where τ is restricted to $\mathbb{S}^1 \cong [0, 2\pi]/\sim$. We now have two symmetries present in our system, namely, translations in x and in τ. Thus, we need to add two phase conditions, which we choose according to Remark 1, and we also need to add the generator $c\partial_x u$ of the spatial translation to (11.29).

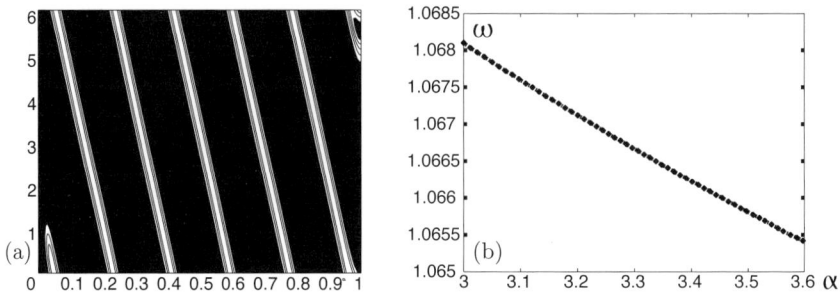

Fig. 11.10. Panel (a) is a space-time contour plot of the third component $u_3(x, \tau)$ of the solution to the Rössler system (space x is horizontal and rescaled time τ is vertical); since the defect is symmetric with respect to the reflection $x \mapsto -x$, the solution is plotted only for $x \geq 0$. Panel (b) shows the dependence of the temporal frequency ω on the parameter α.

We obtain the system

$$\begin{pmatrix} -\omega u_\tau + D u_{xx} + c u_x + f(u) \\ \int_0^{2\pi} \int_{\mathbb{R}} \langle \partial_\tau u_*(x, \tau), u(x, \tau) - u_*(x, \tau) \rangle \, dx \, d\tau \\ \int_0^{2\pi} \int_{\mathbb{R}} \langle \partial_x u_*(x, \tau), u(x, \tau) - u_*(x, \tau) \rangle \, dx \, d\tau \end{pmatrix} = 0, \qquad (11.30)$$

which we wish to solve for (u, ω, c), where u is defined on $\mathbb{R} \times \mathbb{S}^1$. In practice, one truncates the real line $\mathbb{R}$ to a large interval $(-L, L)$ and imposes additional boundary conditions such as Neumann at $x = \pm L$.

As an example, we consider the three-component Rössler system with $u \in \mathbb{R}^3$ where $D = 0.4$ and

$$f(u; \alpha) = (-u_2 - u_3, u_1 + 0.2u_2, 0.2 + u_1 u_3 - \alpha u_3)^T \qquad (11.31)$$

is the nonlinearity which depends on a parameter α. In [35], a defect solution was computed by implementing the system (11.30) with Neumann boundary conditions at $x = \pm L$ in AUTO [14]. The time direction τ is discretized using finite differences, taking the periodic boundary conditions for τ into account. The resulting system of ODEs in the spatial variable x is solved in AUTO as a boundary value problem.

Note that a symmetry condition $x \to -x$ is used at the left-hand boundary in Fig. 11.10(a). Thus the defect is occurring along this edge of the plot. This defect solution is robust and can be continued with the algorithm outlined above. An example of such a numerical computation is shown in Fig. 11.10(b).

11.8 Localized Structures in $\mathbb{R}^2$

Localized planar structures arise in many applications. Examples are light bullets in optical fibres [42], localized roll patterns known as convectons in binary-fluid convection [34], buckling modes in mechanical structures [19], localized micro-structures in solidification [20], and oscillating localized patterns known as oscillons in vertically vibrated sand layers [17, 43] and in the Belousov-Zhabotinsky reaction [44].

Rather than attempting a general discussion of how these patterns could be computed numerically, we focus on localized stationary hexagonal structures in the planar Swift–Hohenberg equation

$$[1 + \Delta]^2 u + \mu u - \nu u^2 + u^3 = 0. \tag{11.32}$$

Three typical localized hexagon structures are shown in Fig. 11.11; note that solutions (a)–(c) are computed for the same values of the parameters μ and ν. There are different ways for attempting to compute such patterns. We may, for instance, discretize in Cartesian coordinates and use three phase conditions to remove translation and rotation symmetries; see, e.g. [27]. However, a discretization in Cartesian coordinates will give preference to patterns with D_4-symmetry, thus possibly leading to spurious solutions when computing hexagonal structures.

An alternative approach is pursued in [8] where the patterns shown in Fig. 11.11 were computed using polar coordinates. We now describe the latter approach in more detail. Written in polar coordinates (r, θ), the Swift–Hohenberg equation for $u(r, \theta)$ is given by

$$[1 + \Delta]^2 u + \mu u - \nu u^2 + u^3 = 0, \tag{11.33}$$

where

$$\Delta_{r,\theta} u = u_{rr} + \frac{1}{r} u_r + \frac{1}{r^2} u_{\theta\theta}.$$

The computational domain is $\theta \in \mathbb{S}^1 \cong [0, 2\pi]/\sim$ and $0 < r < R$ for some $R \gg 1$ with the boundary condition

$$u(R, \theta) = u_r(R, \theta) = 0, \qquad \theta \in \mathbb{S}^1 \cong [0, 2\pi]/\sim \tag{11.34}$$

at $r = R$.

Solutions are now sought using the Fourier ansatz

$$u(r, \theta) = \sum_{n=0}^{N-1} a_n(r) \cos(6n\theta), \tag{11.35}$$

which exploits the D_6-symmetry inherent to hexagons. With this ansatz, the boundary conditions (11.34) become

$$a_n(R) = a'_n(R) = 0, \qquad n = 0, \ldots, N - 1. \tag{11.36}$$

We need to supplement these conditions with pole conditions at $r = 0$ to take care of the singularity of the Laplace operator at the origin. We choose to work with the pole conditions

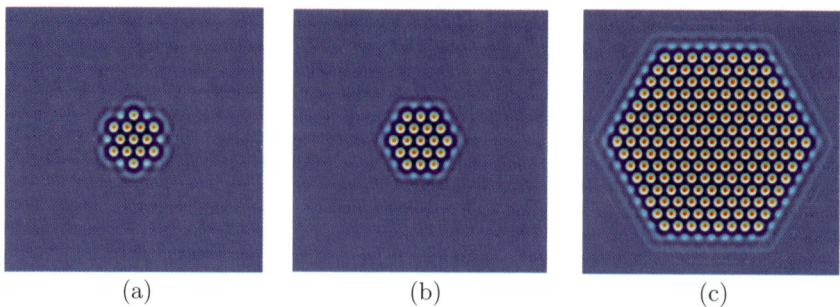

(a)   (b)   (c)

Fig. 11.11. Spatial contour plots of three localized hexagon structures (a)–(c) of the planar Swift–Hohenberg equation (11.32) with $\mu = 0.3$ and $\nu = 1.6$. The computational domain is $R = 100$, and $N = 20$ Fourier modes were used; plotted are the solutions on a 50×50 square.

$$a_0'(0) = a_0'''(0) = a_k(0) = a_k''(0) = 0, \qquad n = 1, \ldots, N-1, \qquad (11.37)$$

which enforce that $u(r, \theta)$ is even in r and make the variational formulation of (11.33) well defined at $r = 0$.

Figure 11.11 shows results of a computation where the resulting ODE boundary value problem has been implemented and solved with AUTO. Note that localized hexagonal structures of arbitrarily wide spatial extent can be computed this way. In fact panels (a)–(c) are three hexagonal patterns on the same continuation branch; details will appear in [8].

11.9 Planar Spiral Waves

As a final example, let us comment briefly on the computation of spiral wave solutions of a planar reaction-diffusion system

$$u_t = D\Delta u + f(u), \qquad x \in \mathbb{R}^2, \ u \in \mathbb{R}^n.$$

Spiral waves are relative equilibria with respect to the rotation symmetry of the plane and, therefore, of the form $u(r, \theta, t) = u_*(r, \theta - \omega t)$ in polar coordinates (r, θ). In particular, the spiral-wave profile $u_*(r, \theta)$ is a solution to the elliptic system

$$D\Delta_{r,\theta} u + \omega \partial_\theta u + f(u) = 0. \qquad (11.38)$$

To factor out the rotation symmetry, we may add the phase condition

$$\int_0^{2\pi} \langle \partial_\theta u_*(R/2, \theta), u(R/2, \theta) - u_*(R/2, \theta) \rangle \, d\theta = 0 \qquad (11.39)$$

posed at $r = R/2$.

Fig. 11.12. A contour plot of the third component u_3 of a spiral-wave solution to the planar Rössler system (11.28) with (11.31) is shown.

To solve (11.38)–(11.39), we truncate to the computational domain $(r, \theta) \in (0, R) \times [0, 2\pi]$ with periodic boundary conditions in θ and Neumann boundary conditions at $r = R$ plus appropriate pole conditions at $r = 0$. Barkley [2] then used finite differences in r and a spectral method in θ to compute spiral waves in the two-dimensional FitzHugh-Nagumo system (11.14).

Figure 11.12 shows the example of one such computation for the Rössler system (11.28) with (11.31) in two spatial dimensions, where we used Barkley's numerical code [2]. Shown is a contour plot of the third component u_3 of a spiral-wave solution at a fixed instant in time.

11.10 Outlook

We have attempted to outline a general approach to computing various kinds of coherent structures on infinite or semi-infinite domains. This procedure involves setting up formally well-posed boundary value problems that factor out the symmetries or degeneracies that arise due to the particular nature of the coherent structure in question. In essence, most of our methods rely on reduction to a two-point boundary value problem, even when solving problems in the plane. Thus, one can harness the full power of software such as AUTO for the continuation of paths of such solutions as parameters vary.

However, to compute more general patterned states, one needs to solve boundary value problems in appropriate two- or three-dimensional spatial domains without the artificial reduction to one-dimensional spatial problems. A reliable package for the continuation of solutions to fully two- and three-dimensional elliptic problems is clearly a pressing requirement. Such a package should be easy to use, yet sufficiently powerful. This would enable the user to set up specific boundary conditions that factor out the symmetries, in the manner described here, in dependence on the underlying properties of the structures that are sought.

Another pressing need, and the subject of much ongoing research, is the derivation of algorithms that are able to compute information on the spectral stability of the coherent structures while these structures themselves are being computed.

Acknowledgments

We thank David Lloyd for his helpful comments and suggestions and, especially, for his help with text and figures in Sects. 11.5 and 11.8. The results reviewed here have been obtained in collaboration over many years, and we would like to thank Jesús Cuevas, Giles Hunt, Panayotis Kevrekidis, Yuri Kuznetsov, David Lloyd, Gabriel Lord, Thomas Melvin, Bart Oldeman, Arnd Scheel, Jean-Mark Vanden-Broek, and Eddie Wilson for their contributions. We are grateful to Elsevier Science, Cambridge University Press, the American Physical Society, and the Society for Industrial and Applied Mathematics for permission to reproduce previously published material.

References

1. F. S. Bai, G. J. Lord, and A. Spence. Numerical computations of connecting orbits in discrete and continuous dynamical systems. *Internat. J. Bifurc. Chaos Appl. Sci. Engrg.*, 6:1281–1293, 1996.
2. D. Barkley. Linear stability analysis of spiral waves in excitable media. *Phys. Rev. Lett.*, 68:2090–2093, 1994.
3. W.-J. Beyn. The numerical computation of connecting orbits in dynamical systems. *IMA J. Num. Anal.*, 9:379–405, 1990.
4. W.-J. Beyn. On well-posed problems for connecting orbits in dynamical systems. *Contemp. Math.*, 172:131–168, 1995.
5. W.-J. Beyn and J. M. Kleinkauf. The numerical computation of homoclinic orbits for maps. *SIAM J. Num. Anal.*, 34:1207–1236, 1997.
6. A. R. Champneys. Homoclinic orbits in reversible systems and their applications in mechanics, fluids and optics. *Physica D*, 112:158–186, 1998.
7. A. R. Champneys, Yu. A. Kuznetsov, and B. Sandstede. A numerical toolbox for homoclinic bifurcation analysis. *Internat. J. Bifur. Chaos Appl. Sci. Engrg.*, 6(5):867–887, 1996.
8. A. R. Champneys, D. J. Lloyd, and B. Sandstede. Localised patterns in the planar Swift-Hohenberg equation. In preparation.
9. A. R. Champneys and G. J. Lord. Computation of homoclinic solutions to periodic orbits in a reduced water-wave problem. *Physica D*, 102:101–124, 1997.
10. A. R. Champneys, J. M. Vanden-Broeck, and G. J. Lord Do true elevation gravity-capillary solitary waves exist? A numerical investigation. *J. Fluid Mech.*, 454:403–417, 2002.

11. C. M. deSterke and J. E. Sipe. Gap solitons. *Progr. Opt.*, 33, 1994.
12. R. L. Devaney. Reversible diffeomorphisms and flows. *Trans. Amer. Math. Soc.*, 218:89–113, 1976.
13. L. Dieci and J. Rebaza. Point-to-point and periodic-to-periodic connections. *BIT Num. Math.*, 44:41–62, 2004.
14. E. J. Doedel, R. C. Paffenroth, A. R. Champneys, T. F. Fairgrieve, Yu. A. Kuznetsov, B. E. Oldeman, B. Sandstede, and X. Wang. AUTO2000: Continuation and bifurcation software for ordinary differential equations (with HOMCONT). Technical report, Concordia University, 2002.
15. E. J. Doedel and M. J. Friedman, Numerical computation of heteroclinic orbits. *J. Comp. Appl. Math.*, 26:317–327, 1989.
16. D. B. Duncan, J. C. Eilbeck, H. Feddersen, and J. A. D. Wattis. Solitons on lattices. *Physica D*, 68:1–11, 1993.
17. J. Fineberg. Physics in a jumping sandbox. *Nature*, 382:793–764, 1996.
18. M. J. Friedman and E. J. Doedel. Numerical computation and continuation of invariant manifolds connecting fixed points. SIAM *J. Numer. Analysis*, 28:789–808, 1991.
19. G. W. Hunt, M. A. Peletier, A. R. Champneys, P. D. Woods, M. Ahmer Wadee, C. J. Budd, and G. J. Lord. Cellular buckling in long structures. *Nonlin. Dyn.*, 21:3–29, 2000.
20. H. Jamgotchian, N. Bergeon, D. Benielle, P. Voge, B. Billia, and R. Guerin. Localised microstructures induced by fluid flow in directional Solidification. *Phys. Rev. Lett.*, 87(16):166105, 2001.
21. J. M. Bergamin, T. Bountis, and M. N. Vrahatis. Homoclinic orbits of invertible maps. *Nonlinearity*, 15:1603–1619, 2002.
22. B. Krauskopf and T. Riess. Numerical analysis of heteroclinic connections between hyperbolic equilibria and hyperbolic periodic orbits. Unpublished, 2007.
23. M. Krupa, B. Sandstede, and P. Szmolyan. Fast and slow waves in the FitzHugh–Nagumo equation. *J. Differ. Eqns.*, 133(1):49–97, 1997.
24. Yu. A. Kuznetsov. *Elements of Applied Bifurcation Theory*, third edition, (Springer-Verlag, New York, USA, 1994).
25. Yu. A. Kuznetsov. Computation of invariant manifold bifurcations. In D. Roose, A. Spence, and B. De Dier, editors, *Continuation and Bifurcations: Numerical Techniques and Applications*, pages 183–195 (Kluwer, Dordrecht, The Netherlands, 1990).
26. L. X. Liu, G. Moore, and R. D. Russell. Computation and continuation of homoclinic and heteroclinic orbits with arclength parameterization. *SIAM J. Sci. Comp.*, 18:69–93, 1997.
27. D. J. B. Lloyd and A. R. Champneys. Efficient numerical continuation and stability analysis of spatiotemporal quadratic optical solitons. *SIAM J. Sci. Comp.*, 27(3):759–773, 2005.
28. D. J. B. Lloyd, A. R. Champneys, and R. E. Wilson. Robust heteroclinic cycles in the one-dimensional complex Ginzburg-Landau equation. *Physica D*, 204:240–268, 2005.
29. G.J . Lord, A. R. Champneys, and G. W. Hunt. Computation of homoclinic orbits in partial differential equations: An application to cylindrical shell buckling. *SIAM J. Sci. Comp*, 21:591–619, 1999.
30. G. J . Lord, D. Peterhof, B. Sandstede, and A. Scheel. Numerical computation of solitary waves in infinite cylindrical domains. *SIAM J. Numer. Anal.*, 37(5):1420–1454, 2000.

31. B. P. Luce. Homoclinic explosions in the complex Ginzburg-Landau equation. *Physica D*, 84(3-4):553–581, 1995.
32. T. R .O. Melvin, A. R. Champneys, P. G. Kevrekidis, and J. Cuevas. Radiationless traveling waves in saturable nonlinear Schrödinger lattices. *Phys. Rev. Lett.*, 97:124101, 2006.
33. F. J. Muñoz Almaraz, E. Freire, J. Galán, E. Doedel, and A. Vanderbauwhede. Continuation of periodic orbits in conservative and hamiltonian systems. *Physica D*, 181:1–38, 2003.
34. J. J. Niemela, G. Ahlers, and D. S. Cannell. Localized travelling-wave states in binary-fluid convection. *Phys. Rev. Lett.*, 1990.
35. B. E. Oldeman and B. Sandstede. Numerical computation of defects in the Rössler model. Unpublished.
36. B. E. Oldeman, B. Krauskopf, and A. R. Champneys. Death of period-doublings: locating the homoclinic-doubling cascade. *Physica D*, 146:100–120, 2000.
37. A. J. Rodríguez-Luis, E. Freire, and E. Ponce. A method for homoclinic and heteroclinic continuation in two and three dimensions. In D. Roose, A. Spence, and B. De Dier, editors, *Continuation and Bifurcations: Numerical Techniques and Applications*, pages 197–210 (Kluwer, Dordrecht, The Netherlands, 1990).
38. B. Sandstede. Convergence estimates for the numerical approximation of homoclinic solutions. *IMA J. Num. Anal.*, 17:437–462, 1997.
39. B. Sandstede, A. Scheel, and C. Wulff. Dynamics of spiral waves on unbounded domains using center-manifold reductions *J. Diff. Eqns.*, 141:122–149, 1997.
40. C. Sparrow. *The Lorenz Equations: Bifurcations, Chaos, and Strange Attractors.* (Springer-Verlag, New York, 1982).
41. S. M. Sun. Non-existence of truly solitary waves in water with small surface tension. *Proc. Royal Soc. A*, 455(1986):2191–2228, 1999.
42. M. Tlidi, A. G. Vladimirov, and P. Mandel. Interaction and stability of periodic and localized structures in optical bistable systems. *IEEE J. of Quant. Electr.*, 39(2):216–226, 2003.
43. P. B. Umbanhowar, F. Melo, and H. L. Swinney. Localized excitations in a vertically vibrated granular layer. *Nature*, 382:793–796, 1996.
44. V. K. Vanag and I. R. Epstein. Stationary and oscillatory localized patterns, and subcritical bifurcations. *Phys. Rev. Lett.*, 92(12):128301, 2004.
45. K. Yagasaki. Numerical detection and continuation of homoclinic points and their bifurcations for maps and periodically forced systems. *Internat. J. Bifurc. Chaos Appl. Sci. Engrg.*, 8:1617–1627, 1998.
46. J. Yang, B. A. Malomed, D. J. Kaup, and A. R. Champneys. Embedded solitons: a new type of solitary wave. *Math. Comp. Simulation*, 55:585–600, 2001.

Continuation and Bifurcation Analysis of Delay Differential Equations

Dirk Roose[1] and Robert Szalai[2]

[1] Department of Computer Science, Katholieke Universiteit Leuven, Belgium
[2] Department of Engineering Mathematics, University of Bristol, United Kingdom

Mathematical modeling with delay differential equations (DDEs) is widely used in various application areas of science and engineering (e.g., in semiconductor lasers with delayed feedback, high-speed machining, communication networks, and control systems) and in the life sciences (e.g., in population dynamics, epidemiology, immunology, and physiology). Delay equations have an infinite-dimensional state space because their solution is unique only when an initial function is specified on a time interval of length equal to the largest delay. Consequently, analytical calculations are more difficult than for ordinary differential equations and numerical methods are generally the only way to achieve a complete analysis, prediction and control of systems with time delays.

Delay differential equations are a special type of functional differential equation (FDE). In FDEs the time evolution of the state variable can depend on the past in an arbitrary way as long as the dependence is a bounded function of the past. However, DDEs impose a constraint on this dependence, namely that the evolution depends only on certain past values of the state at discrete times. (We do not consider here the case of distributed delay.) The delays can be constant or state dependent. The equations can also involve delayed values of the derivative of the state, which leads to equations of neutral type.

In this chapter we mainly discuss the simplest case, namely a finite number of constant delays. Specifically, we consider a nonlinear system of DDEs with constant delays $\tau_j \geqslant 0$, $j = 1, \ldots, m$, of the form

$$x'(t) = f(x(t), x(t - \tau_1), x(t - \tau_2), \ldots, x(t - \tau_m), \eta), \qquad (12.1)$$

where $x(t) \in \mathbb{R}^n$, and $f : \mathbb{R}^{(m+1)n+p} \to \mathbb{R}^n$ is a nonlinear smooth function depending on a number of (time-independent) parameters $\eta \in \mathbb{R}^p$. We assume that the delays are in increasing order and denote the maximal delay by

$$\tau = \tau_m = \max_{i=1,\ldots,m} \tau_i.$$

A solution segment is denoted by $x_t = x_t(\theta) = x(t+\theta) \in C$, $\theta \in [-\tau, 0]$. Here $C = C([-\tau, 0]; \mathbb{R}^n)$ is the space of continuous functions mapping the delay interval into $\mathbb{R}^n$. For a fixed value of the parameter η, a solution $x(t)$ of (12.1) on $t \in [0, \infty)$ is uniquely defined by specifying a function segment x_0 as an initial condition. A discontinuity in the first derivative of $x(t)$ generally appears at $t = 0$ and is propagated in time, even if f and ϕ are infinitely smooth. However, the solution operator of (12.1) smooths the solution, meaning that discontinuities appear in higher and higher derivatives as time increases.

A DDE can be approximated by a system of ordinary differential equations (ODEs) and so standard numerical methods for ODEs could be used. However, to obtain an accurate approximation a high-dimensional system of ODEs is needed, and this leads to expensive numerical procedures. During the last decade, more efficient and more reliable numerical methods have been developed specifically for DDEs. In this chapter we survey those numerical methods for the continuation and bifurcation analysis of DDEs that are implemented in the software packages DDE-BIFTOOL [26, 27] and PDDE-CONT [78]. Where appropriate, we also briefly describe alternative numerical methods. Note that we do not discuss time integration of DDEs; for this topic see, for example, [3] and [7].

The structure of this chapter is as follows. In Sect. 12.1 we discuss numerical methods to compute the right-most characteristic roots of steady-state solutions. In Sect. 12.2 we describe collocation methods for computing periodic solutions and their dominant Floquet multipliers. Section 12.3 presents defining systems for codimension-one bifurcations of periodic solutions that allow one to compute the location of bifurcation points accurately. Computation of connecting orbits is discussed in Sect. 12.4 and of quasiperiodic solutions is discussed in Sect. 12.5. In Sect. 12.6 we briefly discuss how to deal with special types of DDEs, specifically, equations of neutral type and DDEs with state-dependent delays. In Sect. 12.7 we discuss specific details of the software packages DDE-BIFTOOL and PDDE-CONT. Their functionality is illustrated in Sect. 12.8, where we present the bifurcation analysis of several DDE models of practical relevance. Finally, conclusions and an outlook can be found in Sect. 12.9.

12.1 Stability of Steady-State Solutions

In this and the next section we assume that the parameter η is fixed and we omit it from the equations. A steady-state solution $x(t) \equiv x^\star$ of (12.1) satisfies the nonlinear system

$$f(x^\star, x^\star, x^\star, \dots, x^\star) = 0. \tag{12.2}$$

The (local) stability of $x^\star$ is determined by the stability of (the zero solution of) the *linearized equation*

$$y'(t) = A_0 y(t) + \sum_{j=1}^{m} A_j y(t - \tau_j), \tag{12.3}$$

where $A_j \in \mathbb{R}^{n \times n}$ denotes the partial derivative of f with respect to its $(j+1)$th argument, evaluated at x^*. The linearized equation (12.3) is asymptotically stable if all its roots λ of the *characteristic equation*

$$\det(\lambda I - A_0 - \sum_{j=1}^{m} A_j e^{-\lambda \tau_j}) = 0 \qquad (12.4)$$

lie in the open left half-plane (i.e., $\mathrm{Re}(\lambda) < 0$); see, e.g., [40, 62, 75]. Equation (12.4) has an infinite number of roots λ, known as the *characteristic roots*. However, the number of characteristic roots with real part larger than a given threshold is finite. Hence, to analyze the stability of a steady-state solution, one must determine reliably all roots satisfying $\mathrm{Re}(\lambda) \geqslant r$, for a given $r < 0$ close to zero.

Analytical conditions for stability can be found in Stépán [75] and Hassard [44]. These conditions are deduced by using the argument principle of complex analysis, and they give a practical method for determining stability. In recent years, numerical methods have been developed to compute approximations to the right-most (stability-determining) characteristic roots of (12.4), by using a discretization either of the solution operator of (12.3) or of the infinitesimal generator of the semi-group of the solution operator of (12.3). The solution operator $\mathcal{S}(t)$ of the linearized equation (12.3) maps an initial function segment onto the solution segment at time t, i.e.,

$$(\mathcal{S}(t)y(\cdot))(\theta) = y(t + \theta), \quad -\tau \leq \theta \leq 0, \ t \geq 0. \qquad (12.5)$$

This operator has eigenvalues μ, which are related to the characteristic roots via the equation $\mu = e^{\lambda t}$ [67]. To determine the stability, we are interested in the dominant eigenvalues. If t is large then these eigenvalues are well separated, which can be exploited during the eigenvalue computation, but the time integration itself may be costly. In Sect. 12.1.1 we describe a reliable way to compute the dominant eigenvalues of $\mathcal{S}(h)$ where h is the time step of a linear multistep (LMS) method.

Since $\mathcal{S}(t)$ is a strongly continuous semi-group [38, 40], one can define the corresponding infinitesimal generator $\mathcal{A}$ by

$$\mathcal{A}y = \lim_{t \to 0+} \frac{\mathcal{S}(t)y - y}{t}. \qquad (12.6)$$

For (12.3) the infinitesimal generator becomes

$$\begin{aligned} &\mathcal{A}y(\theta) = y'(\theta), y \in \mathcal{D}(\mathcal{A}) \\ &\mathcal{D}(\mathcal{A}) = \{y \in C : y' \in C \quad \text{and} \quad y'(0) = \sum_{j=0}^{m} A_j y(-\tau_j)\}. \end{aligned} \qquad (12.7)$$

Both operators can be discretized by spectral discretizations or time integration methods; this always leads to a representation by some matrix. Eigenvalues of this matrix yield approximations to the right-most characteristic roots. Hence, for computational efficiency it is important that the size of the resulting matrix eigenvalue problem is small, or at least that the stability determining eigenvalues can be computed efficiently by using an iterative method, such as subspace iteration. Accurate characteristic roots can be found by using Newton iterations on

$$\begin{aligned}(\lambda I - A_0 - \sum_{j=1}^{m} A_j e^{-\lambda \tau_j})v &= 0,\\ v_0^T v &= 1,\end{aligned} \tag{12.8}$$

where $v \in \mathbb{R}^n$ and $v_0 \in \mathbb{R}^n$, to obtain accurate characteristic roots λ (and the corresponding eigenfunctions $ve^{\lambda t}$). The difference between the approximate and the corrected roots gives an indication of the accuracy of the approximations.

Below we describe how the characteristic roots can be computed via an approximation of the solution operator by time integration, which is the method that is implemented in DDE-BIFTOOL. We also briefly comment on other approaches.

12.1.1 Approximation of the Solution Operator by a Time Integrator

A natural way to approximate the solution operator is to write the numerical time integration of the linearized equation as a matrix equation. Engelborghs et al. [28] have proposed and analyzed the use of a linear multistep method with constant steplength h to approximate the solution operator $\mathcal{S}(h)$. The delay interval $[-\tau, 0]$ (slightly extended to the left and the right; see below) is discretized by using an equidistant mesh with mesh spacing h, and a solution is represented by a discrete set of points $y_i := y(t_i)$ with $t_i = ih$. A k-step LMS method with steplength h to compute y_k can be written as

$$\sum_{i=0}^{k} \alpha_i y_i = h \sum_{i=0}^{k} \beta_i \left(A_0 y_i + \sum_{j=1}^{m} A_j \tilde{y}(t_i - \tau_j) \right), \tag{12.9}$$

where α_i and β_i are parameters and (in case $t_i - \tau_j$ does not coincide with a mesh point) the approximations $\tilde{y}(t_i - \tau_j)$ are obtained by polynomial interpolation with s_- and s_+ points to the left and the right, respectively.

The discretization of the solution operator is the (linear) map between $[y_{L_{min}}, \ldots, y_{k-1}]^T$ and $[y_{L_{min}+1}, \ldots, y_k]^T$ where $L_{min} = -s_- - \lceil \tau/h \rceil$ and where the mapping is defined by (12.9) for y_k and by a shift for all variables other than y_k. This map is represented by an $N \times N$ matrix, where

$$N = n(-L_{min} + k) \approx n\tau/h. \tag{12.10}$$

Since the time step h is small, the eigenvalues μ of this matrix are not well separated (most eigenvalues lie close to the unit circle). They can be computed by e.g. the QR method, with a computational cost of the order $N^3 \approx n^3(\tau/h)^3$, and so approximations to the characteristic roots can be derived.

To guarantee the reliability of the stability computation, the steplength h in the LMS method (12.9) should be chosen such that all characteristic roots λ with $\text{Re}(\lambda) \geqslant r$ $(r < 0)$ are approximated accurately. Procedures for such a safe choice of h are described in [28, 84] and implemented in DDE-BIFTOOL. They are based on theoretical properties of

(a) the relation between the stability properties of the solution of the linearized equation (12.3) and the stability of the discretized equation (12.9);
(b) an a-priori estimate of the region in the complex plane that includes all characteristic roots λ with $\text{Re}(\lambda) \geqslant r$.

Note that the solution operator can also be discretized by using a Runge-Kutta time integrator [11].

12.1.2 Other Approaches

Breda et al. [12] have developed numerical methods to determine the stability of solutions based on a discretization of the infinitesimal generator. By discretizing the derivative in (12.7) with a Runge-Kutta or an LMS method, a matrix approximation of $\mathcal{A}$ is obtained. The resulting eigenvalue problem is large and sparse, as in the case when the solution operator is discretized by a time integration method. Breda et al. [13] also proposed a pseudo-spectral discretization of the infinitesimal generator. In this approach, an eigenfunction of the infinitesimal generator $ve^{\lambda t}, t \in [-\tau, 0]$, is approximated by a polynomial $P(t)$ of degree p. Collocation for the eigenvalue problem for the infinitesimal generator leads to an equation of the form

$$P'(t_i) = \lambda P(t_i), \tag{12.11}$$

where the collocation points $t_i, i = 1...p$ are chosen as the shifted and scaled roots of an (orthogonal) polynomial of degree p. These equations are augmented with

$$A_0 P(0) + \sum_{j=1}^{m} A_j P(\tau_j) = \lambda P(0), \tag{12.12}$$

which introduces the system-dependent information. The resulting matrix eigenvalue problem has size $n(p+1)$. The first np rows are the Kronecker product of a dense $p \times (p+1)$ matrix and the identity matrix. The last block row consists of a linear combination of the matrices $A_j, j = 0, ..., m$ and the identity. The matrix is full but can be of much smaller size than in the previous case, due to the 'spectral accuracy' convergence, as is shown in the detailed analysis presented in [13].

A pseudo-spectral discretization of the solution operator is proposed in [11, 85]. Here a polynomial approximation $P(t)$ of an eigenfunction, defined on the interval $[-\tau, h]$ has to satisfy p collocation conditions of the form

$$P(t_i + h) = \mu P(t_i),$$

where $\mu = e^{\lambda h}$. These equations are augmented with a condition obtained from integrating the linearized equations over a time interval h. When high accuracy is required, a pseudo-spectral discretization will lead to a more efficient procedure than when a time integration discretization is used, but numerical experiments indicate that for low accuracy requirements both approaches are competitive [85].

However, for the pseudo-spectral approaches no strategy is known that guarantees a priori that all characteristic roots with real part larger than r are computed accurately; this is in contrast to discretization of the solution operator with an LMS method.

12.2 Periodic Solutions

A periodic solution $x^\star(t)$ of an autonomous system of the form (12.1) satisfies

$$x^\star(t + T) = x^\star(t), \ \forall t,$$

where T is the period. An extensive literature exists on the existence, stability and parameter dependence of periodic solutions; see, e.g., [40, §XI.1-2]. These results are essentially analytical in nature and the corresponding methods have different rigorous restrictions and cannot be applied to general nonlinear systems with several delays.

Because of the dependence on the past, periodicity of $x(t)$ at one moment in time, $x(t) = x(t + T)$ for some t, does not imply periodicity for the whole solution. Instead, a complete function segment of length τ has to be repeated. Consequently, a periodic solution to (12.1) can be found as the solution of the following *two-point boundary value problem* (BVP),

$$\begin{cases} x'(t) = f(x(t), x(t - \tau_1), \dots, x(t - \tau_m), \eta), \ t \in [0, T], \\ x_0 = x_T, \\ p(x, T) = 0, \end{cases} \quad (12.13)$$

where x_0 and x_T are function segments on $[-\tau, 0]$ and $[-\tau + T, T]$, respectively, and the period T is an unknown parameter. Furthermore, p represents a phase condition that is needed to remove translational invariance. A well-known example is the classical integral phase condition [21]

$$\int_0^1 \dot{x}^{(0)}(s)(x^{(0)}(s) - x(s)) \, ds = 0, \quad (12.14)$$

where $x^{(0)}$ is a reference solution; see also Chap. 1.

Stable periodic solutions of a DDE can be found by numerical time integration; the convergence of the integration depends on the stability properties of the periodic solution [46]. However, both stable and unstable solutions can be computed by solving the above boundary value problem by either collocation or by a shooting approach. Here we only consider collocation methods.

12.2.1 Collocation

In collocation a periodic solution is computed by using a discrete representation that satisfies the differential equation at a set of collocation points on
$[0, T]$. Doedel and Leung [22] have computed periodic solutions of DDEs using collocation based on a truncated Fourier series; see also [14] for a similar
approach. This Fourier approach has the advantage that periodicity is automatically fulfilled. However, steep gradients in a solution pose problems and
it is not possible to determine the solution stability.

Collocation based on piecewise-polynomial representations is used in AUTO
[18] and CONTENT [52] to compute periodic solutions for systems of ordinary
differential equations; see also Chaps. 1 and 2. We now discuss how piecewise-
polynomial collocation can be used for DDEs. We first rescale time by a factor
$1/T$ such that the period is one in the transformed system

$$\begin{cases} x'(t) = Tf\left(x(t), x(t - \tau_1/T), \ldots, x(t - \tau_m/T), \eta\right), & \text{for } t \in [0, 1], \\ x(\theta + 1) - x(\theta) = 0, & \text{for } \theta \in [-\tau/T, 0], \\ p(x, T) = 0. \end{cases} \quad (12.15)$$

A mesh with $L + 1$ mesh points $\{0 = t_0 < t_1 < \ldots < t_L = 1\}$ is specified.
This mesh is periodically extended to the left with ℓ points to obtain a mesh
on $[-\tau/T, 1]$ with $\ell + L$ intervals. In each interval an approximating polynomial
of degree d is described in terms of the function values at the representation
points (using Lagrange polynomials as basis). These function values are determined by requiring that the approximating collocation solution fulfills the
(time-scaled) differential equations exactly at the collocation points. In each
interval, the collocation points are typically chosen as the (scaled and shifted)
roots of a dth degree orthogonal polynomial.

The approximating polynomial of degree d on each interval $[t_i, t_{i+1}]$, $i = -\ell, \ldots, L - 1$, can be written as

$$u(t) = \sum_{j=0}^{d} u(t_{i+\frac{j}{d}})P_{i,j}(t), \ t \in [t_i, t_{i+1}], \quad (12.16)$$

where $P_{i,j}(t)$ are the Lagrange polynomials through the *representation points*

$$t_{i+\frac{j}{d}} = t_i + \frac{j}{d}(t_{i+1} - t_i), \quad j = 0, \ldots, d.$$

Because polynomials on adjacent intervals share the value at the common
mesh point, this representation is automatically continuous (however, it is
not continuously differentiable at the mesh points).

The approximation $u(t)$ is completely determined by the coefficients

$$u_{i+\frac{j}{d}} := u(t_{i+\frac{j}{d}}), \ i = -\ell, \ldots, L-1, \ j = 0, \ldots, d-1 \text{ and } u_L := u(t_L). \ (12.17)$$

We define the *starting vector* u_s and the *final vector* u_f, both of length $N = n(\ell d + 1)$, as

$$u_s := [u_{-\ell}, \ldots, u_{i+\frac{j}{d}}, \ldots, u_0]^T, \quad u_f := [u_{L-\ell}, \ldots, u_{i+\frac{j}{d}}, \ldots, u_L]^T. \quad (12.18)$$

The collocation points are obtained as

$$c_{i,j} = t_i + c_j(t_{i+1} - t_i), \; i = 0, \ldots, L-1, \; j = 1, \ldots, d,$$

from a set of *collocation parameters* c_j, $j = 1, \ldots, d$, e.g., the shifted and scaled roots of the dth degree Gauss-Legendre polynomial.

A periodic solution for a fixed value of the parameters η is found as the solution of the following $(n((\ell + L)d + 1) + 1)$-dimensional (nonlinear) system in terms of the unknowns (12.17) and T,

$$\begin{cases} \dot{u}(c_{i,j}) = Tf(u(c_{i,j}), u(c_{i,j} - \frac{\tau_1}{T}), \ldots, u(c_{i,j} - \frac{\tau_m}{T})), \eta) = 0, \\ \qquad\qquad\qquad i = 0, \ldots, L-1, \; j = 1, \ldots, d, \\ u_f - u_s = 0, \\ p(u) = 0. \end{cases} \quad (12.19)$$

Here, p again represents a phase condition such as (12.14).

The collocation solution fulfils the time-scaled differential equation exactly at the collocation points. If $c_{i,j}$ coincides with t_i then the right derivative is taken in (12.19), if it coincides with t_{i+1} then the left derivative is taken. Taking into account the periodicity conditions, one can reduce system (12.19) to the following nonlinear system in the unknowns $u = [u_0, \ldots, u_L]^T$ and T,

$$\begin{cases} \dot{u}(c_{i,j}) = Tf(u(c_{i,j}), u((c_{i,j} - \frac{\tau_1}{T}) \bmod 1), \ldots, u((c_{i,j} - \frac{\tau_m}{T}) \bmod 1), \eta) = 0, \\ \qquad\qquad\qquad i = 0, \ldots, L-1, \; j = 1, \ldots, d, \\ u_0 - u_L = 0, \\ p(u) = 0. \end{cases}$$

$$(12.20)$$

Hence, the dimension of the system and the number of unknowns is reduced to $(n(Ld + 1) + 1)$.

When using Newton's method to solve (12.20), the matrix of the linear system to be solved in each iteration is sparse and has a particular structure, as is shown in Fig. 12.1. The matrix consists of a (large) $nLd \times n(Ld + 1)$ matrix filled with two (circular) bands, bordered by one column and $n + 1$ rows. The extra column contains derivatives with respect to the period; n extra rows contain the periodicity condition, and one extra row is due to the phase condition (12.14). The diagonal band is itself a concatenation of $nd \times n(d + 1)$ blocks. The off-diagonal bands are a consequence of the delay terms. When the mesh is equispaced then the off-diagonal band lies at a fixed distance from the diagonal band as is illustrated in Fig. 12.1(a). This is no longer the case when the mesh is non-equispaced; see Fig. 12.1(b).

In the case of collocation for ODEs, the matrix of the linear system has a band structure with a band size proportional to n and d but independent of the number of mesh intervals L; see also Chap. 1. Hence the system can be solved efficiently by a direct band solver. For delay differential equations this

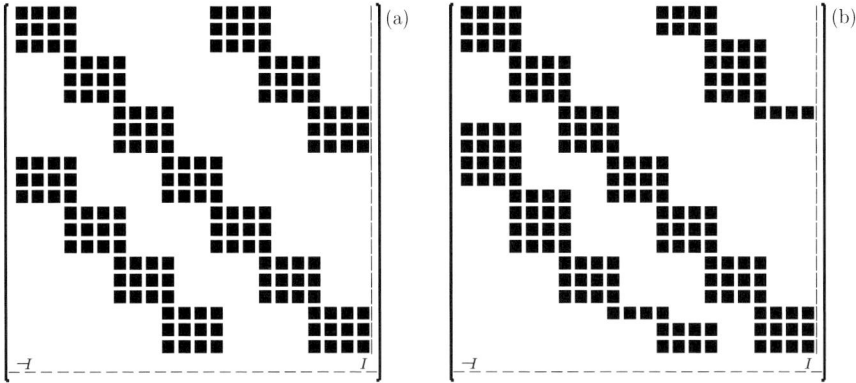

Fig. 12.1. Structure of the matrix arising in the Newton iteration to solve (12.20) for one delay that is smaller than the period T; shown is the case $L = 7$ and collocation polynomials of degree $d = 3$. Panel (a) is for an equispaced mesh and panel (b) for a non-equispaced mesh; each black box represents an $n \times n$ block.

is not possible. Indeed, the structure of the matrix, described above, cannot easily be exploited when using a direct solver, especially in case of several delays and/or a non-equispaced mesh. However, for moderate values of d, n and L the linear system (12.20) can still be solved with a direct method. The efficiency can be increased by using a Newton-chord method, in which case the Jacobian is not recomputed (and factored) in every iteration but remains fixed during a number of iterations. In Sect. 12.2.3 we describe an efficient iterative procedure to solve (12.20).

Furthermore, an adaptive (non-equispaced) mesh can be used to decrease the required number of intervals L for difficult solutions (with steep gradients). For the latter, the interval size $h_i = t_{i+1} - t_i$ is adapted to equidistribute the $(d+1)$th derivative of the solution along the mesh; see [2, 25].

Engelborghs and Doedel [24] have proven that the convergence rate of the maximal continuous error $E = \max_{t \in [0,1]} \|u(t) - u^\star(t)\|$ is $\mathcal{O}(h^d)$ in general and $\mathcal{O}(h^{d+1})$ for Gauss-Legendre collocation points on equispaced and non-equispaced meshes with $h = \max_i h_i$. Special convergence rates at the mesh points (so-called superconvergence) that feature for ODEs are, in general, lost for DDEs.

Note that, in the case of a nonautonomous (or forced) system, the collocation method is essentially the same as in (12.20), except that the phase condition is not needed, since the phase of the solution is determined by the phase of the forcing.

12.2.2 Monodromy Operator and Floquet Multipliers

The stability of a periodic solution is determined by the Floquet multipliers, which are the eigenvalues of the monodromy operator. In the case of autonomous equations there is always a trivial multiplier $+1$, which stems from the fact that the associated linearized equation is always solved by the time derivative of the solution itself. According to Floquet theory, a periodic solution is asymptotically stable if all the multipliers — not counting the trivial one — lie strictly inside the complex unit disk. The main focus of this section is the computation of the monodromy operator using the previously described collocation method.

Denote by $x^\star(t)$ a T-periodic solution of (12.1). As in the previous sections we rescale time by $1/T$. The linearized equation about this periodic solution in rescaled time is

$$\frac{d}{dt}y(t) = T(A_0(t)y(t) + \sum_{j=1}^{m} A_j(t)y(t-\tau)), \qquad (12.21)$$

where $A_j(t)$ denotes the partial derivative of f with respect to its $(j+1)$th argument, evaluated at $x^\star(Tt)$. Also let $U(t,s)$ be the fundamental solution operator of (12.21), which is defined as

$$(U(t,s)\phi_s)(\theta) = y(t+\theta), \quad \theta \in [-\tau/T, 0],$$

where ϕ_s is an initial function and y is the corresponding solution of (12.21). The monodromy operator is defined as

$$\mathcal{M} = U(1,0),$$

that is,

$$\mathcal{M} : C([-\tau/T, 0]; \mathbb{R}^n) \to C([-\tau/T, 0]; \mathbb{R}^n),$$
$$\phi \mapsto y_1,$$

where ϕ is the initial function and y_1 is the solution segment $y_1(\theta) = y(1+\theta)$, $\theta \in [-\tau/T, 0]$. The discretized version of $\mathcal{M}$ is $\mathcal{M}_d : u_s \to u_f$ and its matrix representation can be obtained by solving (12.21) with a collocation method similar to (12.19). This method is used in DDE-BIFTOOL [27].

However, when the maximal delay is larger than the period, u_s and u_f overlap; the computation of $\mathcal{M}_d$ can be improved by exploiting this property. For the sake of generality we use the Riesz representation theorem and write (12.21) in the form

$$\frac{dy(t)}{dt} = T \int_0^{\tau/T} d_\theta \zeta(T\theta, t)y(t-\theta), \qquad (12.22)$$

where ζ is a matrix-valued function of bounded variation that, with (12.21), can be written as

$$\zeta(T\theta, t) = \begin{cases} 0 & \text{if} \quad \theta \leq 0, \\ A_0(t) & \text{if } 0 < \theta < \tau_1, \\ \vdots & \vdots \\ A_0(t) + \sum_{j=1}^{m} A_j(t) \text{ if} & \tau \leq \theta. \end{cases}$$

Notice that (12.22) implicitly depends on the initial function. It can be written explicitly as

$$\frac{dy(t)}{dt} - T \int_0^t d_\theta \zeta(T\theta, t) y(t - \theta) - T \int_t^{\tau/T} d_\theta \zeta(T\theta, t) \phi(t - \theta) = 0. \quad (12.23)$$

We introduce $K = \lceil \tau/T \rceil$ solution segments of $y(t)$ and ϕ as

$$y_1(t) = y(2 - K + t), \ y_2(t) = y(3 - K + t), \dots, y_K(t) = y(1 + t)$$
$$\phi_1(t) = \phi(1 - K + t), \phi_2(t) = \phi(2 - K + t), \dots, \phi_K(t) = \phi(t),$$
$$t \in [-1, 0],$$

such that $\phi_i, y_i \in \mathcal{X} := C([-1, 0]; \mathbb{R}^n)$, and we also define operators on $\mathcal{X}$ as obtained from (12.23) as

$$(\mathcal{A}\phi)(\theta) = \frac{d\phi(\theta)}{dt} - T \int_0^{1+\theta} d_\gamma \zeta(T\gamma, \theta) \phi(\theta - \gamma), \quad D(\mathcal{A}) = C^1([-1, 0], \mathbb{R}^n),$$

$$(\mathcal{B}_i \phi)(\theta) = T \int_{i+\theta}^{i+1+\theta} d_\gamma \zeta(T\gamma, \theta) \phi(i + \theta - \gamma), \quad 1 \leq i \leq N.$$

It is clear that y_K is the only unknown, because all the other y_i can be found from the initial conditions as $y_i = \phi_{i+1}$. Hence, the only equation that has to be solved is

$$\mathcal{A}y_K - \sum_{i=1}^{K} \mathcal{B}_i \phi_i = 0, \quad y_K(-1) = \phi_K(0).$$

In order to eliminate the explicit boundary condition we introduce extended operators on $\hat{\mathcal{X}} = \{(\varphi, c) \in \mathcal{X} \times \mathbb{R}^n : c = \varphi(0)\}$ in the form of

$$\hat{\mathcal{A}} = \begin{pmatrix} \mathcal{A} & 0 \\ \mathcal{L} & 0 \end{pmatrix}, \quad \hat{\mathcal{B}}_i = \begin{pmatrix} \mathcal{B}_i & 0 \\ 0 & 0 \end{pmatrix} \text{ for } i < N \quad \text{and} \quad \hat{\mathcal{B}}_N = \begin{pmatrix} \mathcal{B}_N & 0 \\ 0 & I \end{pmatrix},$$

where $\mathcal{L}\varphi = \varphi(-1)$. The extended monodromy operator is defined on $X = C([-N, 0]; \mathbb{R}^n)$; this space is isomorphic to the further extended

$$\tilde{X} = \left\{ ((\phi_1, c_1), \dots, (\phi_N, c_N)) \in \hat{\mathcal{X}}^N : \phi_k(0) = c_k = \phi_{k+1}(-1), 1 \leq k < N \right\}.$$

In order to obtain stability results it is sufficient to construct the monodromy operator on $\tilde{X}$, which becomes

$$\check{\mathcal{M}} = \begin{pmatrix} 0 & \hat{I} & \cdots & 0 \\ 0 & 0 & \ddots & 0 \\ \vdots & \vdots & & \vdots \\ 0 & 0 & & \hat{I} \\ \hat{A}^{-1}\hat{B}_1 & \hat{A}^{-1}\hat{B}_2 & \cdots & \hat{A}^{-1}\hat{B}_N \end{pmatrix}. \tag{12.24}$$

Because of the identity matrices above the diagonal, the operator $\check{\mathcal{M}}$ is not compact, but its powers $\check{\mathcal{M}}^k$, $k \geqslant K$ are compact.

The operator $\check{\mathcal{M}}$ can be computed by collocation and by inverting the resulting discretized $\hat{A}$ operator. In PDDE-CONT the spectrum of $\check{\mathcal{M}}$ is computed with the iterative Arnoldi-Lanczos method [68], which is implemented in the ARPACK software package [54]. Note that in this iterative process, when the discretized operator is multiplied by a vector, only one solution step with $\hat{A}$ is necessary.

Despite the differences, using either $\mathcal{M}$ or $\check{\mathcal{M}}$ gives the same accuracy of the multiplier calculation [56]. In particular, it was shown in [56] that the computations of the multipliers and of the periodic solution itself have the same accuracy. The exception is the computation of the trivial multiplier $+1$, which was found to be more accurate when using $\check{\mathcal{M}}$. Hence, inferring the accuracy of the computed periodic solution from the accuracy of the trivial multiplier can be deceiving.

12.2.3 Collocation-Newton-Picard

Verheyden and Lust [83] have developed an iterative procedure to solve the linear system arising in Newton's method applied to system (12.19). Consider the unknowns $u_{i+j/d} := u(t_{i+j/d})$ defined in (12.17). Recall the definition of the starting vector u_s and the final vector u_f given in (12.18)

$$u_s := [u_{-\ell}, \ldots, u_{i+\frac{j}{d}}, \ldots, u_0]^T, \quad u_f := [u_{L-\ell}, \ldots, u_{i+\frac{j}{d}}, \ldots, u_L]^T, \tag{12.25}$$

and define the *trajectory vector* as

$$u_t := [u_{\frac{1}{d}}, \ldots, u_{i+\frac{j}{d}} \ldots, u_L]^T, \tag{12.26}$$

where u_s and u_f are of length $N = n(\ell d + 1)$ and u_t is of length $\hat{N} = nLd$ (here ℓ and L denote the number of mesh points in $[-\tau/T, 0]$ and $(0, 1]$, respectively). Note that u_f consists of the last $n(\ell d + 1)$ components of u_t.

The linearization of (12.19) has the following form

$$\begin{aligned} -B\Delta u_s + A\Delta u_t + r_{1,T}\Delta T &= -r_1, \\ -\Delta u_s + \Delta u_f &= -r_2, \\ \alpha_s \Delta u_s + \alpha_t \Delta u_t + \alpha_T \Delta T &= -\alpha, \end{aligned} \tag{12.27}$$

where r_1, r_2 and α denote the residuals of system (12.19) and $-B$, A and $r_{1,T}$ denote the partial derivatives of the collocation conditions with respect to u_s,

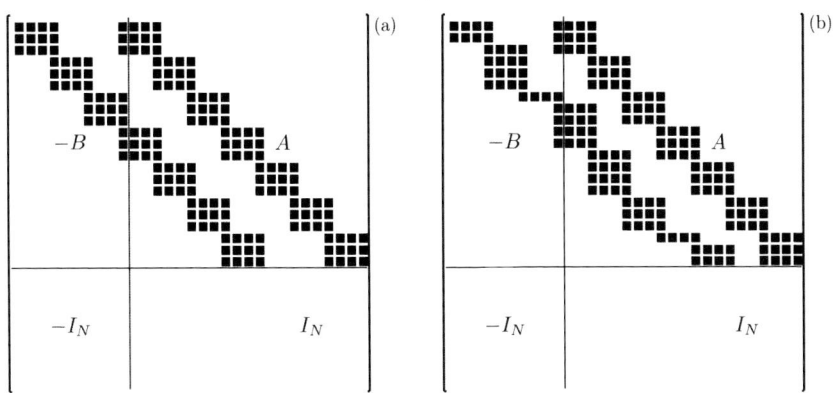

Fig. 12.2. Typical structure of the linearized collocation system for one delay that is smaller than the period T; shown is the case of $\ell + L = 3 + 7$ mesh intervals and collocation polynomials of degree $d = 3$ (the bordering row and column are omitted). Panel (a) is for an equispaced mesh and panel (b) for a non-equispaced mesh; each black box represents an $n \times n$ block.

u_t and T. A typical structure for the matrix of the linearized system in the case of one time delay that is smaller than the period T is shown in Fig. 12.2. Panel (a) is for an equispaced mesh, while in panel (b) a non-equispaced mesh is used. In both cases the mesh with $L = 7$ mesh intervals is extended with $\ell = 3$ additional mesh intervals and the piecewise polynomials have degree $d = 3$.

The linear system can be manipulated and condensed to the form

$$\begin{bmatrix} M - I & b_c \\ \beta_s & \beta_T \end{bmatrix} \begin{bmatrix} \Delta u_s \\ \Delta T \end{bmatrix} = - \begin{bmatrix} r_c \\ \alpha_c \end{bmatrix}. \tag{12.28}$$

Here M is the discretization of the monodromy operator, which can be derived from $M_t = A^{-1}B$. Afterwards, Δu_t can be computed from (12.27). This manipulation is based on the correspondence between the linearization of the collocation scheme and the discretization of the linearized boundary value problem. The condensation is related to the condensation used in AUTO; see Chap. 1. System (12.28) can be solved with the Newton-Picard method [55], which leads to a substantial reduction in the computational cost, especially when only a few Floquet multipliers are larger in modulus than a certain threshold ρ, e.g., $\rho = 0.5$ [83].

12.3 Defining Systems for Codimension-One Bifurcations of Periodic Solutions

Periodic solutions of autonomous DDEs allow three generic codimension-one bifurcations. First, the monodromy operator may have an algebraically double +1 eigenvalue, which corresponds to a limit point (or fold, or saddle-node) bifurcation where the solution ceases to exist. Second, there may be a single −1 multiplier, which gives a period-doubling bifurcation. Third, if two critical complex conjugate multipliers lie on the unit circle of the complex plane, then there is a Neimark-Sacker or torus bifurcation. In this case an invariant torus bifurcates from the periodic solution.

Continuing the bifurcations of periodic solutions in DDEs does not differ substantially from the case of ODEs. In order to compute bifurcations one has to include additional equations to (12.20), which are satisfied by a periodic solution if, and only if, the monodromy operator has a certain kind of singularity. In the period-doubling and the Neimark-Sacker cases, the simplest procedure to construct such a determining system is to require that the monodromy operator has a singular vector. Short algebraic transformations of (12.24) reveals that these bifurcations occur if

$$\left(\hat{A} - \sum_{j=1}^{N} \sigma^j \hat{B}_j\right) v = 0,$$

$$(12.29)$$

$$v^H v = 1,$$

has a unique solution v with the inverse characteristic multiplier $\sigma = \mu^{-1} \neq 1$ on the unit circle. (Throughout, a superscript H denotes the (complex conjugate) transpose.) Because of the appearance of higher powers of σ, this equation is different from the ODE case if the delay is larger than the period. Adding (12.30) to the defining system of the periodic solution (12.20) doubles the size of the problem. The size of (12.30) can be reduced to $n + 1$ by using characteristic matrices that are equivalent to the operator in (12.30) [80]. However, the smallest possible addition would consist of only one additional scalar equation to (12.20) without introducing new variables. This can be achieved by using the bordering theorem [9], which states that the bordered operator

$$\begin{pmatrix} D & \beta \\ \alpha^H & \delta \end{pmatrix} = \begin{pmatrix} A & b \\ c^H & 0 \end{pmatrix}^{-1},$$

exists if both A and A^H have one-dimensional kernels and $b \notin \ker A^H$, $c \notin \ker A$ or A is bijective and $c^H A^{-1} b \neq 0$. Moreover, δ can be used as a test functional of the singularity, because it is zero if, and only if, A is singular. In order to obtain δ it is sufficient to solve the equation

$$\begin{pmatrix} A & b \\ c^H & 0 \end{pmatrix} \begin{pmatrix} \beta \\ \delta \end{pmatrix} = \begin{pmatrix} 0 \\ 1 \end{pmatrix}. \tag{12.30}$$

Hence, using a discretized version of $\hat{A} - \sum_{j=1}^{N} \sigma^j \hat{B}_j$ for the operator A in (12.30) with appropriate choices of b and c^H in the period-doubling and Neimark-Sacker cases, the equation $\delta(x^*, \eta) = 0$ determines the bifurcation point. In a continuation context the resulting β can be re-used as the value of c in the next continuation step. Similarly, by solving the adjoint equation

$$\begin{pmatrix} A^H & c \\ b^H & 0 \end{pmatrix} \begin{pmatrix} \alpha \\ \delta \end{pmatrix} = \begin{pmatrix} 0 \\ 1 \end{pmatrix},$$

the resulting α can be the new value of b in the next continuation step.

In the case of a fold bifurcation in an autonomous system (12.1), because of the algebraically double $+1$ multiplier, the operator A has to be

$$A_{LP} = \begin{pmatrix} \hat{A} - \sum_{j=1}^{N} \sigma^j \hat{B}_j & \phi_0 \\ \mathrm{Int}_{\psi_0} & 0 \end{pmatrix},$$

where

$$\psi_0 = f(x(t), x(t - \tau_1), x(t - \tau_2), \ldots, x(t - \tau_m), \eta),$$

$$\phi_0 = -\sum_{j=1}^{N} j \hat{B}_j \psi_0,$$

and

$$\mathrm{Int}_{\psi_0} \phi = \int_{-1}^{0} \psi_0(\theta) \phi(\theta) \, d\theta.$$

Note that A_{LP} is different from what one would expect by analogy with ODEs; see [20]. Here, ϕ_0 is obtained by computing the Jordan chain of $\hat{A} - \sum_{j=1}^{N} \sigma^j \hat{B}_j$; see, e.g., [48]. The regularity of δ obtained from A_{LP} at the bifurcation point can be proven either by using the equivalence with characteristic matrices [80] or by standard techniques [20].

12.4 Connecting Orbits

A solution $x^*(t)$ of (12.1) at some fixed value of the parameter η is called a *connecting orbit* if the limits

$$\lim_{t \to -\infty} x^*(t) = x^- \quad \text{and} \quad \lim_{t \to +\infty} x^*(t) = x^+ \tag{12.31}$$

exist, where x^- and x^+ are steady states of (12.1). We call the orbit homoclinic when $x^- = x^+$, and heteroclinic otherwise. Orbits of this type exist, for instance, in laser models with optical feedback, which are discussed in Sect. 12.8.1; see also [35]. They also appear naturally when looking for traveling waves in delay partial differential equations [70].

A defining condition for a connecting orbit is that it is contained in both the stable manifold of x^+ and the unstable manifold of x^-. A classical approach in the ODE case is to approximate this condition by truncating the time domain to an interval of length T_c and to apply (so-called) projection boundary conditions [8]: one end point of the connecting orbit is required to lie in the unstable eigenspace of x^- and the other end point in the stable eigenspace of x^+. The projection boundary conditions, therefore, replace the stable and unstable manifolds by their linear approximations near the steady states.

Here, the boundary conditions need to be written in terms of solution segments. Furthermore, x^+ has infinitely many eigenvalues with negative real parts (see Sect. 12.1) and so it is impossible to write the final function segment as a linear combination of all stable eigenfunctions. Instead, it is required that the end function segment is in the orthogonal complement of all unstable left eigenfunctions. We will assume for notational convenience that (12.1) only contains one delay τ; however, the method is implemented in DDE-BIFTOOL for the general case of m fixed delays.

The condition for the initial function segment $x_0(\theta)$ can be written as

$$x_0(\theta) = x^- + \varepsilon \sum_{k=1}^{s^-} \alpha_k v_k^- e^{\lambda_k^- \theta} \qquad \left(\sum_{k=1}^{s^-} |\alpha_k|^2 = 1 \right),$$

where s^- is the number of unstable eigenvalues λ^-, with corresponding eigenvectors v^-. The α_k are unknown coefficients, and ε is a measure for the desired accuracy. An extra condition is added to ensure continuity at $\theta = 0$. Since we cannot write the end conditions for the final function segment in a similar way, a special bilinear form [38] is used to express the fact that the final function segment is in the complement of the unstable eigenspace of x^+. This leads to the s^+ extra conditions of the form

$$w_k^{+H} \left(x(T_c) - x^+ \right) + \int_{-\tau}^0 w_k^{+H} e^{-\lambda_k^+ (\theta+\tau)} A_1(x^+, \eta) \left(x(T_c + \theta) - x^+ \right) d\theta = 0,$$

where $k = 1, \dots, s^+$. Here s^+ is the number of unstable eigenvalues of x^+, w_k^+ are the left eigenvectors corresponding to the eigenvalues λ_k^+, and the matrix A_1 is defined as in (12.3). While this integral condition works well in practice, one slight drawback is that it does not control the distance of the end function segment to the steady state.

As for periodic solutions, connecting orbits arise in one-parameter families and any time-translate is also a connecting orbit. Therefore, a phase condition such as (12.14) needs to be added to select one of these orbits.

For the case of a one-parameter family of connecting orbits a number of free parameters are required to obtain a generically isolated solution. One has to solve (12.1) together with the steady-state equations (12.2) for x^- and x^+ and characteristic equations of the form (12.8) for λ_k^- and v_k^- and λ_k^+ and w_k^+, i.e., a system of n differential equations, supplemented with $(s^- + s^+)(n + 1) + 2n + s^+ + 2$ extra equations, resulting in the need for $s_\eta = s^+ - s^- + 1$ free parameters. This leads to a boundary value problem, which is coupled to a number of algebraic constraints for the equilibria and their stability. The boundary value problem can be solved by a collocation method as in Sect. 12.2.1.

Good starting conditions for Newton's method can be obtained as follows. For a homoclinic orbit, one can start from a nearby periodic solution with a sufficiently large period. Heteroclinic orbits can be approximated by using time integration or by using an extension of the method of successive continuation [19]. Details of the method, including a numerical study of the convergence, are presented in [69].

12.5 Quasiperiodic Tori

In dynamical systems quasiperiodic solutions reside on invariant tori. In this section we describe a method to compute two-dimensional tori as periodic functions on the unit square. In particular we adapt the method of Schilder et al. [72], which uses a finite difference method to discretize the defining equation. Here we use a spectral collocation method that is well suited to delay equations.

A quasiperiodic solution $x(t)$ of (12.1) has two rationally independent frequencies ω_1, ω_2. Hence, there exists a function $y : \mathbb{R}^2 \to \mathbb{R}^n$, which is 2π-periodic in both variables, such that x can be written as $x(t) = y(\omega_1 t_1, \omega_2 t_2)$. Putting u into (12.1) yields a first-order delayed partial differential equation

$$\frac{\partial}{\partial t_1} u(t_1, t_2) + \frac{\omega_2}{\omega_1} \frac{\partial}{\partial t_2} u(t_1, t_2) = \frac{1}{\omega_1} f(u(t_1, t_2), u(t_1 - \omega_1 \tau_1, t_2 - \omega_2 \tau_1), \ldots$$
$$\ldots, u(t_1 - \omega_1 \tau_m, t_2 - \omega_2 \tau_m), \eta), \quad (12.32)$$

where ω_1, ω_2 are unknown frequencies. Because there are translational symmetries in both variables of u, two phase conditions have to be imposed on u in order to fix a unique solution and determine the unknown frequencies. Assuming that we have a reference solution $u^{(0)}$ of (12.32) at η_0, we formulate a condition that minimizes the distance of u at η from $u^{(0)}$, i.e.,

$$\kappa(\theta_1, \theta_2) = \frac{1}{(2\pi)^2} \int_0^{2\pi} \int_0^{2\pi} \|u(t_1 + \theta_1, t_2 + \theta_2) - u^{(0)}(t_1, t_2)\|_2^2 \, dt_1 dt_2.$$

Taking the first derivative of κ with respect to θ_1 and θ_2, the phase conditions become

$$\frac{1}{(2\pi)^2} \int_0^{2\pi} \int_0^{2\pi} \frac{\partial}{\partial t_1} u^{(0)}(t_1, t_2) u(t_1, t_2) \, dt_1 dt_2 = 0,$$

$$\frac{1}{(2\pi)^2} \int_0^{2\pi} \int_0^{2\pi} \frac{\partial}{\partial t_2} u^{(0)}(t_1, t_2) u(t_1, t_2) \, dt_1 dt_2 = 0.$$

In the case of time-periodic systems only the second phase condition is necessary, since the phase in t_1 is fixed by the phase of the forcing. In addition to the phase conditions, we also need boundary conditions that guarantee the periodicity of u, that is,

$$u(0, t_2) = u(2\pi, t_2) \quad \text{and}$$
$$u(t_1, 0) = u(t_1, 2\pi), \quad \forall \, t_1, t_2 \in [0, 2\pi].$$

12.5.1 Spectral Collocation

To obtain an approximation of the quasiperiodic solution the defining sets of equations can be discretized with an appropriate numerical scheme and solved by Newton's method. There are several different spectral collocation methods for partial differential equations that could be used to solve (12.32); for an introduction see Trefethen [81]. Here we use a method that was developed for computationally challenging hyperbolic equations such as the Navier-Stokes equation. The method is a multi-domain spectral collocation method called the staggered grid Chebyshev method, developed by Kopriva and Kolias [50].

The method is similar to the collocation of periodic solutions. It uses piecewise polynomials that are represented by their values at discrete points of a mesh, which is different from the mesh on which the equation is solved. We use a very simple domain subdivision of the area $[0, 2\pi] \times [0, 2\pi]$ that splits it into the rectangles

$$D_{i,j} = [t_1^i, t_1^{i+1}] \times [t_2^j, t_2^{j+1}],$$

where $\{0 = t_l^0 < t_l^1 < \cdots < t_l^{L_l} = 2\pi\}$ with $l \in \{1, 2\}$. On each rectangle $D_{i,j}$ we use the Lobatto points $(t_1^{i,p}, t_2^{j,q}) = (t_1^i + b_1^p(t_1^{i+1} - t_1^i), t_2^j + b_2^q(t_2^{j+1} - t_2^j))$ to represent the solution

$$u(t_1, t_2) = \sum_{p=0}^{d_1} \sum_{q=0}^{d_2} u(t_1^{i,p}, t_2^{j,q}) P^{i,j,p,q}(t_1, t_2), \qquad (12.33)$$

where $P^{i,j,p,q}$ are the Lagrange polynomials through the points $(t_1^{i,p}, t_2^{j,q})$. The function u is now completely determined by the values

$$u_{i,j,p,q} := u(t_1^{i,p}, t_2^{j,q}),$$

which we consider identical if they represent the same point in $[0, 2\pi] \times [0, 2\pi]$. We also need to impose the boundary conditions, which are

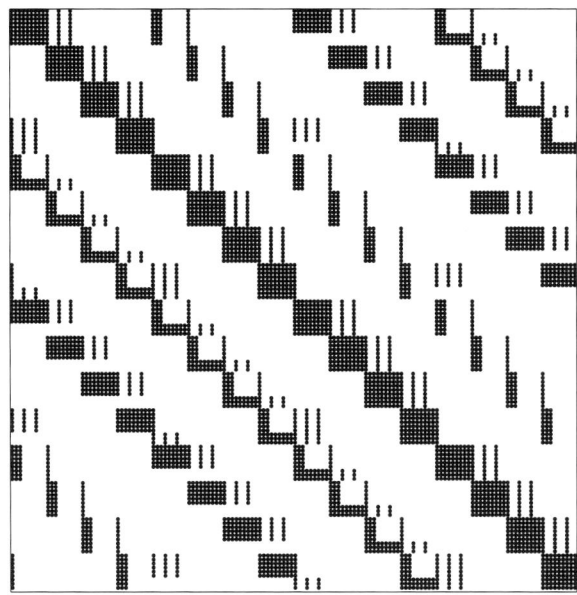

Fig. 12.3. Sparsity structure of the Jacobian of the discretization of (12.32). The seemingly irregular pattern is due to the patching of the rectangles defined by (12.34); parameters are $L_1 = L_2 = 4$, $d_1 = d_2 = 3$ and $n = 1$.

$$u_{0,j,0,q} = u_{L_1-1,j,d_1,q}, \ 0 \leqslant q < d_2, \ 0 \leqslant j < L_2 - 1,$$
$$u_{i,0,p,0} = u_{i,L_2-1,p,d_2}, \ 0 \leqslant p < d_1, \ 0 \leqslant i < L_1 - 1 \text{ and} \qquad (12.34)$$
$$u_{0,0,0,0} = u_{L_1-1,L_2-1,d_1,d_2}.$$

It is also possible to think of the piecewise polynomials as discontinuous in the interfaces and define mortar equations as in spectral penalty methods (see, e.g., Hesthaven [45]) instead of (12.34).

Equation (12.32) is solved on the grid

$$(\hat{t}_1^{i,p}, \hat{t}_2^{j,q}) = (t_1^i + c_1^q(t_1^{i+1} - t_1^i), t_2^j + c_2^q(t_2^{j+1} - t_2^j)),$$
$$0 \leqslant i < L_1, 0 \leqslant j < L_2, 0 \leqslant p < d_1, 0 \leqslant q < d_2,$$

where c_1^p, c_2^q are the Gauss points. Using the polynomial representation (12.33) of u in (12.32) and evaluating at $(\hat{t}_1^{i,p}, \hat{t}_2^{j,q})$ yields a large algebraic system that can be solved by Newton's method. The typical sparsity structure of the Jacobian of this discretized system is shown in Fig. 12.3, but without the borders accounting for the phase conditions.

12.6 Further Classes of Delay Equations

We now briefly review continuation methods for systems that include delays that depend on the state variables or whose right-hand side includes delayed derivatives of the state.

12.6.1 State-Dependent Equations

We briefly describe how a DDE with state-dependent delay (sd-DDE) can be handled; we assume for simplicity that only one delay is present. An sd-DDE can be of the form

$$\begin{cases} \frac{\mathrm{d}}{\mathrm{d}t}x(t) = f_1(x(t), x(t - \tau(x(t))), \eta), \\ \tau(x(t)) = g_1(x(t)), \end{cases} \quad (12.35)$$

where $g_1 : \mathbb{R}^n \to \mathbb{R}$ is a given (explicit) function of the solution $x(t)$, or it can be of the form

$$\begin{cases} \frac{\mathrm{d}}{\mathrm{d}t}x(t) = f_2(x(t), x(t - \tau(t)), \tau(t)), \\ \frac{\mathrm{d}}{\mathrm{d}t}\tau(t) = g_2(x(t), x(t - \tau(t)), \tau(t)), \end{cases} \quad (12.36)$$

where $g_2 : \mathbb{R}^n \times \mathbb{R}^n \times \mathbb{R} \to \mathbb{R}$, and the delay is determined by a differential equation. We assume that all functions in (12.35) and (12.36) are sufficiently smooth and that the delay is bounded, i.e., $0 \le \tau(t) \le r$, $\forall t$. Note that, using $x_1 \equiv x$ and $x_2 \equiv \tau$, (12.36) can be considered as a particular case of (12.35) with the extended state $x \equiv (x_1, x_2)$.

A steady-state solution of an sd-DDE is determined by the state x and the delay τ, i.e., the delay should be considered as a part of the solution. A steady-state solution (x^*, τ^*) of (12.35) or (12.36) can be computed by solving a (nonlinear) algebraic system. The local stability of steady-state solutions of sd-DDEs was studied in [15, 41]. It was shown, under natural assumptions on the right-hand side of the equation and on the delay function τ, that generically the behavior of the state-dependent delay τ, except for its value τ^*, has no effect on the stability, and that in the local linearization τ can be treated as a constant. Hence, to study the local stability of a steady state of (12.35) or (12.36), these equations can be linearized at x^* by setting $\tau \equiv \tau^*$. The resulting linearized equation is a DDE with constant delay, and the numerical procedures discussed in Sect. 12.1 can be used without changes [57].

The existence of periodic solutions for particular cases of sd-DDEs has been studied by several authors, in particular the existence of 'slowly oscillating periodic solutions'. The theory suggests that a Hopf bifurcation theorem holds; see, e.g., [61]. The stability of periodic solutions has only recently been studied; see, e.g., [42] for non-autonomous sd-DDEs. It was proven that the Fréchet derivative of the solution operator of the nonlinear sd-DDE with respect to initial data equals the solution operator of the linearized equation. Based on these results (12.35) and (12.36) can be linearized around a (nonconstant) solution $(x^*(t), \tau^*(t))$ as follows. Let $D_j f_i$ denote the derivative of f_1 with respect to its jth argument, then

$$\tfrac{\mathrm{d}}{\mathrm{d}t}y(t) = D_1 f_1(s)y(t) - D_2 f_1(s)\tfrac{\mathrm{d}}{\mathrm{d}t}x^*(t - \tau(x^*(t)))\tfrac{\partial}{\partial x}\tau(x^*(t))y(t)$$
$$+D_2 f_1(s)y(t - \tau(x^*(t))), \qquad (12.37)$$

with $s = (x^*(t), x^*(t - \tau(x^*(t))))$, respectively, and

$$\begin{cases} \tfrac{\mathrm{d}}{\mathrm{d}t}y_1(t) = D_1 f_2(s)y_1(t) + D_2 f_2(s)y_1(t - \tau^*(t)) - D_2 f_2(s)\tfrac{\mathrm{d}}{\mathrm{d}t}x^*(t - \tau^*(t))y_2(t) \\ \qquad\quad + D_3 f_2(s)y_2(t), \\ \tfrac{\mathrm{d}}{\mathrm{d}t}y_2(t) = D_1 g_2(s)y_1(t) + D_2 g_2(s)y_1(t - \tau^*(t)) - D_2 g_2(s)\tfrac{\mathrm{d}}{\mathrm{d}t}x^*(t - \tau^*(t))y_2(t) \\ \qquad\quad + D_3 g_2(s)y_2(t), \end{cases}$$

with $s = (x^*(t), x^*(t - \tau^*(t)), \tau^*(t))$. These linearized equations contain a time-dependent (no longer state-dependent) delay. If the coefficients in the linear equation are smooth and periodic (with period T) and the delay function is smooth, then the solution operator over the period T (over an interval mT if $\tau_m > T$ and $mT \geq \tau_m$, $\tau_m = \max_{t \in [0,T]} \tau(t)$) is compact [38].

A periodic solution can be computed by solving a two-point boundary value problem in time, similar to (12.13), but in the case of (12.36) the additional equation $\tau(0) = \tau(T)$ must be imposed. The solution of these boundary value problems by collocation and the computation of the Floquet multipliers is conceptually equal to the procedure outlined in Sect. 12.2; see [57].

12.6.2 Collocation Schemes for Equations of Neutral Type

We summarize basic results on two collocation schemes that were proposed in Barton et. al. [6]. Here we consider the simple equation of neutral type

$$\dot{x}(t) = f(x(t), x(t - \tau), \dot{x}(t - \tau), \eta). \qquad (12.38)$$

The collocation scheme of Sect. 12.2.1 discretizes (12.38) by substituting the collocation polynomials and evaluating at the collocation points. In the Jacobian matrix of this discretized system the second derivatives of the polynomials appear. This reduces the accuracy by an order, which is only $O(h^m)$. This drop in the order of convergence is apparent in the examples of [6]. To remedy the situation (12.38) can be transformed into an ODE coupled to a difference equation

$$\dot{x}(t) = y(t) \qquad (12.39)$$
$$y(t) = f(x(t), x(t - \tau), y(t - \tau)); \qquad (12.40)$$

see [6]. Applying the collocation scheme of Sect. 12.2.1 to this system does not introduce second-order derivatives in the Jacobian matrix and, hence, a better convergence can be expected. The numerical experiments in [6] show a convergence rate of $O(h^{m+1})$. In [6] the Gauss-Legendre points were used in the collocation scheme, together with a periodic boundary condition on the algebraic part, but other approaches are possible for delay differential algebraic equations.

12.7 Software Packages

Several software packages exist for simulation (time integration) of delay differential equations, including ARCHI [66], DKLAG6 [16], RADAR5 [36] and XPPAUT [29]. Furthermore, Matlab now contains the solver dde23 [71]. Probably the earliest computer program specifically designed for DDEs has been published by Hassard [43], namely BIFDD which allows a normal-form analysis of Hopf bifurcation points. XPPAUT by Ermentrout [29] allows a limited stability analysis of steady-state solutions of DDEs using the approach described in [58].

By contrast, the software packages DDE-BIFTOOL and PDDE-CONT implement numerical continuation of DDEs as introduced in the previous sections. In this section we describe the functionality of these numerical tools.

12.7.1 DDE-BIFTOOL

The package DDE-BIFTOOL consists of a collection of Matlab-routines for the numerical continuation and bifurcation analysis of systems of DDEs with multiple discrete delays, which may be fixed or state-dependent; for detailed instructions we refer to the user manual [27]. This software allows one to compute branches of steady-state solutions and steady-state fold and Hopf bifurcations with continuation. Given an equilibrium, it allows one to approximate the right-most, stability-determining roots of the characteristic equation, which can be further corrected with Newton's method. Periodic solutions and their Floquet multipliers can also be computed by collocation with adaptive mesh selection. Branches of periodic solutions can be continued starting from a previously computed Hopf point or from an initial guess of a periodic solution profile. For DDEs with constant delays, connecting orbits (both homoclinic and heteroclinic solutions) can also be computed. The numerical methods that are used in the software are as detailed in the previous sections.

DDE-BIFTOOL has no graphical user interface, but a number of routines are provided to plot solution, branch and stability information. Furthermore, automatic detection of bifurcations is not supported. Instead, the evolution of the characteristic roots or the Floquet multipliers can be computed along solution branches, which allows the user to detect and identify bifurcations using an appropriate visualization. Starting points for branch switching at bifurcations on branches of steady-state and periodic solutions can be generated, as well as starting solutions for homoclinic solutions close to periodic solutions.

Several extensions or 'add-ons' have been developed. We mention here a Mathematica program written by Pieroux that allows the automatic generation of the system definition files with symbolically obtained derivatives, software written by Green for the computation of one-dimensional unstable manifolds in DDEs [34], and the extension by Barton for equations of neutral type [6].

12.7.2 PDDE-CONT

PDDE-CONT implements the numerical methods described in Sect. 12.2. It is written in C++ with the use of linear algebra packages UMFPACK [17], LAPACK [1] and ARPACK [54]. The software has a command line interface and a graphical user interface together with a basic plotting facility.

PDDE-CONT can continue periodic solutions of delay equations that are in the form

$$y'(t) = f(t, y(t), y(t - \tau_1(t)), y(t - \tau_2(t)), \ldots, y(t - \tau_m(t)), \eta).$$

The right-hand side f and the delays τ_j can be either T-periodic or time independent. The software does not have any algorithms to continue equilibria apart from the obvious fact that an equilibrium can be considered as a constant periodic solution. Bifurcations of periodic solutions can be continued in two parameters by using test functions as described in Sect. 12.3, but PDDE-CONT cannot switch branches automatically. For detailed instructions see the user manual [78]. Note that PDDE-CONT can be used together with DDE-BIFTOOL by converting the results between the two packages.

Due to the implementation in C++, the performance of PDDE-CONT is significantly better than that of DDE-BIFTOOL (which is implemented in Matlab). Furthermore, PDDE-CONT uses sparse-matrix algorithms that require less memory, so that problems of relatively high dimension can be tackled. The resulting large bordered linear systems (see Sect. 12.2.1) are solved by using bordering techniques from [31, 32]. The large sparse matrix without borders is factorized by UMFPACK and the whole system is solved using the BEMW method [32].

12.8 Examples of Numerical Bifurcation Analysis of DDEs

In this section we illustrate the performance of the numerical techniques described in the previous sections with examples of DDE models of a number of physical and biological phenomena.

12.8.1 DDE-PDE Model of a Laser with Optical Feedback

A longitudinally single-mode semiconductor laser subject to conventional optical feedback and lateral carrier diffusion can be modeled by the hybrid DDE-PDE system

$$\frac{\mathrm{d}A(t)}{\mathrm{d}t} = (1 - \mathrm{i}\alpha)A(t)\zeta(t) + \eta A(t - \tau)e^{-\mathrm{i}\phi} - \mathrm{i}bA(t), \quad (12.41)$$

$$T\frac{\partial Z(x,t)}{\partial t} = d\frac{\partial^2 Z(x,t)}{\partial x^2} - Z(x,t) + P(x)$$
$$- F(x)(1 + 2Z(x,t))|A(t)|^2. \quad (12.42)$$

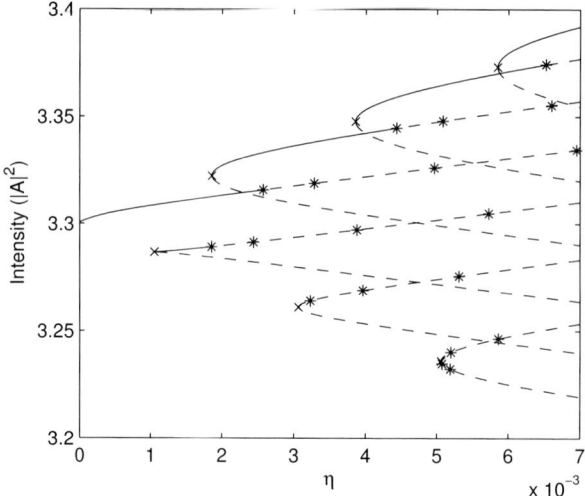

Fig. 12.4. Bifurcation diagram in the plane of intensity $|A|^2$ vs. feedback strength η of steady-state solutions of (12.41)–(12.42) for $\alpha = 3$, $\phi = 0$, $T = 1000$, $d = 1.68 \times 10^{-2}$ and $\tau = 1000$. Stable solutions are drawn as solid curves and unstable solutions as dashed curves; also shown are saddle-node bifurcations ($\times$) and Hopf bifurcations (*). From K. Verheyden, K. Green, and D. Roose, Numerical stability analysis of a large-scale delay system modeling a lateral semiconductor laser subject to optical feedback, *Phys. Rev. Lett.* **69**(3) (2004) 036702 © 2004 by the American Physical Society; reprinted with permission.

Here the complex scalar variable $A(t)$ represents the amplitude of the electrical field $E(t) = A(t)e^{ibt}$, and the real variable $Z(x,t)$, $x \in [-0.5, 0.5]$, represents the carrier density [82]. The functions $\zeta(t)$, $P(x)$ and $F(x)$ are specified in [82]. Continuous-wave solutions, called 'external cavity modes' (ECMs) can be computed as steady-state solutions of (12.41)–(12.42), augmented with a scalar condition for the unknown b and an extra scalar constraint to remove the S^1-symmetry. Zero Neumann boundary conditions for $Z(x,t)$ are imposed at $x = \pm 0.5$. In the computations the time variable is rescaled by a factor of 1000 so that most quantities in the computation are of order one. The symmetry about $x = 0$ is exploited by considering only the interval $[0, 0.5]$. Splitting (12.41) into real and imaginary parts and discretizing (12.42) in space with a second-order central difference formula with constant stepsize $\Delta x = 0.5/128$ results in a DDE system of size $n = 131$.

Figure 12.4 shows the bifurcation diagram of steady-state solutions of (12.41)–(12.42) with $\alpha = 3$, $\phi = 0$, $T = 1000$, $d = 1.68 \times 10^{-2}$ and $\tau = 1000$, obtained by continuation with DDE-BIFTOOL, with the feedback strength η as the parameter. The diagram shows several branches of steady-state solutions arising from saddle-node bifurcations. During continuation the right-

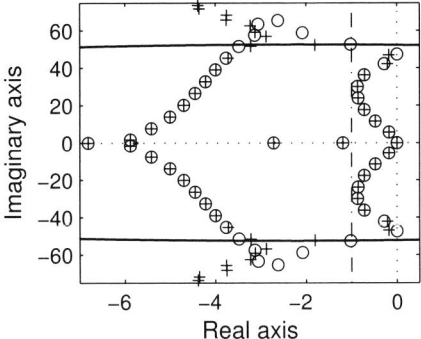

Fig. 12.5. Characteristic roots at the Hopf point of (12.41)–(12.42) $\eta \approx 2.5717 \times 10^{-3}$ from Fig. 12.4. Shown are approximations of characteristic roots with real part larger than $r = -1$ derived from the eigenvalues of the discretization of the solution operator by the sixth-order special-purpose LMS method (+), and their corrections by Newton's method (○). From K. Verheyden, T. Luzyanina, and D. Roose, Efficient and reliable stability analysis of solutions of delay differential equations, *Proceedings of 2006 International Conference on Nonlinear Science and Complexity*, 109–120 © 2007 by World Scientific Publishing; reprinted with permission.

most characteristic roots are computed and monitored, allowing for the detection of Hopf bifurcation points along these branches.

Figure 12.5 shows the characteristic roots at the moment of the first Hopf bifurcation on the middle branch at $\eta \approx 2.5717 \times 10^{-3}$. Since the imaginary part of the right-most pair of characteristic roots is large, the system presents a challenging test case for characteristic root calculation with DDE-BIFTOOL.

Approximations to the characteristic roots were obtained by computing the eigenvalues of the matrix approximation to the solution operator with a sixth-order LMS method, optimized to retain the stability properties of the linearized equation. The steplength h in the LMS method is automatically determined to ensure that all characteristic roots with real part larger than $r = -1$ (threshold specified by the user) are approximated accurately. This leads to the discretization of the delay interval with an equidistant mesh of 27 points. The resulting eigenvalue problem has dimension $131 \times 27 = 3537$, which is large but can still be solved by using the QR-method. These approximations are subsequently corrected by Newton's method applied to (12.8). The approximate characteristic roots shown in Fig. 12.5 were derived from the eigenvalues of the discretization of the solution operator, and their corrections by Newton's method.

For this example a comparison of the computation of the characteristic roots using the pseudo-spectral discretizations of the infinitesimal generator and of the solution operator is presented in [85]. In both cases, a polynomial of degree $p = 32$ is used, so that the linear eigenvalue problems have size

Table 12.1. The computational cost of four algorithms based on a pseudo-spectral discretization of the infinitesimal generator $\mathcal{A}$ and the solution operator $\mathcal{S}(h)$ when using polynomials of degree $p = 32$ to find the right-most characteristic roots shown in Fig. 12.5.

	$\mathcal{A}$		$\mathcal{S}(h)$	
	Right-most	Shift-Invert	Forward	Backward
CPU time (seconds)	106.2	94.1	103.4	55.5
# matrix-vector products	6528	4951	6254	3146

$n(p + 1) = 131 \times 33 = 4323$. Table 12.1 shows the computational cost of the four methods.

To solve the linear eigenvalue problem resulting from the pseudo-spectral discretization of the infinitesimal generator, the Matlab function `eigs` function is used to compute the right-most 30 eigenvalues with a requested tolerance of 10^{-8} (results indicated with 'Right-most'). Note that `eigs` uses Arnoldi's method with implicit restart, and this method does not require the explicit construction of the matrix. For the results indicated with 'Shift-Invert', `eigs` is used in conjunction with the shift-invert technique and returns the eigenvalues λ closest to the shift $\|A_0\| + \|A_1\| \approx 4528.5$, as proposed in [10]. The pseudo-spectral discretization of the solution operator $\mathcal{S}(h)$ leads to two algorithms, called forward and backward variants in [85]. The steplength h is chosen to be 10^{-4} for the forward and 10^{-3} for the backward variant, respectively.

The accuracy of the computed characteristic roots is similar for the four methods. For example, the roots $-0.285 \pm i11.8$ are computed by the four algorithms with a relative error between $6.5\ 10^{-12}$ and $2.4\ 10^{-14}$. The accuracy is lower for the eigenvalues with large imaginary part (the relative error on the purely imaginary eigenvalues $\pm i47.7$ is $\approx 10^{-4}$ for the backward variant, and $\approx 7\ 10^{-6}$ for the three other algorithms. The exponential convergence with respect to the degree p has been confirmed by numerical experiments.

12.8.2 The Mackey-Glass Equation

The equation

$$\dot{x}(t) = ax(t) + b\frac{x(t - \tau)}{1 + x^{10}(t - \tau)} \tag{12.43}$$

models the regeneration of white blood cells [59], and it is today widely known as the Mackey-Glass equation. Although it is a simple equation, not much is known about its solution structure.

The three equilibria of (12.43), i.e., $x_1 = 0$ and $x_{2,3} = \pm \sqrt[10]{-(a + b)/a}$ are connected to each other at $a = -b$ by a supercritical pitchfork bifurcation. The nonzero solutions can lose their stability in a Hopf bifurcation along the curves in parameter space given by

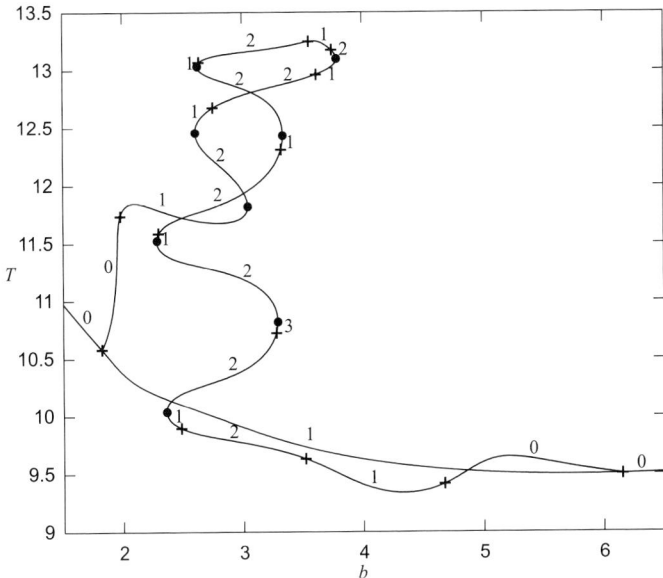

Fig. 12.6. Bifurcation diagram showing period-two solutions of (12.43) for fixed $a = -1.2158$. Fold bifurcations are denoted by dots while period-doubling bifurcations are denoted by +; the numbers of unstable characteristic multipliers are indicated along the different branches.

$$a = -\arccos(-d^{-1})\frac{1}{\tau\sqrt{d^2 - 1}},$$

$$b = \frac{10a}{d - 9},$$

where $|d| \geq 1$. Hopf bifurcations for $d > 1$ are supercritical, so they give rise to stable periodic solutions. These periodic solutions bifurcate further via several period doublings, which then leads to chaotic motion. It was demonstrated in [39] that chaos arises due to the transverse intersection of the two-dimensional unstable and infinite-dimensional stable manifolds of this periodic solution. We remark that some square-shaped solutions of large period can be obtained by singularly perturbing a map to give

$$\varepsilon\dot{x}(t) = ax(t) + b\frac{x(t-1)}{1 + x^c(t-1)},$$

where $\varepsilon \to 0$ and $\varepsilon\tau = 1$; see [60] for details.

Here we analyze the period-two solutions bifurcating from the period-one solutions that in turn can be related to the supercritical Hopf bifurcation of the equilibrium. These solutions form a complicated branch structure that is challenging to compute. In Fig. 12.6 the bifurcation diagram for $a = -1.2158$

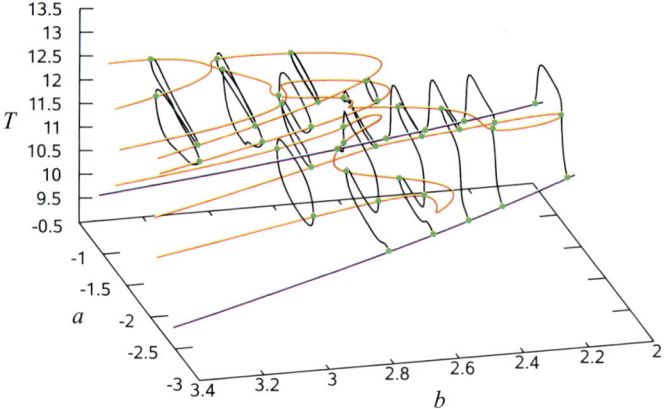

Fig. 12.7. The structure of the period-two solutions of (12.43) over the (a, b)-plane; shown are solution branches (black curves), their fold bifurcations (green dots), period-doubling bifurcations of the period-one solution (blue curves) and fold bifurcations of period-two solutions (red curves).

and $\tau = 2$ is shown. As b varies the solution undergoes several fold and period-doubling bifurcations. These solutions are almost all unstable, and so they cannot be found by simulation; the number of unstable characteristic multipliers is shown along the branches in Fig. 12.6. Furthermore, the period-one solution branch is included, but with twice the period so that it matches up with the branch of period-two solutions.

By investigating the fold bifurcations and computing several branches of solutions we can obtain a fairly complete picture of the structure of periodic solutions. Figure 12.7 shows this structure for the fixed delay $\tau = 2$, where we plot the period T of the solutions as a function of the parameters a and b.

12.8.3 Traffic Model with Driver Reaction Time

The traffic model in Orosz et al. [64, 65] describes the dynamics of N cars on a circular track. Each car has a velocity v_i and an associated headway h_i defined as the distance to the car in front. The headways h_i are calculated from the velocities as

$$\dot{h}_i(t) = v_{i+1}(t) - v_i(t). \tag{12.44}$$

Because of the circular track, we assume that $v_{N+1} = v_1$ and $h_N = L - \sum_{i=1}^{N-1} h_i$. Each car tries to reach its optimal velocity, which is a function of the headway that can be expressed as

$$\dot{v}_i(t) = \beta(V(h_i(t-1) - v_i(t)), \tag{12.45}$$

where β is the sensitivity to velocity differences. Due to the reaction time of the driver, a delayed value of the headway is used in the model. The optimal

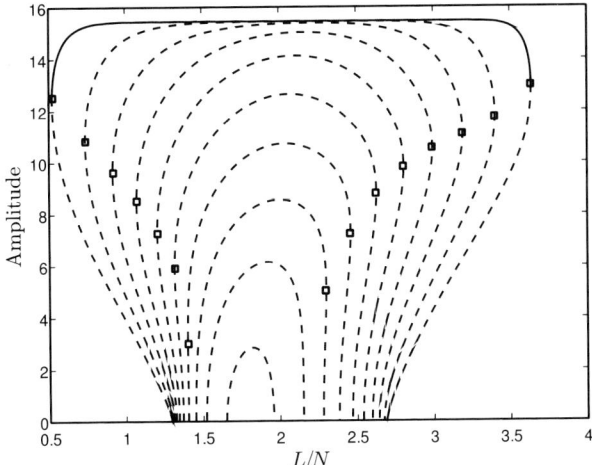

Fig. 12.8. Periodic solution branches of (12.44)-(12.45). Unstable solutions are denoted by dashed lines, continous lines refer to stable solutions and boxes denote fold bifurcations.

velocity is a function of the headway, and it is modeled by the optimal velocity function

$$V(h) = \begin{cases} 0 & 0 \le h \le 1, \\ v^0 \, \frac{(h-1)^3}{1+(h-1)^3} & h > 1. \end{cases}$$

By making use of the algebraic condition for h_N, one can reduce the dimension of system (12.44)–(12.45) by one to $2N - 1$.

In this section we consider $N = 17$ cars, which is the largest number of cars that was considered and (partially) analyzed in [64]. Our starting point is the steady-state solution of the model, which corresponds to equal headways and equal car velocities and so is given by

$$h_i^* = L/N, \quad v_i^* = V(h_i^*).$$

The steady state undergoes several Hopf bifurcations from which branches of periodic solutions arise; they are shown in Fig. 12.8 as computed with PDDE-CONT for typical parameter values of $\beta = 1$ and $\tau = 2$ as a function of the average headway L/N. Note how all branches of periodic solutions feature folds and connect pairs of subcritical Hopf bifurcations. The outer-most branch is stable between the folds, which shows that there is bistability between stable periodic solutions (indicating a traffic jam) and the stable steady state (uniform flow of cars). The other branches of periodic solutions remain unstable throughout, but the outer-most of them has all its unstable Floquet multipliers very close to 1 (for l/N around 2), which means that the

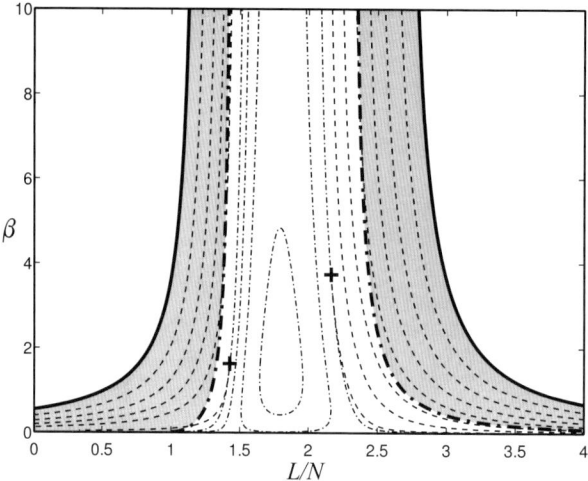

Fig. 12.9. Hopf and fold bifurcation curves of (12.44)–(12.45). The gray regions are bistable, where a stable equilibrium coexists with a stable periodic solution. Between the gray regions the equilibrium is unstable, while outside the gray regions the equilibrium is globally stable.

corresponding periodic solution can be observed as long transients. Physically, these transients give rise to traffic jams, which move towards each other and eventually either merge with the stable traffic jam or disperse [64].

Figure 12.9 shows the curves of fold bifurcations for $N = 17$ cars in the $(L/N, \beta)$-plane. The plot also shows some Hopf bifurcation curves (dash-dotted lines) and points of degenerate Hopf bifurcations ($+$); the regions of bistability are highlighted in gray. In [64] a similar image was computed for $N = 9$ cars with DDE-BIFTOOL by performing one-parameter continuations in L/N for many values of fixed β to find the fold bifurcations. (The locus of Hopf bifurcations is actually known analytically.) As this approach is very time consuming, we used PDDE-CONT instead, which is able to follow the fold bifurcation curves directly in two parameters.

12.8.4 Chatter Motion in Milling

Cutting processes are often subject to the so-called regenerative effect [77], which comes from the fact that a cutting tool always cuts a surface that was produced by the same tool some time ago. The cutting forces nonlinearly depend on the chip geometry, which in turn depends on the current and a delayed tool position. The underlying dynamics of the tool can be considered to be linear and, hence, the nonlinearity comes from the geometry of the chip forming and the cutting force only. There is a vast literature on the dynamics of machining that mainly focuses on the stability of steady cutting; see, e.g.,

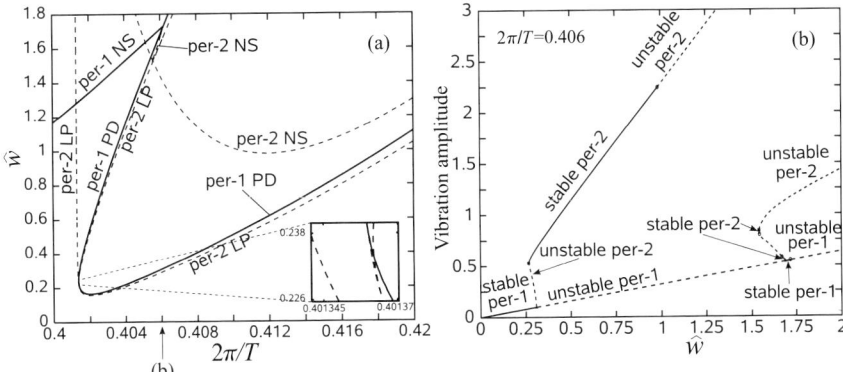

Fig. 12.10. Stability chart of (12.46) in the cutting speed and chip width parameter plane (a); the relative damping is $\zeta = 0.0038$ and the tool cuts continuously for a time of 10.82% of every period. Panel (b) shows a bifurcation diagram for fixed cutting speed; the fold bifurcations of the periodic solutions are due to the non-smooth dynamics of the system.

[47]. However, there are only a few papers on the nonlinear dynamics and they employ either analytical methods [76] or simulation [4].

Machining processes are inherently nonsmooth, because there is the possibility of a loss of cutting force when the tool leaves the work piece. This poses some challenges, although in some cases one can approximate the equations of motion with a smooth system. In the case of turning, which is an autonomous process, DDE-BIFTOOL was used in [23]. Here we summarize the results in [79], where a milling problem was investigated with PDDE-CONT.

The equation of motion of the nonsmooth milling problem reads

$$
\begin{aligned}
\ddot{x}(t) + 2\xi\dot{x}(t) + x(t) &= g(t)\hat{w}(\cos 2\pi t/T + 0.3\sin 2\pi t/T) \\
&\times \big[H(1 + x(t - 2T) - x(t - T))F_c((1 + x(t - T) - x(t))\sin 2\pi t/T) \quad (12.46) \\
&+ H(x(t - T) - x(t - 2T) - 1)F_c((2 + x(t - T) - x(t))\sin 2\pi t/T) \big],
\end{aligned}
$$

where F_c is a nonlinear cutting force function, usually modeled with the power law $F_c(x) = 4\hat{w}/3x^{3/4}$, and H is the Heaviside function. The function g is a T-periodic windowing function that changes its value once in a period between 0 and 1 depending on whether the tool is cutting the material. The two important parameters are the period T, which is inversely proportional to the spindle speed and the dimensionless chip width $\hat{w}$.

In order to conduct a numerical bifurcation analysis of the system with PDDE-CONT, the Heaviside function $H(z)$ is replaced by the smoothed function $(1 + \tanh(Cz))/2$ with a sufficiently large value of C. Equation (12.46) has a unique T-periodic solution which represents steady cutting and can lose its stability either at a Neimark-Sacker or at a period-doubling bifurcation. Figure 12.10(a) shows the bifurcation diagram where these bifurcation curves

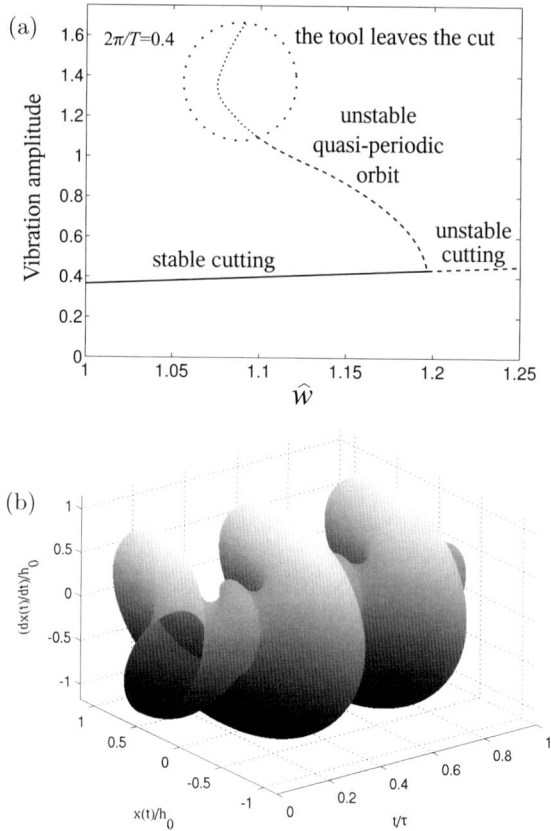

Fig. 12.11. A branch of quasiperiodic solutions for a cutting speed of $2\pi/T = 0.4$ (a) and an invariant torus (b) for a point on the branch just before the tool leaves the work piece and (12.46) becomes invalid.

are shown as solid curves. The period doublings may be subcritical or supercritical; see the bifurcation diagram for a fixed cutting speed in Fig. 12.10(b). Fold and Neimark-Sacker bifurcation curves of the period-two solutions have been continued in two parameters, and they are shown in Fig. 12.10(a) as dashed curves. These numerical results were compared to experimental data in [79].

Quasiperiodic solutions arising at a Neimark-Sacker bifurcation can be computed with the technique described in Sect. 12.5. A branch of invariant quasiperiodic tori was continued with PDDE-CONT until the model loses its physical validity. During the continuation the rotation number $\frac{\omega_2}{\omega_1}$ was kept constant and $\hat{w}$ and T served as free parameters. The resulting curve of quasiperiodic solutions is shown in Fig. 12.11(a). Since T varies only slightly during the continuation, the dependence on the period is not shown in the bifurcation diagram. One of the invariant tori along the branch (near where the model loses its validity) is shown in Fig. 12.11(b). The computation of further quasiperiodic solutions reveals that this system has very narrow Arnol'd tongues in the region above the Neimark-Sacker curve in Fig. 12.10(a).

12.8.5 A Laser with Filtered Optical Feedback

One main objective for studying laser dynamics is to find regions of parameter values where a constant-amplitude coherent light is produced. In many laser systems delay is an important feature. It arises due to the finite travel time of light between components of the system and may lead to different types of dynamic behavior including chaos; see, e.g., [49]. The numerical tools introduced in this chapter are very well suited to the study of nonlinear dynamics in lasers with delayed optical feedback; see also [51].

In this section we summarize some results of Erzgräber et. al. [30], who investigate a DDE model of a semiconductor laser with filtered optical feedback of the form

$$\frac{\mathrm{d}E}{\mathrm{d}t} = (1 + i\alpha)N(t) + \kappa F(t), \tag{12.47}$$

$$T\frac{\mathrm{d}N}{\mathrm{d}t} = P - N(t) - (1 + 2N(t))|E(t)|^2, \tag{12.48}$$

$$\frac{\mathrm{d}F}{\mathrm{d}t} = \Lambda E(t - \tau)\mathrm{e}^{-iC_p} + (i\Delta - \Lambda)F(t), \tag{12.49}$$

where the variable E is the complex optical field, N is the (real-valued) population inversion of the laser, and F is the complex optical field of the filter. The material properties of the laser are given by the linewidth enhancement factor α, the electron lifetime T and the pump rate P. The coupling of the laser with the filter is controlled by the parameter κ, while τ is the time that the light spends in the external feedback loop. The dynamics of the filter is modeled by (12.49). The feedback phase C_p is the phase difference between the laser and the filter fields, and Δ is the detuning between the filter center frequency Ω_F and the solitary frequency Ω_0 of the laser. Hence, $C_p = \Omega_0\tau$ and $\Delta = \Omega_F - \Omega_0$.

The laser equations (12.47)–(12.49) exhibit a rotational symmetry (rotation of both E and F over any angle) that is important for the types of solutions that are supported. It also needs to be dealt with in the continuation to ensure that solutions are isolated. The idea is to consider solutions of the form

$$(E(t), N(t), F(t)) = (A(t)\,\mathrm{e}^{ibt}, N(t), B(t)\,\mathrm{e}^{ibt}).$$

By putting this ansatz into (12.47)–(12.49) we obtain the new system

$$\frac{\mathrm{d}A}{\mathrm{d}t} = (1 + i\alpha)N(t)A(t) - ibA(t) + \kappa B(t), \tag{12.50}$$

$$T\frac{\mathrm{d}N}{\mathrm{d}t} = P - N(t) - (1 + 2N(t))|A(t)|^2, \tag{12.51}$$

$$\frac{\mathrm{d}B}{\mathrm{d}t} = \Lambda A(t - \tau)e^{-i(C_p + b\tau)} + (i\Delta - \Lambda - ib)B(t). \tag{12.52}$$

Note that the stability of this transformed system does not differ from the the stability of (12.47)–(12.49), because the norm of $(E(t), N(t), F(t))$ is the same as the norm of $(A(t), N(t), B(t))$. System (12.50)–(12.52) still has the same rotational symmetry, but the equations are now in a form that can be dealt with in continuation.

The primary interest is in the so-called external filtered modes (EFMs), which are single-frequency periodic solutions of (12.50)–(12.52) that are characterized by fixed $A(t) = A_s$, $N(t) = N_s$ and $B(t) = B_s$. EFMs were extensively studied analytically [63] and with numerical continuation [30]. In order to determine an EFM uniquely one needs to fix the phase, which can be done, for example, by setting $\mathrm{Re}(E_s) = 0$ and treating b as a variable. Figure 12.12(a) shows a bifurcation diagram in the (κ, C_p)-plane that was computed with DDE-BIFTOOL. EFMs are stable in the green region; they are born in saddle-node bifurcations (blue curves) and lose their stability in Hopf bifurcations (red curves).

At Hopf bifurcations periodic solutions arise whose continuation requires a new phase condition [37]. Let us introduce the symmetry group

$$G(\theta) = \begin{pmatrix} e^{i\theta} & 0 & 0 \\ 0 & 1 & 0 \\ 0 & 0 & e^{i\theta} \end{pmatrix}, \quad \mathfrak{G} = \frac{\mathrm{d}G(\theta)}{\mathrm{d}\theta}\Big|_{\theta=0} = \begin{pmatrix} i & 0 & 0 \\ 0 & 0 & 0 \\ 0 & 0 & i \end{pmatrix},$$

which produces a two-parameter family of solutions

$$u(t; \theta_1, \theta_2) = G(\theta_1)(A(t + \theta_2), N(t + \theta_2), B(t + \theta_2))$$

from any solution of (12.50)–(12.52). In a continuation context, when looking for the next solution on a branch of solutions, we want to find the one closest in norm to the previous solution $u^{(0)}$. Hence, the new solution u is chosen to minimize

$$D(\theta_1, \theta_2) = \int_0^1 \|u(t; \theta_1, \theta_2) - u^{(0)}(t)\|_2^2 \, \mathrm{d}t$$

in θ_1, θ_2. Differentiating with respect to both variables and evaluating at $\theta_1 = \theta_2 = 0$ yields

$$\int_0^1 \dot{u}^{(0)}(s)(u^{(0)}(s) - u(s)) \, \mathrm{d}s = 0, \tag{12.53}$$

$$\int_0^1 \mathfrak{G}u^{(0)}(s)(u^{(0)}(s) - u(s)) \, \mathrm{d}s = 0. \tag{12.54}$$

where $\mathfrak{G} = \mathrm{d}/\mathrm{d}\theta G(\theta)|_{\theta=0}$ is the infinitesimal generator of the symmetry group. Note that (12.53) is actually phase condition (12.14), while (12.54) is a new phase condition that fixes the group invariance.

With (12.53) and (12.54) periodic solutions can be continued as isolated solutions. These phase conditions are implemented in both DDE-BIFTOOL and PDDE-CONT. In [30] DDE-BIFTOOL was used to compute the periodic solutions of (12.50)–(12.52) and PDDE-CONT was used to determine their stability boundaries by continuing the Neimark-Sacker bifurcation in two parameters.

The resulting stability regions of the two different types of periodic solutions are colored in Fig. 12.12(a), and examples of typical time series are shown in panels (b)–(d). First, there are the typical relaxation oscillations (not to be confused with relaxation oscillations of slow-fast systems as discussed in Chap. 8), which are a periodic exchange of energy between the electric field E and the inversion N in a semiconductor laser. Relaxation oscillations are fast (on the order of a few GHz) and effectively do not involve the filter; see Fig. 12.12(b). The other type of oscillations are the frequency oscillations, which are slower and oscillate on the time scale given by the external roundtrip time (that is, the delay τ); see Fig. 12.12(c) and (d). These oscillations are unusual for semiconductor lasers because they feature practically constant laser intensity I_L but an oscillating frequency ω_L. Notice that the dynamics of the filter appears to suppress the dynamics of the intensity. Both types of oscillations lose their stability at Neimark-Sacker bifurcations, which are shown as black curves in Fig. 12.12(a).

12.9 Conclusions

We discussed numerical continuation methods for the stability and bifurcation analysis of delay differential equations with constant delays, concentrating on techniques concerning steady-state solutions and periodic solutions. We also described how to compute connecting (homoclinic and heteroclinic) orbits and quasiperiodic solutions. Furthermore, we briefly mentioned how to deal with state-dependent delays and with equations of neutral type. Compared with numerical methods for such tasks in ordinary differential equations the methods we presented are either similar but with a higher computational cost (an example is collocation for computing periodic solutions) or much more complex (as is the case for computing the stability of a steady state or finding a connected orbit). These additional difficulties are due to the infinite-dimensional nature of DDEs.

Rather than trying to give a complete literature survey, we focused on the numerical methods implemented in the software packages DDE-BIFTOOL and PDDE-CONT. Both have about the same functionality as similar packages for ODEs, but with less flexibility and at a higher computational cost. They make continuation and bifurcation analysis for DDEs readily available for scientists dealing with concrete problems arising in applications. We have

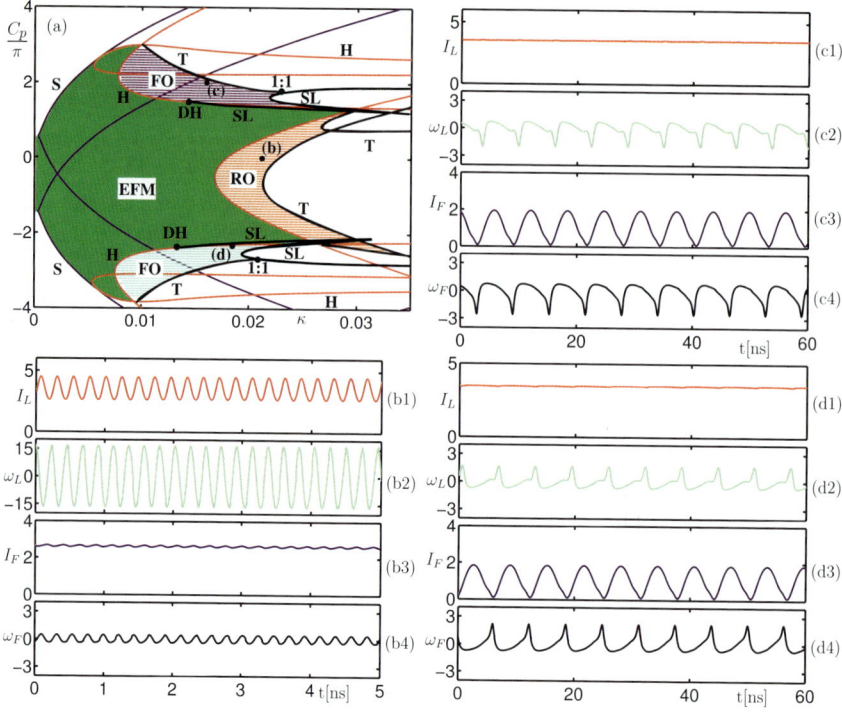

Fig. 12.12. Panel (a) shows the bifurcation diagram in the (κ, C_p)-plane of (12.47)–(12.49) for $\Delta = 0$, $\alpha = 5.0$, $T = 100$, $P = 3.5$, $\tau = 500$ and $\Lambda = 0.007$. EFMs are stable in the green region, which is bounded by curves of saddle-node bifurcations (blue) and Hopf bifurcations (red). Panels (b1)–(b4) show an example of relaxation oscillations, while panels (c1)–(c4) and panels (d1)–(d4) are examples of frequency oscillations; plotted are the laser intensity I_L, its frequency ω_L, the filter intensity I_F, and its frequency ω_F. The stability regions of the different oscillations are shown in panel (a) in orange, purple and light blue, respectively. From H. Erzgräber, B. Krauskopf and D. Lenstra, *SIAM J. Appl. Dyn. Sys.* 6(1) (2007) 1–28 © 2007 by the Society for Industrial and Applied Mathematics; reprinted with permission.

included results on the continuation and bifurcation analysis of several realistic models to illustrate the applicability of the methods.

Numerical developments can also help with the solution of some open theoretical problems. For example, some numerical results on state-dependent DDEs are ahead of the theory and suggest that certain conditions imposed in the theory are rather technical and not fundamental. One of the areas for future work for both theory and numerical methods is that of piecewise-smooth delayed systems, which have important applications, for example, in control theory [5, 73], hybrid testing [53, 74] and machine tooling [23, 77].

Acknowledgements

We thank Bernd Krauskopf, David Barton, Tatyana Luzyanina, Giovanni Samaey and Koen Verheyden for their useful comments and suggestions. We are grateful to the American Physical Society, World Scientific Publishing and the Society for Industrial and Applied Mathematics for permission to reproduce Fig. 12.4, Fig. 12.5 and Fig. 12.12, respectively.

References

1. E. Anderson, Z. Bai, C. Bischof, S. Blackford, J. Demmel, J. Dongarra, J. Du Croz, A. Greenbaum, S. Hammarling, A. McKenney, and D. Sorensen. *LAPACK Users' Guide, third edition*. (SIAM, Philadelphia, 1999).
2. U. M. Ascher, R. M. M. Mattheij, and R. D. Russell. *Numerical Solution of Boundary Value Problems for Ordinary Differential Equations*. (Prentice Hall, 1988).
3. C. T. H. Baker, C. A. H. Paul, and D. R. Willé. A bibliography on the numerical solution of delay differential equations. Technical Report 269, University of Manchester, Manchester Centre for Computational Mathematics, 1995.
4. B. Balachandran. Non-linear dynamics of milling process. *Trans. Royal Soc.*, 359:793–820, 2001.
5. D. A. W. Barton, B. Krauskopf, and R. E. Wilson. Explicit periodic solutions in a model of a relay controller with delay and forcing. *Nonlinearity*, 18(6):2637–2656, 2005.
6. D. A. W. Barton, B. Krauskopf, and R. E. Wilson. Collocation schemes for periodic solutions of neutral delay differential equations. *J. Diff. Eqns. Appl.*, 12(11):1087-1101, 2006.
7. A. Bellen and M. Zennaro. *Numerical Methods for Delay Differential Equations*. (Oxford University Press, 2003).
8. W. J. Beyn. The numerical computation of connecting orbits in dynamical systems. *IMA J. Numer. Analysis*, 9:379–405, 1990.
9. W. J. Beyn, A. Champneys, E. J. Doedel, W. Govaerts, B. Sandstede, and Y. A. Kuznetsov. Numerical continuation and computation of normal forms. In B. Fiedler, editor, *Handbook of Dynamical Systems*, pages 149–219. (Elsevier, 2002).
10. D. Breda. *Numerical computation of characteristic roots for delay differential equations*. PhD thesis (Department of Mathematics and Computer Science, University of Udine, 2004).
11. D. Breda. Solution operator approximation for delay differential equation characteristic roots computation via Runge-Kutta methods. *Appl. Numer. Math.*, 56:305–317, 2006.
12. D. Breda, S. Maset, and R. Vermiglio. Computing the characteristic roots for delay differential equations. *IMA J. Numer. Analysis*, 24:1–19, 2004.
13. D. Breda, S. Maset, and R. Vermiglio. Pseudospectral differencing methods for characteristic roots of delay differential equations. *SIAM J. Sci. Comput.*, 27(2):482–495, 2005.
14. A. M. Castelfranco and H. W. Stech. Periodic solutions in a model of recurrent neural feedback. *SIAM J. Appl. Math.*, 47(3):573–588, 1987.

15. K. L. Cooke and W. Huang. On the problem of linearization for state-dependent delay differential equations. *Proc. Am. Math. Soc.*, 124(5), 1996.

16. S. P. Corwin, D. Sarafyan, and S. Thompson. DKLAG6: A code based on continuously imbedded sixth order Runge-Kutta methods for the solution of state dependent functional differential equations. *Appl. Numer. Math.*, 24(2–3):319–330, 1997.

17. T. A. Davies. UMFPACK Version 4.1 User Guide. Technical report, Department of Computer and Information Science and Engineering, University of Florida, Gainesville, FL, USA, 2003. http://www.cise.ufl.edu/research/sparse/umfpack/.

18. E. J. Doedel, A. R. Champneys, T. F. Fairgrieve, Y. A. Kuznetsov, B. Sandstede, and X. Wang. AUTO97: Continuation and bifurcation software for ordinary differential equations, 1997; available via ftp.cs.concordia.ca in directory pub/doedel/auto.

19. E. J. Doedel, M. J. Friedman, and B. I. Kunin. Successive continuation for locating connecting orbits. *Num. Alg.*, 17:103–124, 1997.

20. E. J. Doedel, W. Govaerts, and Yu. A. Kuznetsov. Continuation of periodic solution bifurcations in odes using bordered systems. *SIAM J. Numer. Analysis*, 42(2):401–435, 2003.

21. E. J. Doedel, H. B. Keller, and J. P. Kernevez. Numerical analysis and control of bifurcation problems (II): bifurcations in infinite dimensions. *Internat. J. Bifur. Chaos Appl. Sci. Engrg.*, 1(4):745–772, 1991.

22. E. J. Doedel and P. P. C. Leung. A numerical technique for bifurcation problems in delay differential equations. *Congr. Numer.*, 34:225–237, 1982. (Proc. 11th Manitoba Conf. Num. Math. Comput., Univ. Manitoba, Winnipeg, Canada).

23. Z. Dombóvári. Bifurcation analysis of a cutting process. Master's thesis (Budapest University of Technology and Economics, 2006).

24. K. Engelborghs and E. Doedel. Stability of piecewise polynomial collocation for computing periodic solutions of delay differential equations. *Numer. Math.*, 91(4):627–648, 2002.

25. K. Engelborghs, T. Luzyanina, K. J. in 't Hout, and D. Roose. Collocation methods for the computation of periodic solutions of delay differential equations. *SIAM J. Sci. Comput.*, 22(5):1593–1609, 2000.

26. K. Engelborghs, T. Luzyanina, and D. Roose. Numerical bifurcation analysis of delay differential equations using DDE-BIFTOOL. *ACM Trans. Math. Software*, 28(1):1–21, 2002.

27. K. Engelborghs, T. Luzyanina, and G. Samaey. DDE-BIFTOOL v. 2.00: a Matlab package for bifurcation analysis of delay differential equations. Technical Report TW-330, Department of Computer Science, K.U.Leuven, Leuven, Belgium, 2001; available at www.cs.kuleuven.ac.be/cwis/research/~twr/research/software/delay/ddebiftool.shtml.

28. K. Engelborghs and D. Roose. On stability of LMS methods and characteristic roots of delay differential equations. *SIAM J. Numer. Analysis*, 40(2):629–650, 2002.

29. B. Ermentrout. XPPAUT3.91 - The differential equations tool. University of Pittsburgh, Pittsburgh, 1998; available at http://www.pitt.edu/~phase/.

30. H. Erzgräber, B. Krauskopf, and D. Lenstra. Bifurcation analysis of a semiconductor laser with filtered optical feedback. *SIAM J. Appl. Dyn. Sys.*, 6(1):1–28, 2007.

31. W. Govaerts. Bordered augmented linear systems in numerical continuation and bifurcation. *Numer. Math.*, 58:353–368, 1990.

32. W. Govaerts and J. D. Pryce. Mixed block elimination for linear systems with wider borders. *IMA J. of Numer. Analysis*, 13:161–180, 1993.

33. K. Green, B. Krauskopf, and K. Engelborghs. Bistability and torus break-up in a semiconductor laser with phase-conjugate feedback. *Physica D*, 173(1-2):114–129, 2002.

34. K. Green, B. Krauskopf, and K. Engelborghs. One-dimensional unstable eigenfunction and manifold computations in delay differential equations. *J. Comp. Phys.*, 197(1):86–98, 2004.

35. K. Green, B. Krauskopf, and G. Samaey. A two-parameter study of the locking region of a semiconductor laser subject to phase-conjugate feedback. *SIAM J. Appl. Dyn. Sys.*, 2(2):254–276, 2003.

36. N. Guglielmi and E. Hairer. Implementing Radau IIA methods for stiff delay differential equations. *Computing*, 67:1–12, 2001

37. B. Haegeman, K. Engelborghs, D. Roose, D. Pieroux, and T. Erneux. Stability and rupture of bifurcation bridges in semiconductor lasers subject to optical feedback. *Phys. Rev. E*, 66:046216 1–11, 2002.

38. J. K. Hale. *Theory of Functional Differential Equations*, volume 3 of *Applied Mathematical Sciences*. (Springer-Verlag, New York, 1977).

39. J. K. Hale and N. Sternberg. Onset of chaos in differential delay eqautions. *J. Comp. Phys.*, 77:221–239, 1988.

40. J. K. Hale and S. M. Verduyn Lunel. *Introduction to Functional Differential Equations*, volume 99 of *Applied Mathematical Sciences*. (Springer-Verlag, New York, 1993).

41. F. Hartung and J. Turi. Stability in a class of functional differential equations with state-dependent delay. In C Corduneanu, editor, *Qualitative problems for differential equations and control theory*, pages 15–31. (World Scientific, Singapore, 1995).

42. F. Hartung and J. Turi. On differentiability of solutions with respect to parameters in state-dependent delay equations. *J. Diff. Eqns.*, 135:192–237, 1997.

43. B. D. Hassard. A code for Hopf bifurcation analysis of autonomous delay-differential systems. In F. V. Atkinson, W. F. Langford, and A. B. Mingarelli, editors, *Oscillations, Bifurcations and Chaos*, volume 8 of *Can. Math. Soc. Conference Proceedings*, pages 447–463 (Amer. Math. Soc., Providence, RI, 1987).

44. B. D. Hassard. Counting roots of the characteristic equation for linear delay-differential systems. *J. Diff. Eqns.*, 136:222–235, 1997.

45. J. S. Hesthaven. Spectral penalty methods. *Appl. Num. Math.*, 33:23–41, 2000.

46. K. J. in 't Hout and Ch. Lubich. Periodic orbits of delay differential equations under discretization. *BIT*, 38(1):72–91, 1998.

47. T. Insperger, B. P. Mann, G. Stépán, and P. V. Bayly. Stability of up-milling and down-milling, part 1: alternative analytical methods. *Int. J. of Mach. Tool Manuf.*, 43(1):25–34, 2003.

48. M. A. Kaashoek and S. M. Verduyn Lunel. Characteristic matrices and spectral properties of evolutionary systems. *Trans. Amer. Math. Soc.*, 334:479–517, 1992.

49. D.M. Kane and K.A. Shore (Eds.), *Unlocking Dynamical Diversity: Optical Feedback Effects on Semiconductor Lasers*, (John Wiley & Sons, 2005).

50. D. A. Kopriva and J. H. Kolias. A conservative staggered-grid chebyshev multidomain method for compressible flows. *J. Comp. Phys.*, 125:244–261, 1996.

51. B. Krauskopf. Bifurcation analysis of lasers with delay. In D.M. Kane and K.A. Shore, editors, *Unlocking Dynamical Diversity: Optical Feedback Effects on Semiconductor Lasers*, pages 147–183. (John Wiley & Sons, 2005).
52. Yu. A. Kuznetsov and V. V. Levitin. CONTENT: A multiplatform environment for analyzing dynamical systems, 1997. Dynamical Systems Laboratory, Centrum voor Wiskunde en Informatica; available via ftp.cwi.nl in directory pub/CONTENT.
53. Y. N. Kyrychko, K. B. Blyuss, A. Gonzalez-Buelga, S. J. Hogan and D. J. Wagg. Real-time dynamic substructuring in a coupled oscillator-pendulum system. *Proc. R. Soc. A* 462:1271–1294, 2006.
54. R. B. Lehoucq, D. C. Sorensen, and C. Yang. *ARPACK Users' Guide: Solution of Large-Scale Eigenvalue Problems with Implicitly Restarted Arnoldi Methods.* (SIAM, Philadelphia, 1998).
55. K. Lust, D. Roose, A. Spence, and A.R. Champneys. An adaptive Newton-Picard algorithm with subspace iteration for computing periodic solutions. *SIAM J. Sci. Comput.*, 19(4):1188–1209, 1998.
56. T. Luzyanina and K. Engelborghs. Computing Floquet multipliers for functional differential equations. *Internat. J. Bif. Chaos Appl. Sci. Engrg.*, 12(12):2977–2989, 2002.
57. T. Luzyanina, K. Engelborghs, and D. Roose. Numerical bifurcation analysis of differential equations with state-dependent delay. *Internat. J. Bif. Chaos Appl. Sci. Engrg.*, 11(3):737–753, 2001.
58. T. Luzyanina and D. Roose. Numerical stability analysis and computation of Hopf bifurcation points for delay differential equations. *J. Comput. Appl. Math.*, 72:379–392, 1996.
59. M. C. Mackey and L. Glass. Oscillation and chaos in physiological control systems. *Science*, 77:287–289, 1977.
60. J. Mallet-Paret and R. D. Nussbaum. A differential delay equation arising in optics an physiology. *SIAM J. Math. Analysis*, 20(2):249–292, 1989.
61. J. Mallet-Paret and R.D. Nussbaum. Boundary layer phenomena for differential-delay equations with state-dependent time lags 1. *Arch. Rational Mech. Analysis*, 120:99–146, 1992.
62. S. I. Niculescu. *Delay Effects on Stability: A Robust Control Approach.* Lecture Notes in Control and Information Science, 269, (Springer-Verlag, New York, 2000).
63. M. Nizette and T. Erneux. Optical frequency dynamics and relaxation oscilations of a semiconductor laser subject to filtered optical feedback. In M. Pessa D. Lenstra and Ian H White, editors, *Semiconductor Lasers and Laser Dynamics II, Proceedings of the SPIE*, pages 6184–32. 2006.
64. G. Orosz, B. Krauskopf, and R. E. Wilson. Bifurcations and multiple traffic jams in a car following model with reaction delay time. *Physica D*, 211:277–293, 2005.
65. G. Orosz, R. E. Wilson, and B. Krauskopf. Global bifurcation investigation of an optimal velocity traffic model with driver reaction time. *Phys. Rev. E*, 70(026207):1–10, 2004.
66. C. A. H. Paul. A user-guide to Archi - an explicit Runge-Kutta code for solving delay and neutral differential equations. Technical Report 283 (The University of Manchester, Manchester Center for Computational Mathematics, 1997).
67. A. Pazy. *Semigroups of Linear Operators and Applications to Partial Differential Equations.* (Springer-Verlag, New York, 1983).

68. Y. Saad. *Iterative Methods for Sparse Systems.* (SIAM, Philadelphia, 2003).
69. G. Samaey, K. Engelborghs, and D. Roose. Numerical computation of connecting orbits in delay differential equations. *Num. Algor.*, 30(3-4):335–352, 2002.
70. G. Samaey and B. Sandstede. Determining stability of pulses for partial differential equations with time delays. *Dyn. Sys.*, 20(2):201–222, 2005.
71. L. F. Shampine and S. Thompson. Solving DDEs in MATLAB. *Appl. Numer. Math.*, 37:441–458, 2001.
72. F. Schilder, H. M. Osinga, and W. Vogt. Continuation of quasiperiodic invariant tori. *SIAM J. Appl. Dyn. Sys.*, 4(3):244–261, 2005.
73. J. Sieber. Dynamics of delayed relay systems. *Nonlinearity*, 19(11):2489–2527, 2006.
74. J. Sieber and B. Krauskopf. Control-based bifurcation analysis for experiments. *Nonlin. Dyn.* in press.
75. G. Stépán. *Retarded Dynamical Systems.* (Longman, London, 1989).
76. G. Stépán and T. Kalmár-Nagy. Nonlinear regenerative machine tool vibrations. In *Proceedings of the 1997 ASME Design Engineering Technical Conferences, Sacramento, California*, September 1997. Paper number DETC97/VIB-4021.
77. G. Stépán, R. Szalai, and T. Insperger. Nonlinear dynamics of high-speed milling subjected to regenerative effect. In G. Radons and R. Neugebauer, editors, *Nonlinear Dynamics of Production Systems*, pages 111–128. (Wiley-VCH, Berlin, 2004).
78. R. Szalai. PDDE-CONT: A continuation and bifurcation software for delay-differential equations, 2005. Department of Applied Mechanics, Budapest University of Technolgy and Economics; available at http://www.mm.bme.hu/~szalai/pdde/.
79. R. Szalai, B. P. Mann, and G. Stépán. Period-two and quasiperiodic vibrations of high-speed milling. In I. Grabec and E. Govekar, editors, *Proceedings of the 9th CIRP International Workshop on Modeling of Machining Operations*, pages 107–114. CIRP, 2006.
80. R. Szalai, G. Stépán, and S. J. Hogan. Continuation of bifurcations in periodic delay-differential equations using characteristic matrices. *SIAM J. Sci. Comp.*, 28(4):1301–1317, 2006.
81. L. N. Trefethen. *Spectral Methods in MATLAB.* (SIAM, Philadelphia, 2000).
82. K. Verheyden, K. Green, and D. Roose. Numerical stability analysis of a large-scale delay system modeling a lateral semiconductor laser subject to optical feedback. *Phys. Rev. E*, 69(3):036702, 2004.
83. K. Verheyden and K. Lust. A Newton-Picard collocation method for periodic solutions of delay differential equations. *BIT Num. Math.*, 45(3):1605–625, 2005.
84. K. Verheyden, T. Luzyanina, and D. Roose. Efficient and reliable stability analysis of solutions of delay differential equations. *Proceedings of 2006 International Conference on Nonlinear Science and Complexity*, pages 109–120. (World Scientific Publishing, Singapore, 2006).
85. K. Verheyden and D. Roose. Efficient numerical stability analysis of delay equations: a spectral method. In D. Roose and W. Michiels, editors, *Time-delay systems 2004*, IFAC Proceedings Volumes, pages 209–214. (Elsevier, Oxford, UK, 2005).